Asheville-Buncombe
Technical Community College
Learning Resources Center
340 Victoria Rd.
Asheville, NC 28801

DISCARDED

JUL - 1 2025

BIODIVERSITY, SCIENCE AND DEVELOPMENT
TOWARDS A NEW PARTNERSHIP

BIODIVERSITY, SCIENCE AND DEVELOPMENT

Towards a New Partnership

Edited by

Francesco di Castri

*Director of Research, CNRS
Centre d'Ecologie Fonctionelle et Evolutive
1919 Route de Mende
B.P. 5051
34033 Montpellier Cedex
France*

and

Talal Younès

*Executive Director
International Union of Biological Sciences
51, Boulevard de Montmorency
75016 Paris
France*

CAB INTERNATIONAL

in association with the

International Union of Biological Sciences

CAB INTERNATIONAL
Wallingford
Oxon OX10 8DE
UK

Tel: +44 (0)1491 832111
Fax: +44 (0)1491 833508
E-mail: cabi@cabi.org
Telex: 847964 (COMAGG G)

©CAB INTERNATIONAL 1996. All rights reserved. No part of this publication may be reproduced in any form or by any means, electronically, mechanically, by photocopying, recording or otherwise, without the prior permission of the copyright owners.

A catalogue record for this book is available from the British Library.

ISBN 0 85198 973 X

Published in association with:

International Union of Biological Sciences (IUBS)
51 Boulevard de Montmorency
75016 Paris
France

Typeset in 10/12 Palatino by Colset Pte Ltd, Singapore
Printed and bound in the UK at the University Press, Cambridge

Contents

Preface   xi

Introduction: Biodiversity, the Emergence of a New Scientific
Field – Its Perspectives and Constraints   1
Francesco di Castri and Talal Younès

Part I: Opening Addresses

Opening Address of UNESCO   12
Federico Mayor, Director General

Opening Address of the Government of France   18
Michel Barnier, Minister of the Environment

Opening Address of ICSU   23
Harold A. Mooney, Vice President

Opening Address of IUBS   26
Francesco di Castri, President

Bridge-building for Biodiversity   30
Reuben J. Olembo, Deputy Executive Director

Written Message from the President of Argentina   35
Carlos Menem

Written Message from the President of Indonesia   38
President Soeharto

Part II: Approaching the Unity of Life Through Biological Diversity

    1. Biological Foundations for Sustainability and Change   42
       C.S. Holling

    2. Molecular Mechanisms Promoting and Limiting Genetic Variation   58
       W. Arber

    3. Genome Dynamics and the Generation of Biodiversity   67
       G. Bernardi

    4. Genes, Memes and Artefacts   80
       M. Gadgil

    5. New Dimensions in the Study of Genetic Diversity   91
       A.P. Ryskov, M.I. Prosniak, V.V. Kalnin, O.V. Kalnina, O.N. Tokarskaya, E.K. Khusnutdinova, G.P. Georgiev and S.A. Limborska

    6. Deep-ocean Biodiversity   104
       J.F. Grassle

    7. Biological Diversity and Genetic Resources   107
       A. Charrier, F. Fridlansky and J.C. Mounolou

Part III: Ecological Functions of Biodiversity

    8. Ecological Functions of Biodiversity: The Human Dimension   114
       P.S. Ramakrishnan

    9. Microorganisms: The Neglected Rivets in Ecosystem Maintenance   130
       D.L. Hawksworth

    10. Biodiversity of Marine Sediments   139
       C. Heip

    11. Linkage between Ecological Complexity and Biodiversity   149
       H. Kawanabe

    12. Biotic Interactions and the Ecosystem Function of Biodiversity   153
       H.A. Mooney

    13. Relations Between Biodiversity and Ecosystem Fluxes of Water Vapour and Carbon Dioxide   162
       E.-D. Schulze

Part IV: Global Inventorying and Monitoring of Biodiversity

    14. Inventorying and Monitoring Biodiversity   166
       P.B. Tinker

15. Monitoring and Inventorying Biodiversity: Collections, Data and Training   171
   N.R. Chalmers

16. Inventories: Preparing Biodiversity for Non-damaging Use   180
   R. Gámez

17. Inventorying and Monitoring Flora in Asia   184
   K. Iwatsuki

18. Monitoring Biodiversity at Global Scales   189
   A.E. Lugo

19. Some Problems of Inventorying and Monitoring Biodiversity in Russia   197
   V.E. Sokolov and B.R. Striganova

20. Inventory and Monitoring for What and for Whom?   208
   P.B. Bridgewater

Part V: Conservation of Biodiversity

21. Biodiversity Conservation and Protected Areas in Tropical Countries   212
   G. Halffter

22. Conservation of Coastal–Marine Biological Diversity   224
   G. Carleton Ray

23. Biological Conservation in a High Beta-diversity Country   246
   J. Sarukhán, J. Soberón and J. Larson-Guerra

24. Conserving Biodiversity: The Key Political, Economic and Social Measures   264
   J.A. McNeely

25. Biodiversity Conservation in the New South Africa   282
   B.J. Huntley

26. Biodiversity in the Developing Countries   304
   T.N. Khoshoo

Part VI: Biodiversity and Agriculture, Grazing and Forestry

27. Biodiversity and Agriculture, Grasslands and Forests   312
   M. Lefort and M. Chauvet

28. Managing Biodiversity in Canada's Public Forests   324
   P.N. Duinker

29. Some Canadian Approaches to Partnership in
    Agricultural Biodiversity   341
    B. Fraleigh

30. Three Levels of Conservation by Local People   347
    A. Gómez-Pompa

31. Some Current Issues in Conserving the Biodiversity of
    Agriculturally Important Species   357
    T. Hodgkin

32. Ecosystem Management: An Approach for Conserving
    Biodiversity   369
    R.C. Szaro, G.D. Lessard and W.T. Sexton

33. Biological Diversity and Agrarian Systems   385
    B. Vissac

Part VII: Biodiversity: Aquaculture and Fisheries

34. Biodiversity, Science and Development: Towards a New
    Partnership in Aquaculture and Fisheries   403
    E.A. Huisman

35. Biodiversity and Aquaculture   409
    R.S.V. Pullin

36. Conservation of Biological Diversity in Hatchery
    Enhancement Programmes   424
    D.M. Bartley

37. Banking Fish Genetic Resources: The Art of the
    Possible   439
    B. Harvey

38. Introduction of Fish Species in Freshwaters: A Major
    Threat to Aquatic Biodiversity?   446
    C. Lévêque

39. Biodiversity and Fisheries and Aquaculture
    Development in Taiwan   452
    I. Chiu Liao and Yew-Hu Chien

Part VIII: Biodiversity and Industry

40. Biodiversity and Marine Biotechnology: A New
    Partnership of Academia, Government and Industry   456
    R.R. Colwell

41. Biotechnologies and the Use of Plant Genetic Resources
    for Industrial Purposes: Benefits and Constraints for
    Developing Countries   469
    A. Sasson

42. A Partnership: Biotechnology, Bio-pharmaceuticals and
    Biodiversity   488
    M. Comer and E. Debus

43. In Defence of Biotechnology   500
    M.F. Cantley

44. Tropical Biodiversity and the Development of
    Pharmaceutical Industries   506
    E.J. Adjanohoun

45. Biodiversity Prospecting: Opportunities and Challenges
    for African Countries   519
    C. Juma and J. Mugabe

Part IX: Biodiversity in Urban and Peri-urban Environments

46. Biodiversity in Urban and Peri-urban Zones   539
    R. Folch

47. The Importance of Urban Environments in Maintaining
    Biodiversity   543
    V.H. Heywood

48. The Role of Biodiversity in Urban Areas and the Role of
    Cities in Biodiversity Conservation   551
    J.-P. Reduron

49. Restoration of Biodiversity in Urban and Peri-urban
    Environments with Native Forests   558
    A. Miyawaki

50. Synurbization of Animals as a Factor Increasing
    Diversity of Urban Fauna   566
    M. Luniak

51. Biodiversity Management in Peri-urban Environments in
    Switzerland   576
    B. Schmid

52. General Considerations on the Biodiversity of Urban and
    Peri-urban Environments   581
    G. Vida

53. The Urban Dimension of Diversity   584
    R. Pesci

Part X: Biodiversity: Culture Values and Ethics

54. Biodiversity in the Twenty-first Century   596
    J. de Rosnay

55. Biodiversity: Cultural and Ethical Aspects   599
    M. Elmandjra

56. The Value of Biological Diversity: Socio-political
    Perspectives   606
    Crispin Tickell

57. The Ethical and Non-economic Rationale for the
    Conservation of Biodiversity   614
    A. Campeau

58. Environmental Ethics and Biodiversity   617
    Ph. Bourdeau

59. Ethics of Biodiversity Conservation   622
    Gian Tommaso Scarascia Mugnozza

60. Reformation Towards a Nature-oriented Culture   630
    Wakako Hironaka

61. Biodiversity and Quality of Life   633
    G. Hauser

Index   639

Preface

During the course of the last years, and especially since the organization of the United Nations Conference on Environment and Development in 1992 in Rio de Janeiro (the 'Rio Summit'), biodiversity has become a very common and fashionable word. It is very widely used by the general public, environmental groups, the scientific community, conservationists and industrialists, and has a very high profile in the political arena. Unfortunately, it is somewhat loosely used, with different meanings, connotations and intentions.

For the general public, at least in the industrialized countries, biodiversity has mostly an emotional connotation. The main concern is for the preservation of some species of animals and plants of a strong charismatic nature or high aesthetic value, such as the panda, elephant, rhinoceros, sequoia, orchids, etc., as well as for stopping the species extinction related to developmental activities taking place, particularly, in the Southern hemisphere. Quite understandably, the main objective of conservationists is to preserve as large a portion as feasible of ecosystems under threat, which are mostly in the tropics, and that they remain as untouched as possible. This perception is exacerbated by the multitude of myths on biodiversity that pervades the media and leads mainly to an overly sensational or catastrophic viewpoint as regards the collapse of biodiversity and the role of humankind therein.

It should be made clear that biodiversity stands at the very foundations of development. This statement applies to the agricultural development *sensu lato* through the selection of crop varieties, landraces and forestry stands, and for a more balanced and diversified rotation of cultures. It is relevant as well to aquaculture and fisheries and also to a great deal of industrial development, particularly in the pharmaceutical sector, the agro-industries and the expanding biotechnological applications.

From a political viewpoint, several recent achievements have been made. Reference should be made to the launching in 1992 of the Convention on

Biological Diversity and to the establishing – within the Global Environmental Facility – of a large-scale funding mechanism. As of February 1995, 168 countries have signed this Convention, while in April 1995, already 118 countries have ratified it. Admittedly, this Convention represents at best a common denominator to reach a consensus among all the member states concerned, each with different priorities in their agenda. It is the result of a few years of intense efforts of *biodiplomacy*, and a long way remains to complete it, but hopefully the implementation processes can be undertaken soon. However, it has the great merit to exist and represents the indispensable basis for further activities on biodiversity. The underlying dissension between countries of the North and of the South lies on the fact that for the former biodiversity is essentially a global issue, supporting their views in a paragraph in the preamble to the Convention which reads as follows: 'Affirming that the conservation of biological diversity is a common concern of humankind.' Conversely, developing countries tend to show a strong 'country-driven' approach. Their understandable concern is to make use of their own national 'intellectual property rights' on biodiversity for economical and developmental benefits, and to ensure equity in the transfer of appropriate technology for applications in agriculture and industry. The consecutive paragraph in the preamble to the Convention, following the one mentioned above, reads indeed 'Reaffirming that States have sovereign rights over their own biological resources'. Both are unobjectionable principles *per se*, and this crucial issue and apparent contradiction are likely to be overcome at a more advanced and mature stage of the implementation of the Convention.

As regards the scientific community, there has been an extraordinary proliferation of papers and articles on biodiversity during the last few years. Some of them are very relevant to comprehend the emerging issues of biodiversity; others are confusing this new approach with the traditional field of species diversity and richness. In addition, some of the scientists concerned have not been exempt from the usual 'preaching to the converted' syndrome, and have been reluctant to establish the necessary contacts and links with all the other actors involved in the biodiversity issue: conservation non-governmental groups, entrepreneurs and industrialists, local populations, politicians and decision-makers. Furthermore, biodiversity does not belong to any single discipline (ecology, zoology, botany, genetics, etc.), and requires a true interdisciplinary approach, which is not easy to adopt given the current institutional patterns in academia, the universities and the research laboratories.

Well aware of the underlying difficulties and constraints of the endeavour, but also of its intrinsic responsibility for clarifying and harmonizing scientific issues, for joining efforts of different biological disciplines, and for providing relevant research results to decision-makers of both industrialized and developing countries, the International Union of Biological Sciences (IUBS), with the joint sponsorship of SCOPE (the Scientific Committee on Problems of the Environment of ICSU) and of UNESCO, launched in 1991 a cooperative scientific programme on biodiversity, called *DIVERSITAS*. Thanks to an earlier three-year feasibility study, only a few particularly relevant themes were selected for the first phase: the ecosystem function of biodiversity, its origin, maintenance and loss, as well as understanding its transformation over both space and time through inventorying and monitoring. Different levels of organization, from the molecular to the population and ecosystems are covered in *DIVERSITAS*. Special attention is given to the rather neglected aspects of microbial diversity and of aquatic biodiversity. The revival of taxonomy as a most relevant and current priority is at the very basis of *DIVERSITAS*.

Three years after the launching of *DIVERSITAS*, and within its conceptual framework, an International Forum called 'Biodiversity, Science and Development. Towards a New Partnership' was held in September 1994 in Paris at the UNESCO headquarters. It was organized by IUBS in conjunction with its General Assembly, with UNESCO, ICSU and the French Government as cosponsors of the Forum.

The main objectives of this event were as follows:

1. To crystallize the results already achieved within *DIVERSITAS* and to explore new avenues and perspectives.

2. Through a truly scientific approach, to demystify current misleading perceptions and attitudes.

3. Having the concept of partnership as a *leitmotiv*, to help build bridges between the basic and the applied aspects of biodiversity, between the scientific and the productive world involved in biodiversity issues, between the biological diversity and the cultural and ethical implications of diversity.

4. To help provide approaches and facts in the process of construction and implementation of the Convention of Biological Diversity.

5. To compare and harmonize viewpoints coming from the governmental and non-governmental side.

In order to achieve these goals, the Forum was organized in a rather unusual way. Instead of the traditional lecture approach most common to scientific congresses, only round tables with a few panellists were organized. After a general framework given by the moderator, panellists made short and provocative interventions stimulating the debate among themselves and with the numerous participants of the floor. Time was then provided for an open discussion that was summarized at the end by the moderator. More than 80 panellists and some 600 participants from different parts of the world, with diverse backgrounds and professional experience, ensured an extremely rich and multifaceted debate.

This volume reflects the peculiar patterns of the Forum; the different chapters and sections take largely into account not only the initial statements of the contributors, but also incorporate the main features of the discussion that took place thereafter.

After a short introduction on a working definition, the perspectives and the constraints of biodiversity, Part I includes the opening messages where the views of governments from both industrialized (France) and developing countries (Argentina and Indonesia), of United Nations organizations

(UNESCO and UNEP) and of the scientific community (ICSU and IUBS) are expressed.

The following three parts deal with the science of biodiversity in three different dimensions, that of mechanisms, function and change. Part II explores the potential of biodiversity to comprehensively approach the unity of life at different levels of integration from the molecular to the organismic and ecosystem; Part III focuses on the still little-explored role of biodiversity in ecosystem dynamics and functioning, especially the impact of global climate change on biodiversity, and conversely the potential role of biodiversity in global change; and Part IV underlines the foundations for global inventorying and monitoring that are necessary to understand the change of biodiversity over both time and space scales.

Part V on the conservation of biodiversity attempts to integrate the basic scientific concepts of conservation biology and the more practical aspects of conservation viewed as a tool of management. It thus provides a bridge with Parts VI, VII and VIII, which are devoted to the management and the rational use of biodiversity, in a crescendo from the more conservative aspects towards the intensification of its utilization in the terrestrial and aquatic environments, and furthermore in the industrial applications.

The last two parts (IX and X) have a more human dimension by addressing biodiversity in the built-in urban environment and emphasizing the cultural values and ethics linked with the concept of biological diversity.

The Editors would like to express their gratitude and deep appreciation to all the institutions that provided the financial support which have made possible the organization of this large Forum. Particular mention should be made to the TOTAL Foundation, the European Commission (DG XII), the French Government, Japan's Committee for IUBS, UNESCO and ICSU.

The Editors would also like to acknowledge the invaluable personal contribution of Robert Barbault, Chairman of the IUBS Committee in France and Pierre Lasserre of UNESCO, and to thank very warmly Colleen Adam for her commitment, sense of responsibility and efficiency during the organization of the Forum, and Amy Freeman who has been instrumental in the editing of this volume and has translated from French a number of chapters.

Francesco di Castri and Talal Younès

Paris and Montpellier
June 1995

Introduction: Biodiversity, the Emergence of a New Scientific Field – Its Perspectives and Constraints

Francesco di Castri and Talal Younès

Since the publication in 1988 of *Biodiversity*, the book edited by Wilson, there has been an exponential rate of articles on biodiversity (Harper and Hawksworth, 1994). They do not refer necessarily to the same aspects, and there is a real need for a better understanding of what biodiversity is in reality.

Jutro (1993) records 14 recent definitions of biodiversity of those most often used. Two of them – largely quoted – are of a more official nature, since they have been approved by most countries in the context of worldwide negotiations, agreements and strategies.

The more extended one is that of the United Nations included in the Convention on Biological Diversity (UNEP, 1992). According to it biodiversity means: *The variability among living organisms from all sources including, inter alia, terrestrial, marine and other aquatic ecosystems and the ecological complexes of which they are part; this includes diversity within species, between species and of ecosystems.* The shortest definition of all is that of the Global Biodiversity Strategy (WRI, IUCN and UNEP, 1992) which regards biodiversity as *The totality of genes, species, and ecosystems in a region.* Despite that participation of the scientific community the elaboration of such definitions has been far from satisfactory; it is interesting to note that both definitions do indeed refer to the three main components of biodiversity: genes, species and ecosystems. Diversity within species is the *genetic* diversity; between species is the *species* or *taxonomic* or *organismal* diversity; and of ecosystems is the *ecological* or *habitat* diversity.

However, according to Harper and Hawksworth (1994), it was Norse *et al.* (1986), in a little-known publication, who first expanded the traditional usage of biological diversity to refer to the three levels of genetic, species and ecological diversity.

Unfortunately, these definitions pay little attention, if any, to the interactions within, between and among the various levels of biodiversity. And

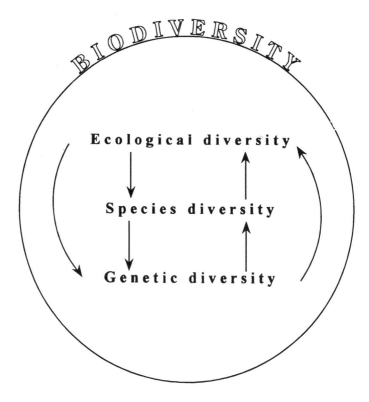

Fig. I.1. A definition of biodiversity based on its components and their interactions.

interaction is the main intrinsic mechanism to shape the characteristics and the functioning of biodiversity. Another shortcoming is that the notion of *scale* seems to have been ignored, while structural and functional attributes of biodiversity can only be determined by a proper consideration of appropriate scales of space and of time.

This is not a simple matter of semantics. Far-reaching consequences depend on resolving this dilemma, i.e. whether biodiversity is a simple umbrella covering a mosaic of heterogeneous activities of genetics, biosystematics and ecology, or it represents an unitary entity shaped by the continuum of all its elements and their interactions.

From its very inception, the *DIVERSITAS* programme has been designed and launched as a multi-level and multi-scale endeavour (di Castri and Younès, 1990; Solbrig, 1991), looking at the interactions and the integration of the different disciplines involved. It intends to provide, therefore, a *leitmotiv* and an unifying principle to all the biological world.

In the light of the above considerations, the simplest operational definition of biodiversity can be formulated as *the ensemble and the interactions of the genetic, the species and the ecological diversity, in a given place and at a given time* (see Fig. I.1; also in di Castri 1995a,b).

It should be stressed that these interactions are of a hierarchical nature.

Fig. I.2. The trilogy of biodiversity presented as a hierarchical zoom.

By interlocking the three diversities, as shown in Fig. I.2 (see also di Castri, 1995a), the classical zooming effect of the hierarchical theory is achieved, so that emerging properties – that is to say, properties that do not exist at a lower level of integration – appear when passing to a higher level, i.e., from the gene to the species up to the ecosystem level, or properties disappearing when moving the opposite way from the ecosystem to the gene level. The emergence or loss of properties of biodiversity depend on the relative position of the three blocks and of the level and intensity of the interactions. Additional information on the hierarchical theory can be found in Allen and Starr (1982), Salthe (1985), Nicolis (1986), O'Neill et al. (1986), di Castri and Hadley (1988), Vrba (1989) and Vrba and Eldredge (1984).

In summary, hierarchy is a central phenomenon of biodiversity, and there needs to be a general theory integrating the hierarchical levels of biodiversity, how they come to be and interact. In this way biodiversity, rather than amalgamating disconnected pieces of scientific research, will become a transdisciplinary scientific field in its own, *the unique trilogy of biodiversity*.

Admittedly, the hierarchy discussed above is not a 'clean' hierarchy *sensu stricto*, since genes, species and ecosystems do not belong all together to the same hierarchical category. This is also implicitly underlined by Harper and

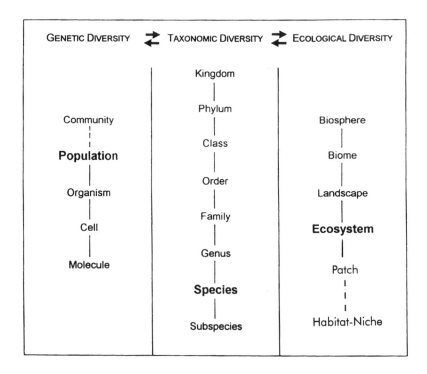

Fig. I.3. Hierarchical patterns and scales of biodiversity (the place of *Community* in the first column and that of *Habitat-Niche* in the third column are admittedly questionable).

Hawksworth (1994). The hierarchical concept is expanded and made more accurate in Fig. I.3 (see also di Castri, 1995a), where the hierarchical patterns of biodiversity are shown as the interactions of three different scales of levels of organization: the genetic, the taxonomic and the ecological. A more sophisticated definition of biodiversity could be, therefore, *the ensemble and the hierarchical interactions of the genetic, taxonomic and ecological scales of organization, at different levels of integration.* Indeed, populations (with their gene pools), species and ecosystems are usually the cornerstone at the intersection of the three scales (Solbrig, 1991). They are also the three main elements considered in conservation biology, and – in practice – they should be viewed all together in their interactions for the conservation of rare species or threatened habitats. Still, this is a too restrictive use of the hierarchical approach.

As regards the genetic scale, for instance, molecular and population genetics should be very closely linked. Concerning the taxonomic scale, an overemphasis on species diversity could lead to biased conclusions. For example, the marine environment is considered to be poorer in species when compared to the terrestrial; however, there are as many as 28 phyla in the marine environment (13 of them being endemic), against only 11 phyla in the terrestrial environment, with just one endemic phylum (Grassle *et al.*, 1991). It must also be noted that not all species are equal when it comes to measuring the biodiversity of a system: a few species can play a keystone role in system functioning, while others may be redundant; some species are dominant and can embrace a very large number of individuals, thus decreasing the equitability of the system, and others, rare species, may be present in a very low number of individuals. Also, a lower number of species can be compensated – to a certain extent – by a very high genetic variability in some populations. Finally, at the ecological scale, the main factor of species extinction may be the fragmentation of landscapes or the disruption of the patch dynamics, and not merely the lack of resilience at the ecosystem level (di Castri, 1995a).

The importance of hierarchical organization of biodiversity, combined with due consideration given to appropriate scales of space and time, cannot be taken as a simple theoretical artefact. From a practical viewpoint, structural and functional attributes of system stability, productivity and sustainability, as well as patterns of ecosystem functioning (di Castri and Younès, 1990), can only be clarified if hierarchies and scales are considered in terms of their interactions. The same applies, in a managerial sense, to the conservation of natural areas as well as to the selection, rotation and 'mixing up' of appropriate crops or forestry stands, or the use of the oceans' biological resources. In addition, the restoration of degraded lands due to extended monoculture or massive deforestation (as is the case in the current crisis of 'economic globalization'), and the redesigning of a more stable and harmonious landscapes should be based on the linkages of the main components of the above-mentioned 'trilogy of biodiversity'.

Perspectives

There are at least six major reasons for which the application of the concept of biodiversity, as understood in this volume and derived from the *DIVERSITAS* programme, can serve as a 'broker' towards a new science, a new development and a new society. Partnership should be the underlying motto and the motor for action.

First of all, biodiversity can become a much needed unifying *leitmotiv* providing a continuum within the biological world. At present, biology is fragmented in a large number of disciplines with little or no connection among them, and too often competing and underestimating each other. In most countries, molecular biology is the most appealing field, while taxonomy is neglected and holds little or no career incentives. Still, both disciplines are equally needed in a close cooperation and interaction if we are to successfully address the problems of biodiversity. The same applies to ecology when it is rightly taken as a scientific discipline with its own logic, and not as an emotional position or a political ideology. Working together towards a common goal, with compatible methods and approaches, can lead to a kind of cultural revolution of the current sectoral mechanisms and institutional patterns, breaking up the existing barriers in research and training in biology. This is not an Utopian dream; as an example of the feasibility of this newly achieved partnership, the CNRS (Centre National de la Recherche Scientifique) of France has decided to establish biodiversity as the core activity of its Department of Life Sciences; the preliminary results already show unexpected openings and a renewed dialogue between previously separated scientists and disciplines. By the way, institutionally speaking, the search for the unity of biology is the very *raison d'être* of an organization such as the International Union of Biological Sciences (IUBS).

Secondly, biodiversity is the backbone for agriculture, animal husbandry and forestry selection, as well as for aquaculture utilization. New varieties have to be constantly discovered, because of a decreased resistance to pests, or in order to reduce the use of pesticides and herbicides. Wild relatives of domestic animal and cultivated plant species are the best basis for breeding new varieties and for biotechnological innovations. Accordingly, a new theme on the conservation of wild ancestors is now operational within the framework of *DIVERSITAS*. In the utilization of the terrestrial and aquatic biological resources, it is particularly important that not only the most visible parts of the ecosystems be studied, but also the frequently neglected biodiversities in the soil and the deep oceans. These hidden environments are very rich in species and play a major role in natural recycling.

Thirdly, the land use and the regional development (*aménagement du territoire*) are now subject to very rapid and dramatic changes because of the globalization of trade and markets. Often this results in a strong intensification of the human impact which sometimes leads to deforestation and desertification, and sometimes to the abandonment of fertile lands, the so called 'human desertification'. In any event, the face of the earth will change in a period of about 20–30 years, even before the likely advent of the climatic global change. The overall biosphere will take on the characteristics and the functioning patterns of an immense ecotone. A major evolutionary responsibility for humankind will be, therefore, the reconstruction of ecosystems and the redesigning of new landscapes including the urban landscape, according to the new rules of the game and the new perturbation regimes. This requires a deep understanding of biodiversity, from the diversity of landscapes down to the diversity within populations.

Fourthly, biodiversity will be fundamental in the new era of industrial applications, during what is sometimes called the 'post-industrial period' in the history of humankind. This new era will be characterized by high international competitiveness, not only noticed through lower prices, but of the remarkable technicality and diversification of the products themselves. It is only by exploring the great potential of the still largely unknown biological world that the entrepreneurs will find new avenues and opportunities for the survival of their own undertakings.

Fifthly, biodiversity is the best tool with which to establish a much-needed bridge within the social and cultural world. The globalization of information and communications, as well as the massive human migrations will be the major features of the next century (migrations from the countryside to the megalopolis, from one to another country, and between continents), and will configure a new picture of the political and social panorama, modifying even more the existing landscapes. Massive migrations imply the integration (and not the exclusion) of new populations through or – more likely – without the assimilation of the different cultural identities. In a similar way, in the biological world there will be an explosion of invasive species (animals, plants, microbes, some of them of a pathogenic nature). Some biological invaders will come from abroad, others will develop *in situ* the capacity to become invaders because of major environmental disruptions. In any event, the evolution of both the biological and the sociocultural relations will be strongly affected by the above-mentioned globalizations (of the economies, of the information, and later on, of the climate), at different hierarchical levels and in their reciprocal interactions (see Fig. I.4, from di Castri 1995c,d). The study of biodiversity cannot be disconnected, therefore, from considering the human dimension. The evolutionary linkages between biological aspects and human culture have already been stressed since 1983 by Dobzhansky *et al.*

Finally, biological diversity is a pillar of human development where a new dialectic synthesis between globalization and diversity is achieved. As said by di Castri (1995e), sustainable development can be metaphorically conceived as a chair with four legs having similar length and strength; these four legs represent the economic, the environmental, the social and the cultural dimensions. Biodiversity taken in a broad sense, is not only instrumental in giving the appropriate weight to the environmental dimension, but also ensures the elements leading to the economic diversification, to the social reconciliation, and to the maintenance of cultural identities.

Constraints

For biodiversity to play an instrumental role in the implementation of the perspectives for action discussed above, a number of constraints should be removed, overcome, or at least clarified.

The current status of taxonomy is perhaps the most difficult constraint to overcome. Only some 1.5 million species are known, and little information is available on the functional attributes of these species. Still, today's conservative estimates of the total number of species range from 5 to 30 million, with other estimates that go even up to 100 million species. In spite of this, there are few new taxonomists in the world because of the lack of career incentives; and for some animal groups there are no specialists at all. Furthermore, the status of several Museums of Natural History and of Herbaria continues to deteriorate year after year, mostly those located in the tropics. One could fairly wonder whether the objective of recognizing all species living in the biosphere is an Utopian one, and whether it would not be more realistic to concentrate on a more comprehensive knowledge of a rather reduced number of taxa for worldwide comparisons of the changes of species diversity in space and time. Admittedly, this would be a very painful and controversial decision

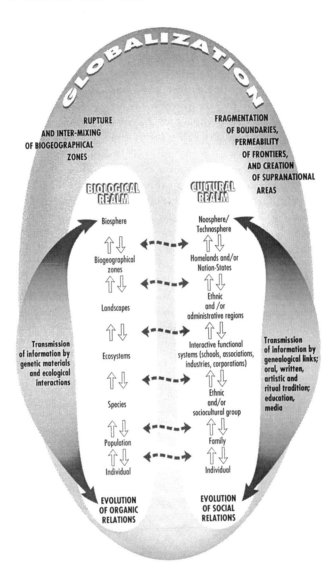

Fig. I.4. Hierarchies with a spatial dimension and a historical trajectory of biological systems (left) and sociocultural systems (right), interacting in a context of globalization. From di Castri (1995c,d).

to be taken. For the time being, only the possibility of estimating species diversity by extrapolation (Harper and Hawksworth, 1994) remains a realistic issue to be envisaged. The importance of establishing at least a small series of sites with an All-Taxon Biodiversity Inventory (ATBI), as proposed by Janzen (1993), could help the process of these extrapolations, but this hypothesis should be rigorously tested and its feasibility discussed in concrete and pragmatic terms.

A second constraint, very much related to the previous one, is that of measuring biodiversity (see Hawksworth, 1994). Genetic diversity is known for certain populations of large species of animals and plants, for several domestic species in view of their breeding, for some laboratory animals and plant species such as *Drosophila*, copepods, mice, *Arabidopsis*, etc. It is also known for a limited number of species that have been used for experimental research on population dynamics and genetics, and for a few species of microbes of medical or agricultural importance. The problem of species diversity has already been discussed above. On the other hand, ecological diversity can be estimated to a certain extent by direct survey, remote sensing or fractal analysis. But the real challenge lies in the possibility of taking into account the emerging properties that appear by the interactions of the three diversities. Progress has been made in understanding the processes of speciation and of species extinction, but there is yet a long way to go before the establishment of the theoretical foundations of biodiversity.

A third constraint is the current dichotomy between the biodiversity agenda and priorities of industrialized countries of the North and developing countries of the South. A reconciliation of national prerogatives with a global interest, based on principles of overall equity, is a *sine qua non* condition for undertaking sensible worldwide actions and endeavours. For instance, inventorying and monitoring of biodiversity are being carried out in numerous countries, sometimes with considerable international funding. However, different taxonomic groups, with different sampling methods and at different intervals of time for monitoring, are usually being chosen by the various countries. It is difficult to see what will be the result of such an exercise, even for those nations currently involved. What is certain is that no useable output will emerge for the understanding of global biodiversity dynamics. Countries should certainly be deeply involved, but only using a pre-designed and previously accepted framework for interchanging results, so that an acceptable level of comparability can be attained (di Castri, 1995a). One of the main goals of the *DIVERSITAS* programme is helping to design such a framework.

A fourth constraint has an educational dimension. A multitude of myths on biodiversity pervades not only the media, but also more and more the textbooks for primary and secondary education. The underlying assumption is that biodiversity is *intrinsically* opposite to economy and development. It is actually poverty which is the main limiting factor for maintaining an acceptable level of biodiversity, at least in the largest part of the world. A booklet prepared by UNESCO (di Castri *et al.*, 1994) mainly addressing planners and decision-makers and two pamphlets designed by TOTAL and IUBS (1994, 1995) mainly addressing the general public are among the examples showing it is possible to achieve a wise popularization of biodiversity, while adhering to scientific foundations and facts. This is an essential task to be faced by UNESCO's environmental education programmes and by the IUBS Commission for Biological Education.

Finally, the last constraint is intrinsic to the very concept of biodiversity, as mentioned above, with the complexity of its hierarchical interactions. Such a complex system can only have a non-linear behaviour with a lower degree of predictability. This does not apply only to biodiversity, but to most of the scientific, managerial and political setting-up in the real world as it is now. Quoting verbatim Holling (1994), *'The future is not just uncertain; it is inherently unpredictable.'* To face this unpredictability, two principles should be adopted: the precautionary principle to keep options open for the future, and the principle of flexibility to maintain an adaptive potential and to choose, in a timely manner, the 'right' way when confronted with the unavoidable and sudden bifurcations and surprises in the years to come. Nothing else is more suitable than biodiversity, with its almost infinite opportunities and interactions, to become the foundation of these two principles.

References

Allen, T. and Starr, T. (1982) *Hierarchy. Perspectives for Ecological Complexity*. University of Chicago Press. Chicago, 310 pp.

di Castri, F. (1995a) Biodiversity. In: *World Science Report*. UNESCO, Paris (in press).

di Castri, F. (1995b) The hierarchical uniqueness of biodiversity. *Biology International, (Special Issue)* 33, 54–57.

di Castri, F. (1995c) Diversité biologique et diversité culturelle. Les fondements de l'universel. In: B. Sitter-Liver (ed.) *Culture within Nature*. Wiese Verlag, Bâle, pp. 395–411.

di Castri, F. (1995d) Facing the globalization of a fractured world. New patterns of international partnership. In: Hadley, M. (ed.) *Integrating Conservation, Development and Research: Scientific Responses to the Environment-Development Challenge*. UNESCO, Paris and Parthenon Publ., Carnforth (in press).

di Castri, F. (1995e) Una silla de cuatro patas. In: *Las Ultimas Noticias*. Santiago, Chile, 29 April 1995.

di Castri, F. and Hadley, M. (1988) Enhancing the credibility of ecology: interacting along and across hierarchical scales. *GeoJournal* 17(1), 5–35.

di Castri, F. and Younès, T. (eds) (1990) Ecosystem function of biological diversity. *Biology International (Special Issue)* 22, 1–20.

di Castri, F., Lasserre, P., Robertson-Vernhes, J., Dogse, P. and Childe, F. (eds) (1994) *Biodiversity. Science, Conservation and Sustainable Use*. Environment and Development Briefs, No. 7, UNESCO, Paris (also available in French and Spanish).

Dobzhansky, T., Boesinger, E. and Wallace, B. (1983) *Human Culture. A Moment in Evolution*. Columbia University Press, New York.

Grassle, J.F., Lasserre, P., McIntyre, A.D. and Ray, G.C. (1991) Marine biodiversity and ecosystem function. *Biology International (Special Issue)* 23, 1–19.

Harper, J.L. and Hawksworth, D.L. (1994) Biodiversity: measurement and estimation. Preface. *Philosophical Transactions of the Royal Society, London* B 345, 5–12.

Hawksworth, D.L. (ed.) (1994) Biodiversity: measurement and estimation. *Philosophical Transactions of the Royal Society, London* B 345, 1–136.

Holling, C.S. (1994) An ecologist view of the Malthusian conflict. In: Lindahl-Kiessling, K. and Landberg, H. (eds) *Population, Economic Development, and the Environment*. Oxford University Press, New York, pp. 79–103.

Janzen, D.H. (1993) Taxonomy: universal and essential infrastructure for development and management of tropical wildland biodiversity. In: Sandland, O.T. and Schei, P.J. (eds) *Proceedings of the Norway/UNEP Expert Conference on Biodiversity, Trondheim, Norway.* NINA, Oslo, pp. 100–112.

Jutro, P.R. (1993) Human influence on ecosystems: dealing with biodiversity. In: McDonnel, M.J. and Pickett S.T.A. (eds) *Humans as Components of Ecosystems.* Springer-Verlag, New York, pp. 246–256.

Nicolis, J.S. (1986) *Dynamics of Hierarchical Systems. An Evolutionary Approach.* Springer-Verlag, Berlin, 397 pp.

Norse, E.A., Rosenbaum, K.L., Wilcove, D.S., Wilcox, B.A., Romme, W.H., Johnston, D.W. and Stout, M.L. (1986) *Conserving Biological Diversity in Our National Forests.* The Wilderness Society, Washington, DC.

O'Neill, R.V., DeAngelis, D.L., Waide, J.B. and Allen, T.F. (1986) *A Hierarchical Concept of Ecosystems.* Princeton University Press, Princeton, 253 pp.

Salthe, S.N. (1985) *Evolving Hierarchical Systems.* Columbia University Press, New York, 343 pp.

Solbrig, O.T. (ed.) (1991) *From Genes to Ecosystems: A Research Agenda for Biodiversity.* IUBS, Paris, 123 pp.

TOTAL-IUBS-UNESCO (1994) *Diversity is Life.* Paris (available also in French)

TOTAL-IUBS (1995) *Cultivons la diversité.* Paris.

UNEP (1992) *Convention on Biological Diversity, June 1992.* United Nations Environment Programme, Nairobi.

Vrba, E.S. (1989) What are the biotic hierarchies of integration and linkage? In: Wake, D.B. and Roth, G. (eds) *Complex Organismal Functions: Integration and Evolution in Vertebrates.* John Wiley & Sons, Dahlem Konferenzen, pp. 379–401.

Vrba, E.S. and Eldredge, N. (1984) Individuals, hierarchies and processes: towards a more complete evolutionary theory. *Paleobiology* 10(2), 146–171.

Wilson, E.O. (ed.) (1988) *Biodiversity.* National Academy Press, Washington, DC.

WRI, IUCN and UNEP (1992) *Global Biodiversity Strategy: Guidelines for Action to Save, Study, and Use Earth's Biotic Wealth Sustainably and Equitably.* World Resources Institute Publications, Baltimore.

Opening Address of UNESCO

Federico Mayor, Director General, United Nations
Educational, Scientific and Cultural Organization

'Life is endless forms', wrote Charles Darwin as he admired the diversity of nature in the Galapagos Islands. He could not have foreseen that this statement on phenotypes would also apply to genetics, physiology and pathology. Endless forms, continuously evolving, with constant transformations and mutations. This formulation would not have surprised Heraclitus, who saw each being as separate and unique at each instant of its existence – endless forms which, through mechanisms of spatial complementarity, meet, unite, and separate, like lock and key, in interactions that reflect the equilibrium we equate with health or the disequilibrium associated with sickness. Endless forms at every level – genetic, cellular and somatic. For human beings, the eyes of the universe, the only creatures who know that they know each individual's uniqueness – reflecting a dynamic and endless diversity – is a corollary of his distinctive freedom. Biodiversity is therefore not only a phenomenon recognized by human beings as occurring outside their own species, but also a biological and spiritual requirement for their own existence as a species. Biodiversity is therefore a key element in the constellation of subjects that vie for our attention today.

It is for me a great pleasure to welcome you today to UNESCO House for the opening of this Forum on Biodiversity, Science and Development: Towards a New Partnership, organized by the International Union of Biological Sciences on the occasion of its 25th General Assembly with the support of the French authorities, the International Council of Scientific Unions and UNESCO.

This Forum is timely, coming at a moment when the debate on biodiversity is at the heart of general concern about environmental issues and is giving rise to much controversy. The very concept of biodiversity is so vast that misunderstandings are frequent. In view of alarming signs of the erosion of biological diversity, of the need to counter it by increased efforts to promote conservation and the sustainable use of resources, and of the economic interests involved in its exploitation, a Convention on biological diversity, was, as you know, drawn up under the auspices of the United Nations Environment Programme and signed at Rio; it came into force with great rapidity and the first Conference of Parties to the Convention will take place this very year in the Bahamas. UNESCO, which was closely associated with the preparation of the Convention, is involved along with UNEP and FAO in the work of the interim secretariat and will help to implement the instrument.

Given what is at stake, it is important at this stage to take stock of our knowledge, to confront frequently diverging viewpoints and to approach this complex issue from different angles in order to facilitate dialogue with decision-makers and give practical application to the findings of science. Your Forum, for example, will be dealing with topics that are crucial to our understanding of biodiversity: cellular and molecular issues related to the concept of the unity and diversity of life, the role of biodiversity in the operation of ecosystems, inventories and long-term surveillance, means of conservation, including conservation in our own everyday urban environment, the use of resources for agriculture, aquaculture and industry, and the cultural and moral dimensions of biodiversity.

The fact that the world of science has responded positively to this initiative shows that it has been understood and welcomed: your presence here is the surest guarantee that the Forum will attain its objectives both in the refining of accumulated knowledge and in addressing the omissions and questions that still remain.

UNESCO is associated with this Forum because the Forum's purpose is in tune with UNESCO's own mission, namely, to further knowledge, facilitate its exchange, promote new partnerships and open the way for science to come up with practical solutions to the problems facing Member States and their populations. Indeed, the science/society interface has always been at the centre of our concerns. Ever since its foundation and the initial impetus given by its first Director General, the biologist Julian Huxley, UNESCO has striven to persuade human societies to adopt a more responsible attitude towards nature and its resources by taking all sorts of initiatives in areas as diverse as geology, soils, water, the oceans and terrestrial ecosystems.

As early as 1948, at Fontainebleau, and with the French authorities, it presided over the founding of the International Union for the Protection of Nature, now IUCN, which was the first non-governmental organization in this field and with which we maintain close ties. In the early 1960s, ICSU, with UNESCO's support, launched an ambitious programme of ecological research, the International Biosphere Programme (IBP), which set out to study the structure and operation of ecosystems and produced a remarkable synthesis of current knowledge. At the same time, IBP revealed the powerful impact of human activities on the functioning of ecosystems. But man and nature were still regarded too much as two separate entities; their mutual relations would need to be clarified before they could be improved. It was with the programme on Man and the Biosphere (MAB), launched by UNESCO after the Conference on the Biosphere in September 1968, that human beings came to be regarded as an integral part of ecosystems. Admittedly, we must seek to protect all species, but first of all the human species in its infinite diversity, that quirk of nature that is 'self-conscious', a maverick, capable of saying 'no', of asserting its independence of biochemical logic and modifying its environment.

At the time the recognition of man's place in nature gradually led to a new approach to the conservation of biodiversity. Instead of keeping local people out of national parks the aim was to involve them in the conservation of the ecosystems in which they lived and from which they drew their livelihood, thus showing that sustainable exploitation and conservation were not contradictory. The biosphere reserves, a practical application of this approach, for example, sought to involve locally active persons in the protection of biodiversity, to promote traditional knowledge and applications and to take account of the cultural and economic interests of the people directly concerned. For this reason they constitute a testing ground for new forms of conservation and they are becoming more and more popular as a very modern approach to nature and sustainable development. The UNESCO Conference to be held in Seville next March, on the invitation of the Spanish authorities, will discuss these new trends. At the same time, the international network constituted by the biosphere reserves within the MAB framework will be strengthened and reorganized on a more formal basis.

Today more than ever the scientific community is being enjoined to represent the interests of populations before decision-makers in its areas of competence. This is why UNESCO is seeking closer ties throughout the world with the scientific community, represented in particular by the International Council of Scientific Unions and its various member unions, especially the International Union of Biological Sciences and the International Union of Microbiological Societies, as well as its scientific committees such as SCOPE (Scientific Committee on Problems of the Environment).

The *DIVERSITAS* programme, with which most of you are familiar, launched jointly by IUBS, SCOPE and UNESCO, illustrates this concern. Its purpose is to combine activities in research, inventory work and surveillance with training and awareness-raising. Regarded by ICSU as one of its most important actions in environmental matters, it is backed by numerous National Committees. Allow me here to pay tribute to the active role played by the French Committee for this programme. I have just referred to inventory work. Would it not be appropriate to conduct an inventory of the planet at the end of this millennium? By bringing together all existing information on all continents we could – indeed should – take stock of our heritage.

It is the duty of all of us to foster international interchange in science and to help UNESCO to facilitate the dissemination of knowledge and access to data, especially for developing countries, and to strengthen research and training institutions. In this connection, UNESCO is particularly anxious to encourage the creation and operation of networks like the MIRCENs (Microbial Resources Centres), biosphere reserves and networks of marine research stations. The research institutes and universities have realized that their own development depends on more effective communication between researchers and teachers from North and South. In this respect computerized networks are turning out to be particularly useful, mainly through the growing coverage of international networks such as Internet, Earn/Bitnet, Rio, etc., and UNESCO is making its contribution.

At present it is also encouraging, under the UNITWIN programme for the twinning of universities in countries of the North and the South, the formation of a network of chairs in ecotechnology. This is because we want more training courses which bring together in close association the most up-to-date information in ecology, economics, the social sciences and technology. These courses are essential not just for decision-makers, who will have to deal with increasingly complex environmental situations, but also to build up pools of experts to meet increasing needs at all levels. At the municipal level in particular, consideration must be given – if we want to tackle environmental problems without further delay – to setting up technical services capable of maintaining water quality, controlling the emission of toxic substances, analysing soils to find out what fertilizers and artificial substances are needed and developing 'outcome–income' industries, etc. These 'eco-jobs' are going to increase in number and we should start preparing the ground now.

In a joint declaration of the chairpersons of our major intergovernmental progammes, biodiversity was treated as a priority field for co-operation between MAB, IOC, COMAR and IHP, particularly in regard to coastal areas and islands.

Yet another paradox – and we have created many such paradoxes in the last few decades – at the very moment when phytogenetic engineering is expanding and cultivation techniques are steadily becoming more sophisticated, the intensive exploitation of the environment and crossing techniques, especially for plants, are impoverishing the gene pool and thus reducing biological diversity. In this context more attention should be paid to the question of property rights and to the social and moral aspects of exploiting biodiversity for commercial and industrial purposes. Here is one field of action in which UNESCO has a duty to work with the other relevant organizations, for example in the implementation of the Convention on Biodiversity.

As I have just said, biodiversity is a vast and complex concept. Its ramifications extend into all spheres of human life and activity. We must take care not to forget, for example, the genetic diversity of the human species. To meet its mortal responsibilities UNESCO has, as you know, recently set up a programme on the human genome.

In this connection, I am delighted that your Forum is to look into the cultural and moral aspects of biodiversity. This concept refers to the complex relations between human beings and nature. Each society has its own response to nature, a sensitivity that is moulded by religion, history, climate, topography and so on. Such representations of the relationship between human beings and nature have been dangerously neglected in the economic models based primarily on Western culture. There lies a particularly important task, the task of rehabilitating and making known the whole range of views of the man-made relationship. The discussions of the forthcoming symposium on 'Science outside the West in the twentieth century', to be organized very shortly by ORSTOM with UNESCO support, will certainly be of great interest in this respect.

As you know, another of UNESCO's roles is to maintain and protect the cultural diversity that constitutes the wealth of humankind and which is threatened by many aspects of the ongoing 'globalization' process. By uniting culture and nature in a single objective, the 1972 Convention on the World Cultural and Natural Heritage makes possible an integrated approach to these two strands of our existence, namely, the creative being and his natural surroundings.

The cultural aspects are closely bound up with moral concepts and the notion of equity, which you are also going to address in your discussions. Full recognition of the role played or playable by the 'users' of biodiversity calls for the development of new partnerships with those users and hence for efforts to raise their awareness and keep them informed. The recognition that biodiversity is essential to human survival implies that man has a responsibility towards future generations. As I remarked when we celebrated the 25th anniversary of the Conference on the Biosphere, 'the population of three quarters of the world is still expanding rapidly while the remaining quarter continues to devour raw materials, space and energy'. How are we to strike a balance between resources and consumption, how are we to harmonize population growth with natural assets and their exploitation, production and consumption with due respect for the distinctiveness of local cultures and with a concern to achieve global equilibrium; that is the real challenge as this century draws to its end. It is also the real subject of the United Nations Conference on Population and Development, which opens today in Cairo. All the inhabitants of the earth must have access to knowledge. All must be educated, that is to say capable of choosing for themselves and of shaping their own future without any outside interference or foreign models. To be free is to be able to decide for oneself. Anything else is merely another form of colonialism.

To conclude, I would remind you that scientific progammes need, more than they have ever needed before, a humanist dimension. The reserach communities that launch them are rather cut off from the rest of society. Students and young researchers increasingly have the most advanced findings of their science at their fingertips. They must be encouraged to have an inquiring mind, to open out to others, and to listen to others. As Saint Bonaventure said, 'Reason will continue to be useful as long as it keeps sight of areas where it is of no use at all'. In areas involving the scientific study of the environment, specialists in the fundamental sciences, engineers, sociologists and economists must be brought into contact with philosophers, writers, linguists and jurists. We are living at an epistemological turning point: we have an extreme need for bridges to link the human with social sciences and the natural and life sciences. Biodiversity is perhaps one of the best examples of the obligatory nature of the multidisciplinary approach today. The supreme function of scientists is to recognize the various dimensions that converge within their own field of study. None of these dimensions should be excluded: to simplify complexity is a sign of irresponsibility and lack of judgement, for it is impossible to master reality without seeing it as a whole, without knowing it through and through. This Forum will do much to improve our knowledge of reality in its complexity and infinite diversity, and thus give us a clearer view of the paths of our common future.

Opening Address of the Government of France

Michel Barnier, Minister of the Environment

I am very happy to be able to participate in this International Forum on Biodiversity. My schedule did not permit me to open the Forum this morning, however I felt it important to be here among you on the opening day of your discussions.

This Forum is taking place at a particularly opportune time: in less than three months the First Conference of the Parties of the Biodiversity Convention will be held in Nassau and your work will provide it with indispensable scientific material. And today, the International Conference on Population and Development opens in Cairo, reminding us that Man and his development are at the heart of our concerns, and that the greatest challenge facing us is to understand and establish an equilibrium between Man and Nature.

From the Invention of the Concept to the Onset of the Biodiversity Convention

Just ten years ago, the word 'biodiversity' did not make up part of our vocabulary. Scientists, who developed this notion, had a particularly pertinent and ambitious vision. The importance that this concept yields today is proof of their insight. In less than a decade, it has resulted in a worldwide Convention on Biodiversity, the first in history.

The idea for this Convention was initiated jointly by the International Union for the Conservation of Nature (IUCN) and ecological and conservationist communities in countries of the North. The Convention's objectives stemmed from concerns related to the acceleration of biodiversity loss, namely in tropical regions. As negotiations were taking place between 1988 and 1992, the magnitude of economic and political stakes surrounding the conservation debate caused its scope to progressively widen. As regards the 'environment-development' perspective highlighted at the Rio Conference, concerns began to extend past species and ecosystem conservation and towards issues of access to genetic resources and transfer of technology.

In the midst of these debates, developing countries raised questions concerning the advantages to be gained primarily by developed countries through selection of cultivated or domestic species and developments in biotechnology, notably in the field of genetic engineering. It is in the poorest countries that the constraints of conservation would weigh most heavily. Developed nations have accepted this line of thinking and generally agree that conservation efforts should go hand in hand with financial aid in order to share costs. Benefits gained from exploiting resources should also be shared by transferring information and technology.

Last Spring I presented a Bill to the Parliament proposing the ratification of the Convention on Biodiversity signed on 13 June 1992 in Rio de Janeiro by the President of France. The Parliament authorized ratification by a unanimous vote from both Chambers on 10 June 1994, and our ratification was sent to the United Nations in New York in August 1994. France will therefore be able to participate fully as a Member of the Convention in the first Conference of the Parties in November 1994.

However, France has never been completely satisfied with the United Nations Biodiversity Convention. In fact, the evening of the adoption of the Final Act in Nairobi, 22 May 1992, France abstained from signing the text to demonstrate its disapproval of the text's unambitious nature. Available scientific grounding was in fact largely underestimated and the global approach to conservation advocated by France was not accepted. It is unfortunate that the Convention does not go farther in terms of conservation, for example by setting up worldwide lists like those established for UNESCO's World Heritage Programme, the Washington Convention or the RAMSAR Convention.

At the same time, the Convention on Biodiversity does leave significant room for national strategies. In the text there is certain flexibility for each country to define the components of biodiversity and its scientific monitoring, to establish conservation programmes and actions to be taken, and to set up protected zones. Furthermore, the text recommends taking into consideration the concept of sustainable use of resources and encourages research, training and education of the general public. This represents considerable progress in realizing that we have to do more, and do it better, in order to conserve our immense treasure of biological life that is in danger of being diminished forever.

France therefore chose to become a Party of the Convention, despite the shortcomings of the text, so that it may participate as an active member and not merely as an observer. This way, France may have a real influence on the application of the Convention, and continue to work towards a global approach to biodiversity.

Numerous International Scientific Programmes, Supported Notably by the GEF

The International Union of Biological Sciences and the French Academy of Sciences should be congratulated on their organization of this ambitious and timely Forum. It is particularly encouraging that it is being held in France, in Paris, and that it is founded on the essential need to establish valid scientific grounding prior to debating issues of biodiversity.

In this respect, it should not be forgotten that France gives particular attention to the *DIVERSITAS* programme, launched by the International Union of Biological Sciences (IUBS) in collaboration with UNESCO and SCOPE (Scientific Committee on Problems of the Environment). The same attention shall be given to all international initiatives aimed at increasing scientific knowledge of biodiversity.

The *DIVERSITAS* Programme has three main points of interest: the origin, maintenance, and loss of biodiversity; the role of biodiversity in ecosystem functioning; and inventorying and monitoring of biodiversity. The French contribution to the international *DIVERSITAS* programme, which the Ministry of Environment participates in significantly (5.3 million francs in 1994), is quite ambitious because it aims to encourage biodiversity research in a number of different ways. At the Ministry of the Environment, we place a priority on the need to find the determining factors influencing biodiversity and biodiversity change. In addition, there needs to be an evaluation of the role that human interventions play and the consequences they have on biodiversity and ecosystem functioning at local and regional levels. In short, we must attempt to understand the importance of biodiversity for human societies, whether it be ecological, economic, ethical, or cultural. One of the originalities of the French approach to biodiversity is this attempt to mobilize the social sciences as much as the biological sciences. I would like in this respect to congratulate Robert Barbault who is in charge of this programme and to give him my encouragement. We are hoping to learn a lot from his work as well as that of his team. He has told me to what extent French scientists are interested in the proposed 'Systematic Agenda 2000' programme, which consists of taking an inventory of all living species on our planet. I welcome this scientific initiative and hope that it will benefit from the support of international organizations.

We also look forward with great interest to the results of other international programmes in progress, such as the Global Biodiversity Assessment. This project, executed by the United Nations Environment Programme (UNEP) and financed by the Global Environment Facility (GEF), was funded by France and Germany. France continues to make substantial contributions to GEF's actions.

In its pilot phase, 43% of GEF's total budget was consecrated to biodiversity projects. Negotiations for reconstituting its resources were successfully completed in Geneva last March and the programme was allotted $2 billion for a four-year period. The French contribution (807 million francs, or 7.3% of total contributions) is coupled with a parallel bilateral contribution of 440 million francs. In other words, France attaches considerable importance to global environmental problems, particularly those related to biodiversity.

France's Long-term Involvement in Biodiversity Conservation Efforts

It should be kept in mind that in Western European countries, concerns regarding biodiversity go back many years, long before the invention of the concept of biodiversity. At the international level, this is demonstrated by UNESCO's excellent Convention on World Heritage – this first of its kind. At the national level, in France, we can look to examples such as the law of 1930 on natural sites, landscapes, and monuments, complemented in 1957 by regulations allowing the creating of the first natural reserves. In addition, we can recall the law of 1960 concerning national parks and the law of 10 July 1976, the keystone of France's nature protection system.

As far as conservation of biodiversity is concerned, the United Nations Biodiversity Convention does not pose France any additional legal constraints, whether it be in relation to its national nature protection policy, its international activities resulting from its more specific conventions, or its involvement as a member of the European Union.

Within the European framework, France is interested in strengthening scientific studies, based on inventorying natural zones with an ecological interest, in order to designate sites which will become part of a European Ecological Network of special conservation zones, called 'NATURA 2000'. In fact, out of the 205 types of natural habitats to be protected that have already been inventoried, 140 of them are in France. Taking into account its overseas departments, territories and its austral lands, France has almost an entire sampling of the ecosystem diversity found on the planet.

In 1978, the French Ministry of Environment created jointly with the Museum of Natural History a national data bank of wild species and natural environments, called the Flora and Fauna Secretariat. The Bureau of Genetic Resources, created in 1983 to coordinate France's efforts in this domain, is housed in the Museum as well. Furthermore, France is a candidate for welcoming a new Nature Centre, a project of the recently established European Environmental Agency.

Finally, France has made significant efforts at the international level to protect endangered species such as the African elephant and rhinoceros, as well as whales. In Mexico last May, the International Whaling Commission accepted France's initiative for creating a sanctuary for marine mammals around the Antarctic.

Ethical and Cultural Dimensions

I am particularly interested in this Forum's initiative to include in the topics to be addressed a discussion of ethical and cultural aspects of biodiversity. In reality, to justify nature conservation actions, we often rely on utilitarian justifications such as 'wild species need to be saved because they might be useful if one day we discover such and such a molecule . . .'

In this respect, let me say that I find particularly unfortunate the tendency today to try to claim ownership of certain elements of biodiversity through patents (from genes to complete living organisms). Economists have even tried to find ways to give a monetary value to wild species and natural environments in an attempt to justify their conservation. This is undoubtedly an important, although limited approach.

I agree with those who believe that wild species and natural environments are part of our heritage. Heritage, by definition, is what we have inherited and what we will pass on to future generations. It is not simply economic in nature, but cultural as well. Today, Man has the power to choose whether to diminish this heritage by using it in an unsustainable fashion, or to make use of it like a 'good father' – as an old expression from French law would say. There is a moral dimension here which I feel we must not lose sight of.

I would like to sincerely thank UNESCO for hosting this International Forum which has brought together the best scientists in the field, and I would especially like to thank the organizers for having conceived of the Forum keeping in mind veritable dialogue and openness. This Forum is the kind of initiative which will give other countries the momentum necessary to overcome shortcomings of the Biodiversity Convention, and to succeed in truly protecting humanity's common heritage: the treasure of all living beings.

Opening Address of ICSU

Harold A. Mooney, Vice President,
International Council of Scientific Unions

This international forum on 'biodiversity - science and development' is timely and extraordinary for a number of reasons.

It has brought together an impressive star-filled cast to address a major environmental issue in an unusually comprehensive context and in a truly unique interactive manner.

The timing is right in that the nations of the world that have signed the Biodiversity Convention are now beginning to put into actions the historic and bold provisions of the Convention. However, the scientific basis for many of the Convention requirements are not yet established. This symposium can aid in this important process.

This symposium is being held in association with the General Assembly of the International Union of Biological Sciences (IUBS). IUBS has put such importance on this symposium that they have devoted virtually the entire assembly period to this event, giving an exemplary ratio of science to business, giving substance to the statement that this is not a time for business as usual.

This event is an important and complex undertaking that has taken vision and unusual dedication as has been given by the President of IUBS, Francesco di Castri, the Executive Director, Talal Younès, and Colleen Adam, the Executive Assistant.

The Grand Issue of Our Time

The reduction of the Earth's carrying capacity is the central concern of all of us. We are seeing ample evidence of this reduction in what has been termed global change in its broadest context, that is land degradation (depletion of soil stocks, soil toxification, etc.), biotic change (biotic impoverishment and breakdown of biogeographic barriers), atmospheric change (ozone increase in the troposphere and decrease in the stratosphere, etc.), and the inevitability of climate change.

The driving force of all these changes is human population growth and non-sustainable resource use. It is these topics that were the central focus of two UN Conferences, the 1992 Rio Conference on sustainable use and the 1994 Cairo Conference on population. At this meeting we will focus on understanding and managing biodiversity-key elements in the sustainable use of this poorly understood keystone to the earth's functioning and human well-being.

What is ICSU?

1. ICSU represents 80 of the world's scientific academies or research councils, with 50 more nations involved by national representation in ICSU's Scientific Unions. Many of these nations are signatories to the Biodiversity Convention and hence are committed to assessment and protection of their biotic resources.

2. ICSU is composed of 20 scientific unions ranging from the history and philosophy of science (IUHPS) to astronomy (IAU) and to numerous unions representing the biological sciences:

- IUBS, International Union of Biological Sciences
- IUBMB, International Union of Biochemistry and Molecular Biology
- IUPAB, International Union of Pure and Applied Biophysics
- IUIS, International Union of Immunological Societies,
- IUMS, International Union of Microbiological Societies
- IUNS, International Union of Nutritional Sciences
- IUPHAR, International Union of Pharmacology
- IUPS, International Union of Physiological Sciences

A number of these unions have direct interest in biodiversity including, foremost, IUBS but also IUBMB and IUMS.

3. ICSU also supports 19 bodies that are interdisciplinary. Many of these are concerned with biodiversity including:

SCOPE, Scientific Committee on Problems of the Environment
IGBP, International Geosphere-Biosphere Progam
SCOR, Scientific Committee on Oceanic Research
COGENE, Committee on Genetic Experimentation
CASAFA, Application of Science to Agriculture, Forestry, and Aquaculture

Programmes with explict responsibilities in biodiversity are: *DIVERSITAS* a coalition of programmes of IUBS, SCOPE, UNESCO. In addition, in the International Geosphere Biosphere Programme there is a research element within the core project on Global Change and Terrestrial Ecosystems that explicitly addresses ecological complexity.

All of these efforts are small in the face of the great need for solid scientific information on the nature, status, and management of biodiversity. Fortunately, there are many other groups, in addition to those named above, such as the IUCN as well as environmental organizations that have biodiversity as a top priority. The efforts of all of these programmes are small in face of the needs of what is rightly called the biodiversity crisis – the irrevocable loss of the Earth's biotic capital. This Forum can help mobilize the scientific community and partners in business and government to bring new understanding, but more importantly new means to conserve and sustainably utilize the Earth's biotic richness.

Opening Address of IUBS

Francesco di Castri, President International Union of Biological Sciences

I would like, above all, to thank very sincerely the organizations which have given their sponsorship to this Forum on Biodiversity, namely UNESCO, the French Government and ICSU.

I would also like to express my appreciation to the various institutions that have given financial support to the International Union of Biological Sciences in order to successfully carry out this endeavour, particularly, the European Commission, the TOTAL Foundation, and numerous French institutions such as CNRS, CIRAD, INRA, ORSTOM, IFREMER and the Ministries of Environment and of Research.

All of my gratitude equally goes to those who accepted to participate in the Forum, and to the delegates of the General Assembly of the International Union of Biological Sciences, to which this Forum is closely associated.

The French Academy of Sciences and the French Committee of the International Union of Biological Sciences deserve special mention, because they have been our closest and friendliest partners throughout the organization of these activities.

I believe that in contemporary science there is no concept as fundamental, yet at the same time as controversial and as little understood, as the concept of biodiversity.

It is a fundamental concept, because biodiversity is life, it is the variety of living organisms at different levels of integration. The diversity of genes which make up the germplasm of a population (genetic diversity); the diversity of species, which all together shape a given ecosystem (specific diversity); and the diversity of ecosystems, which in their fragmentation or interpenetration constitute a regional landscape (ecological diversity). The new concept of biodiversity emerges precisely from the interlocking of and the interactions among these three diversities.

Biodiversity is thus the leitmotif unifying all of the biological sciences, from the molecular level to individuals, species, ecosystems and the entire biosphere. In connection with the similar concept of complexity, a bridge is formed with the physical, chemical and mathematical sciences as well.

Biodiversity gives us a framework for managing renewable natural resources, whether it is in view of conserving biodiversity or putting terrestrial (agriculture, forests, breeding) or aquatic resources to use. In our daily activities, in our daily lives, we are in contact with and choose different levels of biodiversity as regards our food, leisure activities, work and conviviality. There are, therefore, many connections between biological diversity and cultural diversity, hence the diverse ways of life and value systems that govern human societies.

In this regard, and to begin right away by putting an end to one of the myths pertaining to biodiversity, it is not true that man is implacably destructive and intrinsically in opposition with biodiversity. Quite the contrary, with the early emergence of agriculture and the diversification of landscapes surrounding us (which are all almost entirely human constructs), man has laid the groundwork for greater genetic and ecological diversification. Over time, man has become the dominant evolutionary factor and has acquired a kind of evolutionary responsibility of which – admittedly – he is not always aware.

I said earlier that biodiversity is one of the concepts we know the least about. Nevertheless, we have an obligation to defend the biological unknown around us, even the genetic unknown that we carry in us, and which should facilitate partnership with other species.

Why this unknown? Partly, because we have identified only a minute number of existing species (i.e. 1.5 million out of 10, 30 or even 50–100 million species), and we know even less about the genetic potential they carry within them. We know only with absurd imprecision the current extinction rate. Most of the figures estimated in the press, even specialized journals (for instance that 50,000 species disappear each year), are founded on incomplete, often highly biased, data.

We also do not know what role biodiversity plays in ecosystem functioning, or the redundancy level or thresholds for diversity beyond which irreversible changes in a system's structure or even its collapse would occur.

The enormity of this unknown which makes up our most precious asset must, of course, be reduced through scientific research. Is this possible, however, considering the moderate appreciation and support – and I use here a euphemism – given to the science which is instrumental for appreciating this diversity, that is to say, plant, animal and particularly microbial taxonomy? This is precisely one of the matters which will be addressed during the discussions throughout the Forum.

I also mentioned, finally, that no other concept is as controversial as biodiversity – whether our impressions of it have come from media presentations or through our own diverse perceptions.

Biodiversity is on a razor's edge between emotions and scientific data. It lies at the border between fixist preservation, which seems to encourage life-saving measures for any single species, and dynamic conservation which aims to give all species the possibility of following their evolutionary path, which intrinsically includes notions of change and extinction.

As an ecologist, I also have to admit that biodiversity is the demarcation line between scientific ecology and deep ecology, the latter being rooted more in the dogma of ideology than in experimental approach.

There are two other controversies which can lead to a true confrontation. One occurs in all countries. The real battle to conserve biodiversity should be carried out on the entire territory of a given country, and not just in limited areas designated for preservation (national parks, reserves). Concentrating solely on these areas and on the conservation of species would be to lose the battle before it started. Rational management of areas and habitats (whether protected or not) is the only way to ensure that biodiversity can take on its true dimensions.

The other controversy, which has already been the topic of endless debates in the United Nations and elsewhere, is the concept of equity in the distribution and use of biodiversity between countries in the North and South, especially in tropical countries. Equality in the distribution of biodiversity has no meaning from a biological and biogeographical standpoint. Nevertheless, equitable use of biodiversity by the various countries is an indispensable concept, but one that still has a way to go before its wide-ranging consequences are accepted.

A Convention on Biodiversity was approved at the Rio Summit in 1992. This Convention represents the best possible common denominator, in view of the different approaches of the North and South and the gaps in our knowledge which I mentioned earlier. It is commendable that the Convention was launched, even in conditions that were not, and probably will never be, very favourable. However, the weight of unknown elements will have to progressively lessen, and the controversies be clarified, if this Convention is to have a chance to function with all the impact necessary. I think that this Forum can also introduce some elements in this sense.

Now, some words concerning the role of the Forum and the *DIVERSITAS* Programme, from which this Forum emanated.

It is, above all, a matter of clearly highlighting three key scientific points: scales and interactions of biological diversity, the role of biodiversity in ecosystem functioning, and the ways and means of inventorying biodiversity in space and over time.

We should also try to grasp the meaning of biodiversity, in view of conserving it, but also because of its fundamental role for using terrestrial and aquatic ecosystems, and everything else it can offer in terms of industry and biotechnology.

There are two final aspects, often overlooked, which we will address directly: biodiversity in anthropic environments – urban and peri-urban areas – and the links between biological and cultural diversity, including ethical, legislative, and social aspects.

We know very well that the International Union of Biological Sciences, although its primary objective is to study the diversity of life, cannot take on this task alone. Through the *DIVERSITAS* programme, partnership has been established with UNESCO and SCOPE. This partnership, however, must be extended to include not only scientists, but also users – from farmers, to fishermen and industrialists, decision-makers and civil society. This is why this Forum is so open, with an emphasis on discussion of problems, rather than on long individual presentations.

It is my great hope that these discussions show the way towards new kinds of operational partnership, and we will be very interested to work on this with all the various actors concerned with biodiversity.

The range of experiences and approaches which have come together here for this Forum, with an exceptional participation of the scientific community, including that of the General Assembly of the International Union of Biological Sciences, makes me very optimistic about what will come out of this Forum. I see it more as a starting point than as an end in and of itself.

Undoubtedly, some aspects concerning biodiversity will be demystified, but through this demythification, biodiversity's importance can only become greater, ready to take on its vital dimension for all of us.

Bridge-building for Biodiversity

Reuben J. Olembo, Deputy Executive Director, United Nations Environment Programme

On behalf of the Executive Director of UNEP, Ms Elizabeth Dowdeswell, I wish to express UNEP's appreciation and gratitude for the opportunity to address this Forum of distinguished scientists assembled here today. I would also like to convey to you Ms Dowdeswell's greetings and best wishes for a successful forum. Her absence today with you should not in any way suggest a diminished sense of commitment to the scientific community on her part or that of UNEP.

On behalf of UNEP, I wish to reaffirm our commitment and our strong desire to continue our already well-established partnerships in seeking scientific underpinnings to our activities in support of the worldwide quest for sustainable development.

Achieving sustainable development and clearly delineating the role of biodiversity therein will not be easy. Intrinsic, ethical and religious values long espoused by conservationists must now be complemented with direct economic benefits now demanded by developers and financial prospectors. We also must be prepared to promote due processes of equity and universal justice in the sharing of the benefits of biodiversity in so far as it is now recognized as a common good to the global humankind. These are now matters enshrined in the Biodiversity convention, a consensus political statement which now unites the world on the biodiversity issue.

We are sure that the deliberations and decisions of the Convention's COP and its subsidiary bodies, in particular of organs such as the Subsidiary Body for Scientific, Technical and Technological Advice (SBSTTA) would greatly benefit from results and outputs of deliberations arising from scientific discussions, such as this Forum.

It is UNEP's view that the development and adoption of due consultations required in the resolution of biodiversity issues may be best achieved through the meticulous application of the scientific method to the results management approach adopted by UNEP for its Corporate Programme framework for the 1994-95 biennium. Under this framework, the elaboration of the biodiversity issues will be undertaken and systematically pursued under one or a combination of any of the following sequential critical paths or processes:

1. identification of the problems;
2. identification of solutions or options;
3. effective implementation or identified solutions as per options.

There are a number of points for intervention by the scientific community into these processes. Here I would like to draw your particular attention to the Biodiversity Country Studies, the Global Biodiversity Assessment (GBA) Project, the work of the Convention's subsidiary scientific body, the SBSTTA and that of GEF's Scientific and Technical Advisory Panel (STAP). They all require critical inputs from the scientific community.

The past 20 years have seen the fruition of some of UNEP's synergistic as well as catalytic and cooperative efforts, some of which have led to several seminal publications and documents. They have served to bring into a sharp focus pertinent issues of biodiversity conservation and its sustainable use. I will cite some examples, such as the World Conservation Strategy, the World Charter for Nature, the Global Biodiversity Strategy, the Global Marine Biodiversity Strategy, Caring for the Earth: A Strategy for Sustainable Living, the Global Biodiversity Status Report, the Global Biogeochemical Cycles and the State of the World's Lakes Inventory.

It is our earnest desire to collaborate with more partners, in a more intensive and systematic fashion. I am sure this path will eventually lead us towards a closer and more fruitful partnership with scientific communities such as yours.

A fair and effective implementation of the Biodiversity Convention would require the attainment of its threefold objectives. These are biodiversity conservation, sustainable use of biodiversity and its related components, and the equitable sharing of benefits so derived.

The Biodiversity Convention contain specific articles that promote international technical and scientific cooperation among contracting parties, in particular between developed and developing countries.

I mention some of these provisions: development and implementation of national policies; development and enhancement of institutional capacity building, establishment of joint research programmes and joint ventures for the development or transfer of technologies relevant to the objectives of the Convention.

The Convention also provides for the establishment of the Subsidiary Body for Scientific, Technical, and Technological Advice (the SBSTTA). This multi-disciplinary, government expert body will provide advice to the conferences of the parties and, as appropriate, their other subsidiary bodies relating to the implementation of the convention.

In particular, the SBSTTA will provide assessments of the state of scientific and technical knowledge, and the effects of various types of measures taken through efficient and state-of-the-art technologies for the conservation and sustainable use of biological diversity, the mitigation and adaptation to climate change, and the reduction of the emissions of ozone-depleting substances. It will also provide advice on scientific programmes and international cooperation in research and development.

The Convention requires that scientific, technical, and economic assessments be undertaken to support the implementation of proposed measures. Most developing countries face difficulties in conducting national, let alone participating in international, assessments that provide the scientific, technical, and economic basis for national, regional, and international decision-making.

UNEP appreciates fora such as this to explore the interfaces between ethics and law, environment and economics, refining the definitions of sustainable development in the context of biodiversity conservation and sustainable use, exploring ways and means of formulating viable, implementable and fully fundable biodiversity strategies and action plans.

I am pleased to learn that you intend to address aspects of these issues seriously and systematically.

I would like to share with you the three key limiting factors which contribute to the inability of developing countries to make timely informed decisions about the means available to conserve and sustainably use their share of the world's biological diversity.

The first is the inability to synthesize existing data in meaningful and digestible ways. The second is the absolute lack of relevant and necessary data on critical aspects of biodiversity such as the magnitude and distribution of biological diversity, the direct and indirect threats to biological diversity, the structure and functioning of ecological systems, and the proper valuation of biological diversity. The Biodiversity Country Studies programme and the Global Biodiversity Assessment (GBA) Report, of which I shall say more later, are part of the effort directed at this limitation.

The third is the failure of existing economic systems and policies to value the ecosystems and the ecological services of the environment and the existing inequity in the ownership, use, management and flow of benefits derived from both the use and conservation of biological resources.

It must be stressed that even when the scientific knowledge of the structure and functioning of a particular ecosystem may be adequate to design a conservation project, there may still be insufficient knowledge necessary for the design of, say, a severely damaged habitat ecosystem restoration or some sustainable use projects. To address these constraints in the area of biodiversity, more investments in research need to be made and the research be action oriented.

To date there has been no comprehensive international assessment of the status of biological diversity. However, provisions of Article 25 of the Biodiversity Convention establish and outline the functions of the subsidiary scientific body (SBSTTA) which include providing the Conference of the Parties with a scientific and technical assessment of the status of biological diversity. As a major contribution and means of mobilizing the scientific community worldwide to provide a solid basis for decision-making by the COP in their implementation of the Convention and Agenda 21, UNEP has initiated a number of undertakings that may prove to be among the most important scientific exercises in the field of biodiversity.

Special reference may be made here to three of these exercises. The first is the preparation of national *Biodiversity Country Studies, Strategies and Action Plans* to underpin and re-enforce in-country biodiversity assessment and planning processes, to identify national priorities for action, and to provide a baseline for monitoring effectiveness of national actions, policies and programmes on conservation and sustainable use of genetic resources.

The second exercise relates to the *Biodiversity Data Management (BDM) Project* for capacity building in developing countries and improved in-country networking of biodiversity information to empower the countries to effectively manage their biodiversity data and information and to underpin the sustainable use of their genetic resources. The BDM undertaken in partnership with the World Conservation Monitoring Centre (WCMC), will build on the country study process and strengthen the associated National Biodiversity Units (NBUs) and other national institutions generating and maintaining biodiversity data. The BDM will also contribute to the Environmental and Natural Resource Information Network (ERIN) under the UNEP – GEMS/GRID Earthwatch programme.

The third exercise is the *Global Biodiversity Assessment (GBA) Project*. GBA, an independent, critical and peer-reviewed scientific analysis, will examine the current state of knowledge, identify gaps in knowledge, and draw attention to scientific issues where scientists have reached a consensus and those where uncertainty has led to conflicting points of view. Some of the issues currently under scrutiny include:

- characterization, origins, dynamics and distribution;
- ecosystems functioning;
- multiple values of biodiversity;
- human influences on biodiversity and socioeconomic issues;
- biodiversity and biotechnology;
- data and information management and communication;
- legal measures and consensus-building.

GBA will not prepare any recommendations, but be a pure source of knowledge that can be drawn upon by decision-makers and scientists.

The preparation of the assessment has so far involved several hundred scientists from over 40 countries. Many more scientists will be involved through the extensive peer review process which will take place in the coming months. We would welcome the participation of as many scientists as possible among this audience as peer reviewers of the Assessment. The GBA report, planned for printing in the second quarter of 1995, will be distributed later that year.

Much of the scientific research needed to generate up-to-date and reliable information must be performed at the local and regional level in the biodiversity-rich developing countries. Considering that so much research is needed to improve our understanding of ecosystems, we cannot any longer afford to underutilize any latent scientific and technical talent available worldwide.

I refer in particular to the utilization of indigenous and local knowledge. This will be essential for the protection and sustainable use of biological diversity and for soil conservation, agroecosystems management and habitat/species restoration ecology. A robust scientific and technical infrastructure and institutional capacity in developing countries will not only result in national and international policies and regulations based on the best possible scientific understanding, but it will also contribute to the design and implementation of sustainable and cost-effective measures. It would also enhance effective participation and representation by developing countries in regional and global environmental fora. A higher percentage of the world's research and development activities should accordingly be undertaken within and by developing countries by appropriate boosts to their current inadequate technical, financial and other resources.

Institutional Building and Networking

It is evident that there are major problems in the development of viable scientific capacity and infrastructures to meet the biodiversity development needs of developing countries. These include: lack of financial or economic resources; deficiencies in education systems; scarcity of scholarship programmes; scarcity of adequately trained human resources; relative isolation of their scientists and researchers in certain fields; limited modalities for regional and international scientific cooperation; poor or no relations between science and industry; low compensation for the research and development staff; and undue reliance on technology trade or turn-key projects. Taken together, these factors have resulted in the inhibited growth of indigenous capacity for sound science.

Bridge-building can be a short-cut strategy for institutional and scientific capacity building. Originating in the South, the INBIO programme of Costa Rica presents an excellent approach to biodiversity institution-building that can be replicated in many of the biodiversity-rich countries of the South. More than two decades ago, UNESCO and UNEP joined hands to establish Microbiological Resources Centres (MIRCENS) in strategically placed institutions in several developing countries to tap the underutilized and unexplored potential of microorganisms for scientific research and development, coupling these developing country centres with advanced developed country institutions. This programme has already proved its worth and it should be further developed and expanded. Finally, the private sector, a reservoir of some of the best applied research, should be coopted and enlisted in institutional strengthening and capacity building. In the Rio spirit of Agenda 21, building bridges and securing partnerships is at the heart of the new directions in the global quest for sustainable development.

Written Message from the President of Argentina

Carlos Menem

As President of Argentina, I feel greatly honoured to share with you at this International Forum some of my thoughts on biodiversity – a subject of such great importance to humankind.

Each passing day makes it clearer to us that biological diversity, the natural heritage of our world, is endangered.

The accelerating loss of biodiversity is due to socioeconomic and environmental problems such as the alteration and destruction of habitats, the expansion of farming land, the impact of the extractive industries, pollution and the introduction of alien species.

Biodiversity is fundamental to each country's socioeconomic development and if we fail to take the appropriate measures for its preservation and its rational use, we will put our future at risk as well as that of our descendants.

Our country is no exception. Its borders enclose a great variety of natural environments with significant climatic variations. This creates an environmental mosaic harbouring a great diversity of species that can adapt to such particular and extreme conditions.

Argentina's greatest cultural and economic asset is its land. It is this natural resource that provided the conditions for a farming system based on crops suitable for temperate zones and for which there is a strong demand in international markets.

In order to gather more scientific knowledge about our biological diversity, several studies are being carried out in our country to establish an inventory of flora and fauna. These studies are carried out with the participation of institutions affiliated to national universities, the National Council of Scientific and Technical Research (CONICET), the National Parks Administration and the National Institute of Farming Technology (INTA). The inventory will provide a basis for future activities concerning the conservation and the sustainable use of these resources.

As regards *in situ* conservation, Argentina depends on a national system for protected areas headed by the National Parks Administration, which was initially concerned with protection and tourism. Now research is being carried out and data recorded on the components of the natural environment and progress has also been made in defining management and sustainable use objectives.

As for *ex situ* conservation, INTA heads a National Programme for the conservation of genetic resources in germplasm banks. These banks store species currently used in commercial agriculture.

In addition to systematic recording of data, we must also coordinate our efforts to evaluate the importance of biological resources for industrial and pharmaceutical use.

If the conservation of biological diversity is to be effective, it must be based on action at a number of levels involving scientists, politicians, non-governmental organizations, administrators of protected areas and local communities, and efforts must be made to bring in organizations working in the development field.

Resources should be used in a sustainable way to satisfy present and future needs. The aim is to improve the quality of life by maximizing the productive potential of ecosystems through the use of appropriate technology and active public participation. A new outlook is necessary here, encouraging the introduction of the environmental variable in both global and sectoral planning.

Science should provide the basis for all measures aimed at protecting biological diversity.

Scientific research programmes should focus on: identifying, inventorying and gathering documentation on the status and the distribution of biological diversity; monitoring changes; studying the functions which biological diversity performs in relation to the maintenance and operation of ecosystems; studying means of recovery; establishing criteria and techniques for the sustainable use of biological resources and researching the safe manipulation of genetically altered organisms.

International cooperation represents a decisive step in the area of conservation – whether it takes the form of exchanges of information or technology transfers.

The developing countries possess most of the world's genetic resources but they lack the necessary technology to transform them. On the other hand, those countries need the resources in question for their activities in the field of biotechnology. Consequently, concerted international action is needed to create locally the capacity for selecting, adapting, and developing the appropriate technology.

The Convention on Biological Diversity provides a suitable framework for dealing with and negotiating on this issue. For the first time, the conservation and sustainable use of biodiversity is acknowledged as a global and cross-sectoral problem. Likewise, it is accepted that the benefits arising from the utilization of genetic resources should be shared on a fair and equitable basis.

I am sure that this large group of distinguished scientists and experts gathered here in Paris will make a significant contribution to the understanding of the diverse and complex issues raised by biodiversity. And as President of Argentina, I wish you every success in your endeavours.

Written Message from the President of Indonesia

President Soeharto

First of all, I would like to express my warmest congratulations to the International Union of Biological Sciences (IUBS) for the initiative it has taken to hold this international forum on biodiveristy. I would like also to extend my highest appreciation to UNESCO, the International Council of Scientific Unions (ICSU), and the French Government which have initiated this forum. I should also like to avail myself of this opportunity to thank all parties concerned that have accorded a warm welcome and provided facilities to the Indonesian Delegation attending the current meeting.

At the UN Summit Conference on the Environment and Development held in Rio de Janeiro two years ago, I emphasized the importance of establishing a New Global Partnership in dealing with the environment issue of development, especially economic development and the enhancement of the quality of human resources.

At this august forum, I would like to reaffirm the importance of a new global partnership in dealing with biodiversity based on mutually beneficial cooperation and shared interest aimed at improving the prosperity and securing the survival of humankind on earth.

I am aware that Indonesia is today one of the primary centres of biodiveristy. Indonesia is endowed with potential genetic resources which are important for the development of superior biological sources for agriculture and industiral development in the future. These biodiversity resources are scattered in almost all parts of the country, both *in situ* and *ex situ*, whether they are used traditionally by the community or in cultured forms. These are all superior varieties of biological resources that play an important role in agricultural and industrial development.

These potential biodiversity resources provide hope for the discovery of various basic needs of life, not only for our nation but also for the international community. They also serve to ensure the continuity of sustainable development and consolidation of the function and quality of the environment on earth. The effort to transform the potential of these biodiversity resources certainly requires both science and technology. Indonesia's capabilities in exploiting these potential resources are still very limited. This is the reason why we are extending our hands and opening up the possibility for mutually beneficial cooperation with the international community.

With such a view, we attach very great importance to the Forum that discusses biodiversity while focusing its attention on the scientific and technological aspects of the new partnership. This Forum becomes increasingly important when we see it as the follow-up of the Earth Summit in Rio de Janeiro and the initial realization for the biodiversity convention as was agreed upon by many nations, including Indonesia. Indonesia ratified the convention on biodiversity last 1 August 1994.

The International Forum aims to deliberate on a number of issues that become the primary concern of the Earth Summit. This is the reason why the goals of the Forum are global in nature and have a far-reaching scope. Consequently, it also needs the participation of those who have primary biodiversity centres, who are highly-skilled to conduct research, scientific, and technological development and application, who have the decision-making powers, and those who are able to translate biodiversity potentials in concrete economic activities for the sake of the well-being of humankind.

All parties must have the concern, determination, and commitment to put into reality the commonly agreed-upon conclusions and recommendations. All must also be willing to cooperate in the planning of the protection, preservation, and utilization of the resources of biological diversity in sustainable development, including plans for the development of technology, human resources, and financial resources.

In the beginning, the relationship between man and biodiversity resources was well-balanced and supportable. However, with the increased number of economic activities and population, there has been an excessive drain on biological resources that has led to the degradation of the function of biodiversity accompanied by degradation of genetic resources and declining quality of human life.

Indonesia has closely followed the evolution of global environment that affects the national environment, including biodiversity and population. Indonesia is also actively involved in international cooperation on the environment, conservation of biological natural resources and international genetic resources.

In the field of population, we are trying hard to control the growth rate and to improve the people's well-being through a Family Planning Campaign by adopting the motto 'A Small and Happy Prosperous Family'. This effort has succeeded in reducing the population growth rate which was 2.5% in 1970 to around 1.6% today. The decline in the population growth rate is also accompanied by a lower infant mortality rate and longer life expectancy.

I should like to state here that Indonesia's success in these achievements has been due to the cooperation with various international organizations and other countries. Based on our experience, we offer the possibility of establishing cooperation with the international community in the field of biodiversity.

I have mentioned earlier that our ability to deal with biodiversity is still limited, both in terms of the exploitation of the potential, its utilization and conservation. We have a small number of experts in basic biology in comparison to the diverse biodiversity resources that are scattered all over the Indonesian archipelago which has more than 17,000 islands. Therefore, we are hoping for the participation and cooperation of international experts in the field of basic research, particularly in discovering traits that are useful for human life.

Mutually beneficial cooperation on the subject of biodiversity can hopefully be established, which does not harm others' interests. The effort to prevent the trading of products originating from biological resources from the tropical forests, for example, is mutually damaging. Such a trade policy pattern will not solve the global environmental issue. But on the contrary, it will only create tensions and conflicts preventing closer cooperation.

The issue of biodiversity cannot be separated from the issues of population and economic development. The degradation of biodiversity must be overcome through economic and social development, as well as control of the population growth rate.

The poor population relies heavily on local lands and biological resources. Therefore, if no efforts are made to improve its quantity and quality, the environment and biodiversity resources will be harmed. Efforts to free them from poverty are not only aimed at enhancing social justice but also concretely stemming the depletion of biodiversity and degradation of genetic resources. These are both challenges and opportunities for the experts to discover scientific information inherent in the biological resources so that their benefits for mankind will be recognized and thus spare them from possible extinction due to continuous utilization. Therefore, it is indeed a wise step if the poor and the less fortunate can also enjoy the fruits of economic development and improve their standard of living, so that eventually they too can participate in providing constructive contributions to development.

Without doubt the most important issue of biodiversity confronting us is population, development, and its equitable distribution, as well as a harmonious and mutually beneficial international cooperation. Indonesia has tried to harmonize equitable distribution of development in all sectors, population control and preservation of development, including the management of biodiversity.

In the development of the function and quality of life, including biodiversity, Indonesia has determined to set aside 18.6 million hectares of forest on its territory which contains a variety of biodiversity resources. With such an extensive area it is hoped that biodiversity resources can be protected and function as providers of genetic resources, for both national and international needs. This extensive area is hopefully sufficient to allow biodiversity resources to evolve in line with the development on the Earth's surface. In order to secure the survival of *ex situ* biodiversity resources and those used by the community, protective measures have also been adopted. We are intensifying *ex situ* conservation in the forms of botanical gardens, genetic reserves and developments of backyards as conservation sites and, at the same time, the utilization of genetic resources. Efforts are also made to improve on biodiversity such as high-yielding varieties, both in types and kinds, in order to produce various agricultural products to meet the demands for food, medicine, fibres, and various manufactured goods.

As a developing country, we are well aware that the huge natural resources and advanced technological mastery have provided benefits and comparative advantages to the industrialized countries, so that the latter can still set terms in their relations with the developing countries. We have to address this very disadvantageous condition of the developing countries if we want to develop a new and mutually beneficial cooperation.

At present, we must all try together on a global scale to secure the overall function of biodiversity resources. The terms and conditions of joint efforts must be commonly agreed upon without any coercion that may prove to be harmful to other parties. Common goals must be set to guarantee the survival of humankind on this earth and the preservation and ever-lasting utilization of biodiversity. As a result, we must all learn to enhance the practices and traditions of a global relationship.

1 Biological Foundations for Sustainability and Change

C.S. Holling
Arthur R. Marshall Jr Chair in Ecological Sciences, Department of Zoology, University of Florida, 111 Bertram Hall, PO Box 118525, Gainesville, FL 32611-8525, USA

Introduction

Perceptions, problems and policies are formed as much by our images of the world as by the world itself. Those images provide for each of us the sense of order required for action in a world of the uncertain and the unknown. That is as true for communities of scientists, as it is for those of businessmen, politicians and citizens. The more the visions of these different communities diverge, the more uncertain is action, for we are a social beast and communal decisions are the norm. Divergence of visions occurs during times of momentous change, where old visions falter and our perceptions of the unknown rise. We lose track of who and perhaps why we are in this world of the uncertain and unknown. At those times, both opportunities for the nimble and dangers for the society seem to dominate what earlier seemed secure and knowable.

Science can only partly help in rediscovering a lost sense of place and purpose, but the more its own concepts and visions are in harmony with a deeper attachment to and wonder with the mystery of life, the more science can contribute to the foundations for a sustainable and creative future. And that seems to me to be happening now in a process of discovery, rediscovery and enrichment of scientific theories of change.

In the community of science, there is a growing recognition of the existence of alternative models of change and of their strengths and limitations, together with a hesitant emergence of an adaptive, evolutionary image that places each of the alternatives in perspective. It embraces change in any complex adaptive systems whether biological, economic or social. That has occurred because understanding has emerged independently in a variety of disciplines in the last half century, as hypotheses and theories were challenged by unexpected surprises from nature not easily comprehended by traditional theory. As a consequence, although the scientific community has historically been less of a community than a collection of separate disciplines, now the ground-breaking areas of the natural sciences are acting more like a community of integrative disciplines that are also reaching out to the social sciences.

At the same time as these images and theories of change have ripened, the scientific community has organized, through the International Council of Scientific Unions and national bodies, the largest international interdisciplinary effort that environmental science has seen. Its goal is to understand the biophysical functions of the planet as a whole, the dynamics and structure of its constituent ecosystems and biota and the role of expanding human influences. The principal organizing forces for much of this biological, physical, and social science is given by a set of new phrases, agendas and programmes: Global Change, Biodiversity and Ecosystem Functions, Sustainable Development.

It is an interesting time. First, a deeper understanding of change is emerging. Second, a new class of global problems, international policies and national actions are evolving in response to a global transformation that has somewhat familiar political, economic and social dimensions but quite novel biophysical ones. Third, the scientific community is now unravelling the way the planet functions in its physical, biological and human interactions. Driving all that is a transformation in the views of the roles of diversity of biological life and human cultures in providing a sustainable basis for creative change.

This chapter explores that emerging evolutionary view of diversity in three sets of arguments and their conclusions. First, alternative images of change exist, each of which leads to different definitions of problems and solutions. Second, a broadly evolutionary view is emerging, founded on three axioms of biology – exponential growth, limits, and variety. Third, the interactions among those axioms generate, for living systems, an adaptive cycle of four phases of birth, growth, death, and renewal that is a useful heuristic for interpreting change and responses to it from scales of genes to societies.

Policy and Science

Fundamental science and policy action are uneasy partners. Consider biology and the biodiversity issue.

Biology exists as a series of interrelated subdisciplines, each functioning within a narrow range of scales. The molecular geneticist deals with fast and small molecular events, the cell biologist with sets of biochemical interactions operating within seconds to days, the comparative physiologist with organ systems to whole individuals over minutes to months, the population biologist with interacting individuals over days to years, the ecosystem ecologist with leaves to landscapes over weeks to centuries and the evolutionary biologist with lineages of organisms over millennia.

Each of these subdisciplines focuses on objects and defines public issues over very specific ranges of sizes and speeds of processes. But in combination, they cover an enormous range of scale of objects from molecules to the planet, and of scale of problems from dynamics of genetic diversity to transformation of global atmosphere.

Political and economic issues, on the other hand, drive policy action and inaction at very different scales. Human social groups in national and international communities are the objects of attention – local communities, businesses, ethnic goups, nations, communities of nations. And a narrow window of time defines problems and responses – typically months to years. As a consequence, the very fast processes underlying the loss of genetic diversity seem like inconsequential noise for any policy or political relevance. The very slow process underlying the formation of taxonomic and ecosystem diversity seems like a constant background requiring no response until impacts occur that force politics, policy and people to adapt. In short there is a disconnect between the scales of politics and the scales of transformations occurring in nature.

Two conditions could more easily harmonize the science and the policy. One is to discover a common arena where both the science for diversity and policies for sustainability interact. The second is for a unified sense of change and its character that has relevance for both biological and social systems.

Images of Change

So much presently seems uncertain or unknown that many of the calls for action or inaction, however well supported by technical argument, are largely determined by images people hold of the way change takes place or should take place. Because each image is partially relevant, impressive and convincing technical arguments can be mobilized for each one no matter how opposite the resulting calls for action or inaction.

Four belief systems, and an emerging fifth one, are driving present debate and public confusion. Each reflects different implicit assumptions concerning stability and change, as I have suggested elsewhere (Holling, 1987). Alternatively they can be labelled (albeit unfairly) by a caricature of their causal assumptions, as I shall do here.

The first is a view of smooth exponential growth where resources are never scarce because human ingenuity always invents substitutes. It was the basic view of Herman Kahn and is the foundation for Julian Simon's arguments (Simon and Kahn, 1984). It assumes that humans have an infinite capacity to innovate and that nature changes gradually – fast enough to be detected yet slow enough to be managed. It is a view of Nature Cornucopian. It leads to libertarian kinds of policies designed to free individual enterprise. It is the view that says loss of biological diversity, or accumulation of greenhouse gases requires no draconian national or international action because the effects will accumulate slowly enough to allow people to discover substitutes and create opportunity.

The second is a hyperbolic view where increase is inevitably followed by decrease. It is a view of fundamental instability where persistence is only possible in a decentralized system where there are minimal demands on nature. It is the 'Small – is – beautiful' economic view of Schumacher (1973) or some extreme environmentalists. If the previous view assumes that infinitely ingenious humans do not need to learn anything different, this view assumes that humans are incapable of learning how to deal with the technology they unleash. It is a view of Nature Anarchic. It leads to policies of decentralization with odd similarities to the policies of Nature Cornucopian in its desire for freedom from external centralized control. Where it differs, however, is in the perception that retreat from present patterns of overconsumption is so essential that authoritarian regulation and control is necessary.

The third is a view of logistic growth where the issue is how to navigate a looming and turbulent transition – demographic, economic, social and environmental – to a sustained plateau. This is the view of several institutions with a mandate for reforming global resource and environmental policy: the Bruntland Commission, the World Resources Institute, the International Institute of Applied Systems Analysis, and the International Institute for Sustainable Development, for example. Many individuals are contributing skilful scholarship and policy innovation. They are among some of the most effective forces for change. It is a view of Nature Balanced. It leads to policies of national and international collaboration for regulations and controls over such issues as biodiversity loss and greenhouse gas accumulation.

The fourth is a view of nested cycles organized by fundamentally discontinuous events and processes. That is, there are periods of exponential change, periods of growing stasis and brittleness, periods of readjustment or collapse and periods of reorganization for renewal. Instabilities organize the behaviours as much as do stabilities. That was the view of Schumpeter's (1950) economics and it has more recently been the focus of fruitful scholarship in a wide range of fields – ecological, social, economic and technical. That has formed the body of my own ecological research for the past 20 years. I find striking similarities in Harvey Brook's (1986) view of technology, Brian Arthur's (1989) and Kenneth Arrow's view of the economics of innovation and competition (Waldrop, 1992), Mary Douglas (1978) and Mike Thompson's (1983) view of cultures, Don Michael's (1984) view of human psychology and Barbara Tuchman's (1978) and William McNeill's (1979) view of history. It is a view of Nature Resilient. It leads to policies that attempt to restore or enhance the resilience of natural ecological systems so that they can experience wide change and still maintain the integrity of their functions. Since diversity is one major source of resilience, the incentives needed to protect, restore and enhance diversity come to the fore.

The emerging fifth view is evolutionary and adaptive. It has been given recent impetus by the paradoxes that have emerged in successfully applying the previous more limited views. Complex systems behaviour, discontinuous change, chaos and order, self-organization, non-linear system behaviour and adaptive evolving systems are all the present code words characterizing the more recent activities. It is leading to integrative studies that combine insights and people from developmental biology and genetics, evolutionary biology, physics, economics, ecology, and computer science. The Santa Fe Institute is an interesting experiment (Waldrop, 1992) in applying collaborative approaches to explore the insights and opportunities opened by such an evolutionary paradigm. It is a view of Nature Evolving. It leads to policies for sustainable opportunity, where investments are made to support or enhance the foundations underlying ecological, social and economic function (Holling, 1994).

The point is not that one of these beliefs is correct and the others wrong. Each is true but each is a partial truth. Because we are only now beginning to come to grips with understanding the changing reality, there is therefore no limit to the ability of a good scientist to invent compelling lines of causal explanation that inexorably support his or her particular beliefs. How can even the best intentioned politician possibly be expected to deal with that? How can even the most reflective of the public? With every issue having supporting evidence and explanation and denying counter-evidence and counter-explanation – all legitimate – the issues seem to be ones for which there is no independent reality of nature, only moral issues that can be dealt with by social debate. Can we ever separate belief from fact?

A Synthesis

The source of a synthesis that can give direction is coming from a rediscovery and an enrichment of an essentially biological paradigm of evolution. Here I will explore that development from a foundation of my own roots in biological ecology.

The Three Axioms of Biology

For much of this century, ecology has been more a biological tradition than an environmental one. Just as Darwin's theories of species change were shaped by Malthus, so ecologists' theories have been rooted in the foundations of Darwinian natural selection and evolution. The environment has been seen principally as a fixed and exhaustible backdrop to the actions of and interactions among the biota in their confrontation with the physical environment. It was a demographer, Raymond Pearl (1927) who was one of the first to encapsulate the essence of this perception in a simple, and for that reason influential, representation of population growth. He showed that the populations of many organisms in the laboratory grew over time as an S-shaped curve to a plateau that was sustained if food resources were continually maintained. The logistic equation fitted many of those patterns of population growth, of yeast for example, or of *Drosophila*, with remarkably close fits, and the two parameters of the logistic began to be seen as reflecting two of three axioms of ecology.

The parameter 'r', or the instantaneous rate of growth, represented the universal axiom that populations have the inherent propensity to grow exponentially, inevitably exceeding or reaching limits of the external environment. The parameter 'K', or the plateau to which the logistic growth approached, in turn, represented the axiom that the environment sets ultimate limits to growth.

Those two fundamental properties are viewed as being essentially self-evident and incontrovertible axioms by most biologists. In biologically based ecology, they appear in a number of forms. For example, one of the pioneers in ecology, Robert MacArthur, proposed that species of organisms can be designated as following one of two principal strategies as represented by the symbols r and K. The r-strategists are the pioneer species, the opportunists that can deal with the wide range of physical variability often found in recently disturbed habitats, but cannot compete well with other species. They are the 'weeds' of the biota. They have short lifetimes, small size, and small, widely dispersed and abundant propagules. In contrast, the K-species are the conservative species with strong competitive abilities. They are able to outcompete the r-strategists, persist for long periods, and have long life-times, large size and fewer but larger propagules.

In summary, two of the fundamental axioms of ecological and evolutionary biology are that organisms are exuberantly overproductive and that limits set by space, time and energy are inevitably encountered. The foundations for all modern ecology and evolutionary biology rest in part upon the consequences of those two axioms. But alone, they reinforce or produce a highly determined, static and equilibrium-centred view not at all incongruent with the way turbulent nature seems to behave. They seem to lead to a picture of Nature Balanced – harmonious but fixed.

But there is a third element to this foundation that gradually has become seen as a third axiom. This concerns processes that generate variability and novelty. It was that third axiom that Darwin saw could transform a static view of a given fixed set of species on the planet to one of evolution of species. Exuberant overproduction, interacting with limits set the conditions for Darwinian 'survival of the fittest'. But it is the continual propagation of variation in species' attributes that provides the source for continued experimentation in an environment that itself changes. That is where one major impact of biodiversity lies, in providing options for different futures. It is the result of the interaction of that third axiom of variability with the axioms of overproduction and limits that provides the answer to what are the roles for biodiversity.

The addition of the third axiom turns a static image of Nature Balanced to one of Nature Evolving. The conceptual foundations to establish this re-emergence of an evolutionary paradigm are being drawn from growing experience in understanding the operation of complex, non-linear systems where discontinuous behaviour and structural change is the norm. Scholars in an unusual variety of disciplines have contributed to the development of these theories – from thermodynamics (Nicolis and Prigogine, 1977), oceanography (Broecker, 1987), climatology (Lorenz, 1963), atmospheric chemistry (Crutzen and Arnold, 1986), evolutionary and developmental biology

(Kauffman, 1992) and ecology (May, 1977; Levin, 1992; Holling, 1992). All deal with the reality of abrupt changes organized by several equilibria, of the existence of multistable states and of the interplay between order and disorder in evolving self-organizing systems. That is what our world is and that is what is at the heart of making sustainable development feasible.

Social scientists are also major contributors. Historians, like William McNeill (1979), have long argued a view of history as being a sequence of discontinuous events and of human response to them. Wildavsky and Douglas (1982) argue for the inevitability and need for risk and surprise in any human development. Mary Douglas (1978) and Mike Thompson (1983) use their background in cultural anthropology to characterize institutions as being driven by similar interplay's between stability and instability. And when someone of the stature of Kenneth Arrow suggests the need for transforming economics by non-linear theory (as quoted in Waldrop, 1992), a revolution in thought may be occurring in that field, as well. In every instance, these theories owe their force to the resolution of puzzles that appear when earlier incomplete or inadequate concepts encounter surprising reality.

The way key subcultures in the natural and social sciences view the world is converging on these theories of change. Those theories rationalize the paradoxes of stability and instability, of order and disorder or of stasis and evolutionary change. Since those are the same paradoxes inherent in the goal of sustainability and development, an avenue opens for directly relevant cooperation between critical parts of the social and natural sciences and between science and politics.

As a start, I shall describe that view of change as it applies to ecological systems. I do that not to force an inappropriate analogy on the way social and economic systems function, but to search for the common foundations for change that underlie the operation of any complex living system.

The Adaptive Renewal Cycle

In ecology, the notion of a unique optimal path to a sustained optimal climax has increasingly become seen to be a static and unrealistic view – a version of Nature Balanced. In contrast, a view has emerged from the study of a variety of ecosystems that emphasizes cycles of establishment, growth, disturbance, and reorganization that are controlled by small sets of variables, each organizing the time and space structures over a different range of scales (Holling et al., 1994). It has led to one version of asynthesis that emphasizes four primary stages in an ecosystem cycle (Holling, 1986).

The traditional view of ecosystem succession has been usefully seen as being controlled by two functions: *exploitation*, in which rapid colonization of recently disturbed areas is emphasized and *conservation*, in which slow accumulation and storage of energy and material is emphasized. In ecology the species in the exploitive phase have been characterized as r-strategists and in the conservation phase as K-strategists, drawing from the traditional

Biological Foundations for Sustainability and Change 49

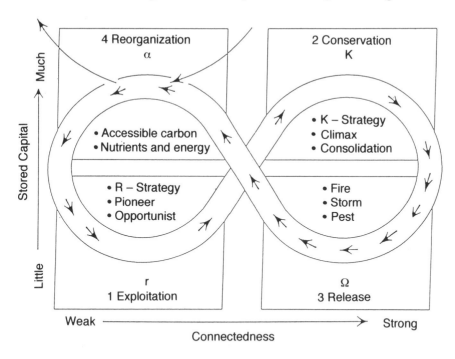

Fig. 1.1. The four ecosystem functions (r, K, Ω ,α) and the flow of events among them. The arrows show the speed of that flow in the cycle, where arrows close to each other indicate a rapidly changing situation and arrows far from each other indicate a slowly changing situation. The cycle reflects changes in two attributes: (i) y axis: the amount of accumulated capital (nutrients, carbon) stored in variables that are the dominant structuring variables at that moment and (ii) x axis: the degree of connectedness among variables. The arrows entering and leaving the α phase suggest where the system is most sensitive to external influence.

designation of parameters of the logistic equation and two of the three axioms of biology. To an economist or organization theorist, those functions could be seen as equivalent to the entrepreneurial market for the exploitation phase and the bureaucratic hierarchy for the conservation phase.

But the revisions in ecological understanding indicate that two additional functions are needed, as summarized in Fig. 1.1. One is that of *release*, or 'creative destruction', a term borrowed from the economist Schumpeter (as reviewed in Elliott, 1980), in which the tightly bound accumulation of biomass and nutrients becomes increasingly fragile (overconnected in systems terms) until it is suddenly released by agents such as forest fires, insect pests or intense pulses of grazing. We designate that the Ω phase.

The second is one of *reorganization*, in which soil processes minimize nutrient loss and reorganize nutrients so that they become available for the next phase of exploitation. This last phase is essentially equivalent to processes of innovation and restructuring in an industry or in a society – the kinds of economic processes and policies that come to practical attention at times of economic recession or social transformation. We designate that the α phase.

During this cycle, biological time flows unevenly. The progression in the ecosystem cycle proceeds from the exploitation phase (r, Box 1, Fig. 1.1) slowly to conservation (K, Box 2), very rapidly to release (Ω, Box 3), rapidly to reorganization (α, Box 4) and rapidly back to exploitation. During the slow sequence from exploitation to conservation, connectedness and stability increase and a 'capital' of nutrients and biomass is slowly accumulated. For an economic or social system the accumulating capital could as well be infrastructure capital, levels in the organization or standard operating techniques that are incrementally refined and improved.

As the progression to K, Box 2 proceeds, the nutrient capital becomes more and more tightly bound within existing vegetation, preventing other competitors from utilizing the accumulated capital until the system eventually becomes so overconnected that rapid change is triggered. The agents of disturbance might be wind, fire, disease, insect outbreak or a combination of these. The stored capital is then suddenly released and the tight organization is lost allowing the released capital to be reorganized to initiate the cycle again. Human enterprises can have similar behaviour, as, for example, when corporations such as IBM or General Motors accumulate rigidities to the point of crisis, followed by efforts to restructure.

Instabilities trigger the release or Ω phase, which then proceeds to the reorganization or α phase where the weak connections allow unrestricted chaotic behaviour and the unpredictable consequences that can result. Stability begins to be re-established in the r phase of Box 1. In short, chaos emerges from order, and order emerges from chaos. I find this description similar to the 'order through fluctuation' pattern of complex system dynamics described by Prigogine (1980).

Resilience and recovery are determined by the fast release and reorganization sequence, whereas stability and productivity are determined by the slow exploitation and conservation sequence.

Moreover, there is a nested set of such cycles, each with its own range of scales. In the typical boreal forest for example, fresh needles cycle yearly, the crown of foliage cycles with a decadal period and trees, gaps and stands cycle at close to a century or longer periods. The result is an ecosystem hierarchy, in which each level has its own distinct spatial and temporal attributes (Fig. 1.2).

These hierarchies are not static structures. Levels are transitory structures maintained by processes which are changing over time and space. The cycles are all operating concurrently, influencing one another. They are rhythms within rhythms, providing not the static structures of a well-oiled machine shop clanking and vibrating at a myriad of frequencies, but rather those of a jazz band, building rhythms and rifts around each other, coalescing into both short and long rhythmic structures around islands of rhythmic discord.

A critical feature of such hierarchies is the asymmetric interaction between levels (Allen and Starr, 1982; O'Neill *et al.*, 1986). In particular, the

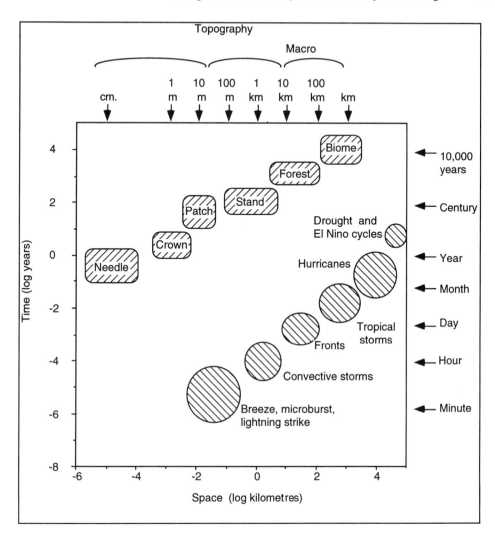

Fig. 1.2. Hierarchies of biotic structures and atmospheric processes in the boreal forest.

larger, slower levels maintain constraints within which faster levels operate. In that sense, therefore, slower levels control faster ones. If that was the only asymmetry, however, then hierarchies would be static structures and it would be impossible for organisms to exert control over slower environmental variables.

However, it is not broadly recognized that the birth, growth, death and renewal cycle, shown in Fig. 1.1, transforms hierarchies from fixed static structures to dynamic adaptive entities whose levels are vulnerable to small disturbances at the transition from gowth to collapse (the Ω phase), and the transition from reorganization to rapid growth (the α phase). During other times, the slower and larger processes are stable and resilient, constraining the lower levels and immune to the buzz of noise from small and faster processes. It is at the two phase transitions between gradual and rapid change that the large and slow entities become sensitive to change from the small and fast ones.

When the system is reaching the limits to its conservative growth, it becomes increasingly brittle. The system is very locally stable, but that stability extends over a small domain. A small disturbance can push it out of that stable domain into abrupt and dramatic change. The nature and timing of the collapse-initiating disturbance determines, within some bounds, the future trajectory of the system. Therefore this brittle state presents the opportunity for a small-scale change to rapidly cascade through the overconnected system, bringing about its rapid transformation. This is the 'revolt of the slave variable' (Diener and Poston, 1984). The path to collapse can be initiated by both internal conditions or external events, but typically it is internally induced brittleness that sets the conditions for collapse.

The second opportunity for small-scale processes to cause system change is during the transition from reorganization to exploitation – from α to r. During this reorganization phase the system is in a state opposite to the one previously described. There is little local regulation and stability, so that the system can easily be moved from one state to another. Resources for growth are present, but they are disconnected from the processes that facilitate and control growth. In such a weakly connected state, a small-scale change can nucleate a novel structure amidst the sea of chaos. This structure can then use the available resources to grow explosively and establish a new exploitative path along which the system develops and then locks into. This transition occurs as small-scale changes sow seeds of order in the larger and slower chaos within which they are embedded. That represents a transient but critically important bottom-up asymmetry that provides the opening for evolutionary change.

It is this stage where the third axiom of biology most manifests itself. For it is at this disconnected phase where the greatest amount of reshuffling of existing entities occurs and where novel assemblages might emerge. It turns the image of Nature Resilient, where cycles ever repeat themselves into Nature Evolving. The role of this α phase in ecology is the same as that phase in cell biology where cell division and fusions allow rearrangement of genes. It is where experiments are initiated to be tested in later phases. It is where variety and diversity fulfils an evolutionary role.

Sustainability and Diversity

To this point I have emphasized the evolutionary or creative role for diversity. But diversity also provides a foundation for sustainability as well. That role helps maintain the resilience of existing structures whether ecological, social or economic. I shall turn again to ecology for example.

Any ecosystem – a forest, a lake, a grassland, or a wetland – contains hundreds to thousands of species interacting among themselves and their physical and chemical environment. But not all those interactions have the same strength or the same direction. That is, although everything might be ultimately connected to everything else if the web of connections are followed far enough, the first order interactions that structure the system, increasingly seem to be confined to a small number of biotic and abiotic variables whose interactions form the 'template' (Southwood, 1977), or niches, that allow a great diversity of living things, to, in a sense, 'go along for the ride' (Carpenter and Leavitt, 1991; Cohen, 1991; Holling, 1992). Those species are affected by the ecosystem but do not, in turn, notably affect the ecosystem, at least in ways that our relatively crude methods of measurement can detect. Hence at the extremes, species can be regarded either as 'drivers' or as 'passengers' (Walker, 1992), although this distinction needs to be treated cautiously. The driver role of a species may only become apparent every now and then under particular conditions that trigger their key structuring function.

There are typically a variety of species engaged in each set of 'drivers'. Examples include the set of herbivore grazing antelopes that structure the vegetation of the savannas of East Africa at intermediate scales from metres to kilometres; or the suite of 35 species of insectivorous birds that set the timing for insect outbreaks in the forests of eastern Canada (Holling, 1988). In these sets of species, each one performs its actions somewhat differently from others, and each responds differently to external variability. The result of their joint action is an overlapping set of reinforcing influences that are less like the redundancy of engineered devices and more like portfolio diversity strategies of investors. The risks and benefits are spread widely to retain overall consistency in performance independent of wide fluctuations in the individual species. That is at the heart of the role of functional diversity in maintaining the resilience of ecosystem structure and function.

We chose to term this functional diversity following the terms suggested by Schindler (1988, 1990) and by Holling *et al.* (1994). Such diversity provides great robustness to the process and, as a consequence, great resilience to the system behaviour. Moreover, this seems the way many biological processes are regulated – overlapping influences by multiple processes, each one of which is inefficient in its individual effect but together operating in a robust manner. For example, those are the features of the multiple mechanisms controlling body temperature regulation in endotherms, depth perception in animals with binocular vision and direction in bird migration.

Because of the robustness of this functional diversity, and the non-linear way behaviour suddenly flips from one pattern to another and one set of controls to another, gradual loss of species involved in controlling structure initially would have little perceived effect over a wide range of loss of species. Then as loss of those species continued, different behaviour would suddenly emerge more and more frequently in more and more places. To the observer, it would appear as if only the few remaining species were critical when in fact all add to the resilience. Although behaviour would change suddenly, resilience (measured as the size of stability domains, sensu Holling, 1973) would gradually contract. The system, in gradually losing resilience, would become increasingly vulnerable to perturbations that earlier could be absorbed without change in function, pattern and controls.

The fundamental point is that only a small set of self-organizing processes made up of biotic and physical processes are critical in forming the structure and overall behaviour of ecosystems, and that these establish sets of relationships, each of which dominates over a definable range of scales in space and time. Each set includes several species of plants or animals, each species having similar but overlapping influence to give functional redundancy. It is those sets of species that control ecosystem structure and function. Protecting, conserving or restoring those species and their conditions for existence, at the same time will provide the niches for all others – those 'going along for the ride', and those that potentially can assume novel controlling influence as an evolutionary phenomenon.

Conclusion

The way key subcultures in the natural and social sciences view the world is converging on these theories of change. Those theories rationalize the paradoxes of stability and instability, of order and disorder or of stasis and evolutionary change. The image of Nature Resilient as one in opposition to that of Nature Evolving is only apparent. The appropriate image is of Nature Resilient and Evolving, since both attributes of flexible persistence and of experimentation interact. Since those are the same apparent paradoxes inherent in the goal of sustainability and development, an avenue opens for directly relevant cooperation between critical parts of science and policy.

Complex living systems need both stability and change. The first is provided by resilience – properties that allow a system to be subjected to external change and still persist in its basic operations and functions. Diversity of number of controlling species or entities, and diversity of those entities in space provide the enduring foundations for resilience. But if that was all, however, such persistence would be stasis – systems doomed to repeat identical cycles on a wheel of determinism.

Opportunity for novelty emerges because such cycles inexorably go through brief periods of disorganization and turbulence where unexpected events and novel combinations can seed a different future. Such experiments can be ever-repeated without persistent chaos for the whole because of the integrity provided by the hierarchical structure. So long as simple information is transferred between levels, repeated experimentation is possible within levels without collapse of the whole. The more those localized vulnerabilities at one level start to connect with similar vulnerabilities at neighbouring levels, the more turbulent and sweeping the possible transformations. That is, perhaps, what our global society is experiencing now. Local industries are restructuring, a revolution in communication and computers is transforming the scale of interactions among people, an ideological empire has collapsed and the enduring values that sustain us are weakened.

Václav Havel, President of the Czech Republic said it well on the occasion of receiving the Liberty Medal in Philadelphia, 4 July 1994. Our deeper scientific theories of the world and of our place in it

> remind us, in modern language, of what we have long suspected, of what we have long projected into our forgotten myths and perhaps what has long lain dormant within us as archetypes. That is the awareness of our being anchored in the Earth and the universe, the awareness of that we are not here alone nor for ourselves alone, but that we are an integral part of higher, mysterious entities against whom it is not advisable to blaspheme. This forgotten awareness is encoded in all religions. All cultures anticipate it in various forms. It is one of the things that forms the basis of man's understanding of himself, of his place in the world, and ultimately of the world as such.

Protecting and restoring the diversity of life is one of the foundations for sustainability. But sustainability is not just for ecological or social or economic reasons. It is for all three. Sustainability as a single goal and a single focus could lead to stasis and rigidity unless the opportunities are available for evolutionary change, i.e. for human development in its deepest sense of physical, intellectual and spiritual well-being.

References

Allen, T.F.H. and Starr, T.B. (1982) *Hierarchy: Perspectives for Ecological Complexity*. The University of Chicago Press, Chicago.

Arthur, B.W. (1989) Competing technologies increasing returns, and lock-in by historical events. *The Economic Journal* 99, 116–131.

Broecker, W.S. (1987) Unpleasant surprises in the greenhouse? *Nature* 328, 123–126.

Brooks, H. (1986) The typology of surprises in technology, institutions and development. In: Clark, W.C. and Munn, R.E. (eds) *Sustainable Development of the Biosphere*. ILASA, Laxenburg, Austria, pp. 325–347.

Carpenter, S.R. and Leavitt, P.R. (1991) Temporal variation in paleoimnological record arising from a tropic cascade. *Ecology* 72, 277–285.

Cohen, J. (1991) Tropic topology. *Science* 251, 686–687.

Crutzen, P.J. and Arnold, F. (1986) Nitric acid cloud formation in the cold Antarctic stratosphere: a major cause for the springtime 'ozone hole'. *Nature* 324, 651–655.

Diener, M. and Poston, T. (1984) On the perfect delay convention or the revolt of the slaved variables. In: Haken, H. (ed.) *Chaos and Order in Nature*. Springer-Verlag, Berlin, pp. 249–268.

Douglas, M. (1978) Cultural bias. Paper for the Royal Anthropological Institute, London No. 35. Royal Anthropological Institute.

Elliott, J.E. (1980) Marx and Schumpeter on capitalism's creative destruction: a comparative restatement. *Quarterly Journal of Economics* XCV(1), 45–68.

Holling, C.S. (1973) Resilience and stability of ecological systems. *Annual Review of Ecology and Systematics* 4, 1–23.

Holling, C.S. (1986) The resilience of ecosystems; local surprise and global change. In: Clark, W.C. and Munn, R.E. (eds) *Sustainable Development of the Biosphere*. Cambridge University Press, Cambridge, pp. 292–317.

Holling, C.S. (1987) Simplifying the complex: the paradigms of ecological function and structure. *European Journal of Operational Research* 30, 139–146.

Holling, C.S. (1988) Temperate forest insect outbreaks, tropical deforestation and migratory birds. *Memoirs of the Entomological Society of Canada* 146, 21–32.

Holling, C.S. (1992) Cross-scale morphology, geometry and dynamics of ecosystems. *Ecological Monographs* 62, 447–502.

Holling, C.S. (1994) New science and new investments for a sustainable biosphere. In: Jansson, A.M., Hammer, M., Folke, C. and Costanza, R. (eds) *Investing in Natural Capital*. Island Press, Washington, DC, pp. 57–73.

Holling, C.S., Schindler, D.W., Walker, B. and Roughgarden, J. (1994) Biodiversity in the functioning of ecosystems: an ecological primer and synthesis. In: Perrings, C., Molder, K.-G., Folke, C., Holling, C.S. and Jansson, B.-O. (eds) *Biodiversity: Ecological and Economic Foundations*. Cambridge University Press, Cambridge.

Kauffman, S.A. (1992) *Origins of Order: Self-Organization and Selection in Evolution*. Oxford University Press, Oxford.

Levin, S.A. (1992) The problem of pattern and scale in ecology. *Ecology* 73, 1943–1967.

Lorenz, E. (1963) Deterministic nonperiodic flow. *Journal of the Atmospheric Sciences* 20, 130–141.

May, R.M. (1977) Thresholds and breakpoints in ecosystems with a multiplicity of stable states. *Nature* 269, 471–477.

McNeill, W.H. (1979) *The Human Condition: An Ecological and Historical View*. Princeton University Press, Princeton.

Michael, D.N. (1984) Reason's shadow: notes on the psycho dynamics of obstruction. *Technological Forecasting and Social Change* 26, 149–153.

Nicolis, G. and Prigogine, I. (1977) *Self-Organization in Non-equilibrium Systems*. John Wiley, New York.

O'Neill, R.V., DeAngelis, D.L., Waide, J.B. and Allen, T.F.H. (1986) *A Hierarchical Concept of Ecosystems*. Princeton University Press, Princeton.

Pearl, R. (1927) The growth of populations. *Quarterly Review* 2, 532–548.

Prigogine, I. (1980) *From Being to Becoming, Time and Complexity in the Physical Sciences*. W.H. Freeman, New York.

Schindler, D.W. (1988) Experimental studies of chemical stressors on whole lake ecosystems. Baldi Lecture. *Verf. Internat. Verein.* 23, 11–41.

Schindler, D.W. (1990) Experimental perturbations of whole lakes as tests of hypotheses concerning ecosystem structure and function. *Oikos* 57, 25–41.

Schumacher, E.F. (1973) *Small is Beautiful: Economics as if People Mattered*. Harper & Row, New York.

Schumpeter, J.A. (1950) *Capitalism, Socialism and Democracy*. Harper, New York.

Simon, J.L. and Kahn, H. (1984) *The Resourceful Earth: A Response to Global 2000*. Basil Blackwell, Oxford.

Southwood, T.R.E. (1977) Habitat, the templet for ecological strategies? *Journal of Animal Ecology* 46, 337–365.

Thompson, M. (1983) A cultural bias for comparison. In: Kunreuther, H.C. and Linnerooth, J. (eds) *Risk Analysis and Decision Processes: The Siting of Liquid Energy Facilities in Four Countries*. Springer, Berlin.

Tuchman, B.W. (1978) *A Distant Mirror*. Ballantine, New York.

Waldrop, M.M. (1992) *Complexity*. Simon and Schuster, New York.

Walker, B.H. (1992) Biological diversity and ecological redundancy. *Conservation Biology* 6, 18–23.

Wildavsky, A.B. and Douglas, M. (1982) *Risk in Culture: An Essay on the Selection of Technical and Environmental Dangers*. University of California, Berkeley.

2 Molecular Mechanisms Promoting and Limiting Genetic Variation

W. Arber
Department of Microbiology, Biozentrum, University of Basel, Klingelbergstrasse 70, CH-4056 Basel, Switzerland

Summary

A large body of knowledge on microbial evolution has been obtained in studies on microbial genetics during the last 50 years. In particular, many different molecular mechanisms each contributing in its specific way to spontaneous mutagenesis and genetic recombination on the one hand and to the limitation of genetic variation to tolerable rates on the other hand have been characterized in considerable detail. Whereas many of these often genetically encoded mechanisms act within genomes (intragenomic level) to bring about genetic variation in microbial populations, DNA acquisition (horizontal gene transfer) acts between genomes (intergenomic level). DNA acquisition by either transformation, conjugation or virus-mediated transduction is a quite frequently encountered event which can stepwise improve microbial fitness. On this view, the evolution of microbial strains should be considered as the result of both the steady divergence of the genetic information and the occasional acquisition of biological functions 'developed' by other branches of the evolutionary tree, which should thus be drawn with horizontal connectors between its branches. The intragenomic development and the acquisition of genetic information as well as the control over the frequency of genetic variation largely depend on genetically encoded enzyme systems and organelles. Nevertheless, biological evolution is not directed towards a specific goal. Rather, variety generator functions provide populations of organisms with many different genetic variants, the fate of which depends on the action of steadily exerted natural selection.

Growth and Evolution of Haploid Microorganisms

Bacteria are unicellular haploid microorganisms which propagate vegetatively by cell fission. The prototype of genetic studies, *Escherichia coli* K-12, carries the bulk of its genetic information on a single chromosome of about 4.7×10^6 base pairs length. Occasionally, bacteria contain accessory genetic information on one or more small DNA molecules called plasmids. Upon cell division, each daughter cell receives usually a full set of the cellular genetic information called the bacterial genome. Genome duplication must thus precede cell division. DNA replication is one of the sources of genetic variation. Other sources are actions of both external and internal mutagens and different kinds of DNA rearrangement or recombination processes. Genetic variation occurring by any of these processes brings about a DNA sequence alteration in individual cells of a bacterial population. This alteration, or spontaneous mutation, may or may not become manifest as a phenotypic alteration. In the extreme situation, a single mutation may be lethal for the cell. In all other cases, the mutant cell will initiate the formation of a mutant subclone in the growing microbial population.

Because of the haploid nature of the bacterial genome, phenotypic manifestation of mutations is rapid and natural selection becomes quickly manifest and can provide selective disadvantage to less fit mutants and selective advantage to occasionally occurring better fit mutants. In growing cultures of *E. coli*, one can statistically observe one new mutant per a few hundred cells in each generation. Therefore, in an exponentially propagating bacterial culture it is not possible to maintain a genetically pure clone for a long time. Rather, the growing bacterial population originating from a single cell will contain an increasing proportion of mutants of different kinds. Molecular genetic methodology allows one in many cases to identify and characterize new mutations and to conclude on the mutagenesis mechanism involved and on its evolutionary role.

Classical Microbial Genetics Revealed Natural Strategies for Gene Acquisition

The first identification of bacterial mutants about 50 years ago indicated that bacteria have genes. Attempts to explore bacterial genetics experimentally were quite successful and rapidly revealed three distinct natural processes by which genetic information from a donor strain becomes horizontally transferred to a recipient strain. These processes are: (i) transformation in which DNA fragments liberated from the donor bacterium are taken up from the environment by a recipient bacterium, (ii) conjugation in which a conjugative plasmid mediates a direct contact between the donor and the recipient cells as well as the directed transfer of genetic information from the former to the latter, and (iii) transduction, in which a bacterial virus particle acts as a vector for the transfer of donor DNA to a recipient bacterium being infected by the transducing phage particle.

In each case, a partial diploidy can result from the gene transfer which is often followed by recombinational integration of the acquired genes into the recipient genome, usually by general recombination between homologous sequences. Occasionally, other recombination processes such as site-specific recombination or transposition can also establish a permanent association of the acquired genes with the recipient genome. Molecular mechanisms involved in these processes of horizontal gene transfer and in the stable establishment of transferred genetic information in the recipient bacteria have been thoroughly explored in the past decades and are mostly quite well understood. As a rule, enzymes catalyse these processes and gene products are also engaged in the formation of organelles (such as conjugative pili and of viral particles) promoting horizontal DNA transfer.

Prerequisites for Success of DNA Acquisition

Despite an involvement of gene products in the natural strategies for horizontal DNA transfer, successful gene acquisition is statistically relatively rare in nature, particularly if donor and recipient cells do not belong to the same strain of bacteria. This is in line with the evolutionary role of the process which should also ensure a certain degree of genetic stability of the evolving population. Gene transfer can occasionally occur between relatively unrelated species of microorganisms, e.g. between Gram-positive and Gram-negative bacteria. The low frequencies result from a number of natural barriers against horizontal gene transfer and against the stable establishment of acquired foreign genetic information.

Prerequisites for the success of DNA acquisition include the conditions:

1. that the foreign DNA is taken up by the recipient cell which does require either competence for transformation, surface compatibility for conjugation or susceptibility to viral infection in the case of transduction;

2. that invading foreign DNA withstands the degradation by restriction systems of the recipient bacteria;

3. that acquired genes become stably inherited, which requires either recombination with the recipient genome or autonomous replication in the recipient cell and adequate partition upon cell division; and

4. that acquired genetic information is expressed and that the resulting gene products do not disturb the functional harmony of the recipient cell.

It is at this last level that natural selection exerts its action in analogy to what happens after any other mutagenesis event altering the genetic potential of an individual.

One can observe that gene acquisition is statistically most successful if it occurs in small steps. In this regard it is interesting to note that restriction systems, while reducing the overall frequency of DNA acquisition, also stimulate the establishment of short segments of genetic information in the recipient genome, because linear DNA fragments with free ends are recombinogenic. In addition, DNA acquisition in small steps reduces the possibility that the functional harmony of the resulting hybrid genome becomes disturbed. In view of the universality of the genetic code, biological functions developed in one species of microorganisms have a good chance to carry out the same functions after horizontal transfer also in cells of another species, although differences in codon usage may affect the efficiencies of acquired functions.

Intragenomic DNA Rearrangements as Sources of Genetic Variation

Although bacteria reproduce asexually and are thought not to depend on recombination for their propagation, they encode a number of different recombination systems. General recombination requires relatively extended regions of sequence homology at which the recombination exchange can occur. This enzyme-mediated 'classical' recombination process, which is well known for its generation of genetic variety at the level of chromosomes in diploid organisms, cannot carry out this same function in haploid organisms. However, these enzymes are nevertheless present in bacteria, where they fulfil repair functions on DNA damaged by various mutagens. General recombination also gives rise to unequal crossing over by acting at homologous DNA segments carried at different locations in the genome, e.g. gene duplications and higher amplifications can result from this process, as well as replicon fusion, deletion formation and DNA inversion.

Cointegration of different DNA molecules, deletion formation and DNA inversion can also be mediated by site-specific recombination systems. Such systems are quite widespread and a number of them are mechanistically well understood on the basis of both *in vivo* and *in vitro* explorations. Here, extended sequence homology is not required for efficient recombination. Rather, specific DNA sequences are chosen by the involved enzymes to site-specifically carry out DNA strand cleavage and religation. For some such enzymatic systems it was shown that recombination can also occur, although with decreasing efficiencies, at DNA sites which increasingly diverge from the preferred site. It is this property to occasionally carry out a recombination at an unusual site which offers the possibility to create, once in a while, a novel DNA arrangement. This has, for example, been studied for DNA inversion, by which the rare use of a DNA site deviating from the consensus site can bring about new gene fusions, in which parts of the reading frames of two genes are fused together. The same recombination process can also bring a gene under another expression control. Hence, the evolutionary relevance of the activities of such DNA rearrangement systems resides in the production of novel DNA arrangements at low frequencies. Since many different sequences can occasionally serve for such recombination, the system can be considered as a variety generator, by which some individuals in large populations undergo sequence alterations which are then submitted to natural evolutionary selection.

Transposition of mobile genetic elements is another widely encountered recombination process which functions as variety generator. *E. coli* carries several mobile genetic elements in its genome, often in several copies at different locations for a specific element. At low frequencies, these elements undergo transposition, i.e. a copy can become inserted at a newly selected site in the genome. This process is mediated by transposase, an enzyme encoded by the mobile genetic element itself. Criteria for the selection of insertion targets by the transposase vary from one element to another. Some elements highly prefer specific DNA sequences as targets, but occasionally also use other sequences. Other elements display a regional preference for insertion, but can use many different non-homologous sites within the preferred region or occasionally outside of this region. Therefore, a very large number of potential integration sites for mobile genetic elements exists in the bacterial genome. Often biological functions become altered by such a transposition event. This alteration may more often affect a function negatively than positively. The latter condition can, for example, be found if transposition brings a gene under a different expression control. In addition to simple transposition, as has been described here, transposition can go along with more extensive DNA rearrangements such as DNA inversion, deletion or cointegration. Finally, a segment with chromosomal genetic information may sometimes become mobilized if this information is carried between two identical mobile genetic elements. This can be highly relevant in DNA acquisition processes both for the mobilization of chromosomal genes and for their stable establishment after transfer into a recipient bacterium. Much knowledge has been gained in recent years on the molecular interactions of the described transposition processes, and families of mechanistically related processes have been defined. These families often have ramifications into analogous mobile genetic elements carried in the chromosomes of higher organisms.

Synopsis of Molecular Processes Involved in the Generation and Limitation of Genetic Diversity

Biological evolution is a highly dynamic process. Genetic diversity reflects the present state of biological evolution. It thus gives a static picture of the evolutionary process. The left side of Fig. 2.1 shows four categories of sources of genetic variation, each containing a number of different molecular processes. DNA acquisition and DNA rearrangements have already been discussed. The effects of environmental mutagens such as some chemical substances or radiations and the consequences of infidelities of DNA replication are quite well known and have been widely described in the literature. Processes of each of these categories contribute in parallel to the formation of genetic variations, each in its specific way and with characteristic frequency. On the other hand, as has been described, natural selection is a

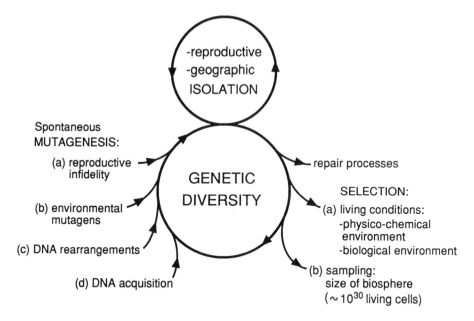

Fig. 2.1. Genetic diversity reflects the actual state of biological evolution in which the four categories of processes drawn to the left generate diversity. Individual sequences are withdrawn from the pool of diversity by repair processes and by natural contraselection. Note that mutagenesis is defined here as any spontaneous alteration of nucleotide sequences of the genome, independently of the molecular mechanism involved and of the resulting phenotype.

major force steadily reducing genetic diversity, since only a minority of novel DNA sequence variations can provide a selective advantage to mutated cells and their progeny. In addition, enzymatically mediated repair systems rapidly remove a majority of sequence alterations resulting both upon DNA replication and from the action of environmental mutagens. Since living conditions are in no way stable but show temporal and local variations, selection does not always give preference to the same genotypes. Selection is an important doorway to the impact of the environment on the evolution of life, although in most situations so far explored selection has no immediate directional influence on the production of specific mutants; it normally just acts selectively on the populations of the different phenotypes available.

The parallel action of several mechanistically different mutagenesis processes may have its deeper relevance, since the products resulting from different mutation mechanisms may show qualitative differences. For example, whereas nucleotide substitution, which often results from replication infidelity, may cause the replacement of one amino acid of a protein by another, DNA inversion may bring about the fusion of two DNA domains encoding different functions. The consequence of such gene fusion may be a novel biological function.

A certain degree of genetic stability is essential for the long-term maintenance of various forms of life. Therefore, mutagenesis mechanisms must be relatively inefficient. Only a small proportion of cells of a population should be affected by mutation per generation. This goal is achieved by a combination of quite different strategies such as control of the expression of genetic functions involved in mutagenesis, translational and post-translational regulations, but also antagonistic effects, such as systems for the repair of DNA damage. Similar antagonistic effects have already been discussed to keep DNA acquisition in a tolerable and useful range of rates. These findings are consistent with the well-known fact that a labile equilibrium is best maintained if promoting and limiting forces act antagonistically on the system in question.

The independent actions of mutagenesis mechanisms, as defined by the first three categories shown in Fig. 2.1, are steadily exerted on the entire genome of bacteria as genetic information is transmitted from generation to generation. Some of the mechanisms also act on DNA of resting cells which have no opportunity to grow. The sum of these mutagenic events can explain the evolutionary tree which one can draw in the classical way with an aleatoric branching as a function of time and with an occasional ending of a branch by extinction. However, in view of the gene acquisition strategy which is quite efficient in bacteria, horizontal connectors between branches should be added to the classical tree. Such connectors can temporarily function between many possible pairs of branches. In most cases of horizontal gene transfer, rather small parts of the donor genome are successfully incorporated into the recipient genome, following the strategy of acquisition in small steps. It is to be noted that wide ranges of microbial strains form a unity of life in view of these evolutionary mechanisms. They have not only their common root back at an origin far in the past, they can also profit at any time of successful developments accomplished by other branches. However, acquisition is in no way a directed process. DNA transfer just occurs once in a while. Its usefulness is decided upon by the steadily acting selection, hence by the living conditions encountered by the organisms. In view of these possibilities, DNA acquisition also projects the unity of life into a far future.

Genes Encoding Evolutionary Functions are Carried on the Genome

Scientific publications often refer to spontaneous mutations as 'errors' committed by enzymes or they interpret mobile genetic elements as being 'selfish'. Is evolution indeed the result of accumulated errors? On the basis of an overwhelming amount of indications obtained in microbial genetics during the last 50 years, I rather promote the view that evolution is heavily influenced by the action of a large number of gene products, many of which may have no other role than to contribute to evolution. Examples of enzymes acting as variety generators and of organelles promoting horizontal gene transfer have been given. Their actions are often counteracted by other gene products, such as DNA repair enzymes or restriction enzymes, whose role may be to balance promoters of genetic variation into a range of genetic plasticity both tolerable and useful for the evolving microbial strain in cause.

In conclusion, we postulate that besides a large number of housekeeping genes and genes for accessory functions which may be required for all individuals of a population, the bacterial genome also carries other genes, the functions of which are not relevant for the accomplishment of individual lives. These genetic determinants rather ensure a progressive evolution by providing means for a steady adaptation to changing living conditions. Interestingly, the genes for these evolutionary functions are carried and inherited on the same chromosome which carries also the genes essential for the lives of all individuals. This statement may appear obvious, since the only way to inherit genes is to carry them on the chromosome. However, from another point of view, the presence of genes, the actions of which are more often destructive than useful is puzzling and might not point to the 'altruistic' nature which I would like to attribute to evolutionary functions. Variety generators generally cause more deleterious mutations than beneficial ones. Because of this situation, individual evolutionary genes are probably not selected for directly, i.e. by their own contribution to individual fitness. Rather, genes with evolutionary functions are indirectly selected by their occasional production of better fit mutants of housekeeping and accessory genes, the products of which act to the benefit of individuals.

It will be an interesting scientific task to explore how far the conclusions drawn from data of microbial genetics are also relevant for higher organisms.

Acknowledgement

The author acknowledges the financial support received from the Swiss National Science Foundation, grants 31-30040.90 and 5001-035228.

References

Arber, W. (1990) Mechanisms of microbial evolution. *Journal of Structural Biology* 104, 107–111.
Arber, W. (1991) Elements in microbial evolution. *Journal of Molecular Evolution* 33, 4–12.
Arber, W. (1993) Evolution of prokaryotic genomes. *Gene* 135, 49–56.
Arber, W. (1995) The generation of variation in bacterial genomes. *Journal of Molecular Evolution* 40, 7–12.
Berg, D.E. and Howe, M.M. (eds) (1989) *Mobile DNA*. American Society of Microbiology, Washington, DC.

Joset, F. and Guespin-Michel, J. (1993) *Prokaryotic Genetics: Genome Organisation, Transfer and Plasticity*. Blackwell Scientific Publications, Oxford.

Kuperlapati, R. and Smith, G.R. (eds) (1988) *Genetic Recombination*. American Society of Microbiology, Washington, DC.

Moses, R.E. and Summers, W.C. (eds) (1988) *DNA Replication and Mutagenesis*. American Society of Microbiology, Washington, DC.

Neidhardt, F.C., Ingraham, J.L., Low, K.B., Magasanik, B., Schaechter, M. and Umbarger, H.E. (eds) (1987) *Escherichia coli and Salmonella typhimurium*, Vol. 2. American Society of Microbiology, Washington, DC.

Genome Dynamics and the Generation of Biodiversity

G. Bernardi
Laboratoire de Génétique Moléculaire, Institut Jacques Monod, 2 Place Jussieu, 75005 Paris, France

Summary

Some years ago Wilson and co-workers proposed that the higher rates of karyotypic change and species formation of mammals compared to cold-blooded vertebrates are due to the formation of small demes, as favoured by the social structuring and brain development of the former. Here, evidence is reviewed which indicates that mammals are more prone to karyotypic change and species formation than cold-blooded vertebrates because of their different genome organization. A similar evidence has also recently become available for birds. Although this different organization appears to be a necessary and, in all likelihood, a sufficient condition for the increased rates of karyotypic change and species formation found in mammals, it is still possible that social structuring and brain development may have played an additional accelerating role.

Introduction

Two basic issues in biodiversity are the formation and the extinction of species. As pointed out recently (Benton, 1995) the present-day 5–50 million species were reached by a process of massive diversification, which was not, apparently, a stochastic process, substantial bursts of multiplication of major taxonomic groups coinciding with their invasion of new habitats, which was often associated with the acquisition of new adaptations. This diversification was interrupted by mass extinctions, the largest of which have been identified.

In contrast with species extinction, species formation raises very special problems. The difficulty of producing discontinuities, namely groups (species), from a continuous process (evolution) was realized by Darwin who considered that speciation was 'the mystery of mysteries' and by Bateson ('the origin and nature of species remains utterly mysterious'). Needless to say, discussions on the nature of speciation assume that species have an objective existence, a point on which not everybody is in agreement. We will accept here Mayr's definition of species as groups of actually or potentially interbreeding natural populations which are reproductively isolated from other such groups.

It should be briefly recalled here, first, that isolating factors may be prezygotic (e.g. seasonal or habitat isolation) or postzygotic (hybrid inviability and sterility); and, second, that the major speciation models comprise: (i) *allopatry*, physical isolation inevitably leading to evolutionary divergence through natural selection and genetic drift; extensive biogeographical evidence exists in favour of this model; (ii) *sympatry*, involving no physical isolation, but genetic similarity except at the few loci responsible for reproductive isolation and ecological divergence; this model is supported by the cichlids of Lake Victoria; and (iii) *stasipatry*, chromosomal rearrangements reducing fecundity when heterozygous.

We will concentrate on the last model because the view that karyotypic change and species formation are closely linked appears to be supported in the case of the speciation of vertebrates (which will be discussed here) by the work of Wilson and co-workers.

According to White (1968) and Grant (1973), fixation of karyotypic mutations can facilitate species formation and adaptive evolution at the organismal level (see Bush et al., 1977) by acting: (i) as a sterility barrier (stasipatric species formation; White, 1968, 1978), the mutant karyotype functioning at the population level as a cytogenetic reproductive isolating mechanism; (ii) as a regulatory mutation, producing an altered pattern of gene expression that results in an organism with a new and fitter phenotype (see, for example, Zieg et al., 1977); (iii) as a linker of loci that previously were far apart in the genome, thereby creating a particular combination of alleles (Dobzhansky, 1970).

These points were reconsidered by Wilson et al. (1974, 1975), Levin and Wilson (1976) and Bush et al. (1977) when they found that cold-blooded vertebrates exhibit a species formation rate which is, on the average, 20% that of mammals, and a karyotypic change rate which is close to 10% that of mammals. Bush et al. (1977) proposed that

> The propensity to form small demes is attributable to several factors, one of which may be especially important for understanding how mammals have achieved remarkably high rates of adaptative evolution. This factor is social structuring. If it were not for this social factor, mammals, because of their high dispersal power, might have evolved at the same rate or more slowly than most lower vertebrates.

More recent work has stressed the importance of brain development in organismic evolution (Wyles et al., 1983; Wilson, 1985).

I will propose here a different explanation for the different karyotypic change rate of cold-blooded vertebrates and mammals. The basic idea,

Genome Dynamics and the Generation of Biodiversity 69

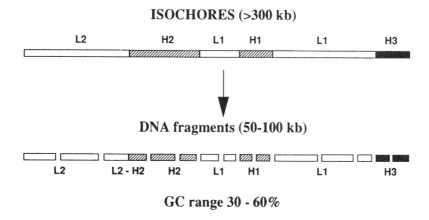

Fig. 3.1. Scheme of the isochore organization of the human genome. This genome, which is a typical mammalian genome, is a mosaic of large (>300 kb) DNA segments, the isochores. These are compositionally homogeneous (above a size of 3 kb) and can be subdivided into a small number of families, GC-poor (L1 and L2), GC-rich (H1 and H2), and very GC-rich (H3). The GC-range of the isochores from the human genome is 30-60% (from Bernardi, 1993b).

first mentioned six years ago (Bernardi, 1989) and further developed later (Bernardi, 1993a), is that the genomes of mammals are more prone to karyotypic changes than those of cold-blooded vertebrates, because of their different organization. The evidence in favour of this explanation will be reviewed. Although the genome organization of mammals appears to be a necessary and, in all likelihood, a sufficient condition for their increased rates of karyotypic change and species formation, the factors discussed by Wilson and co-workers may have played an additional accelerating role.

Genome Organization is Different in Mammals and Cold-blooded Vertebrates

The vertebrate genomes are mosaics of *isochores* (Fig. 3.1), comprising long DNA segments that are homogeneous in base composition (Macaya *et al.*, 1976; Thiery *et al.*, 1976). In warm-blooded vertebrates, isochores (or, more precisely, the 100–200 kb DNA fragments derived from them; see Fig. 3.1) cover a very wide GC range, and attain very high GC values. In the human genome, isochores can be assigned to two GC-poor families (L1 and L2) representing two thirds of the genome, and to three GC-rich families (H1, H2 and H3) forming the remaining one third (Fig. 3.2). In contrast, in cold-blooded vertebrates, isochores are much more uniform in composition and never attain the highest GC levels of the genomes of mammals (Fig. 3.5), as shown by Thiéry *et al.* (1976), Hudson *et al.* (1980) and of Bernardi and Bernardi (1990a,b).

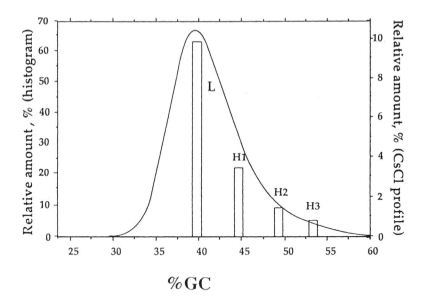

Fig. 3.2. Histogram of the isochore families from the human genome. The relative amounts of major DNA components derived from isochore families L (i.e., L1 + L2), H1, H2, H3 are superimposed on the CsCl profile of human DNA (from Mouchiroud et al., 1991).

On the other hand, histograms of GC levels of third codon positions of genes from mammals are very highly biased towards very high GC levels. In cold-blooded vertebrates, in contrast, the distribution covers a lower GC range and is more symmetrical (Bernardi et al., 1985, 1988; Bernardi and Bernardi, 1991) (Fig. 3.3). In conclusion, the compositional patterns of mammals and cold-blooded vertebrates are very different, at both the DNA and the coding sequence levels.

Gene distribution is similar in all vertebrates

Compositional correlations (Bernardi et al., 1985) exist between exons (and their codon positions) and isochores (Fig. 3.4), as well as between exons and introns (Aïssani et al., 1991). These correlations concern, therefore, coding and non-coding sequences and are not trivial since coding sequences only make up about 3% of the genome, whereas non-coding sequences correspond to 97% of the genome. The compositional correlations represent a *genomic code* (Bernardi, 1993b). It should be noted that a **universal correlation** (Fig. 3.4) holds among GC levels of codon positions (third positions against first and/or second positions). This is apparently due to compositional constraints working in the same direction (towards GC or AT), although to different extents, on different codon positions, as well as on the isochores.

The compositional correlations between GC_3 (the GC level of third codon

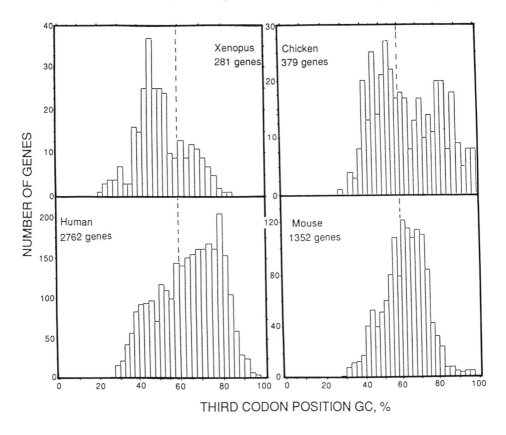

Fig. 3.3. Compositional distribution of third codon positions from vertebrate genes. The number of genes taken into account is indicated. A 2.5% GC window was used. The broken line at 60% GC is shown to provide a reference (from Bernardi, 1993b).

positions) and isochore GC have a practical interest in that they allowed us to position the coding sequence histogram of Fig. 3.3 relative to the CsCl profile of Fig. 3.5 and to assess the *gene distribution* in the human genome (Mouchiroud et al., 1991; Bernardi, 1993b). In fact, if one divides the relative number of genes per histogram bar by the corresponding relative amount of DNA, one can see that gene concentration is low and constant in GC-poor isochores, increases with increasing GC in isochore families H1 and H2, and reaches a maximum in isochore family H3, which exhibits at least a 20-fold higher gene concentration compared to GC-poor isochores (Fig. 3.6).

Recent results (mentioned in Bernardi, 1993b) have shown that the gene distribution is similar in all vertebrate genomes in that the isochore family H3 hybridizes preferentially on the GC-richest DNA fractions in all cases.

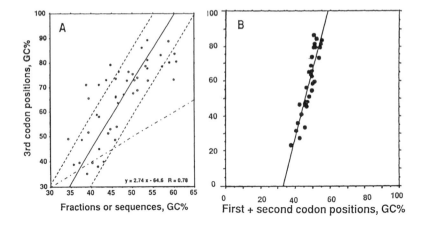

Fig. 3.4. GC levels of third codon positions from human genes are plotted against the GC levels of DNA fractions (dots) or extended sequences (circles) in which the genes are located. The correlation coefficient and slope are indicated. The dash-and-point line is the diagonal line (slope = 1). GC levels of third codon positions would fall on this line if they were identical to GC levels of surrounding DNA. The broken lines indicate a ± 5% GC range around the slope (from Mouchiroud et al., 1991).

Chromosome organization is different in cold- and warm-blooded vertebrates

Differences related to those described at the DNA and coding sequence level also exist at the chromosomal level. In human metaphase chromosomes T(elomeric) bands, the subset of R(everse) bands that is most thermal-denaturation resistant, are essentially formed by GC-rich isochores (mainly of the H2 and H3 families). In contrast, R'-bands, namely the R-bands exclusive of T-bands, comprise both GC-rich isochores (of the H1 family) and GC-poor isochores. Finally, G(iemsa) bands are formed almost exclusively by GC-poor isochores (Saccone et al., 1992, 1993; see Fig. 3.7). The difference in GC level between G-bands and T-bands is about 15%. About 20% of genes are present in G-bands and about 80% in R-bands (60% of them in T-bands). The location of a majority of genes in T-bands is of interest in view of the association of telomeres with the nuclear matrix and envelope (de Lange, 1992).

Whereas the organization of metaphase chromosomes just described is essentially valid for all mammals and also for birds, cold-blooded vertebrates are characterized by metaphase chromosomes in which an R banding cannot be elicited (Schmid and Guttenbach, 1988).

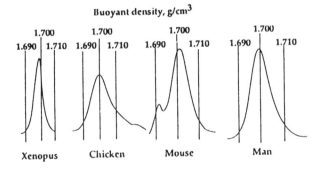

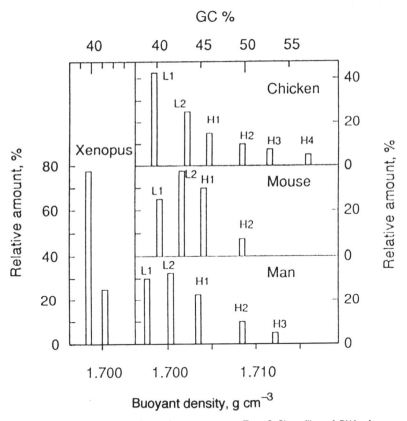

Fig. 3.5. Compositional patterns of vertebrate genomes. Top: CsCl profiles of DNAs from *Xenopus*, chicken, mouse, and man (from Thiery et al., 1976). Bottom: Histograms showing the relative amounts, modal buoyant densities and GC levels of the major DNA components from *Xenopus*, chicken, mouse, and man, as estimated after fractionation of DNA by preparative density gradient in the presence of a sequence-specific DNA ligand (Ag^+ or BAMD; BAMD is bis (acetatomercurimethyl) dioxane). The major DNA components are the families of large DNA fragments (see Fig. 3.1) derived from different isochore families. Satellite and minor DNA components (such as rDNA) are not shown in these histograms (from Bernardi, 1993b).

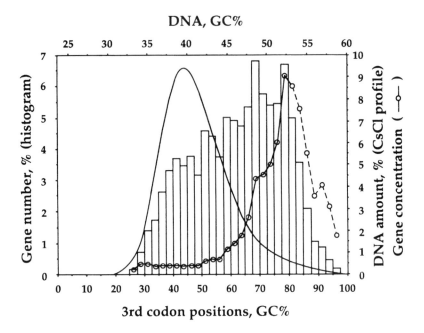

Fig. 3.6. Profile of gene concentration in the human genome as obtained by dividing the relative amounts of genes in each 2.5% GC interval of the histogram by the corresponding relative amounts of DNA deduced from the CsCl profile. The apparent decrease in gene concentration for very high GC values (broken line) is due to the presence of rDNA in that region. The last concentration values are uncertain because they correspond to very low amounts of DNA (from Bernardi, 1993b).

Rearrangements are most frequent at compositional discontinuities

The genomes of warm-blooded vertebrates differ compositionally from those of cold-blooded vertebrates in that about one third of the mammalian genome underwent GC increases, whereas the remaining two-thirds did not. These two compartments were called the neogenome and the paleogenome, respectively (Bernardi, 1989; Fig. 3.8). Such changes took place through the fixation of directional point mutations, as shown by sequence comparisons of homologous genes (Bernardi et al., 1988). The compositional differences in the isochore pattern of warm- compared to cold-blooded vertebrates are paralleled by chromosomal changes, namely the formation of R-bands (and their subset of T-bands).

The crucial point now is that recombinational events in mammals are much more frequent in the Reverse bands and at the Reverse-Giemsa border than in Giemsa bands or than in the genomes of cold-blooded vertebrates.

Detailed information, mainly concerning the human genome, shows that translocation breakpoints are not randomly located on chromosomes (Sutherland and Hecht, 1985; see also Fig. 3.9). Indeed, R-bands and G/R borders are the predominant sites of exchange processes, including

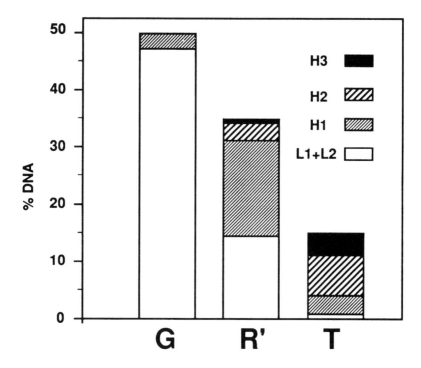

Fig. 3.7. A scheme of the relative amounts of isochore families L1 + L2, H1, H2 and H3 in G-bands, R'-bands and T-bands; R'-bands are R-bands exclusive of T-bands (from Saccone et al., 1993).

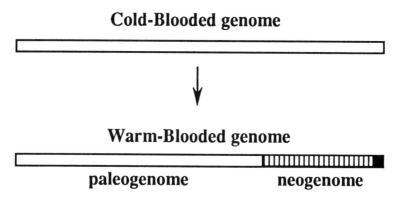

Fig. 3.8. Scheme of the compositional genome transition accompanying the emergence of warm-blooded from cold-blooded vertebrates. The compositionally fairly homogeneous genomes of cold-blooded vertebrates are changed into the compositionally heterogeneous genomes of warm-blooded vertebrates. The latter comprise a paleogenome (corresponding to about two thirds of the genome) which did not undergo any large compositional change and a neogenome (corresponding to the remaining third of the genome, with the GC-richest part only representing 3% of the genome). In the scheme, GC-poor isochores (open bar), GC-rich isochores (hatched bar) and GC-richest isochores (black bar) are represented as three contiguous regions, neglecting the mosaic structure of the isochores in the warm-blooded vertebrate genome (see Fig. 3.1).

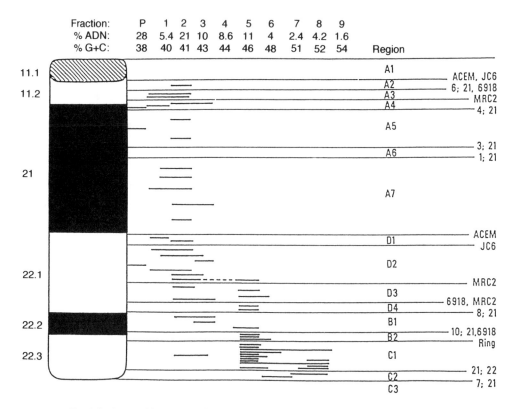

Fig. 3.9. Compositional map of the long arm of human chromosome 21. Long horizontal lines indicate positions of the breakpoints associated with the rearranged chromosomes listed at the right of the figure. Short horizontal lines indicate the compositional DNA fractions hybridizing single-copy probes localized in different Giemsa positive (dark) and Giemsa negative (light) or Reverse bands (from Gardiner et al., 1990).

spontaneous translocations, spontaneous and induced sister-chromatid exchanges, and the chromosomal abnormalities seen after X-ray and chemical damage. They also include the *hot spots* for the occurrence of mitotic chiasmata (Morgan and Crossen, 1977; Sutherland and Hecht, 1985; Kuhn and Therman, 1986; Hecht, 1988). Likewise, fragile sites tend to be more frequent in R-bands or near the border of R- and G-bands (Aurias et al., 1978; Yunis and Soreng, 1984; Hecht, 1988). It has been suggested that fragile sites could be the points at which chromosomes break to form rearrangements (Hecht and Hecht, 1984a,b). These observations indicate that R-bands and G/R borders are particularly prone to recombination and suggest that these phenomena are associated with the compositional discontinuities at G/R borders and within R bands. Cancer-associated chromosomal aberrations are also non-random, with a limited number of genomic sites consistently involved and frequently associated with cellular oncogenes and fragile sites (Mitelman and Heim, 1988). Incidentally, chromosomal rearrangements may have as an important consequence the activation of oncogenes by strong promotors that have been put upstream of them by the rearrangement (Klein, 1983).

At this point, the question should be raised as to why recombination phenomena are so frequent in the neogenome of warm-blooded vertebrates. The high GC level *per se* is unlikely to be responsible because, although rare, genomes which are relatively high in GC (but in a uniform way) are also found among cold- blooded vertebrates (Bernardi and Bernardi, 1990a,b). What seems to matter is, therefore, the presence of compositional discontinuities which exist not only at the borders of G- and R- (or T-) bands, but also within R-and T-bands (Gardiner *et al.*, 1990). The latter are due to the presence of thin G-bands (which can be detected at high resolution; Yunis, 1981) and also to the compositional heterogeneity of the GC-rich isochores present in R- and T-bands.

Concerning the molecular basis for the association between chromosome breakpoints and compositional discontinuities, one possibility is that chromatin structure is more 'open' at compositional discontinuities (as was shown to be the case for CpG islands and the very GC-rich genes which associated with them (Tazi and Bird, 1990; see also Aïssani and Bernardi, 1991a,b)) so allowing recombination to take place, possibly using the many repeated sequences concentrated in those regions. Indeed, the regions which are most susceptible to recombination correspond to high concentrations of Alu sequences (Zerial *et al.*, 1986), CpG islands (Aïssani and Bernardi, 1991a,b) and minisatellites (Jeffreys *et al.*, 1985).

In conclusion, what is proposed here is that the two major compositional shifts which occurred in the genomes of vertebrates, namely those accompanying the emergence of birds and mammals, led to the formation of strongly compositionally compartmentalized genomes which are characterized by a remarkable degree of instability. It should be stressed that, although we do not believe that 'if it were not for social structuring, mammals, because of their high dispersal power, might have evolved at the same rate or more slowly than most cold-blooded vertebrate' (Bush *et al.*, 1977), we can accept that social structuring and brain development may have played an additional accelerating role.

References

Aïssani, B. and Bernardi, G. (199la) CpG islands: features and distribution in the genome of vertebrates. *Gene* 106, 173–183.

Aïssani, B. and Bernardi, G. (1991b) CpG islands, genes and isochores in the genome of vertebrates. *Gene* 106, 185–195.

Aïssani, B., D'Onofrio, G., Mouchiroud, D., Gardiner, K., Gautier, C. and Bernardi, G. (199I) The compositional properties of human genes. *Journal of Molecular Evolution* 32, 497–503.

Aurias, A., Prieur, M., Dutrillaux, B. and Lejeune, J. (1978) Systematic analysis of 95 reciprocal translocations of autosomes. *Human Genetics* 45, 259–282.

Benton, M.J. (1995) Diversification and extinction in the history of life. *Science* 268, 52–58.

Bernardi, G. (1989) The isochore organization of the human genome. *Annual Review of Genetics* 23, 637–661.

Bernardi, G. (1933a) Genome organization and species formation in vertebrates. *Journal of Molecular Evolution* 37, 331–337.

Bernardi, G. (1993b) The isochore organization of the human genome and its evolutionary history – a review. *Gene* 135, 57–66.

Bernardi, G. and Bernardi, G. (1990a) Compositional patterns in the nuclear genomes of cold-blooded vertebrates. *Journal of Molecular Evolution* 31, 265–281.

Bernardi, G. and Bernardi, G. (1990b) Compositional transitions in the nuclear genomes of cold-blooded vertebrates. *Journal of Molecular Evolution* 31, 282–293.

Bernardi, G. and Bernardi, G. (1991) Compositional properties of nuclear genes from cold-blooded vertebrates. *Journal of Molecular Evolution* 33, 57–67.

Bernardi, G., Mouchiroud, D., Gautier, C. and Bernardi, G. (1988) Compositional patterns in vertebrate genomes: conservation and change in evolution. *Journal of Molecular Evolution* 28, 7–18.

Bernardi, G., Olofsson, B., Filipski, J., Zerial, M., Salinas, J., Cuny, G., Meunier-Rotival, M. and Rodier, F. (1985) The mosaic genome of warm-blooded vertebrates. *Science* 228, 953–958.

Bush, G.L., Case, S.M., Wilson, A.C. and Patton, J.L. (1977) Rapid species formation and chromosomal evolution in mammals. *Proceedings of the National Academy of Sciences, USA* 74, 3942–3946.

de Lange, T. (1992) Human telomeres are attached to the nuclear matrix. *EMBO Journal* 11, 717–724.

Dobzhansky, T. (1970) *Genetics of the Evolutionary Process*, 3rd edn. Columbia University Press, New York.

Gardiner, K., Aissani, B. and Bernardi, G. (1990) A compositional map of human chromosome 21. *EMBO Journal* 9, 1853–1858.

Grant, V. (1973) *Plant Species Formation*. Columbia University Press, New York.

Hecht, F. (1988) Enigmatic fragile sites on human chromosomes. *Trends in Genetics* 4, 121.

Hecht, F. and Hecht, B.K. (1984a) Fragile sites and chromosome breakpoints in constitutional rearrangements. I. Amniocentesis. *Clinical Genetics* 26, 169–173.

Hecht, F. and Hecht, B.K. (1984b) Fragile sites and chromosome breakpoints in constitutional rearrangements. II. Spontaneous abortions, stillbirths and newborns. *Clinical Genetics* 26, 174–177.

Hudson, A.P., Cuny, G., Cortadas, J. Haschemeyer, A.E.V. and Bernardi, G. (1980) An analysis of fish genomes by density gradient centrifugation. *European Journal of Biochemistry* 112, 203–210.

Jeffreys, A.J., Wilson, V. and Thein, S.L. (1985), Hypervariable 'minisatellite' regions in human DNA. *Nature* 314, 67–73.

Klein, G. (1983) Specific chromosomal translocations and the genesis of B-cell-derived tumors in mice and men. *Cell* 32, 311–315.

Kuhn, E.M. and Therman, E. (1986) Cytogenetics of Bloom's syndrome. *Cancer Genetics and Cytogenetics* 22, 1–18.

Levin, D.A. and Wilson, A.C. (1976) Rates of evolution in seed plants: net increase in diversity of chromosome numbers and species numbers through time. *Proceedings of the National Academy of Sciences USA* 73, 2086–2090.

Macaya, G., Thiery, J.P. and Bernardi, G. (1976) An approach to the organization of eukaryotic genomes at a macromolecular level. *Journal of Molecular Biology* 108, 237-254.

Mitelman, F. and Heim, S. (1988) Consistent involvement of only 71 of the 329 chromosomal bands of the human genome in primary neoplasia-associated rearrangements. *Cancer Research* 48, 7115-7119.

Morgan, W.F. and Crossen, P.E. (1977) The frequency and distribution of sister chromatid exchanges in human chromosomes. *Human Genetics* 38, 271-278.

Mouchiroud, D., D'Onofrio, G., Aïssani, B., Macaya, G., Gautier, C. and Bernardi, G. (1991) The distribution of genes in the human genome. *Gene* 100, 181-187.

Saccone, C., De Sario, A., Della Valle, G. and Bernardi, G. (1992) The highest gene concentration in the human genome are in T-bands of metaphase chromosomes. *Proceedings of the National Academy of Sciences USA* 89, 4913-4917.

Saccone, C., De Sario, A., Wiegant, J., Rap, A.K., Della Valle, G. and Bernardi, G. (1993) Correlations between isochores and chromosomal bands in the human genome. *Proceedings of the National Academy of Sciences USA* 90, 11929-11933.

Schmid, M. and Guttenbach, M. (1988) Evolutionary diversity of reverse® fluorescent chromosome bands in vertebrates. *Chromosoma* 97, 110-114.

Sutherland, G.R. and Hecht, F. (1985) *Fragile Sites on Human Chromosomes*. Oxford University Press, Oxford.

Tazi, J, and Bird, A.P. (1990) Alternative chromatin structure at CpG islands. *Cell* 60, 909-920.

Thiery, J.P., Macaya, G. and Bernardi, G. (1976) An analysis of eukaryotic genomes by density gradient centrifugation. *Journal of Molecular Biology* 108, 219-235.

White, M.J.D. (1968) Models of species formation. *Science* 159, 1065-1070.

White, M.J.D. (1978) *Modes of Species Formation*. Freeman & Co, San Francisco.

Wilson, A.C. (1985) The molecular basis of evolution. *Scientific American* 253, 148-157.

Wilson, A.C., Sarich, V.M. and Maxson, L.R. (1974) The importance of gene rearrangement in evolution: evidence from studies on rates of chromosomal, protein, and anatomical evolution. *Proceedings of the National Academy of Sciences USA* 71, 3028-3030.

Wilson, A.C., Busch, G.L. and King, M.C. (1975) Social structuring of mammalian populations and rate of chromosomal evolution. *Proceedings of the National Academy of Sciences USA* 72, 5061-5065.

Wyles, J.S., Kunkel, J.G. and Wilson, A.C. (1983) Birds, behavior and anatomical evolution. *Proceedings of the National Academy of Sciences USA* 80, 4394-4397.

Yunis, J.J. (1981) Mid-prophase human chromosome. The attainment of 2900 bands. *Human Genetics* 56, 291-298.

Yunis, J.J. and Soreng, A.M. (1984) Constitutive fragile sites and cancer. *Science* 226, 1199-1204.

Zerial, M., Salinas, J., Filipski, J. and Bernardi, G. (1986) Gene distribution and nucleotide sequence organization in the human genome. *European Journal of Biochemistry* 160, 479-485.

Zieg, J., Silverman, M., Hilmen, M. and Simon, M. (1977) Recombinational switch for gene expression. *Science* 196, 170-172.

Genes, Memes and Artefacts

M. Gadgil
Centre for Ecological Sciences, Indian Institute of Science, Bangalore 560012 and Biodiversity Unit, Jawaharal Nehru Centre for Advanced Scientific Research, Jakkur, Bangalore 560064, India

Summary

The ever growing diversity of life over evolutionary time has led to the evolution of an increasingly complex network of biotic interactions. This, in turn, has favoured the evolution of more complex forms of life that could harness other forms in increasingly sophisticated ways, grounded in flexible patterns of behaviour and elaborate systems of social communication. This process has led to the domination of terrestrial ecosystems by social animals, whose behavioural repertoire includes memes, or elements of behaviour transmitted through social imitation. Meme-based cultural evolution has led to intelligent life; thus, it seems appropriate to suggest that diversity of life forms was a necessary precondition for the evolution of intelligence. This cultural evolution has further led to the elaboration of artefacts, another form of replicating entities. Like living organisms, artefacts have been multiplying, becoming ever more diverse and complex, till now they are beginning to decimate the diversity of life. This process may be contained only if humanity accepts an environmental ethic that treasures the diversity of life which constituted the matrix in which our species was born.

Introduction

> An enduring environmental ethic will aim to preserve not only the health and freedom of our species, but access to the world in which the human spirit was born.
>
> Edward O. Wilson (1992)

> True from a low materialistic point of view, it would seem that those thrive best who use machinery wherever its use is possible with profit; but this is

the art of the machines – they serve that they may rule. They bear no malice towards man for destroying a whole race of them, provided he creates a better instead; on the contrary, they reward him liberally for having hastened their development. It is for neglecting them that he incurs their wrath, or for using inferior machines, or for not making sufficient exertions to invent new ones or for destroying them without replacing them; yet these are the very things we ought to do, and do quickly; for though our rebellion against their infant power will cause infinite suffering, what will things come to, if that rebellion is delayed?

Samuel Butler (1872)

We live in a world rich in diversity of life, of human cultures, of material objects fabricated by people. Humans were shaped in a variegated community of beings. But as people have successfully elaborated a complex world of non-living objects around them, the diversity of life forms no longer appears essential to their continuing persistence. Indeed human-made artefacts, in the course of their exuberant multiplication, are today threatening to decimate the diversity of the very community of beings in which our spirit was born (Gadgil, 1993). Of course, we may still be able to maintain the health and perhaps even the freedom of our species in this community of non-beings. But one hopes with Edward Wilson (1992) that humans will come to nurture an environmental ethic which would treasure the rich diversity of life which was so essential to the birth of our spirit.

Replicating Entities

The ecological play in the evolutionary theatre we dwell in is today staged by three kinds of replicating entities; genes, memes and artefacts. All replicating entities tend to multiply and spread to newer and newer environments. Since the process of replication is imperfect, newer forms of replicating entities constantly surface. Many of these variants perish, but others come to persist by fitting into newer niches. Invading newer niches often poses difficult challenges, challenges that can be met only by resorting to more complex structures and behaviour patterns. So replicating entities tend to become ever more complex with time. Bonner (1988) has masterfully analysed this process of exuberant multiplication, ever-increasing diversification, and growing complexity of living organisms. Living organisms are carriers of genes, four billion year old replicating entities that are fundamental to all life processes. Similar evolutionary processes are at work in the case of two other kinds of replicating entities, memes, or behaviour patterns transmitted through social imitation and material artefacts fabricated especially by humans making more and more copies of prototypes that fulfil some defined purpose. These processes of evolution of memes and artefacts have by now assumed serious implications that need to be thought through with great care.

Evolving Complexity

Complexity of construction and along with it flexibility of behaviour have flourished particularly since living organisms moved on to land 400 million years ago. Life on land is much more difficult; organisms are full of water and must guard against desiccation; air does not support and permit them to move around as easily as water does. To overcome these constraints organisms have elaborated many intricate relationships. Of particular interest among these are mutualistic ones, such as among algae and fungi constituting lichens, or insect pollinators and mammalian seed dispersers of flowering plants. The sugary nectar and fruit pulp that many flowering plants produce as a reward for services in pollination, and the pollen and seed that may be usurped are significant food sources on land and animals have evolved a variety of fascinating adaptations to take advantage of these. Among the most notable of these are group cooperation in locating and harvesting such food sources, be they trails laid by ants gathering grass seeds or the honey bee dance communicating the direction and distance from a nectar and pollen source (Wilson, 1990).

Memes as Replicators

The honeybee dance is particularly intriguing, as it is an elaboration of intentional movements to instigate a fellow group member to join in an activity in less advanced forms of bees and wasps. These are memes, instances of replication through social imitation of behaviour patterns (Dawkins, 1976). Like genes, memes too can multiply, change, and become more complex and diverse over evolutionary time. In honeybees the dance form and the pattern of flight induced by it does vary a great deal, but this variation is within rather narrow genetically prescribed limits. Other animals at that level of organization have an even narrower repertoire of memes. So in the world of 100 million years ago, while the diversity of genes and life forms was already enormous, that of memes must have been rather limited.

But the potential for flourishing of memes has grown as organisms evolved ever more intricate interactions. Honeybees, with their social communication, are efficient harvesters of nectar and pollen, moving over kilometres to bring food back to their hives. Here it is stored as a reserve in the form of honey creating a most attractive cache of food. Honeybees not only cooperate in foraging; they also defend their hives as a group, inflicting suicidal stings on the predators. So it takes large, complex mammals like honey badgers with their thick skin and stiff fur to tap this food resource. And evolution has produced blackthroated honeyguides, birds which lead honey badgers to honeybee hives, and feed on the wax after the badgers are finished.

Birds, like honeyguides, and mammals, like badgers, have well-developed nervous systems and elaborate behaviour patterns. Many bird and mammalian species live in social groups, so that conditions are favourable for the elaboration of memes. Among the most fascinating of memes that have evolved in these groups are those of bower birds. Males of these birds build bowers of twigs, decorated with objects such as pebbles, shells and feathers to court females. Diamond (1988) offered them coloured poker chips and found that fashions developed in the colour of chips selected by males. In one valley, the males tended to go in for blue chips, whereas in another the popular choice was pink. Thus developed behaviour patterns recalling cultural traditions of ornamentation among our own species.

Material Artefacts

Bowers are material artefacts fabricated by animals. Since male bower birds apparently copy each other's constructions, these artefacts too are replicating entities like genes or memes. But use of artefacts really started two million years ago, when our hominid ancestors began to use specially fashioned pieces of stone, wood, and horn to deal more effectively with their food. This allowed them to greatly expand their range of food items, for now they could dig for tubers as well as warthogs or cut through the hide of rhinoceros with a facility greater than that of a lion. This widened food niche might in turn have generated selection pressures favouring a greater role for memes. For it would be of considerable advantage to pass on information on what is edible and what is not, how a prey species is best ambushed and a predator avoided, as learnt behaviour patterns among members of a social group. Indeed it has been speculated that a skeleton of *Homo erectus* found at Kobi Fora in East Africa has deformities caused by excessive intake of vitamin A from consumption of animal fat (Leakey, 1981). Information that it is prudent to limit fat consumption would best be propagated in the group as learnt behaviour – as memes. Increasing facility at getting at a greater and greater range of food sources through the use of artefacts may then have increased the value of adaptation to specific local conditions as *Homo erectus* spread from East Africa through large areas of Asia with very varied environmental regimes. This in turn must have favoured greater flexibility of behaviour and proliferation of a great variety of local cultural traditions. Conditions were thus particularly favourable for the multiplication and diversification of both memes and artefacts among our group-living hominid ancestors (Boyd and Richerson, 1985; Lumsden and Wilson, 1981).

However, the number of artefacts employed by hominids remained limited until anatomically modern *Homo sapiens* began to bury their dead and to use ornaments some 45 thousand years ago. Since ornaments have no function with respect to the non-human world in the way that a hand axe has, their form and numbers can vary tremendously. Use of ornaments must from the beginning have served to enhance social prestige. In that case, prestige would depend on the quality, novelty, and quantity of ornaments, not in any absolute, but only in a relative sense. That would create possibilities of open-ended competition, under which numbers and variety of artefacts could grow without any limit (Gadgil, 1993). Planning and fabrication of such artefacts would in turn favour more flexible behaviour, and a greater role for memes or cultural traditions, accompanied by the development of a rich system of symbolic communication.

Of Biosphere and Noosphere

Over the last 2 million years hominids have been living in a world of three kinds of replicators: genes, memes and artefacts. Each of these vary in the course of replication. The variation in genes is largely random, the process of natural selection prunes out much of it, favouring the occasional variant that enhances the rate of survival and replication of its carrier. So new genes come to spread very slowly. With hominids, variation in memes and artefacts is much more often deliberate, directed towards forms which in some sense serve better the interest of carriers. The rate of increase as well as change of memes and artefacts can therefore be far more rapid.

This potential for rapid change will be realized to the extent that the carriers can visualize the consequences of bearing particular memes, or having access to particular artefacts. The advantage of thus visualizing the future and of planning strategies of dealing with it might have played a significant role in favouring the development of symbolic language which probably arose some 50,000 ybp in our own species. It is this language that permits humans to deal with not just the here and now as all other animals do, but with elsewhere and in the past or future as well. Using this facility people make models of the working of the world and devise ways of modifying it to their own liking. Mental models of the working of the world are memes, and people have come to develop a great variety of increasingly complex ones over time. A whole range of non-living artefacts, clothes, houses, knives and aeroplanes, synthetic drugs, and chemical pesticides have been devised to modify the environment to human purpose. And a whole range of artificially created biological communities, corn fields and pine plantations, carp ponds and cattle ranches are also devised to meet human demands. Indeed, in the words of Vernadsky, memes and artefacts have transformed the biosphere into a noosphere (Chardin, 1956).

Patterns of Artefact Evolution

There are notable parallels in the evolution of life forms and artefacts. Consider, as an example, means of transport. Living organisms have colonized newer and newer environmental regimes over their evolutionary history. So have transport vehicles. Beginning with logs of wood carrying people across short stretches of water, they now cover all types of terrain on land, fly in the air, dive deep under water, and even move through outer space. In the process their populations have been growing rapidly. Ten thousand years ago there were no transport vehicles on land. Three thousand years ago

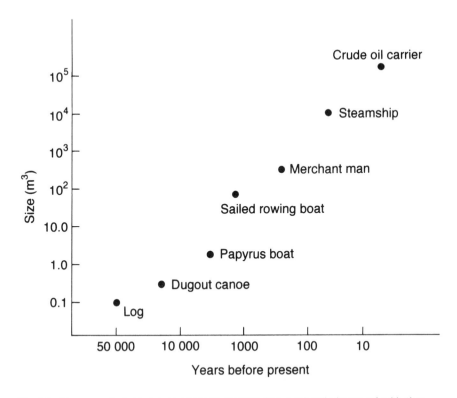

Fig. 4.1. The upper limit of size of transport vehicles has continually increased with time.

domesticated oxen and horses and carts had become the most prevalent means of transport on land. At the time there may then have been one for every ten persons in parts of the Old World, and none in the New World. Today motorbikes and automobiles dominate the scene. There is one for every five or so people in a population that has grown a hundredfold. So the population of transport vehicles has grown by a factor of more than 200 in these last few thousand years. Just as the upper limit of size of living organisms has gone on increasing, the upper limit of size of transport vehicles too has been constantly going up. From a log of about $0.2 \, m^3$ in size 50,000 years ago, the biggest ships of today, the giant oil tankers are larger in size by six orders of magnitude (Fig. 4.1). Just as the upper limit of the complexity of organization of living creatures has gone on increasing with time, so has that of means of transport. A log floating over water is a simple, single object only marginally changed from its natural form. An ox-cart has several parts composing it, but one carpenter can fabricate it all by himself using relatively simple tools. A modern aeroplane has thousands of different parts and only a team of highly trained people can put it together working in a coordinated fashion for years. Finally just as the diversity of life has gone on increasing over evolutionary time, so has that of transport vehicles. A few thousand years ago there were just a few types, dug out canoes, rafts of bamboos or reeds, kayaks of

86   M. Gadgil

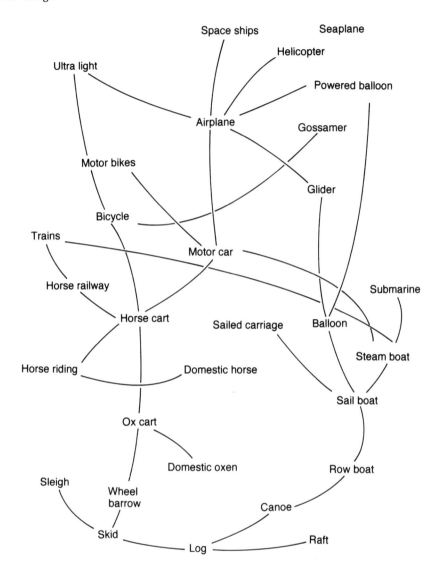

Fig. 4.2. The evolutionary history of artefacts constitutes a web rather than a bush.

animal skin, simple two-wheeled carts. Today we have a mindboggling variety not only of canoes, catamarans, kayaks, and animal carts but many designs of sail boats, power boats, submarines, bicycles, motorbikes, cars, trains, balloons, gliders, aeroplane and spaceships (Finn, 1987).

The tremendously greater rate of change of form and creation of diversity of artefacts has been made possible by the transfer of techniques, the memes underlying them among very different forms of artefacts. This means that the evolutionary lineages of artefacts are complex webs rather than the arborescent bushes of lineages of living organisms. Fig. 4.2 indicates only a miniscule fraction of the complexity of this web. The simple sailboats of today are fabricated out of synthetic materials, carry on board electronic navigational aids and radiocommunication systems all of which were first developed in a variety of contexts that had nothing to do with sailing over water.

For this reason whole communities of artefacts mutually reinforce each other and multiply, change, and diversify together. Thus the network of highways in the United States encourages production of more and more automobiles and the production of automobiles creates demands for more highways. The proliferation of automobiles encourages the development of mobile phones. Consumption of fossil fuels by the automobiles promotes the development of gigantic oil tankers transporting oil over vast stretches of the sea. And oil spills on the sea encouraged the development of the bacterium capable of metabolizing it, leading to the first ever patenting of a new kind of artefact, the genetically engineered living organism.

Positive Feedback

Indeed the last few centuries have witnessed the gradual elaboration of a system of positive feedbacks that today promotes the production of ever larger numbers of more and more complex artefacts (Fig. 4.3). Groups of people each with a highly specialized set of memes are today organized in conjunction with complex sets of artefacts as industries, dedicated to the production of a range of other artefacts. This industrial production is supported by demands from individual consumers, other industries as well as by groups of people dedicated to subjugating and destroying other groups of people – the military. These demands are stimulated by propagation of special memes – memes that attach social prestige to the possession of more and more artefacts. These are promoted by specialized groups involved in commercial advertising or political propaganda. There are other organized groups of people in the research and development establishments who are continually engaged in producing better and better models of how nature works and applying them to the production of artefacts – satisfying the demands of consumers and military machines, enhancing the capabilities of industrial production and the extent and power of advertising and propaganda establishments. With a high degree of mutualism, the community of artefacts is achieving greater and greater intensities of replication of their members. Humans are of course intimately involved in and apparently in charge of this process. But as Samuel Butler (1872) foresaw so long ago in his classic *Erewhon*, human control of this process might increasingly be an illusion. It is said, in recognition of the central role of genes in controlling life processes, that a chicken is an egg's way of making more eggs. Perhaps, today people have become artefacts way of making more artefacts.

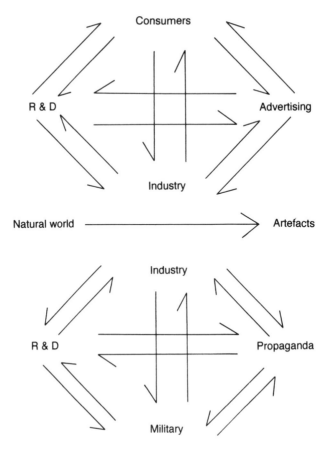

Fig. 4.3. A system of positive feedbacks currently promotes the production of more and more artefacts at the cost of the natural world.

Choices Before Us

Fig. 4.4 attempts to summarize the history of the interaction of two types of replicating entities, life forms and human-made artefacts on the earth. It is clearly a story of a losing battle for the diversity of life against that of artefacts. The rich diversity of life forms set in complex ecosystems provided the milieu in which the human spirit was born. Indeed one may speculate that human fascination for ever-more complex artefacts has its roots in the human fascination for complex living organisms, be they trees, birds or mammals. Humans were motivated to design flying machines by their fascination for birds and bats. In fact one of the earliest, the Avion designed in 1890 by Clement Ander, was clearly inspired by the shape of bat wings. But once effective flying machines came on the scene, they have very rapidly grown larger and larger, more and more complex and acquired a whole diversity of forms.

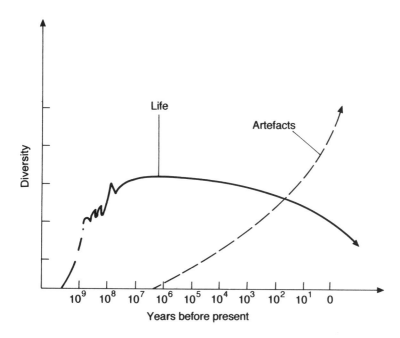

Fig. 4.4. A schematic depiction of the history of diversity of life and of material artefacts on the earth.

Today they outdo birds and bats in many ways, flying much faster, and over far greater distances, though perhaps birds and bats still have an edge in manoeuverability. But these flying artefacts are competing with birds for air space; and when gulls or vultures are hit by planes, attempts are launched not to reduce competition by aeroplanes, but to get rid of the birds. This is because many well-organized groups of humans stand to acquire economic, political, and military power by manufacturing and flying aeroplanes, whereas nobody acquires such power by ensuring that the world remains full of birds and bats.

But perhaps it is inappropriate to believe that the process will indeed run its full course with the near complete elimination of the diversity of the natural by the artificial. In the past humans have at least in part successfully sought to embrace new ways of life friendly to the community of beings around them. This was the message of Buddha, himself born in a sacred grove of *Shorea robusta*, a dipterocarp. That message still effectively protects thousands of temple groves in Southeast Asia. One may then hope with Edward Wilson that humanity will come to embrace a new environmental ethic, that will preserve not only the health of our species, which the artificial may achieve, but also our freedom and access to the world in which the human spirit was born, the world of natural diversity.

References

Bonner, J.T. (1988) *The Evolution of Complexity by Means of Natural Selection*. Princeton University Press, Princeton, New Jersey.

Boyd, R. and Richerson, P.J. (1985) *Culture and the Evolutionary Process*. Chicago Press, Chicago.

Butler, S. (1872) *Erewhon: or Over the Range*. Jonathan Cape, London.

Chardin, P.T. de (1956) The antiquity and world expansion of human culture. In: Thomas, W.L. Jr (ed.) *Man's Role in Changing the Face of the Earth*, Vol. 1. The University of Chicago Press, Chicago, pp. 103-112.

Dawkins, R. (1976) *The Selfish Gene*. Oxford University Press, Oxford.

Diamond, J. (1988) Experimental study of bower decoration by the bowerbird *Amblyornis inornatus* using colored poker chips. *American Naturalist* 131, 631-653.

Finn, S. (1987) *The Big Book of Transport*. Octopus Books, London.

Gadgil, M. (1993) Of life and artifacts. In: Kellert, S.R. and Wilson, E.O. (eds) *The Biophilia Hypothesis*. Island Press, Washington, DC, pp. 365-377.

Leakey, R.E. (1981) *The Making of Mankind*. Abacus, New York.

Lumsden, C.J. and Wilson, E.O. (1981) *Genes, Mind and Culture*. Harvard University Press, Cambridge, Massachusetts.

Wilson, E.O. (1990) *Success and Dominance in Ecosystems: The Case of the Social Insects*. Ecology Institute, Germany.

Wilson, E.O. (1992) *The Diversity of Life*. The Belknap Press of Harvard University Press, Cambridge, Massachusetts.

New Dimensions in the Study of Genetic Diversity

A.P. Ryskov[1], M.I. Prosniak[2], V.V. Kalnin[1], O.V. Kalnina[1], O.N. Tokarskaya[1], E.K. Khusnutdinova[3], G.P. Georgiev[1], and S.A. Limborska[2]*

[1]*Institute of Gene Biology, Russian Academy of Sciences, Moscow 117334, Russia;* [2]*Institute of Molecular Genetics, Russian Academy of Sciences, Moscow 123182, Russia; and* [3]*Department of Biochemistry and Cytochemistry, Bashkir Science Centre, Ufa 450054, Russia*

Mobile Repetitive Elements and Genetic Diversity

It is obvious that biodiversity is a result of genetic diversity among different living organisms. Genes are subject to changes, i.e. mutagenesis, and many different mutations are accumulated within genomes. In the eukaryotic organisms, most of them are not phenotypically expressed due to the presence of another, non-mutated, copy of the same gene in the nucleus. Frequently, mutations within genes do not change the functioning of a protein encoded by the gene and thus cannot be detected. However, sometimes mutations result in the appearance of genetic diseases and at least some of them are eliminated from the population by natural selection.

Many more changes seem to occur not in genes encoding proteins but in DNA sequences surrounding them. These sequences may contain elements for the control of gene expression. They may change the level and tissue specificity of gene expression. For many other DNA sequences, no function has yet been established, but it cannot be ruled out that they also modulate gene expression to some extent. Experiments to test this idea are difficult and have not yet been done.

Changes in such genomic areas do not create dramatic phenotypic changes, but they may induce some quantitative variations in the content of corresponding proteins and lead to differences between individuals belonging to the same species or even to the same small group.

It is very difficult to detect such changes as they occur with a low probability and do not give a clear phenotype. Therefore, it is easier to detect such changes when simultaneously studying the properties of a number of

---

* Corresponding author

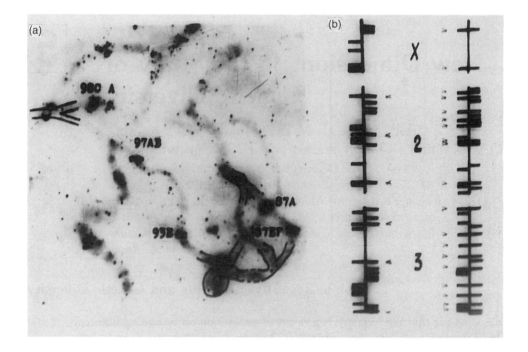

Fig. 5.1 Varying location of mdg1. (a) *In situ* hybridization 15w3 of mdg1 to salivary gland chromosomes of *D. melanogaster* gtx13z gt. One can see that in the unpaired region of the third chromosome (indicated) the localization of mdg1 in the chromosomes originating from different parents is quite different. (b) Comparison of mdg1 distribution among two 3 13z representatives of different strains (gtw and gt) and two representatives of the same strain (Oregon RC). Horizontal lines = mdg1 sites; symbol > indicates cases of mdg1 location at the same site in two compared larvae.

different genome elements. This can be done in studying repetitive sequences of the genome, i.e. sequences represented by many copies. Except for a few cases of rRNA genes (histone genes etc.), these repetitive sequences do not fulfil any obvious function.

First, such an approach was used in the studies of some cloned repetitive sequences of *Drosophila melanogaster* by Ilyin et al. (1977) who demonstrated the mobility of these sequences. One such sequence, mdg1, was cloned and its DNA was hybridized with preparations of polytene chromosomes where the location of hybridization sites could be accurately determined. It was found that the location of mdg1 in individuals taken from different strains was completely different. Less than one third of 20–30 mdg1 copies shared the same location. Moreover, if the individuals were taken from the same strain, about one-third of all sites of mdg1 location were still different. Only homozygous organisms with exactly the same genotype had the same pattern of mdg1 localization (Fig. 5.1).

Later on, many other repetitive elements of *D. melanogaster* were found to have varying distribution throughout the genome. These repetitive sequences represent mobile elements, or elements able to transpose from one site of the genome to another. These elements can influence the activity of genes near which they are located. They may be an important factor determining genetic differences among individual flies. Each fly has a characteristic pattern of mobile elements distribution.

Mini- and Microsatellites as a Source of Genetic Variability

All living organisms have mobile elements in their genomes. That is a general feature. However, the properties of mobile elements and their number strongly vary from one species to another. In many species including vertebrates, the distribution of mobile elements is more constant than in *D. melanogaster*; they transpose with a lower frequency. On the other hand, they are sometimes too numerous for a careful study of their distribution.

Therefore, some other repetitive sequences are usually used to follow genomic differences. These are so-called 'minisatellite' and 'microsatellite' sequences (ms, mcs). They are represented by several dozen to several hundred copies spread throughout the genome. Each copy consists of a series of tandem repeats of a core consensus sequence, 2–4 bp (mcs) or 10–60 bp (ms) long. At a given locus, numerous alleles differing in the number of core repeats may occur. Such sequences are not mobile elements and cannot transpose. However, the reiteration of a short simple sequence increases the chances of incorrect replication of such elements: the number of repetitive simple sequences within the element may change during DNA synthesis. Also, their number may change as a result of unequal recombination. Such changes are still very rare, but they do occur sometimes, producing changes in the length of one of the minisatellite copies. The change, if it has happened in a germ cell, may be inherited. As a result of such events, different individuals have different sets of elements differing in their size.

The easiest way to visualize the variability among individuals is as follows. DNA isolated from a blood sample (or other tissue) of several individuals is digested by one of the restriction endonucleases, i.e. the enzymes cutting DNA only at specific sites containing a certain sequence of 4–8 base pairs long. As a rule, a restriction endonuclease is used which cuts DNA in many different places but not within the minisatellite sequence leaving the internal repeats intact. As a result, the whole minisatellite sequence is not cut whereas the flanking sequences are cut not far from the minisatellite boundaries. After electrophoretic separation of DNA fragments and hybridization with a minisatellite probe, one can see the pattern of size distribution of individual minisatellite copies. Usually two different representatives of the same species strongly differ in this pattern of distribution.

As ms- and mcs-bands are inherited, the method is widely used to determine the parenthood in animals and humans. It is also a very good approach for individual identification in forensic medicine.

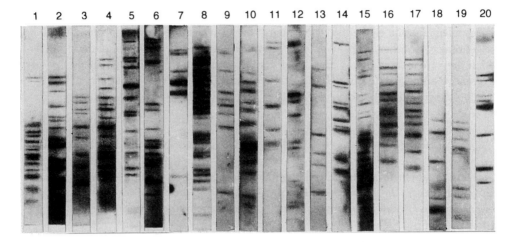

Fig. 5.2 DNA fingerprinting of animal, plant and bacterial genomes. ^{32}P-labelled M13 DNA was hybridized with Southern blots of restriction endonuclease digests of DNA from mouse (1), cow (2), pig (3), rabbit (4), chicken (5), sheep (6), loach (7), *Drosophila melanogaster* (8), *Bombyx mori* (9), *Blatta orientalis* (10), *Chrysomela gamellata* (11), *Anodonta celensis* (12), *Lumbricus terrestris* (13), *Echinococcus granulosus* (14), cotton (14), barley (15), soybean (16), orange (17), *Poncirus trifoliata* (18), yeast (19), and *Vibrio cholerae* (20) Samples (1-5 mg) of DNA were digested with *Bsp*RI (1-7,9,14,18), *Eco*RI+*Hin*dIII (8), *Bsp*RI+*Alu*I (10-13,16), *Hin*dIII (14-17,19,20), *Hin*dIII+*Bsp*RI (16).

There are some other methods of genome analysis generally designated as DNA typing approaches to solve the mentioned problems. Combining several methods of analysis, Gill *et al.* (1994) could identify the remnants recently found near Ekaterinburg with those of Tsar Nicholas II and his family. Thus, the distribution of mini- and microsatellite bands is a sort of genetic portrait of a given organism. Therefore the method is called 'DNA fingerprinting', and the picture of distribution of the bands a DNA fingerprint.

The method was first developed by Jeffreys in the UK, who described the minisatellite families referred to as Jeffreys' minisatellites (Jeffreys *et al.*, 1985). Later several other minisatellites were described. One of them is M13 phage minisatellite which consists of a repeating 15-bp sequence GAGGGTGGXGGXTCT. It was discovered independently in Belgium (Vassart *et al.*, 1987) and in Russia (Jincharadze *et al.*, 1987; Ryskov *et al.*, 1988). Ryskov *et al.* have tested this sequence for hybridization with DNA from many different species and in all cases the sets of hybridizing bands have been determind (Ryskov *et al.*, 1988, 1990; Ginzburg *et al.*, 1989; Monastyrskii *et al.*, 1990, 1991).

This DNA sequence, designated as the M13 minisatellite (abbr. M13-ms), is really ubiquitous among all tested living organisms. It was found in man, different mammals, birds, insects, fungi, plants, bacteria, etc. Any of the species tested always contained some M13-ms sets. This is a unique case of a minisatellite sequence which can be detected everywhere (Fig. 5.2).

This peculiar characteristic of M13-ms, i.e. its presence in genomes of practically all organisms, allows one to apply its analysis for solving a number of problems. These include the establishment of parenthood, individual identification, identification of cell lines and bacterial strains, epidemiological analysis, applications in population biology and ecology, etc.

For example, using blot-hybridization with M13-msDNA, one can distinguish between pathogenic and non-pathogenic strains of bacteria (Ginzburg et al., 1989). In particular, different epidemiological forms of cholera vibrions were found to depend not on the pathogenicity acquired by new strains but on the spreading of one and the same pathogenic strain to different places. The method was also successfully used for the identification of some helminthose agents and different phytopathogenic fungus isolates differed in their toxin spectra (Ryskov et al., 1990; Monastyrskii et al., 1990, 1991). It may be important for efficient protection of crop plants and rice from diseases induced by fungus parasites.

Population Studies Based on DNA Fingerprinting Techniques

The question arises as to whether M13-ms fingerprinting can only be applied for differentiation between individuals, or if it can also be used for differentiation of large or small populations. This is not obvious because DNA fingerprints are rather complex and the population-dependent pattern may not be marked by numerous individual differences.

The most obvious area of application of hybridization with M13-ms is the study of dynamics of genetic processes in populations of plants and animals. It allows (i) the determination of parentage in free-ranging populations for ecological studies of complex mating systems, (ii) estimation of the extent and character of genetic diversity in related populations of wild and domestic animals, and (iii) genetic differentiation and evaluation of the relationship among individuals in natural populations (Wetton et al., 1987; Jeffreys and Morton, 1987; Burke et al., 1989; Reeve et al., 1990; Tokarskaya et al., 1990; Gilbert et al., 1990, 1991). It also has a wide application in breeding and conservation programmes concerning economically important and domestic species of animals (Dunnington et al., 1991; Haberfeld et al., 1992) as well as for the reproduction of endangered species in captivity where there is a special problem of maintenance of genetic diversity (Tokarskaya et al., 1990, 1994). The possibilities and capacity of DNA fingerprinting in genetic differentiation studies were tested by using the artificial populations of the endangered crane species from the Oka State National Reserve nursery as a model system for demonstrating an adequate approach to fingerprinting data analysis. The system consisted of birds from two species, two populations, interspecies and interpopulation hybrids. In our experiments, genetic heterogeneity of artificial populations was studied by DNA fingerprinting combined with a special mathematical method, multivariate analysis described in Jambu (1978). The latter allows

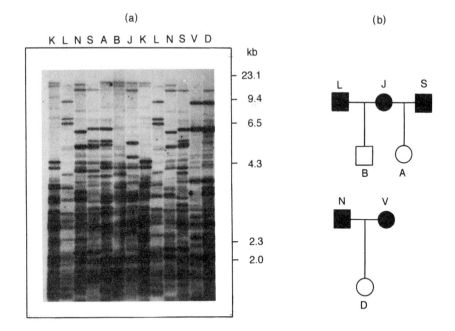

Fig. 5.3 DNA fingerprinting analysis of individuals from the crane captive population. Capital letters stand for the first letter of the bird name. Nine birds were analysed: five Siberian cranes - Julia (female), Kunovat (male) and males, Sergei, Nazar, Libbi from natural populations of Ob and Yakutia, respectively. One female (Vashin) is the Sandhill crane; offspring, Beliy, Agidel and Dakota, were bred in captivity and represent interpopulation hybrids and one interspecies hybrid, respectively. The band-sharing analysis of the individual DNA fingerprints permitted making the right choice among putative fathers for all three offspring and reconstructing the parent-offspring relationships in the captive colony.

a combined characterization of all bands present on fingerprints in the form of several factors.

Fig. 5.3 shows the DNA fingerprints obtained with DNA of cranes from two Siberian populations of the same species (*Grus leucogeranus* Pall) and cranes from another species (Sandhill cranes, *Grus canadensis pratensis* L) and their hybrids. The pictures are very complex and harbour information on genetic relationships in this group. Fig. 5.4 presents the distribution of the individuals studied in the system of coordinates for the first two factors obtained by a computer analysis. The first factor (abscissa) can be associated with interspecies differences, that is the symbols corresponding to the Siberian cranes are on the left-hand side of the graph, and the symbol corresponding to the Sandhill crane (Vashin) is on the right-hand side. Naturally, the interspecies hybrid (Dakota) occupies an intermediate position between the species and its parents (Vashin–Nazar).

The second factor (ordinate) appears to be explicitly associated with inter-population differences. The symbols corresponding to individuals from the Ob

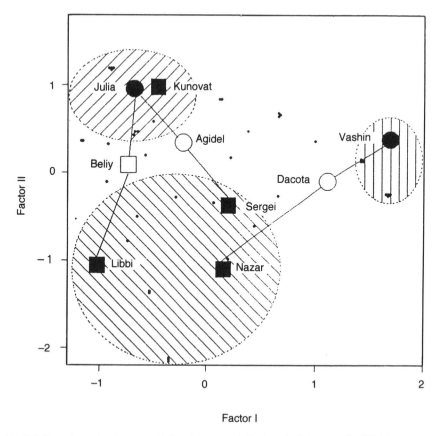

Fig. 5.4 Two-dimensional representation in space of the two first factors of artificial crane population structure recognized by means of factor correspondence analysis method. Symbols (open and dark circles and squares) correspond to individuals and dots correspond to characters, measured on equal scales and can be plotted in the same system of coordinates for visualization.

river population occupy the upper part of the graph, whereas those from the Yakut population are located in its lower portion. The interpopulation hybrids occupy the space just between the Ob and Yakut groups. The method permits revealing characters associated with genetic differentiation of species and populations.

Fingerprinting data analysis by means of a mathematical method applied here can be used for monitoring the species gene pools and breeding of the endangered species in captivity. This approach also offers an opportunity for using a small number of specimens for obtaining information on genetic differentiation of populations and species of interest. This could be useful in analogous studies of wild populations of many other endangered species for which DNA samples are limited either due to the number of individuals or due to difficulties in obtaining biological probes. On the other hand,

DNA Fingerprinting Reveals Ethnogeographical Diversity of Human Populations

More extensive were the studies performed in human populations. DNA fingerprints of 278 individuals were obtained and the same methods of computer analysis were applied to them. One of the populations, Talish, is from Southern Azerbaijan and belongs to the Indo-Iranian group. The representatives of the four other populations live in the middle of Russia (The Urals). The first of these, Komi population, represents the Finno-Ugric ethnic group. The three others are Bashkirs, representatives of the mixed origin group (Turkish and Finno-Ugric). The Bashkir samples were taken from three regions that are referred to different tribal subdivisions on the basis of archaeological, ethnographic and anthropological evidence.

Fig. 5.5 shows a typical pattern of hybridization of human genomic DNA samples with M13 phage DNA. Restriction fragment frequency profiles for each molecular weight in each population are shown in Fig. 5.6. The restriction fragments of 7.5–14 kb are characterized by relatively low band frequencies (mean number 0.012). Below 7.5 kb the mean band frequency rises to 0.131. One band (3.4 kb) is notable by its high frequency: 100% in three populations (a, c, d), 0.986 and 0.894 in the two others (b, e). Obviously these typically used histograms are not informative enough for comparisons of populations.

The distribution of individuals in the coordinate system of the first two factors is presented in Fig. 5.7. Different symbols mark different populations. One of the distinctive features of this method is the use of individuals, not samples, as is customary in traditional population studies. Therefore, each point corresponds to one individual. The Talish population is well distinguished from the Ural group by the first factor (Fig. 5.7a). This conforms to the Talish population history and geography. It is the most remote from the others ethnically and geographically. It can be suggested that the first factor characterizes the Caucasian component here. As shown in Fig. 5.7a, the four Ural populations overlap in this factor space. Figure 5.7b represents the data in the coordinate system for the second and third factors (without the Talish population). The Ural populations can be well distinguished. One of the Bashkir subpopulations (Abzelilovsky) overlaps with the Komi population to a great extent. This was expected from the well known fact that the Bashkir nation takes its origin from the Finno-Ugric group (Kuseev, 1979). In the same overlapping zone, there is another Bashkir population (Arkhangelsky) that has an intermediate geographical location between the Komi and Abzelilovsky populations. The third of the Bashkir populations (Ilishevsky) has the smallest

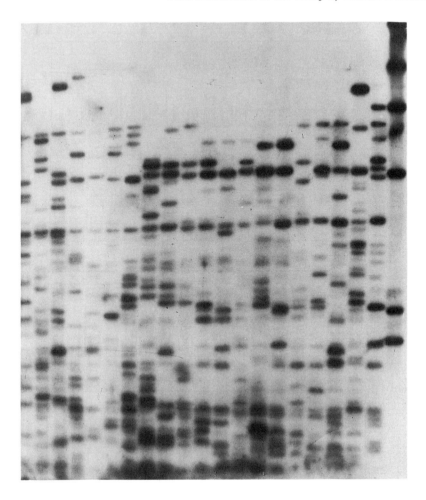

Fig. 5.5 Example of a typical hybridization pattern of human genomic DNA samples with M13 phage DNA. On the right of autoradiograph there is marker, lambda-DNA restricted by *Hind*III.

degree of overlapping with the others and lives near Tatarstan. Interpretation of the overlapping zones is very important because they might be a result of interethnic crosses. This process seems to play a key role in the formation of the Bashkir nation (Kuseev, 1979). The results demonstrate some new possibilities of multilocus DNA fingerprint probes in population studies and in the area of ethnogenesis historical investigations. It is also possible to solve the inverse problem: to identify an individual belonging to a certain preliminary studied population.

Another important application of such population analysis is the evaluation of the genetic diversity of a small ethnic group, which may be used for making recommendations regarding the survival of this group. Of course, in

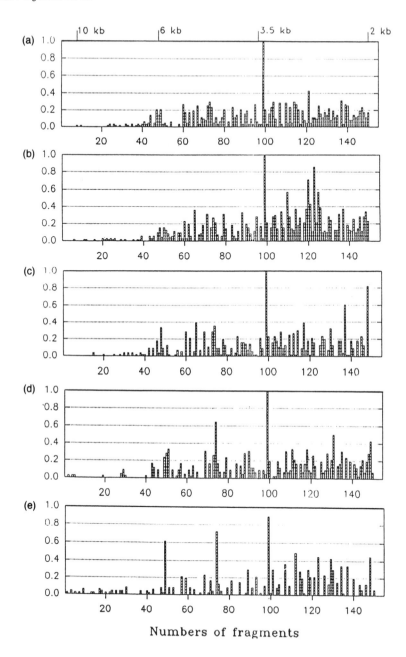

Fig. 5.6 Restriction fragment frequencies profiles of populations investigated. Frequencies are given for each of 150 fragments ranging from 2 to 14 kb. (a) Abzelilovsky, (b) Arkhangelsky, (c) Ilishevsky, (d) Komi, (e) Talish.

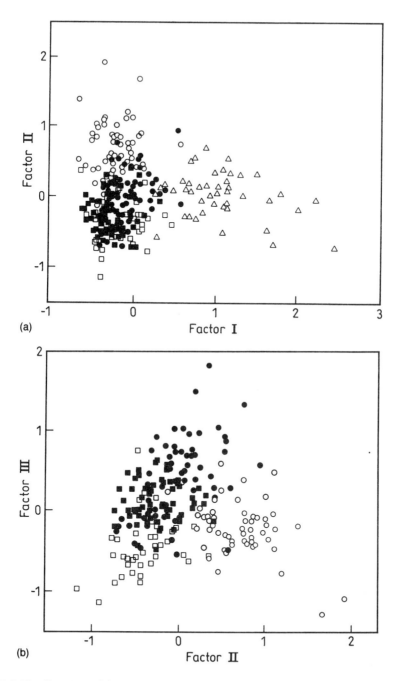

Fig. 5.7 The disposition of five human populations studied in the space of three first factors obtained by means of factor correspondence analysis. (a) All populations studied are in the space of first and second factors; each point in these pictures corresponds to one individual. Different symbols designate individuals from different populations: △, Talish population; □, Komi; ●, Abzelilovsky; ■, Arkhangelski; ○, Ilishevsky. (b) Ural populations are in the space of second and third factors.

the case of the Bashkirs, such a problem does not exist, but it may be very important for some small Northern nations consisting of a few dozen or a few hundred individuals.

Acknowledgements

This work was partly supported by the Russian State Programmes 'Funds of Fundamental Researchers', 'Biodiversity' and 'Frontiers in Genetics'.

References

Burke, T., Davies, N.B. and Hatchwell, B.J. (1989) Parental care and mating behaviour of polyandrous dunnocks *Prunella modularis* related to paternity by DNA fingerprinting. *Nature* 338, 249–251.

Dunnington, E.A., Gal, O., Siegel, P.B., Haberfield, A., Cahaner, A., Lavi, U., Plotsky, Y. and Hillel, J. (1991) Deoxyribonucleic acid fingerprint comparisons between selected populations of chickens. *Poultry Science* 70, 463–467.

Gilbert, D.A., Lehman, N., O'Brien, S.L. and Wayne, R.K. (1990) Genetic fingerprinting reflects population differentiation in the California Channel Island fox. *Nature* 344, 764–767.

Gilbert, D.A., Packer, C., Pusey, A.E., Stephens, J.C. and O'Brien, S.J. (1991) Analytical DNA fingerprinting in lions: parentage, genetic diversity, and kinship. *Journal of Heredity* 82, 378–386.

Gill, P., Ivanov, P.L., Kimpton, C., Piercy, R., Benson, N., Tully, G., Evett, I., Hagelberg, E. and Sullivan, K. (1994) Identification of the remains of the Romanov family by DNA analysis. *Nature Genetics* 6, 130–135.

Ginzburg, A.L., Grzhibovsky, G.M., Tokarskaya, O.N., Ryskov, A.P., Brukhanov, A.F., Shaginyan, I.A., Yanishevski, N.V. and Smirnov, G.B. (1989) Study of epidemiological importance of *Vibrio cholerae* atoxigenic strains. *Genetika* 25, 1320–1324.

Haberfeld, A., Dunnington, E.A. and Siegel, P.B. (1992) Genetic distances estimated from DNA fingerprints in crosses of White Plymouth Rock chickens. *Animal Genetics* 23, 159–165.

Ilyin, Y.V., Tchurikov, N.A., Ananiev, E.V., Ryskov, A.P., Yenikolopov, G.N., Limborska, S.A., Maleeva, N.E., Gvozdev, V.A. and Georgiev, G.P. (1977) Studies on the DNA fragments of mammals and *Drosophila* containing structural genes and adjacent sequences. *Cold Spring Harbor Symposia on Quantitative Biology* 42, 959–969.

Jambu, M. (1978) *Classification Automatique Pour l'Analyse des Données. Méthodes et algorithmes*. Bordas, Paris.

Jeffreys, A.J. and Morton, D.B. (1987) DNA fingerprinting of dogs and cats. *Animal Genetics* 18, 1–15.

Jeffreys, A.J., Wilson, V. and Thein, S.L. (1985) Individual specific fingerprints of human DNA. *Nature* 316, 76–79.

Jincharadze, A.G., Ivanov, P.L. and Ryskov, A.P. (1987) *Doklady Akademii Nauk USSR* 295, 230–232.

Kuseev, R.G. (1979) *The Origin of Bashkir People.* Nauka, Moscow.

Monastyrskii, O.A., Ruban, D.N., Tokarskaya, O.N. and Ryskov, A.P. (1990) DNA fingerprinting of some *Fusarium* isolates differentiated toxinogenically. *Genetika* 26, 374-377.

Monastyrskii, O.A., Ruban, D.N., Tokarskaya, O.N. and Ryskov, A.P. (1991) Identification of *Pyricularia oryzae* cav. isolates differentiated by a formula of virulence with the help of DNA restriction and DNA fingerprinting methods. *Genetika* 27, 1662-1667.

Reeve, H.K., Westneat, D.F., Noon, W.A., Sherman, P.W. and Aquadro, C.G. (1990) DNA 'fingerprinting' reveals high levels of inbreeding in colonies of the eusocial naked mole-rat *Proceedings of the National Academy Sciences USA* 87, 2496-2500.

Ryskov, A.P., Jincharadze, A.G., Prosnyak, M.I., Ivanov, P.L. and Limborska, S.A. (1988) M13 phage DNA as a universal marker for DNA fingerprinting of animals, plants and microorganisms. *FEBS Letters* 233, 388-392.

Ryskov, A.P., Romanova, E.A., Benediktov, I.I. and Penkova, R.A. (1990) Analysis of genetic polymorphism of helminthose agents by the DNA fingerprinting. *Vestnik S.-H. Nauki* 6, 147-149.

Tokarskaya, O.N., Zooyev, A.V., Sorokin, A.G., Panchenko, V.G. and Ryskov, A.P. (1990) DNA fingerprinting in cranes: a new approach to species typing. *Genetika* 26, 599-606.

Tokarskaya, O.N., Kalnin, V.V., Panchenko, V.G. and Ryskov, A.P. (1994) Genetic differentiation in a captive population of the endangered crane species. *Molecular and General Genetics* 245, 658-660.

Vassart, G., Georges, M., Monsieur, R., Brocas, H., Lequarre, A.-S. and Christophe, D. (1987) A sequence in M13 phage detects hypervariable minisatellites in human and animal DNA. *Science* 235, 683-684.

Wetton, J.H., Carter, R.E., Parkin, D.T. and Walters, D. (1987) Demographic study of a wild house sparrow population by DNA fingerprinting. *Nature* 327, 147-149.

6 Deep-ocean Biodiversity

J.F. Grassle
Director, Institute of Marine and Coastal Science, Rutgers University, PO Box 231, Marine Science Building, Dudley Road, New Brunswick, NJ 08903-0213, USA

The ocean is the cradle of life, yet our folklore, and perhaps our instincts, suggest a hostile, if not terrifying, environment. Until the major deep-sea expeditions in the latter part of the last century, life was thought to be absent in the deep ocean. An unexpectedly high diversity of life in the deep sea was discovered in the 1960s from qualitative trawl samples (Hessler and Sanders, 1967). With the identification and enumeration of the entire macrofauna from a large number of quantitative box core samples (Grassle and Maciolek, 1992), there is now even stronger evidence for a richness of species that is among the highest on the planet. Eighteen samples from a single deep-sea station with an aggregate surface area of less than $2\,m^2$, from off the east coast of the United States, yielded 436 species, most of which were undescribed (Grassle and Maciolek, 1992). In similar samples from the sea floor beneath the Mid-Pacific Gyre approximately every other individual collected was a different species (Poore and Wilson, 1993). This extraordinary diversity is not yet reflected in most textbook accounts of the deep sea, and the deep-ocean floor is frequently called a desert, where high pressures, cold and darkness prevent most animals from surviving. Although the high diversity of species in the deep-sea is now established, the sum of all quantitative seafloor samples covers less than $100\,m^2$. Considering that the deep ocean floor covers about 60% of the globe our knowledge of its fauna is based on a minuscule fraction of its vast extent.

All of the animal phyla (the major divisions of life) have been found in the ocean (marine Onychophora are known only from fossils) and 13 are found solely in the ocean. Only a small number of taxonomic specialists can properly identify species in the most species diverse marine environments. The number of systematists, working on marine invertebrates must be increased and the systematists presently working should be better supported. With greater effort, specialized samplers, lengthy periods of processing, and better techniques for identifying small organisms, the known diversity of life in all marine environments is certain to increase sharply over present estimates. When all of the invertebrate groups are quantitatively studied, coral reefs are likely to prove to be at least as rich in species as the deep sea. The assumption that there are more species on land than in the ocean can no longer be taken as a certainty (Grassle and Maciolek, 1992).

At the same time as we learn about the diversity of life forms in the ocean, there is evidence that oceanic ecosystems are endangered. There is no place in the ocean free from pollutants (Grassle *et al.*, 1990), coral reefs are in decline (Hughes, 1994), and fishing practices have resulted in the collapse of major fisheries. A few years ago it would have seemed incredible that the seemingly inexhaustible Grand Banks would be closed to fishing for commercial species, such as cod, haddock, and yellowtail, but such is now the case.

In order to better understand the functioning of marine ecosystems, to distinguish between the consequences of natural and anthropogenic change, and to better manage aquatic ecosystems, new methodologies are needed. Marine and terrestrial systems differ in fundamental ways that will affect approaches to management and plans for conservation (Norse, 1993). They also differ in their knowledge bases: we know comparatively little about life in the marine environment, and lack the kind of common sense understanding of marine ecosystems, which comes readily from walking through a forest.

Major advances in understanding the marine environment have occurred when scientists have become a more continuous presence in the environment, e.g. when ecologists began using study sites in the intertidal zone as a laboratory for *in situ* experiments. The use of diving apparatus and research submersibles has also caused a revolution in our ability to study and understand oceanic life, for example, when coral reef scientists spent long enough watching the behaviour of individual coral reef fish throughout their life history, or were present to witness the initiation of events such as the spreading catastrophic mortality of sea urchin populations and the bleaching of corals in the Caribbean. Our view of the deep-sea floor as a homogeneous ooze changed radically during the first visits with submersibles. Research submersibles also allowed the first deep-sea *in situ* experiments to be conducted and the discovery of unique hydrothermal vent ecosystems. These major discoveries have all occurred in just the last two decades. Even so, we remain occasional, outside observers and, in most cases, can only speculate on the ways in which oceanic life forms interact with the forces that control their survival.

Marine ecosystems must be described in terms of single coherent biological units: genes, populations, species, and communities and we need to understand the processes affecting the interactions of these units to form functional patterns at each of the levels of biological organization. We can now routinely receive at our desks continuous information on physical properties of the ocean by using satellites and ocean observing systems. These data will be used to develop better models of the physical environment comparable to weather prediction models on land. *In situ* microscopes and high resolution acoustic imaging systems are beginning to provide information on interactions among marine organisms. These technologies will enable us to see, in real time, what fish see, or what worms in the mud see. The level of understanding afforded by continuous, landscape-scale, real-time observations is taken for granted in terrestrial environments. With new acoustic and laser survey methods, we will soon have the first organism-scale topographic and sediment maps of the ocean. Continuous high-resolution sediment maps are seldom available, yet they are a prerequisite for designing biological surveys.

References

Grassle, J.F. and Maciolek, N.J. (1992) Deep-sea species richness: regional and local diversity estimates from quantitative bottom samples. *American Naturalist* 139, 313-341.

Grassle, J.F., Maciolek, N.J. and Blake, J.A-(1990) Are deep-sea communities resilient? In: Woodwell, G.M. (ed.) *The Earth in Transition*. Cambridge University Press, New York, pp. 464-469.

Hessler, R.R. and Sanders, H.L. (1967) Faunal diversity in the deep sea. *Deep-Sea Research* 14, 65-78.

Hughes, T.P. (1994) Catastrophes, phase shifts, and large-scale degradation of Caribbean coral reef. *Science* 265, 1547-1551.

Norse, E.A. (ed.) (1993) *Global Marine Biological Diversity* Island Press, Washington, DC, pp. 37-46.

Poore, G.C.B. and Wilson, G.D.F. (1993) Marine species richness. *Nature* 361, 597-598.

Biological Diversity and Genetic Resources

A. Charrier[1], F. Fridlansky[2] and J.C. Mounolou[2]
[1]ENSAM, Place Viala, 34060 Montpellier, and
[2]CGM, CNRS, 91198 Gif-sur-Yvette, France

For millenia Man has drawn the various resources he needed from the diversity of the living world and since the Azilian period he has developed remarkable efforts to master the surrounding nature, to grow better crops and to breed specific animals. Until the end of this century the idea that natural resources were unlimited was generally accepted and justified the systematic development of human populations and activities (Levêque, 1994). A change occurred when it was realized that this assumption was not true forever. Limits to human demographic expansion are now foreseen both in terms of space and of natural resources. Present human activities threaten the very existence of other living species and possibly of his own. A reappraisal of the relationship between society and biodiversity follows and presses the need for deeper knowledge and better management of biological resources.

Biodiversity integrates diversity at various levels of the living world organization: from molecules and genes to cells and organisms, from populations to specific ecosystems and the biosphere in general. On a given site or in a large area the number of species, their relative abundance, the intensity of their activities as well as the rate of their reproduction are used as indicators of biological wealth and its adaptive potential. They depend on the background of intraspecific genetic diversity as well as the environmental pressures that introduce specific constraints (Solbrig, 1991). How the constant evolution of the stock of genetic information is integrated in the organization of biodiversity in space and time and how man interferes in this process is the topic of present scientific debates.

Genetics Teaches Lessons and Opens Questions

From the beginning of this century until the 1960s, genetics told us that hereditary material (DNA) is composed of genes. Genes, in general, are replicated faithfully during cell proliferations and this is the fundamental basis of heredity. Genes code for the synthesis of proteins that are the true actors of cellular life. These molecules contribute to the elaboration of cell structures, they enact the synthesis of all other cellular components and catalyse cell functions. Mutations, through the modification of DNA sequences are an escape from the biological uniformity that would result from the conservative reproduction of genes. This variety of gene forms in a population is shuffled on the occasion of genetic exchanges between individuals at the time of sexual reproduction or through horizontal transfers. At any given time the genetic diversity in a population is the result of the random occurrence of conservative replication of preexisting gene forms and combinations, the mutations and genetic exchanges that have just appeared, the positive or negative selection enforced by external pressures and the random events that have placed specific individuals at the origin of the group (drift, migrations and foundations).

This executive view was questioned in the 1970s and in the early 1980s when it was discovered that many mutations (if not all) were not involved in the selection game. The neutralist theory supports a different interpretation of the contribution of genetic variations to the observed biodiversity. Mutations are considered as an inherent property of DNA metabolism. Genetic information is constantly evolving and only random processes are responsible for the generation of the biological pattern observed at any time and in any location.

By the end of the 1980s two factors induced a major reappraisal of the current views. On one hand, evidence for selection had not been rebutted by neutralism and the impacts of human activity on the evolution of genetic diversity through domestication, pollution, or direct exploitation, were clear examples of determined orientations and of choices bearing on specific genes. On the other hand, the formidable impulse given to biology by sequencing techniques and other molecular approaches changed our view about DNA. First, the genome is no longer considered as a set of stable DNA molecules that may rarely get mutated. It is constantly susceptible to processes that demonstrate its fluidity: amplifications, transpositions, insertions and exclusions of sequences. Such events have been amply used during evolution to generate tactics of differentiation and organ specializations (gene families), strategies of reproductive incompatibilities and speciation. They are also involved in interindividual and interspecific exchanges different from the ones enacted by sexual reproduction (retroposons and retroviruses). In parallel the second major output of the genome sequencing programmes is the discovery of genes that had not been revealed by the former approaches. In some cases similitudes with already known sequences suggest functions for these genes; in others their roles remain to be elucidated. Of course one obvious idea is to postulate that they escaped classical analysis because they are not essential and are involved in the ability of the organism to respond to environmental changes (Oliver et al., 1992). This however remains to be demonstrated.

These lessons of modern genetics generate three questions about the origin and the organization of present genetic diversity. The first one bears on the very abilities of DNA molecules *in vivo* to mutate and to exchange genes: how has this property evolved? how is it related to environmental changes? how has it contributed to the history of biological diversity in the past?

The true relation between genetic diversity and biodiversity is the second subject of interrogation. It is generally accepted that genetic diversity is an insurance for a species or a community to face changes due to time, climate or other environmental modifications. If all genetic structures are not able to persist when new selection pressures occur, some of them do and constitute the genetic core from which a new step in evolution emerges in the long run. Today in the tropics (and up until the turn of this century in Europe) farmers use composite seeds of various genotypes, varieties or even species, which are differently adapted to local environments, in order to yield a crop and fulfill short-term needs whatever the circumstances are. That genetic diversity plays such a role in general is not definitely proven, however. Counter examples exist: some species, like bees, appear poorly polymorphic and exhibit a high adaptive potential, whereas others that are very diverse such as some molluscs seem to have been restricted to specific and constant habitats for a million years. Thus, the relationship between genetic diversity and biodiversity still needs to be thoroughly examined.

Lastly, the critical question of how humans interfere with genetic diversity in general is still pending. In some cases, as in the ones that are presented later in this chapter, humans have modified the frequencies of different types of genes when the individuals chosen for domestication were multiplied. The overall gene pools of some species have indeed been modified in some animals to the point where individuals can no longer live without their breeders. In plants, gene flows between species have also been manipulated. Some biological forms have suffered considerable regression, and others have been created that did not develop spontaneously. In consequence, we have difficulties today in evaluating the true long-term adaptive potential of present gene pools and a more critical examination is necessary if sustainable development is to be envisaged (Cauderon, 1980, 1992). In the meantime, two conflicting possibilities have been proposed: develop systematic and general measures of nature preservation with a long-term perspective, or set up the artificial conditions of conservation for the part of biodiversity that is useful now. The two examples that follow show how such a dilemma can be overcome.

The African Rice and the European Rabbit
Two Examples of an Approach to Genetic Diversity

African rice

The African rice, *Oryza glaberrima*, is a traditional crop grown from Senegal to Chad and down to the southern coast of Western Africa. This culture is steadily declining under the pressure of social changes and of the expansion of the Asian rice, *Oryza sativa*. However, it is recognized that *O. glaberrima* has specific qualities that allow growth and yield under conditions that are marginal for the Asian plant. In addition, the African rice is specifically involved in some cultural events of African societies and this also acts in its favour.

A systematic analysis of *O. glaberrima* genetic diversity was conducted some years ago by several national research institutions in Western Africa, two French ones (ORSTOM and CIRAD) and two international ones International Plant Genetic Resources Institute (IPGRI) and the International Institute of Tropical Agriculture (IITA). The survey was extended to a wild relative species, *Oryza breviligulata*, which spreads eastwards to Tanzania. This species lives in temporarily flooded swamps and also occurs as a weed in and around the rice fields. It was hypothesized that the two species might still exchange genes. Material collected in Africa was both conserved in gene banks (about 10,000 accessions) and analysed using identification criteria including agronomic and morphometric characters, physiological properties, protein polymorphism, and DNA sequences (Pernes, 1984; Miezam and Ghesquiere, 1986).

Results are extremely enlightening. The examination of so called 'neutral' characters or genes shows that the wild species, *O. breviligulata*, has the largest diversity. Estimations for 19 polymorphic enzymatic loci and 49 alleles are for wild and weedy *O. breviligulata* respectively 0.131 and 0.064, versus 0.042 for the cultivated species *O. glaberrima* (Second, 1982). Diversities found in *O. glaberrima* or in the plants of *O. breviligulata* that live in the field associated with the rice are the same and constitute a subset of that of the wild plant (the seven multiallelic combinations representing 90% of the local cultivars exist in wild populations). These neutral markers exhibit a broad geographic and ecological distribution. In *O. breviligulata* a clinal variation for two loci is found from Western to Eastern Africa. On the contrary, *O. glaberrima* is differentiated into various ecotypes infeodated to Northern or Southern regions, to dry or humid conditions and to types of rice cultivation (dry land or flooded rice).

However, when emphasis is placed on the characters that are required by humans to obtain a crop of *O. glaberrima* the effect of selection is clear. Differentiations of the wild and cultivated rices concern the life cycle in relation to photoperiodism, grain yield, dispersal of the spikelets associated with natural shedding and dormancy. For the genes involved in domestication humans have retained and multiplied very specific alleles. They constitute a subset that is not entirely included in the diversity of the wild plants. Numerous new cultivars (19 multiallelic combinations) arising from recombination of existing genes were selected by farmers for various agroecological conditions. Even more interesting is the fact that this whole subset is present among the plants of *O. breviligulata* found as weeds in rice fields. In brief, genetic exchanges in both directions are demonstrated between the African rice and its wild relative (Bezançon, 1993).

Three dramatic pieces of information are provided by these studies: (i) domestication was driven by humans from only two or three wild types among hundreds of other possibilities; (ii) the discovery of gene flows between related species is essential for elaborating conservation rules *in situ* and *ex situ* and for managing collections that would favour the dynamic conservation of diversity; and (iii) the generation by farmers of new recombinant genotypes adapted to various agroecological environments is a guide for the elaboration of synthetic populations in plant breeding.

European rabbit

The case of the European rabbit, *Oryctolagus cuniculus*, also provides an original insight into similar genetic diversity problems because of a different approach. Interest in this animal stems from the fact that it is the only domesticated mammal originating from Western Europe and more precisely from Spain (and Southern France). This species evolved at glaciation times and started its expansion when the ice retreated. Actually it did so both spontaneously and with the efficient partnership of humans who considered it first as easy game, later as a source of meat and as a nuisance for crops. A survey of genetic diversity in *O. cuniculus* leads to the conclusion that diversity is maximal in Spain and is gradually declining northward. Taxonomists even distinguish two subspecies, one in the South Western part of the Iberian Peninsula and in Northern Africa, and one in Northern Spain and the rest of the world. This distinction is consistent with genetic observations of enzyme or immunoglobin polymorphism and with the identification of two general maternal lineages (one in the South Western part of the Iberian Pensinsula, one in the rest of the world; Biju-Duval *et al.*, 1991). Maternal inheritance is due to the transmission of mitochondrial DNA by the female gametes only from one generation to the other. As females are territorial, mitochondrial DNA molecules stay where the mothers live and tend not to move around. It was consequently inferred that the appearance of a new mitochondrial DNA sequence in one place could be the result of a mutational event or of the artificial introduction of a female from a different location. In order to distinguish between these two possibilities an examination of archaeological items and present day animals was carried out in the same locations. Results clearly demonstrate the occurrence of stable populations with specific mitochondrial sequences in Southern France from 10,000 BP to 2000 BP. These lineages have even given progenies that can be found today on the same sites. In Roman times humans started to move rabbits around. The appearance in one place of a new mitochondrial sequence that cannot be easily related by mutation to the local ones but is identical to those present in a different region indicates such human-directed introduction (Monnerot *et al.*, 1994; Hardy *et al.* 1995). Not only does this approach give an indication of the origin of the animals but it yields a precise dating of the events involved through ^{14}C measurements and archaeological information. In particular this shows that although a wide genetic pool was accessible, domestication attempts that have led to present day animals and races have all been made in only one specific maternal lineage. The expansion of this original material through Western Europe took place during the Middle Ages in parallel with the transportation of wild genotypes that were not domesticated.

As in the case of the African rice, humans have sampled and carried out selection efforts on a limited genetic stock, although much more diversity was present in wild rabbits. Recently attempts have been made in some ways to benefit from genetic exchanges between domesticated animals and the others but with limited success. Indeed managers of game reserves and hunting societies know that introduced animals do not easily mix with local ones. Adaptive, behavioural and genetic barriers may explain present difficulties. This means that more efforts must be devoted to the understanding of relations and exchanges between genetically different animals and populations.

In the meantime, conservation measures, both *in situ* and *ex situ*, take into account the extent of existing genetic diversity. Gene banks that have developed for domestic animals (sperm conservation and artificial insemination programmes) will have to open their doors to wild genetic resources and optimize diversity.

Conclusion

For the last 25 years international organizations concerned with genetic resources (FAO/IPGRI) have efficiently focused their activities on cultivated plants and domestic animals. However, the conviction that this would not suffice gradually emerged and culminated at the Rio Conference. From all horizons claims are made in favour of a global reappraisal of the relationship between humans and biological diversity in general. The concern has come to the point where conservation is proposed as the major action to continue in the near future. However, there is controversy about what should be conserved: all species and ecosystems or only the biological items that have (or may have) a direct utility?

Both the lessons of modern genetics and the two examples presented above may help to clarify the debate and to fund the principles of future actions. Indeed from the case of the rice or that of the rabbit we learn that humans are intimately associated with the recent evolution of these species. New biological forms of rice have even been created. Humans have had profound impact on their genetic diversity. But this was imposed rather slowly through centuries and it is clear that the genetic wealth of the two organisms has not disappeared. Our present concern is that, under economic pressures and demographic expansion, the human interference becomes so potent and pressing that the living species do not have the opportunity to accmmodate the change of speed and gradually lose their genetic diversity. Unable to conserve everything as it is, we must face the risk of choices. The question is not to throw away our future because of immediate needs or benefits, hoping that progress in biological engineering will one day enable any living form to be reconstructed. The responsibility of preparing the genetic resources our societies will seek can be assumed if advancing knowledge is continuously integrated in the procedures of conservation and management of core biological diversity. What has to be conserved and how to do it need systematic and recurrent reappraisal in the light of four factors: (i) present diversity, (ii) its ability to evolve, (iii) environmental changes, and (iv) human needs transformations.

References

Bezançon, G. (1993) Le riz cultivé d'origine africaine *Oryza glaberrima* et les formes sauvages et adventices apparentées. Thèse de doctorat es Sciences. Université de Paris XI Orsay. 232 pp.

Biju-Duval, C., Ennafaa, H., Dennebouy, N., Monnerot, M., Mignotte, F., Soriguer, R., El Gaied, A., El Hili, A. and Mounolou, J.-C. (1991) Mitochondrial DNA evolution in lagomorphs: origin of systematic heteroplasmy, organisation of diversity in European rabbits. *Journal of Molecular Evolution* 33, 92–102.

Cauderon, A. (1980) Sur la protection des ressources génétiques en relation avec leur surveillance, leur modelage et leur utilisation. *Comptes Rendus de l'Académie d'Agriculture* 66, 1051–1068.

Cauderon, A. (1992) Complexes d'espèces, flux de gènes et ressources génétigues des plantes. Actes du colloque international en hommage à Jean Pernès, XXIX–XXXV, Paris, 8–10 Janvier.

Hardy, C., Callou, C., Vigne, J-D., Casane, D., Dennebouy, N., Mounolou, J.-C. and Monnerot, M. (1995) Rabbit mitochondrial DNA diversity from prehistoric to modern times. *Journal of Molecular Evolution* 40, 227–237.

Levêque, C. (1994) Environnement et diversité du vivant. Presses Pocket-Cité des Sciences et de l'Industrie (coll. Explora).

Miezam, K. and Ghesquiere, A. (1986) Genetic structure of African traditional rice cultivars. *Rice Genetics Proceedings*. Rice Genetics Symposium, IRRI, Los Banos, Philippines, pp. 91–107.

Monnerot, M., Vigne, J-D., Biju-Duval, C., Casañe, D., Callou, C., Hardy, C., Mougel, F., Soriguer, R., Dennebouy, N. and Mounolou, J.-C (1994) Genetic and historic approach. *Genetics Selection and Evolution* 26, 167s–182s.

Pernes, J. (1984) Gestion des ressources génétiques des plantes. Technique et Documentation Lavoisier (Paris). Vol. 1, 212 pp. Vol. 2, 346 pp.

Oliver, S.G., Van der Aart, Q.J.M., Agostini-Carbone, M.L. *et al.* (1992) The complete DNA sequence of yeast chromosome II. *Nature* 357, 38–46.

Second, G. (1982) Origin of the genetic diversity of cultivated rice (*Oryza* spp). Study of the polymorphism scored at 40 loci. *Japanese Journal of Genetics* 57, 25–57.

Solbrig, O.T. (1991) *From Genes to Ecosystems: a Research Agenda for Biodiversity*. Monograph No. 7, International Union of Biological Sciences, Paris.

8 Ecological Functions of Biodiversity: The Human Dimension

P.S. Ramakrishnan
Professor of Ecology, School of Environmental Sciences, Jawaharlal Nehru University, New Delhi 110067, India

Summary

While considering natural ecosystems in a purely biological sense, addition or deletion of species, functional or structural groups, and indeed ecosystem components, are sometimes shown to be of no/limited consequence in altering a given ecosystem level process. However, it is often forgotten that different functional groups control/govern different processes in an ecosystem, with a certain degree of redundancy attached with them, in such a manner that others could take over similar function, although to a limited extent. Biodiversity becomes crucial for the system as a whole and indeed at the landscape level, as a buffer against perturbations, and for maintaining overall ecological integrity.

Humans, however, are often kept out of the definition of the ecosystem boundary. If humans are included as an integral component of ecosystem function, their implications for the role of biodiversity would change drastically. There are many examples where humans, through species inputs that are intentional or otherwise, have contributed to biodiversity and ecosystem function alterations. The parallelism that has often been recognized by this author between ecologically significant keystone species and socially selected key species, and its value for enhanced/altered ecosystem level functions including their value to the humans is a case in point. Nowhere else is this linkage between ecological and social processes more obvious than in agroecosystems and traditional village ecosystems, since the relationship between biodiversity and function in these situations is to a significant extent determined by human intervention. Thus, for example, biodiversity is not only an important regulator of agroecosystem function in a strictly biological sense of its impact on production and other ecological processes, but is valuable for satisfying a variety of other needs of the farmer and society at large.

© 1996 CAB INTERNATIONAL. *Biodiversity, Science and Development: Towards a New Partnership* (eds F. di Castri and T. Younès)

The implications of this approach in a consideration of biodiversity issues, where humans are considered within the ecosystem boundary rather than outside it, for ecosystem function and sustainable management of natural resources with peoples' participation is discussed using appropriate examples.

Introduction

Although biodiversity versus ecosystem stability and resilience has been a matter of academic concern among ecologists for some time now (Odum, 1971), it is only recently that attempts have been made to clearly delimit the role of biodiversity in ecosystem function (Mooney et al., 1994). However, our understanding of this is still limited by lack of appropriately designed experimental designs that would account for spatial and temporal issues, at the same time accounting for species redundancy problems and extraneous complications such as perturbations, confounding biodiversity-related functional aspects. The recent interest in the role of biodiversity on ecosystem function arises out of a concern for utilization of natural resources in a sustainable manner that satisfies current needs without compromising the needs or options of future generations.

In recent times, activities initiated in pursuit of an improved quality of life, through the production of food, fibre, shelter and consumer goods, recreation, etc. have led to unintended results (Fig. 8.1). In this biodiversity has been a major casualty. It is in this context of biodiversity depletion that varied aspects of disrupting biotic interactions on ecosystem function have been discussed by the different panellists. Since humans play a key role in altering ecosystem level processes, integrating humans as part of the ecosystem rather than keeping them outside it is important. This becomes even more crucial in the tropical context, because biodiversity concentrations are largely tied up with regions where traditional societies live; these traditional societies by the very nature of their livelihood activities are part of the ecosystem/landscape functions, rather than being external players (Ramakrishnan, 1992a,b). Biodiversity is not only an important regulator of ecosystem function in a biological sense, but is also crucial for satisfying a variety of societal functions. The following discussion encompassing humans as part of the ecosystem function is placed in this context.

Humans Integrated within Ecosystem Boundaries

Ecologists generally tend to view an ecosystem strictly in a biological sense, keeping humans outside its structural/functional attributes. In this view of things, humans sitting outside the biological ecosystem boundary bring about ecosystem alterations through perturbations. The impact of human activities on the ecosystem and the effect of altered ecosystem properties on the humans are not viewed as an integrated whole, in the context of biodiversity influencing ecosystem function. However, contrary to the traditional ecologist's view

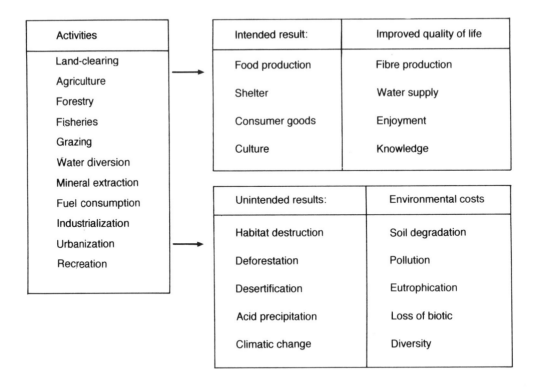

Fig. 8.1. Human activities affecting the sustainability of the biosphere (from Lubchenco et al., 1991).

point, if we take a view that humans form an integral component of the ecosystem function, and that this integrative view point is even more obvious at a landscape level, then the role of biodiversity in ecosystem function would become very obvious. Indeed, this is the way in which many traditional/tribal societies perceive the immediate environment around them (Table 8.1). The concept of village as an ecosystem, with all its ramifications involving agriculture, animal husbandry and the domestic sector enmeshed with the forest and forest-related activities such as hunting and gathering of food, fodder, fuel wood, and medicine and forest farming as done under shifting agriculture, is an example of integrating humans within the ecosystem boundaries and for evaluating the role of biodiversity in ecosystem function in a broader context (Ramakrishnan, 1992a,b).

A total of 171 plant species, and over a dozen animal species are used by four tribes, the Nishis, the Hill Miris, the Sulungs and the Apatanis living in a cluster of villages, in the Lower Subansiri District of Arunachal Pradesh of north-eastern India (Gangwar and Ramakrishnan, 1990). The plant species belong to a wide variety of categories – for leafy vegetables, edible fruits, a variety of other food items, cereals, tubers, liquor, medicine, household goods, traditional dresses, dyes, tattoos, fish, animal poisons, or salt extrac-

Table 8.1. Shifting agriculture (Jhum) in north-eastern India and social disruption (from Ramakrishnan, 1992b).

North-eastern India has over 100 different tribals, linguistically and culturally distinct from one another; the tribes often change over very short distances, a few kilometres in some cases. Shifting agriculture or jhum as the tribals call it, is the major economic activity. This highly organized agroecosystem was based on empirical knowledge accumulated through centuries and was in harmony with the environment as long as the jhum cycle (the fallow length intervening between two successive croppings) was long enough to allow the forest and the soil fertility lost during the cropping phase to recover.

Supplementing the jhum system is the valley system of wet rice cultivation and home gardens. The valley system is sustainable on a regular basis year after year because the wash-out from the hill slopes provides the needed soil fertility for rice cropping without any external inputs. Home gardens extensively found in the region have economically valuable trees, shrubs, herbs and vines and form a compact multi-storied system of fruit crops, vegetables, medicinal plants and many cash crops; the system in its structure and function imitates a natural forest ecosystem. The number of species in a small area of less than a hectare may be 30 or 40; it therefore represents a highly intensive system of farming in harmony with the environment.

Linked on this land-use are the animal husbandry systems centred traditionally around pigs and poultry. The advantage here is primarily that they are detritus-based or based on the recycling of food from the agroecosystem unfit for human consumption.

Increased human population pressure and decline in land area resulting from extensive deforestation for timber for use for industrial man and jhum has brought down the jhum cycle to 4-5 years or less. Where population densities are high, as around urban centres, burning of slash is dispensed with, leading to rotational/sedentary systems of agriculture. These are often below subsistence level, though the attempt is to maximize output under rapidly depleted soil fertility. Inappropriate animal husbandry practices introduced into the area, such as goat or cattle husbandry, could lead to rapid site deterioration through indiscriminate grazing/browsing and fodder removal, as has happened elsewhere in the Himalayas. The serious social disruption caused demands an integrated approach to managing the forest-human interface.

---

tion. The Sulungs, for example, use the starch extracted from the pith of *Metroxylon sagu* as their staple food.

The Nagaland Department of Agriculture, through their well organized grassroots level Village Development Boards (VDBs) have recently been able to catalogue over a thousand plant species used by the villagers, as perceived by the local communities (A.M. Gokhale, Agriculture Development Commissioner, Nagaland, personal communication). This is based on a quick evaluation done during a short 6-month period; a detailed study would lead to many more! Such an extensive use of biodiversity suggests that the tribes of north-eastern India, and traditional farmers elsewhere (Brookfield and Padoch, 1994), rely heavily on the forest ecosystem to meet their varied needs. This implies that conserving this biodiversity is crucial for their survival, as a component of the rainforest ecosystem.

It is not only the mere presence of biodiversity and the functional role it has for tribal humans that is significant here. The manner in which tribals manipulate this biodiversity for ecosystem functional integrity and for their own function within the landscape is interesting. Under a 60-year shifting agriculture cycle, more than 40 crop species are found. Here the emphasis is on cereals which are largely placed towards the base of the slope as they are less nutrient-use efficient, whereas the more nutrient-use efficient tuber crops are placed towards the top of the slope where soil fertility levels are low. Under shorter 10- or 5-year cycles, the cropping pattern shifts with emphasis on tuber crops (Ramakrishnan, 1992b). This is an elegant example of adaptation towards optimization of resource use and risk coverage, through manipulation of biodiversity, by the humans within the ecosystem.

Through mixed cropping involving a large number of species, and traditional weed management strategies, shifting agriculture farmers of north-east India ensure an effective check on nutrient loss during the cropping phase. The traditional weed management practice where about 20% of the weed biomass is left undisturbed in the plot by the shifting agriculture farmer (Ramakrishnan, 1988) is also a practice common to the Mayan agriculture in Mexico (Chacon and Gliessman, 1982).

Within a given landscape, the tribal farmer of north-east India also has a variety of land use systems contributing towards biodiversity at all levels ranging from the subspecific, through the species, population, and the ecosystem (Ramakrishnan, 1992b). Thus, apart from the diversity in cropping patterns within the shifting agriculture systems that he maintains, he may have fallow systems, sedentary systems on hill slopes, wet rice cultivation on valley lands involving a variety of rice cultivars, and a whole variety of tightly packed home gardens resembling a forest where perennial trees and shrubs of economic value are grown along with herbs and vines. This mosaic of ecosystem types of the landscape performs a variety of functions towards the integrity of the system as a whole, while having a variety of service functions for the humans. The valley lands capture water and nutrients washed out from the hill slopes useful for rice production. The home garden imitating a natural forest is self-sustaining through tight recycling of nutrients within and is of value to the humans as a source of food supplement, medicines, fodder, fuel wood, fibre, spices and condiments, for their use and for cash income (Gliessman, 1989; Ramakrishnan, 1992b). These and the secondary successional vegetation on the hill slopes effectively capture the nutrients which could otherwise be lost through hydrology (Ramakrishnan, 1992b). Indeed, evidence now suggests that the diversity and adaptability of indigenous farming practices have much to offer in explaining why some are successful at conserving resources and others are not (Brookfield and Padoch, 1994).

The linkages that exist between the various subsystems (agriculture with further subdivisions within, animal husbandry and the domestic) of the village ecosystem and its further linkages with the forest ecosystem at the landscape level (Fig. 8.2) illustrate how humans form part of the ecosystem function at the village level, and indeed form part of the landscape functional unit itself.

Ecological Functions of Biodiversity   119

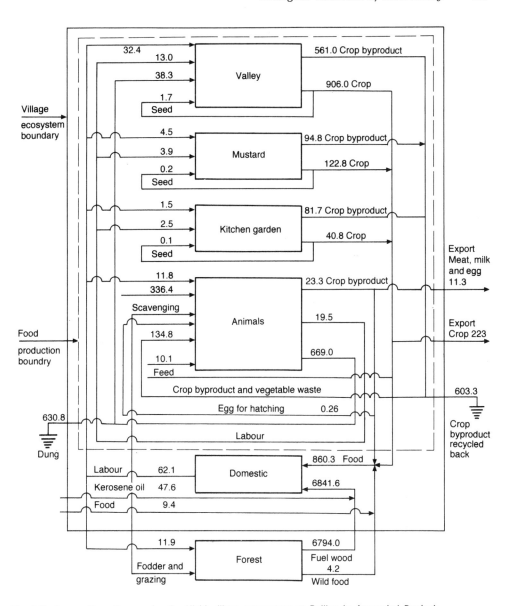

Fig. 8.2. Energy flow diagram for the Nishi village ecosystem at Balijan in Arunachal Pradesh in north-eastern India. Unit = MJ × 103 (from Ramakrishnan, 1992b).

Biodiversity in Agroecosystem Function

Through biodiversity in the agroecosystem, a number of ecological functions over a range of scales in space and time, and in relation to different components and functional groups are ensured (Swift et al. 1994). Thus the functional properties of the domains of species or functional groups can be considered as an hierarchy of nested interacting systems. The following gives examples at different scales.

At a subspecific level, ecotype populations adapted to a given soil type could determine weed–crop competition. In a study of ecotype differentiation in *Cynodon dactylon*, in the context of the calcicole–calcifuge problem in the Indo-Gangetic alluvial soils of north-western India, specially adapted ecotypes suited to a given soil type were shown to be more aggressive than others that were not (Ramakrishnan and Gupta, 1972).

At a species to species level, in north-western Indian wheat plots, *Trigonella polycerata*, a leguminous weed enhanced crop yield at low densities of the weed due to improved nitrogen nutrition through this legume (Kapoor and Ramakrishnan, 1975). The adverse effect of this weed was not realized until the weed density went beyond 3200 plants per m^2.

At population/functional group level, the shifting agriculture farmer of north-east India controls sediment and nutrient loss from plots by his traditional weed management practice (Ramakrishnan, 1988), mentioned in the earlier section. The farmer deals with it in two different ways. In all cases of weed management the farmer leaves about 20% of the weed biomass in his plot. The remaining 80% biomass is weeded and put directly back into the plot as mulch. With retention of about 20% of the weed biomass, the loss of sediment and a labile element such as potassium through run-off is reduced by about 20% when compared with total weeding.

Although losses would be least in the unweeded plots, this would, to a large extent, be negated through weed–crop competition and the consequent reduced crop yield. Further study has shown that if it was not for integrated weed management practices of the traditional farmer, the highly shortened 4–5 year cycles now prevalent would have resulted in even more distorted ecosystem function than at present, through adverse effects on nutrient cycling. The 80% weed biomass put back into the plot, rather than being discarded, helps in nutrient cycling through rapid decomposition. The integrated weed management concept of the traditional jhum farmer indicates that the farmer knows precisely how intense the weeding should be so that the weed stops interfering with crop yield, and yet the beneficial effects are manifest. Such a subtle distinction between 'weed' and 'non-weed' status is widespread for many traditional societies (e.g., *mal monte* and *buen monte* which means bad weeds and good weeds of the Mayan region of southern Mexico and the Guatemalan highland farmers, Gliessman, 1988).

At an ecosystem level, mixed cropping systems, very common to traditional tropical agroecosystems, are suggested to promote enhanced nutrient use, tighter nutrient cycling, higher crop yield, pest and disease

Table 8.2. Shifting agriculture (jhum) and sustainable development for north-eastern India (from Ramakrishnan, 1992b).

For improving the system of land use and resource management in north-eastern India, the following strategies suggested by Ramakrishnan and his co-workers are based on a multidisciplinary analysis. Many of these proposals have already been put into practice.

- With wide variations in cropping and yield patterns under jhum practised by over a hundred tribes under diverse ecological situations, where transfer of technology from one tribe/area to another alone could improve the jhum, valleyland and home garden ecosystems. Thus, for example, emphasis on potato at higher elevations compared to rice at lower elevations has led to a manifold increase in economic yield despite low fertility of the more acid soils at higher elevations.

- Maintain a jhum cycle of minimum 10 years, (this cycle length was found critical for sustainability when jhum was evaluated using money, energy, soil fertility biomass productivity, biodiversity, and water quality, as currencies) by greater emphasis on other land use systems such as the traditional valley cultivation or home gardens.

- Where the jhum cycle length cannot be increased beyond the five-year period that is prevalent in the region, redesign and strengthen this agroforestry system incorporating ecological insights on tree architecture (e.g. the canopy form of trees should be compatible with crop species at ground level so as to permit sufficient light penetration and provide fast recycling of nutrients through fast leaf turnover rates).

- Improve the nitrogen economy of jhum at the cropping and fallow phases by introduction of nitrogen-fixing legumes and non-legumes. A species such as the Nepalese alder (*Alnus nepalensis*) is readily taken in because it is based on the principle of adaptation of traditional knowledge to meet modern needs. Another such example is the lesser known food crop legume *Flemingia vestita*.

- Some of the important bamboo species, highly valued by the tribals, can concentrate and conserve important nutrient elements such as N, P and K. They could also be used as windbreaks to check wind-blow loss of ash and nutrient losses in water.

- Condense the time-span of forest succession and accelerate restoration of degraded lands based on an understanding of tree growth strategies and architecture, by adjusting the species mix in time and space.

- Improve animal husbandry through improved breeds of swine and poultry.

- Redevelop village ecosystems through the introduction of appropriate technology to relieve drudgery and improve energy efficiency (cooking stoves, agricultural implements, biogas generation, small hydroelectric projects, etc.). Promote crafts such as smithying and products based on leather, bamboo and other woods.

- Strengthen conservation measures based upon the traditional knowledge and value system which the tribal communities could identify, e.g. redevelopment of traditional agroecosystems to conserve agroecosystem biodiversity; the revival of the sacred grove concept based on cultural tradition which once enabled each village to have a protected forest although few are now left.

control, and weed suppression (Altieri and Liebman, 1988; Gliessman, 1989; Ramakrishnan, 1992b). Although broad generalizations are difficult, as with other ecosystem types, biodiversity's function is system-dependent. By emphasizing internal controls of ecosystem function, rather than promoting dependence upon external subsidies, these agroecosystem models are accessible to all sections of the society, and if properly managed and redeveloped, could promote equity and some degree of social justice. Apart from the high energy efficiency of these systems, very often the well managed systems could be as productive as more modern agriculture.

With a multilayered canopy and with a high leaf area index for efficient light capture and a layered root mass distribution in the soil profile for optimal nutrient use, crop productivity could be high (Table 8.2), indeed comparable to those of the secondary successional fallows up to 20 years in the same area (Ramakrishnan, 1992b). A characteristic feature of traditional systems is the high rate of biomass accumulation in relation to economic output. Without this high organic matter production it would become necessary to constantly import costly inorganic fertilizers which are often hard to come by and whose effectiveness in the face of high temperature and rainfall under many tropical situations is questionable (Ramakrishnan, 1994a).

Sequential harvesting of crops is an effective way of managing up to 30 or 40 crop species over both space and time, as shown for the north-east Indian shifting agricultural systems (Ramakrishnan, 1992b). Successive harvesting of crops creates more space for the next crop species at the peak of its growth, and the biomass put in after each harvest also ensures efficient recycling of organic crop residues.

Mixed cropping is also receiving considerable attention from modern agricultural scientists as a biological pest suppressant (Litsinger and Moody, 1976). The use of native varieties would probably ensure that a high degree of natural chemical defences are maintained (Janzen, 1973). Further, under mixed cropping, it is unlikely that any one of the pest populations of insects, fungi, bacteria or nematodes would reach epidemic levels due to high genetic diversity.

On a hilly terrain, mixed cropping systems are particularly important for soil and nutrient conservation. As already noted earlier, the shifting agriculture farmer in the north-eastern hill region of India ensures effective use of the nutrient gradient on the steep hill slope by concentrating species that have a high nutrient-use efficiency along the nutrient-poor top of the slope and by organizing less use-efficient species along the nutrient-rich base of the slope. By this means he is also able to achieve a high leaf area index for optimizing photosynthesis, checking erosive losses and maximizing production.

At the landscape level, the land use system in a situation such as north-eastern India, has a variety of interdependent ecosystem types (Ramakrishnan, 1992b). Apart from plots that may be under different shifting agricultural cycles, the hill slopes have forest fallows under varied development phases. In this, some of the herbaceous communities of the early successional phases conserve phosphorus within the system; a bamboo species such as *Dendrocalamus hamiltonii* coming later in the succession in a fallow system is known to conserve potassium within the system. Along the hill slopes, the farmer may develop home gardens, with an assemblage of trees, shrubs, herbs and climbers that are economically valuable, but tightly packed together as though imitating a forest. All these ecosystem types on the hill slopes in turn may affect water and nutrient flows for valley land wet rice cultivation down below. These natural and human-made ecosystem types are closely interlinked with the village ecosystem function.

Keystone Species, Biodiversity and Ecosystem Function

The role of keystone species in conserving and enhancing biodiversity, and indeed in manipulating ecosystem function is a critical area which has not been adequately explored. Keystone species play a crucial role in biodiversity conservation, through key functions that they perform in an ecosystem. Therefore, they could be used for not only managing pristine ecosystems (Ramakrishnan, 1992a,b), but also for building up biodiversity in both natural and managed ecosystems, through appropriately conceived rehabilitation strategies (Ramakrishnan et al., 1994a).

In many areas in north-east India where local population pressures are more intense, the shifting agricultural cycle has come down to less than 5 years, leading in extreme cases to continuous cropping year after year, through a rotational fallow system of agriculture. The fallow period could be a short 1 or 2 years, or it may even be just the fallow regrowth of the previous winter season. In these mixed cropping systems, the farmer uses a number of lesser known plants of food value, of which *Flemingia vestita* (Gangwar and Ramakrishnan, 1989) is significant for this discussion. Selected by the farmer on the basis of experience, the tuber of this plant, rich in protein and stored in gunny bags placed below-ground, serves as a food supplement during the lean season when traditional food supply is limited. Our studies suggest that through the mixed cropping system and by raising a pure crop of this species once in 3 or 4 years, the farmer is able to add as much as 250 kg of nitrogen per hectare per year into his plot. The farmer, on the basis of intuitive experience knows how to adapt traditionally evolved technology to fit in with his present day requirements. This lesser known crop legume is a keystone species for the farmer's agroecosystem integrity, under conditions of extreme pressure on the land (Ramakrishnan et al., 1994b).

While working in the sacred grove forests of Cherrapunji in Meghalaya, we came across four dominant tree species, namely, *Engelhardtia spicata*, *Echinocarpus dasycarpus*, *Sysygium cuminii* and *Drimycarpus racemosus* that contain a high level of nitrogen, phosphorus and potassium in the leaf tissue, in spite of the fact that these species grow in highly infertile soils (Khiewtam and Ramakrishnan, 1993). These are keystone species in an ecological sense that perform key functions of nutrient conservation in this protected ecosystem. Through their role in ecosystem function, they contribute towards supporting biodiversity in these relict forests, often protected by local people for religious and cultural reasons. However, many groves in the region have been degraded because of human impacts resulting from a gradual decline in the traditional value system that has occurred during the last few decades. Therefore, manipulating these keystone species is a simpler way of managing a whole variety of biodiversity in the ecosystem.

In the successional forests of the north-eastern hills of India, a variety of species could be categorized as keystone species. Nepalese alder (*Alnus nepalensis*) conserves nitrogen in the system through nitrogen fixation (Ramakrishnan, 1992b). We found that many species of bamboo (*Dendrocalamus hamiltonii, Bambusa tulda,* and *B. khasiana*) play a key role in conservation of nitrogen, phosphorus and potassium in the jhum fallows. These keystone species not only determine ecosystem function in that particular serial stage of succession where it appears, but even determine the very process of succession itself due to their impact on nutrient cycling. Indeed, they modulate ecosystem function, both in space and time. Such species could be of value in rehabilitating ecosystems, following a successional pathway.

Linkages between Ecological and Social Processes

Linking ecological and social processes is crucial for appreciating the relationship between biodiversity and ecosystem function and to utilize this relationship for human welfare through sustainable management of resources. The linkages could be at two levels: at the process level or at the ecosystem/landscape level.

The whole variety of patterns of crop organization as adopted by the shifting agriculture farmer of north-east India in response to soil fertility gradient on the hill slope or in response to shortening of the shifting agricultural cycle illustrate synchrony of nutrient availability with nutrient use by crop species, both in space and time (Ramakrishnan, 1994). Further, sequential harvesting of crops in the multispecies cropping system is an elegant way in which the shifting agriculture farmer achieves synchrony between nutrient release and uptake at different phases of the cropping season. Depending upon differences in litter quality and microenvironmental conditions prevailing at the time, nutrient release gets regulated.

While working in north-east India, we soon realized that species which are ecologically valuable keystone species performing key functions in the ecosystem and thereby contributing to support/enhance biodiversity are also species that are socially valued by the local community, for cultural or religious reasons. Thus the Nepalese alder coming up in the jhum plots are protected from slashing and burning during jhum operations. Based on intuitive experience, the shifting agriculture farmer realizes that this species is beneficial to their cropping system; indeed our studies suggest that this species may conserve up to about 120 kg of nitrogen per hectare per year through nitrogen fixation (Ramakrishnan, 1992b). Similarly, the bamboo species, mentioned above, which conserve nitrogen, phosphorus or potassium in the system are also highly valued by all communities; they are often grown along the margin of agricultural plots and as part of the home garden system, or used for a variety of traditional activities, such as, house construction, making household utensils, tubing for water transportation, thatching material, etc. These and other ecologically important keystone species are also socially selected keystone species. A broader survey in the Indian context also suggests such a close parallelism between ecological and social processes. The implication is that the linking of ecological and social processes is significant for enhancing biodiversity in the ecosystem, building up of biodiversity based on ecological integrity of the system (because keystone species identified from a given ecosystem type would promote biodiversity that is compatible within that system), and indeed for ensuring rural peoples' participation as an integral part of the ecosystem function, rather than being a manipulator from outside (Ramakrishnan, 1993). This last point is extremely important for rehabilitation of degraded ecosystems and for rural development (Ramakrishnan et al., 1994a). Being able to identify themselves with a value system that they cherish through socially selected keystone species, the local communities would be able to participate in the rehabilitation programme.

At the ecosystem/landscape level, linkages between ecological and social dimensions could be highly complicated. While working on sustainable development for the tribals of north-eastern India, operating under an untenable shifting agricultural cycle of 5 years or less, we soon realized that building upon a traditional knowledge base and technology provides an opportunity for involving community participation. In doing so, a holistic approach was found to be appropriate in order to appreciate process level linkages within the ecosystem/landscape (Table 8.3).

While linking ecological with social processes for the purpose of sustainable development with peoples' participation, one may end up by identifying one or two critical driving factors that would trigger the developmental process. To cite one such example, in the Himalayan region, the present author while working at the G.B. Pant Institute of Himalayan Environment and Development together with the scientists of this Institute, identified water as the key factor for land use development in the entire Himalayan belt – from north-west to the north-east. Water is a scarce commodity outside the monsoon season even in areas that may receive up to 24 m of rainfall in an exceptional year, as at Cherrapunji. Let alone water use for agricultural development, people often find it difficult to procure drinking water. Therefore it was not surprising when local communities consistently identified water as the key resource in short supply. By harvesting surface runoff water of the rainy season and by diverting subsurface seepage water through cheap rainwater harvesting tanks (Kothyari et al., 1991), we could link it with a variety of ecosystem rehabilitation efforts which elicited enthusiastic community participation. Agroecosystem redevelopment in the Garhwal and the Kumaon hills, mixed plantation forestry and Ringal bamboo-related forestry activities

Table 8.3. Economic yield and crop community traits for agriculture under different cycles for the Garos at Burnihat in Meghalaya (from Ramakrishnan, 1992a).

	Follow period (years)		
	5	10	30
Economic yields (tons ha^{-1} yr^{-1})			
Seeds	0.107	1.153	2.180
Leaf fruit	0.129	0.074	0.024
Tubers			0.192
Total	0.556	1.842	3.296
ANP	14.060	11.576	15.213
Growth rate (g m^{-2} day^{-1})	3.8	3.2	4.2
NPP	18.461	14.709	17.746
Growth rate (g m^{-2} day^{-1})	5.1	4.0	4.9
Number of cultivars	8	12	14
LAI	0.59	1.49	3.20
HT	0.030	0.125	0.135
HI (grain + seed)	0.184	0.230	0.182
Labour (days ha^{-1} yr^{-1})	149	305	436

ANP, annual net production; NPP, net primary productivity; LAI, leaf area index; HI, crop yield/NPP because seeds, leaves and tubers are harvested; HI (grain + seed), above ground crop yield/ANP.

in the Kumaon hills, whole watershed development in the Sikkim region and redevelopment of shifting agriculture in the north-east are some of the activities which had a strong community participation. Because the critical factor was provided, namely water outside the monsoon season, rehabilitation costs were minimized and with distinct economic benefits accruing to the local communities rather quickly. Through such an approach, biodiversity could be built up by strengthening traditional agroecosystems and by accelerating forest succession. This again works on the principle that in rural environments, it is appropriate to consider humans as an integral component of ecosystem function, for effectively managing biodiversity.

Conclusions

Sustainable development and effective rehabilitation/management of ecosystems with adequate concern for biodiversity represent two sides of the same coin (Ramakrishnan, 1994a; Ramakrishnan et al., 1994). To be effective it is crucial to have a good appreciation of the role of biodiversity in ecosystem function. In this view, ecological issues are tied up with social, economic, anthropological and cultural dimensions, since the guiding principles of sustainable development (Table 8.4) cut across these very disciplinary realms, with obvious trade-offs.

This implies that we have to make a series of compromises to achieve

Table 8.4. The guiding principles of ecological sustainable resource management (adapted from Hare *et al.*, 1990).

- International equity: providing for today while retaining resources and options for tomorrow.
- Conservation of culture and diversity and ecological integrity.
- Constant natural capital and 'sustainable income'.
- Anticipatory and precautionary policy approach to resource use, erring on the side of caution.
- Resource use in manner that contributes to equity of the environment to supply renewable resources and assimilate wastes.
- Qualitative rather than quantitative development of human well-being.
- Pricing of environmental values and natural perspectives to cover full environmental and social costs.
- Global rather than regional or national perspectives of environmental issues.
- Efficiency of resource use by all societies.
- Strong community participation in policy and practice in the process of transition to an ecologically sustainable society.

sustainable development in such a way that we do not lose track of the ultimate objective, namely, rehabilitation and management of natural resources in a manner that, while conserving biodiversity, satisfies current needs, and at the same time allows for a variety of options for the future (Ramakrishnan *et al.*, 1994a). Although an ecosystem type (artificial ecosystems such as agriculture, a fish pond in a village or village itself visualized as an ecosystem; or natural ecosystems such as grazing land, forests or rivers) may be the appropriate unit for convenient handling, a cluster of interacting ecosystem types (a 'landscape') may be the most effective for a holistic treatment. A watershed is one such landscape unit. Further, from a sustainable development point of view, whereas one may bear in mind a long-term ideal objective to be achieved, ecological, social economic or cultural constraints may necessitate designing short-term strategies, for enabling peoples' participation in the developmental process (Ramakrishnan, 1994b). Sustainable development of the rural landscape is the obvious pathway for integrating humans into ecosystem function, with biodiversity considerations.

References

Altieri, M.A. and Liebman, M. (eds) (1988) *Weed Management in Agroecosystems: Ecological Approaches.* CRC Press, Boca Raton, Florida.

Brookfield, H. and Padoch, C. (1994) Appreciating biodiversity: a look at the dynamism and diversity of indigenous farming practices. *Environment* 36, 6–11; 37–45.

Chacon, J.C. and Gliessman, S.R. (1982) Use of the 'non-weed' concept in traditional tropical agroecosystems of south eastern Mexico. *Agro-Ecosystems* 8, 1-10.

Gangwar, A.K. and Ramakrishnan, P.S. (1989) Cultivation and use of lesser-known plants of food value by tribals of north east India. *Agricultural Ecosystems and Environment* 25, 253-267.

Gangwar, A.K. and Ramakrishnan, P.S. (1990) Ethnobiological notes on some tribes of north-eastern India. *Economic Botany* 44, 94-105.

Gliessman, S.R. (1988) Ecology and management of weeds in traditional agroecosystems. In: Altieri, M.A. and Liebman, M. (eds) *Weed Management in Agroecosystems: Ecological Approaches*. Boca Raton, Florida, pp. 237-244.

Gliessman, S.R. (ed.) (1989) *Agroecology: Research in the Ecological Basis for Sustainable Agriculture*. Ecological Studies 78, Springer-Verlag, New York.

Hare, W.L., Morlowe, J.P., Rae, M.L., Gray, F., Humphries, R. and Ledgar, R. (1990) *Ecologically Sustainable Development*. Australian Conservation Foundation, Fitzroy, Victoria.

Janzen, D.H. (1973) Tropical agro-ecosystems. *Science* 182, 1212-1219.

Kapoor, P. and Ramakrishnan, P.S. (1975) Studies on crop legume behaviour in pure and mixed stands. *Agro-ecosystems* 2, 61-74.

Khiewtam, R. and Ramakrishnan, P.S. (1993) Litter and fine root dynamics of relict sacred grove forest of Cherrapunji in north-eastern India. *Forest Ecology and Management* 60, 327-344.

Kothyari, B.P., Rao, K.S., Saxena, K.G., Kumar, T. and Ramakrishnan, P.S. (1991) Institutional approaches in development and transfer of water harvest technology in the Himalaya. In: Tskiris, G. (ed.) *Proceedings of the European Conference On Advances in Water Resources Technology*, Athens, 20-22 March 1991. A.A. Balkema, Rotterdam, The Netherlands, pp. 673-78.

Litsinger, J.A. and Moody, K. (1976) Integrated pest management in multiple cropping systems. In: Stelly, M. (ed.) *Multiple Cropping*. American Society of Agronomy Madison, Wisconsin, pp. 293-316.

Lubchenco, J., Olson, A.M., Brubaker, L.B., Carpenter, S.R., Holland, M.M., Hubbal, S.P., Levin, S.A., MacMahon, J.A., Matson, P.A., Melillo, J.M., Mooney, H.A., Peterson, C.H., Pulliam, H.R., Real, L.A., Regal, P.J. and Risser, P.G. (1991) The sustainable biosphere initiative: an ecological research agenda. *Ecology* 72, 371-412.

Mooney, H.A. et al. (1994) *Biodiversity and Ecosystem Function*. SCOPE, John Wiley, Chichester (in press).

Odum, E.P. (1971) *Fundamentals of Ecology*. W.B. Saunders, Philadelphia.

Ramakrishnan, P.S. (1988) Successional theory: implications for weed management in shifting agriculture, mixed cropping and agroforestry systems. In: Altieri, M.A. and Liebman, M. (eds) *Weed Management in Agroecosystems: Ecological Approaches*. CRC Press, Boca Raton, Florida, pp. 183-196.

Ramakrishnan, P.S. (1992a) Tropical forests. Exploitation, conservation and management. *Impact of Science on Society*, 42, no. 166, 149-162.

Ramakrishnan, P.S. (1992b) *Shifting Agriculture and Sustainable Development of North-Eastern India*. UNESCOMAB Series, Paris, Parthenon Publ., Carnforth, Lancs, UK. 424 pp. (Republished by Oxford University Press, New Delhi, 1993.)

Ramakrishnan, P.S. (1993) Evaluating sustainable development with peoples' participation. In: Moser, F. (ed.) *Sustainability – Where Do We Stand?* Proceedings of the International Symposium. Technische Universitat, Graz, Austria, pp. 165-182.

Ramakrishnan, P.S. (1994) The jhum agroecosystem in north eastern India: a case study of the biological management of soils in a shifting agricultural system. In: Woomer, P.L. and Swift, M.J. (eds) *The Biological Management of Tropical Soil Fertility*. TSBF, Nairobi and Wiley-Sayce, Chichester, pp. 189-207.

Ramakrishnan, P.S. and Gupta, U. (1972) Ecotypic differences in *Cynodon dactylon* (L.) Pers. related to weed-crop interference. *Journal of Applied Ecology* 9, 333-339.

Ramakrishnan, P.S., Cambell, J., Demierre, L., Gyi, A., Malhotra, K.C., Mehndiratta, S., Rai, S.N. and Sashidharan, E.M. (1994a) *Ecosystem Rehabilitation of the Rural Landscape in South and Central Asia: An Analysis of Issues*. Special Publication, Hadley, M. (ed.) UNESCO (ROSTCA), New Delhi.

Ramakrishnan, P.S., Purohit, A.N., Saxena, K.G. and Rao, K.S. (1994b) *Himalayan Environment and Sustainable Development*. Diamond Jubilee Publications, Indian National Science Academy, New Delhi.

Swift, M.J., Vandermeer, J., Ramakrishnan, P.S., Anderson, J.M., Ong, C.K. and Hawkins, B. (1994) Biodiversity and agroecosystem function. In: Mooney, H.A. *et al.* (eds) *Biodiversity and Ecosystem Properties: A Global Perspective*. SCOPE Series. John Wiley, Chichester, UK (in press).

Toky, O.P. and Ramakrishnan, P.S. (1981) Cropping and yields in agricultural systems of the north-eastern hill region of India. *Agro-Ecosystems* 7, 11-25.

9 Microorganisms: The Neglected Rivets in Ecosystem Maintenance

D.L. Hawksworth
International Mycological Institute, Bakeham Lane, Egham, Surrey TW20 9TY, UK

Summary

Microbes have existed for over 3.5 billion years; they shaped the Earth's atmosphere, first colonized land, combined to make our cells, recycle elements, and mediate key life processes. Current knowledge gaps are immense; less than 5% of the fungi, bacteria and viruses on Earth have been named, and little is known of the range, ecological requirements, and roles of those that have. Although some ecological processes depend on a few microbial species, the extent to which there is real duplication of function is obscure; in any case, replication in any function increases resilience to perturbations. The roles of microorganisms in different ecosystems are diverse, but the communities most at risk from reductions in microbial diversity are those in extreme environments, the boreal and temperate forests, and the oceans. The current level of ignorance of most microorganism groups has major implications for biodiversity conservation and sustainable use: microbially mediated ecological processes should be monitored, the protection of undisturbed habitats is essential to microbial conservation, and consolidated international action is needed to provide the microbial data pertinent to ecosystem maintenance. We need to bridge the microbial knowledge gap.

Introduction

The term 'microorganism' is most commonly used as a synonym of 'bacterium', but is nonsensical in any systematic sense. As used in this chapter, a microorganism can be defined as an organism belonging to a phylum either many members of which cannot be seen by the unaided eye, or where microscopic examination, and in many cases growth in pure culture, is essential for their

Table 9.1. Examples of extreme environments inhabitated by microorganisms.

- High temperatures: hot springs and fumaroles, supporting thermophilic cyanobacterial biofilms growing at about 74°C in hot springs, and archaebacterial hypothermophilic chemolithoautotrophs and heterotrophs at up to 112°C (*Pyrodictium* species have an optimal growth temperature of 105°C.)
- High salinity: salt lakes support anaerobic phototrophs, heterotrophs, sulphate reducers, acetogens and methanogens; also aerobic photosynthetic bacteria such as *Acidiphilium*, *Erythrobacter* and *Roseobacter*; *Halomonas* can grow at up to 33% salt.
- High pressure: in the deep sea, including bacteria associated with deep-sea animals and around deep-ocean hydrothermal vents, withstanding pressures and thriving to over 500 atmospheres.
- High pH: alkaline environments, including soda lakes, with alkalophilie bacteria occurring above pH 8, with a *Plectonema* reported to still grow at pH 13.
- Low pH: strongly acid environments, including acid crater-lakes and acid mine waters; *Thermoplasma* and *Thiobacillus* species can grow down to pH 2.
- Low water activity: particularly caused by high sugar concentrations; the fungus *Xeromyces bisporus* can grow at less than 0.65 a_w.
- Low temperatures: arctic, antarctic and alpine habitats, with microbial activity continuing down to −12°C.

For further information see Edwards (1990), Guerro and Pedrós-Alió (1993) and Postgate (1994).

identification. In this sense, the term encompasses: algae, bacteria (including archaebacteria, cyanobacteria, mollicutes and mycoplasmas), fungi (including yeasts and lichen-forming species), protozoa (including slime moulds), and viruses (including phages and viroids).

Some groups of microorganisms, notably cyanobacteria, were already diverse 3.5 billion years ago (Schopf, 1993). During this unparalleled period of evolution, they have been able to occupy a wider range of habitats on Earth than other organisms (Price, 1988), and to develop groups with adaptations to extreme environments where no other life can occur (Table 9.1). They shaped the Earth's atmosphere, first colonized land, combined to make our cells, recycle elements, and mediate key life processes. Measured in terms of genetic differences within particular molecular sequences, they represent the greatest diversity of Life on Earth (Embley et al., 1994). It follows that microbial groups cannot be ignored, but merit being closer to centre-stage in discussions of biodiversity maintenance.

The significance of microorganism biodiversity has received increased recognition over the last five years, and several overviews have been provided (Hawksworth, 1991, 1994; Hawksworth and Colwell; Allsopp et al., 1995). In this chapter, I will draw attention to particular aspects, the pertinence of which is not yet as widely recognized as they merit: the knowledge gap in our understanding of microorganisms, their roles in ecosystem processes, the application of the redundancy concept, and the ecosystems most at risk through changes in microbial diversity.

The Knowledge Gap

In considering what is known of microorganisms, particularly in relation to ecological processes and the maintenance of ecosystems, the knowledge gaps are immense.

- Less than 5% of the Earth's species are known and able to be identified.
- There are no internationally accepted methods of measuring the size, diversity and activity of functional groups *in situ*, and identical protocols have rarely been employed.
- The precise ecological and biogeochemical processes in which most microorganisms are involved are unclear.
- Microbial ecologists generally study microbial systems in isolation, micro- and macrobiologists rarely participating in integrated programmes.
- The range and habitat requirements of most microorganisms is poorly known.

It is a common misconception that microorganisms do not have distributions, but rather that they are everywhere and that the environmental conditions determine which grow (Gams, 1992). Although this may be true for certain soil bacteria and 'weedy' opportunistic saprobic fungi, the hypothesis cannot be upheld as a general rule. The majority of bacteria, fungi, and viruses are either linked to particular macroorganisms as mutualists, commensals or pathogens, or confined to precise ecological niches.

The distributions of certain ancient microorganism groups, especially soil bacteria and cyanobacteria, are not surprisingly much wider than more recently evolved taxa. Many lichenized fungi, polypores and slime moulds also tend to have very wide ranges compared to vascular plants for parallel reasons. However, it is the more cosmopolitan species that are the exception rather than the rule.

Microorganisms in Ecosystem Processes

The ecosystem processes in which different groups of microorganisms are involved include the production of trace gasses (e.g. methane, dimethlysulphide), population regulation (i.e. biocontrol by pests and pathogens), mutualisms essential for macroorganism function (e.g. mycorrhizas, nitrogen-fixers, gut microbiotas), soil stability (cyanobacterial and lichen crusts) and structure (e.g. fungal mycelia), decomposition of plant and animal remains and products, carbon cycling (i.e. all photosynthetic microbial and lichenized groups in the sea and on land), rock weathering (e.g. bacteria, lichenized fungi), and as components of food webs and chains (e.g. in termite colonies).

Extreme environments

In extreme environments (Table 9.1), particularly those in which only microorganisms can exist, all the ecological processes which enable the communities to persist must be performed by microbes. Particularly spectacular are cryptoendolithic communities in Antarctica, comprising mixtures of fungi and cyanobacteria able to photosynthesize at −10°C with an optimum at 0°C, and remaining frozen for 9–10 months of the year (Friedman et al., 1988). In regions with permafrost, bacteria can be recovered in large numbers from permanently frozen situations. In only slightly less extreme situations, cyanobacteria and fungi lichenized with algae are the dominant organisms on rock surfaces and the ground. In these situations they are the key producers, and also provide food or shelter for a variety of other micro- and macroorganisms.

In arid lands, cyanobacteria, green algae, and crustose and squamulose lichens reduce soil erosion, trap moisture, and add nutrients (especially nitrogen) so contributing to the survival of both vascular plants and animals. Maximal photosynthesis in biological soil crusts in the Negev desert can occur at a water content equivalent to precipitation of 0.2–0.3 mm, imbibing from dew and fog providing enough for activity and growth (Lange et al., 1992). They fix nitrogen, and improve soil stability, water infiltration, and seed germination (Warren, 1995).

Terrestrial ecosystems

In terrestrial ecosystems, other than those in extreme enviroments, and although less conspicuous to the observer, microorganisms also play crucial roles in ecological processes.

In boreal forest ecosystems, in addition to the trees being ectomycorrhizal, lichenized fungi form a major component of the ground vegetation (often with a limited number of species dominating); the lichens on trees and on the ground are important as food and(or) shelter for vertebrates and invertebrates. Further, the gut microbes in large ruminant mammals which browse in these forest systems, for example moose and reindeer, are crucial to their existence. Disruption of these various mutualistic systems will have consequences for the whole community.

Ectomycorrhizas are a particularly key component of the functioning of boreal and temperate forests, providing a mechanism for the trees to obtain nutrients that would otherwise limit their growth, especially phosphorus (Read, 1991; Read et al., 1992). Their significance is evident from the observation that declines in the fruiting of ectomycorrhizal fungi correlate with forest deterioration in Europe (Fellner, 1993).

Knowledge of the microbiota in tropical forests is especially poor, to the extent that it is difficult to assess the functional uniqueness of particular forest microbes. However, groups of microorganisms play crucial roles in diverse ways in the maintenance of tropical forest systems, from serving as nitrogen fixers and as mutualists of plants and invertebrates, to playing key roles in decomposition – including the entrapment of litter above ground (Lodge et al., 1995). Certain species of endomycorrhizal and litter-trapping fungi may have keystone roles in tropical forests. Further, the loss of macrofungi

Marine and freshwater ecosystems

Halophilic and marine microorganisms, including fungi, have been the object of considerable research into their identity, ecology, and physiology (Moss, 1986; Rodriguez-Valera, 1991), but the relationship between their diversity and ecosystem function has rarely been considered.

The significance of, and diversity among, minute microorganisms in the open oceans has only recently started to be appreciated; 10^3–10^8 living particles of each of marine viruses, cyanobacteria, prochlorophytes and bacterioplankton can occur in every millilitre of water in the upper ocean (Koike et al., 1993). The green sulphur bacterium *Chlorobium* can photosynthesize at depths of 80 m in the Red Sea where the light intensity is only 0.01 μEinst m^{-2}s^{-1} (Ormerod et al., 1993), contributing to primary productivity at depths where this process is otherwise impossible.

The diversity of the microorganisms associated with corals, especially mutualists, is a vital component in the maintenance of the health of coral reef ecosystems. Short-term variations in those associated with mucus may provide early warnings of stress to the system caused by disease, pollution, or other factors (Santavy, 1995).

Nutrient imbalance in lakes and rivers, such as that caused by human interference, has profound effects on the abundance of different microorganisms, including those characteristic of eutrophication and toxin-producing dinoflagellates and cyanobacteria. Entire microbial recycling pathways may be lost (Schindler, 1990), the cycling rates of limiting nutrients changing substantially as community structure changes (Carpenter and Kitchell, 1993). Certain chytrids may limit the explosion of certain diatoms and other algae (Dix and Webster, 1995) and so play important roles in ecosystem maintenance.

Redundancy in Microbial Biodiversity

At our current level of ignorance as to the organisms present in ecosystems and their physiologies, it is hazardous to generalize as to either the extent to which particular species have identical roles, or to speculate that losses would not impact on ecological processes. Some ecologically significant processes involve limited numbers of species, such as the production of dimethylsulphide from the oceans and nitrogen fixation. Other processes *appear* to be much less dependent on small numbers of species, for example denitrification, methanogenesis, and decomposition. The word 'appear' must be emphasized, due to our current ignorance. For example, although we know that a large number of basidiomycete fungi are associated with the decomposition of wood, closer scrutiny reveals successional patterns suggesting differences in

Table 9.2. Ecosystems most at risk from a loss of microorganism biodiversity.

- Highest risk: extreme environments, open oceans, arid lands, and arctic-alpine ecosystems.
- High risk: boreal and temperate forests, coral reefs, and agroecosystems.
- Medium risk: savannas, temperate grasslands, and Mediterranean ecosystems.
- Low risk: tropical forests, coastal systems and estuaries.

the chemical breakdowns they mediate (Dix and Webster, 1995). Similarly, whereas some boreal and temperate forest trees can form ectomycorrhizas with 100 or more basidiomycete fungi, as these associations are investigated in more detail successional patterns related to factors such as root age start to be recognized (Read et al., 1992).

In cases where there is true replication in an ecosystem process sense, that replication in itself may be significant in ecosystem maintenance by increasing the resilience of a community to species loss and environmental or human perturbations (Perry et al., 1989).

In the case of the numerous microorganisms dependent on particular macroorganisms for their existence, as obligate host-specific mutualists, commensals, or pathogens (parasites), the loss of any macroorganism inevitably involves the concurrent extinction of suites of microorganism species. Such losses of host-specific microorganisms perhaps merit more emphasis in discussions of extinction rates, but are unlikely to have as significant effects on ecosystem function as the disappearance of the macroorganisms on which they depend.

Whether there is truly significant duplication in function of microorganisms in particular ecological processes or not, we can be confident that there is little redundancy in microbially mediated processes. The loss of functional groups of microorganisms performing fundamental ecological roles will inevitably lead to ecosystem modification or even collapse.

Ecosystems at Risk from Microbial Biodiversity Loss

The extent to which different ecosystems are at risk from the loss of microbial diversity varies according to the overall species richness of the systems. Those which have relatively few microorganisms in key functional areas, or which consist only or predominantly of microorganisms, will be the most at risk (Table 9.2).

The high risk categorization for agroecosystems is not always appreciated, yet sustainable systems can depend on the exploitation of desired microbial processes. Continuous harvesting of the same crop, the introduction of nitrogen-fixing bacteria (e.g. legumes with *Rhizobium* nodules), and plant, soil, and pest management practices, will all lower the diversity of plants and invertebrates, and so the overall diversity of microbes. Tillage and the mechanical disruption of soil breaks mycelial systems and so the nutrient-capturing effectiveness of mycorrhizal fungi. And the application of fertilizers, pesticides and other agrochemicals, can alter the diversity and abundance of the microbiota, including species with roles in the control of soil-borne diseases and the maintenance of soil structure.

Awareness of changes in functional microbial groups in an ecosystem has the potential to provide an early warning that an ecosystem process is starting to be modified or be lost before damage is reflected in the macro-vegetation, by which time it may be impractical to reverse or contain. This can be approached by monitoring microbially mediated processes in soil (Anderson and Ingram, 1993), the use of lichens as bioindicators of air pollution (Richardson, 1992) and forest disturbance (Rose, 1992), ectomycorrhizal fungal fruit bodies as an indicator of acid rain (Fellner, 1993), diatoms in freshwater quality (Round, 1991), and bacteria in coral systems (Santavy, 1995).

Implications for Biodiversity Conservation

The microbial groups drive the systems that support all life on Earth. They are interwoven into the food chains and food-webs on which all macroorganisms depend, and play unique roles in the circulation of matter. They deserve to be more than ignored or cursorily noted in studies of ecosystem function and maintenance:

Three major policy implications for biodiversity conservation arise from this overview.

1. Our level of ignorance is such that, with respect to microorganisms, we rarely know which, if any, are the crucial rivets keeping an ecosystem intact. The comparison of the effects of biodiversity loss to failing rivets binding an aeroplane or ship's structure together (Ehrlich and Ehrlich, 1981) is not totally inappropriate from a microbiological perspective. It follows that the monitoring of microbially mediated ecological processes should be a component of the prudent management of biodiversity, especially within protected areas.

2. The microbial component of an ecosystem is so complex and inadequately understood that the 'restoration' or 'rehabilitation' of an ecosystem is unlikely to ensure the protection of microbial diversity to its original extent. The emphasis for conservation strategies from the microbiological perspective, therefore should be on undisturbed sites.

3. Consolidated international action is needed to bridge the microbial knowledge gap in aspects crucial to ecosystem maintenance and sustainable use. A Committee on Microbial Diversity was established by IUBS and IUMS in 1994 to prioritize and seek support for such actions, and also to provide inputs to *DIVERSITAS* and other initiatives. If the aspirations of this Committee can start to be realized, even in a modest way, the most critical rivets will be identified, and microbiologists will be able increasingly to provide the microbial data needed to understand ecosystem maintenance, and so to contribute more fully to the conservation and sustainable use of the Earth's resources.

Acknowledgements

This contribution has benefited from discussions on the issues raised with various colleagues contributing to the IUBS/SCOPE/UNESCO Ecosystem Function of Biodiversity theme of DIVERSITAS, particularly Drs S.S. Dhillion, D.U. Hooper and D.J. Lodge.

References

Allsopp, D., Colwell, R.R. and Hawksworth, D.L. (eds) (1995) *Microbial Diversity and Ecosystem Function*. CAB International, Wallingford.

Anderson, J.M. and Ingram, J.S.I. (1993) *Tropical Soil Biology and Fertility. A Handbook of Methods*, 2nd edn. CAB International, Wallingford.

Carpenter, S.R. and Kitchell, J.F. (eds) (1993) *The Trophic Cascade in Lakes*. Cambridge University Press, Cambridge.

Dix, N.J. and Webster, J. (1995) *Fungal Ecology*. Chapman and Hall, London.

Edwards, C. (ed.) (1990) *Microbioloy of Extreme Environments*. Open University Press, Milton Keynes.

Ehrlich, P.R. and Ehrlich, A.H. (1981) *Extinction: The Causes and Consequences of the Disappearance of Species*. Random House, New York.

Embley, T.M., Hirt, R.P. and Williams, D.M. (1994) Biodiversity at the molecular level: the domains, kingdoms and phyla of life. *Philosophical Transactions of the Royal Society of London, Biological Sciences* 345, 21–33.

Fellner, R. (1993) Air pollution and mycorrhizal fungi in central Europe. In: Pegler, D.N., Boddy, L., Ing, B. and Kirk, P.M. (eds) *Fungi of Europe: Investigation, Recording and Mapping*. Royal Botanic Gardens, Kew, pp. 239–250.

Friedman, E.I., Hua, M. and Ooamp-Friedman, R. (1988) Cryptoendolithic lichen and cyanobacterial communities of the Ross Desert, Antarctica. *Polarforschung* 58, 251–259.

Gams, W. (1992) The analysis of communities of saprophytic fungi with special reference to soil. In: Winterhoff, W. (ed.) *Fungi in Vegetation Sciences*. Kluwer Academic, Dordrecht, pp. 183–223.

Guerreo, R. and Pedrós-Alió, C. (eds) (1993) *Trends in Microbial Ecology*. Spanish Society for Microbiology, Barcelona.

Hawksworth, D.L. (ed.) (1991) *The Biodiversity of Microorganisms and Invertebrates: Its Role in Sustainable Agriculture*. CAB International, Wallingford.

Hawksworth, D.L. (1994) Biodiversity in microorganisms and its role in ecosystem function. In: Solbrig, O.T., van Emden, H.M. and van Oordt, P.G.W.J. (eds) *Biodiversity and Global Change*. Revised edn. CAB International, Wallingford, pp. 83–89.

Hawksworth, D.L. and Colwell, R.R. (eds) (1992) Microbial diversity amongst microorganisms and its relevance. *Biodiversity and Conservation* 1, 219–345.

Koike, I., Hara, S., Terauchi, K., Shibata, A. and Kogure, K. (1993) Marine viruses – their role in upper ocean dissolved organic matter (DOM) dynamics. In: Guerreo, R. and Pedrós-Alió, C. (eds) *Trends in Microbial Ecology*. Spanish Society for Microbiology, Barcelona, pp. 311–318.

Lange, O.L., Kidron, G.J., Büdel, B., Meyer, A., Kilian, E. and Abeliovich, A. (1992) Taxonomic composition and photosynthetic characteristics of the 'biological soil crusts' covering sand dunes in the western Negev Desert. *Functional Ecology* 6, 519–527.

Lodge, D.J., Hawksworth, D.L. and Ritchie, B.J. (1995) Microbial diversity and tropical forest functioning. In: Orians, G.H., Dirzo, R. and Cushman, J.H. (eds) *Biodiversity and Ecosystem Processes in Tropical Forests*. Springer-Verlag, New York, in press.

Moss, S.T. (ed.) (1986) *The Biology of Marine Fungi*. Cambridge University Press, Cambridge.

Ormerod, J.G., Aukrust, T.W. and Johnsen, I.J. (1993) Frugal *Chlorobium*: the ultimate phototroph. In: Guerreo, R. and Pedrós-Alió, C. (eds) *Trends in Microbial Ecology*. Spanish Society for Microbiology, Barcelona, pp. 59–62.

Perry, D.A., Amaranthus, M.P., Borchers, J.G., Borchers, S.L. and Brainerd, R.E. (1989) Bootstrapping in ecosystems. *BioScience* 39, 230–237.

Postgate, J. (1994) *The Outer Reaches of Life*. Cambridge University Press, Cambridge.

Price, D. (1988) An overview of organismal interactions in ecosystems in evolutionary and ecological time. *Agriculture, Ecosystems and Environment* 24, 369–377.

Read, D.J. (1991) Mycorrhizas in ecosystems – nature's response to the 'Law of the Minimum'. In: Hawksworth, D.L. (ed.) *Frontiers in Mycology*. CAB International, Wallingford, pp. 101–130.

Read, D.J., Lewis, D.H., Fitter, A.H. and Alexander, I.J. (eds) (1992) *Mycorrhizas in Ecosystems*. CAB International, Wallingford.

Richardson, D.H.S. (1992) *Pollution Monitoring with Lichens*. Richmond Publishing, Slough.

Rodriguez-Valera, F. (ed.) (1991) *General and Applied Aspects of Halophilic Microorganisms*. Plenum Press, New York.

Rose, F. (1992) Temperate forest management: its effects on bryophyte and lichen floras and habitats. In: Bates, J.W. and Farmer, A.M. (eds) *Bryophytes and Lichens in a Changing Environment*. Clarendon Press, Oxford, pp. 211–233.

Round, F.E. (1991) Diatoms in river water-monitoring studies. *Journal of Applied Phycology* 3, 129–145.

Santavy, D. (1995) The diversity of microorganisms associated with marine invertebrates and their roles in the maintenance of ecosystems. In: Allsopp, D., Colwell, R.R. and Hawksworth, D.L. (eds) *Microbial Diversity and Ecosystem Function*. CAB International, Wallingford, pp. 211–229.

Schindler, D. (1990) Experimental perturbations of whole lakes as tests of hypotheses concerning ecosystem structure and function. *Oikos* 57, 25–41.

Schopf, J.W. (1993) Microfossils of the Early Archean apex chert: new evidence of the antiquity of life. *Science* 260, 640–646.

Warren, S.D. (1995) Ecological role of microphytic soil crusts in and ecosystems. In: Allsopp, D., Colwell, R.R. and Hawksworth, D.L. (eds) *Microbial Diversity and Ecosystem Function*. CAB International, Wallingford, pp. 199–209.

Biodiversity of Marine Sediments [10]

C. Heip
Centre for Estuarine and Coastal Ecology, Netherlands Institute of Ecology, Vierstraat 28, 4401 EA Yerseke, The Netherlands

Introduction

Biodiversity of the marine environment in general and of marine sediments in particular is poorly known, both in descriptive terms of species richness and its distribution along latitudinal and depth gradients, and in terms of the ecological and evolutionary processes that regulate it. In a recent, otherwise excellent account on species diversity in ecological communities (Ricklefs and Schluter, 1993), the marine benthos is not even mentioned and only two chapters deal with the marine environment at all (Underwood and Petraitis, 1993; McGowan and Walker, 1993). This ignorance or lack of interest is not justified as the marine benthos occupies 70% of the Earth's surface and is an important agent in major global biogeochemical cycles.

Overall, at our present state of knowledge, marine biodiversity appears to be low: about 200,000 marine animal species, perhaps 20,000 marine plant species and an even much lower number of marine viruses and microorganisms (bacteria, fungi, protozoans, microalgae) have been scientifically described. Of the 200,000 described animal species, only a few thousand are planktonic, about 130,000 are from hard substrates and about 60,000 from sediments. The marine plant inventory is probably more complete than that of the animals since there are no deep-sea plants. Even if the true number of marine species was an order of magnitude lower than that of the land, on a higher taxon level the marine environment is certainly much richer than the land. Of the 33 animal phyla, only five do not occur in the seas and 13 are endemic to it (Grassle *et al.*, 1991). This implies that genetic, biochemical, and physiological diversity is also much higher in the oceans than on land (Lasserre, 1992). This chapter will deal exclusively with the fauna of marine sediments. Marine sediments cover most of the Earth's surface and although they resemble terrestrial soils in some respects, they also have many unique characteristics. In the ocean they are covered by about 4 km of water, most of which is cold, about 2°C. The labile organic material that is the food of marine benthos is produced in the surface water layers far away and only a very small percentage of the primary produced material reaches the ocean floor. Animal densities are therefore very low in large parts of the oceans, and deep-sea animals are often very small compared to related species from shallower water. However, this low abundance, low biomass environment has a surprisingly high diversity (Sanders, 1968).

Scales and Patterns

Following the developments in terrestrial ecology, marine ecologists described diversity patterns on different scales. One important restriction is the scale of observation. Marine sediment samples are discrete units, with a square centimetre or at best square decimetre order of magnitude, designed for the study of transects or areas that are very often in the tens to hundreds of kilometres or hundreds to thousands of square kilometre range. A sampling strategy based on a device covering less than one tenth of a square metre once every 10 km must appear absurd to a terrestrial ecologist, but this is exactly what happens in many marine studies. Methodological problems in deep-sea research are numerous and make it difficult to obtain good quantitative information from all but the shallowest sediments. Samples from the seas and *a fortiori* from the deep-sea floor are very expensive to get (one box-corer from 4000 m depth, covering an area of $0.25 \, m^2$, takes about 4 hours of shipping time at a cost of about 3000 ecu).

It is, therefore, logical that studies and explanations of patterns in marine benthic biodiversity have focused on the small (local) and on the large (global) scales and that the field of landscape (or seascape) marine benthic ecology is not really flourishing. Scale interactions may be more important than in terrestrial systems. Movement of adult individuals may be very restricted in many sessile or burrowing species but individuals may cover enormous distances as planktonic larvae, and, in contrast to the land, many marine benthic species cover very large areas such as entire ocean basins. Restrictions to dispersal, such as the continents and the mid-ocean ridges, or sills such as between the Mediterrananean and the Atlantic, do, however, exist and may exist for long periods of time. Such barriers to dispersal are now very frequently overcome with human help, often with catastrophic results.

Local Biodiversity

Local biodiversity may be high in marine sediments and increases from shallow to deep water. Intertidal areas typically have a very low species number. On the continental shelves and slopes, local species numbers are of the order of tens of meiofauna species per $10 \, cm^2$ and macrofauna species per $0.1 \, m^2$. Nematodes and polychaetes in particular may display high local

Table 10.1. Total number of macrofaunal species on the specified total surface sampled in different deep-sea areas.

Number of species	Surface sampled	Area	Author
278–351	1.62 m^2	W. Atlantic, 2100 m	Grassle and Maciolek (1992)
324–363	1.62 m^2	W. Atlantic, 1500 m	Grassle and Maciolek (1992)
315	1.25 m^2	San Diego Trough, 1230 m	Jumars (1976)
146	0.25 m^2	Rockall Trough, 1800–2900 m	Gage (1979)
130	0.32 m^2	NW Atlantic, 2800 m	Rowe et al. (1982)

diversity. As an example, the nematode fauna from the continental slope on the west coast of Corsica had a species richness between of 101–148 per sample of 10 cm^2 at depth of 160–1000 m (Soetaert et al., 1991), with many cogeneric species. For macrofauna, Grassle and Maciolek (1992) calculated that the number of species in deep waters typically numbers many hundreds in total sampled areas of 1–2 m^2 (Table 10.1).

Local diversity in terrestrial communities is often explained in terms of species interactions, i.e. competition, predation. Related species living together may coexist by partitioning resources (Lack, 1944) and niches can be described as part of a multidimensional space that is defined by resource axes which represent the biotic and abiotic factors along which resources are partitioned (Hutchinson, 1959). This partitioning may be part of a process leading to increased species diversity (McArthur, 1958) because of specialization but there is a limit to this and thus to the number of species in a community. This is the principle of limiting similarity (McArthur and Levins, 1967) and a large body of theory was developed in the 1970s that essentially was based on the notion that equilibrium properties of communities, including the number of coexisting species, were determined largely by species interactions (Schluter and Ricklefs, 1993).

Most of the evidence for the importance of species interactions in creating and maintaining local diversity comes from terrestrial studies. The marine studies, mostly from intertidal environments (Connell, 1983; Reise, 1985: Underwood and Petraitis, 1993), are all small-scale. The evidence for competitive interactions in deeper water benthos is circumstantial. One interesting example is the spatial segregation of nematode species in the 5 cm top sediment layer of a station on the continental slope off Corsica. Six species of the genus *Sabatieria* co-occur that all have the same mouth structure and therefore probably the same food. Also six species of the genus *Acantholaimus* cooccur, but they have different mouth structures and presumably different food. The *Sabatieria* species potentially competing for the same food source live at different depths in the sediment whereas the *Acantholaimus* species, that already evolved towards different morphologies, do not (Fig. 10.1) (Soetaert et al., 1995). The surprising consequence of such small-scale segregation is that nematode communities from the same depth layer in different

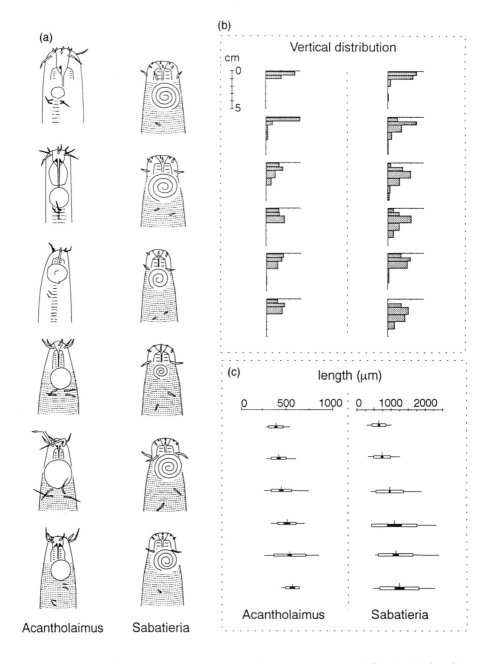

Fig. 10.1. Local diversity of nematodes on a continental slope station off Corsica. (a) Anterior end with buccal structure, (b) vertical distribution into the sediment (percentage of total number in two 0.5 cm surface layers and four 1 cm subsurface layers) and (c) length distribution (average individual length with standard error (black), standard deviation (white) and range (bar)) of six co-occurring species of both the genera *Acantholaimus* and *Sabatieria*.

stations, tens of kilometres apart in water depths differing hundreds of metres, are more similar than communities living at the same station in two adjacent depth-layers one centimetre from each other.

Besides species interactions, disturbance is often invoked as a diversity-regulating mechanism in terrestrial environments. Again the importance of regular or irregular disturbances to explain patterns of diversity will be different on local and on large scales. Increased habitat heterogeneity is created by the activity of the biota themselves. Patches consist of small-scale biogenic structures, such as burrows, tubes, feeding pits, trails etc. On the mesoscale, habitats are often not discrete but continuously changing due to some overriding physical factor such as wave energy, current velocity, fronts etc. Grassle and Maciolek (1992) suggest that high overall diversity in the deep sea is maintained by the input of small patches of ephemeral resources and the disturbance that results from the activities of individual animals. In this respect bottom mounds, burrows, feeding pits, other animals such as sponges or xenophyophoreans have been mentioned. Food input is patchy and rarer as depth increases. Food inputs may be pulsed and tend to accumulate in depressions and burrows made by animals on the deep-sea floor. Rare species tend to be associated with these rare resources.

Regional and Global Biodiversity Patterns

Trends in diversity on a truly global scale have been documented mainly from the terrestrial environment, but why these trends exist is not really understood (Clarke, 1992). From the marine environment, Thorson (1957) showed a pronounced increase in the species richness of epifauna from hard substrates towards the tropics, but the number of macrofaunal species in sediments appeared to be roughly the same for arctic, temperate and tropical areas. On the other hand, Stehli et al. (1967) clearly demonstrated a diversity trend of bivalve molluscs at species, genus and family level from the tropics to the poles and in a later study (Stehli et al., 1975, cited in Clarke, 1992) demonstrated the same for foraminiferans. Since both groups form calcareous skeletons, Clarke (1992) suggested that perhaps such trends may not exist in other taxa. However, recently, global-scale latitudinal trends in deep-sea epifauna (Gastropoda, Bivalvia and the non-calcareous Isopoda) have been described by Rex et al. (1993) from the Atlantic Ocean. All three taxa showed highly significant latitudinal gradients in the North Atlantic with elevated tropical diversity and depressed diversity in the Norwegian Sea.

On a somewhat smaller, regional scale, trends in benthic biodiversity have been described from the whole North Sea (Heip et al., 1992). A clear trend in biodiversity with latitude was found, but this trend was opposite for macrofauna, where diversity increases toward the north, and copepods, where diversity increases toward the south (Fig. 10.2). The higher diversity of copepods in the south (38 species per sample) is easily explained by the presence of sands with a median grain size larger than 200 μm, permitting

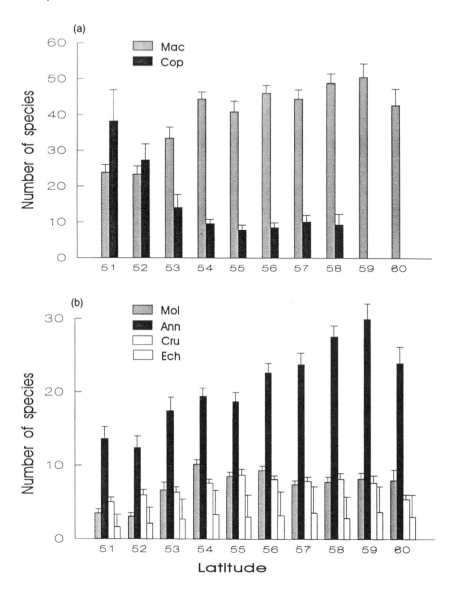

Fig. 10.2. Latitudinal trends in number of species per sample for meiobenthic copepods and macrofauna (a) and main taxonomic groups (molluscs, annelids, crustaceans and echinoderms) within the macrofauna (b) in the North Sea as sampled during the 1986 North Sea Benthos Survey. Sample size is 10 cm^2 for the meiofauna and 0.1 m^2 for the macrofauna.

the existence of numerous small and slender interstitial (living between sand grains) species. For the macrofauna there is a regular increase of diversity at least between 51 and 58°N, mainly due to polychaetes, but which also exists to a certain degree within the other three main macrofaunal groups (Fig. 10.2). It is more difficult to explain these patterns for the macrofauna than for copepods. Historically, the southern part of the North Sea was dry land until 4000 years ago and colonization of this area occurred both from the south and the north. However, the deeper areas to the north are much richer in species that would easily colonize the newly available North Sea habitats. In the modern situation the existing current and productivity and sedimentation patterns are all important aspects to explain biodiversity patterns, and human activity, especially the fisheries, is becoming a more and more dominant factor as well.

Regional patterns in deep sea biodiversity are also most often explained in ecological and historical terms. In ecological terms the congruity in global-scale patterns of diversity may be explained by coupling between surface and sedimentary processes. Such coupling is probably much more intense than anticipated even a few years ago. It has been shown that the sinking rates of particles are much higher than previously thought and that a response between benthic activity to accumulation on the sediments of detritus derived from primary production at the surface is possible and indeed exists. Surface productivity increases poleward as does benthic biomass and perhaps the often observed inverse relationship between productivity and diversity also holds for the marine benthos. Another reason may be that the carbon flux is more variable in the north. Historical factors may also be important. The low diversity in the Norwegian Sea has been attributed to the Quaternary glaciation and the effects of the sea ice cover. In prosobranch gastropods local diversity is correlated with regional diversity and the dispersal potential of the regional species pool (Stuart and Rex, cited in Rex et al., 1993). Local diversity thus appears to be regulated by colonization from the regional species pool and reflects the historical evolutionary build-up of regional diversity.

How Many Species Exist in Marine Sediments?

Grassle and Maciolek (1992) published a very extensive study of deep-sea diversity from ten stations along 176 km of the 2100 m isobath (depth contour) off New Jersey and Delaware in the US and four additional stations at 1500 m and 2500 m depth. In these stations a total of 798 species representing 171 families and 14 phyla were identified on a total sampled surface of 21 m^2. Of these species, 460 (58%) were new to science. About 20% of the species were found at all ten 2100 m stations and 34% occurred at only one station. Of the total soft-sediment fauna 28% of species occurred only once and 11% only twice.

The number of species found rises continuously as more samples and individuals are collected. At a single station species were added at a rate of about 25 per $0.5 \, m^2$. When samples are added along the 176 km transect the rate of increase is about 100 species per 100 km. The rate of increase across depth contours is even greater. Since the deep sea at depths greater than 1000 m occupies about $3.10^8 \, km^2$ of the Earth's surface, if a linear addition of one species per km is extrapolated to 1 species per km^2, the global deep-sea macrofauna would contain on the order of 10^8 species. Using the same reasoning, a similar number could be found for the deep-sea meiofauna. Taking into account the fact that the deepest, oligotrophic areas of the ocean floor have densities of macrofauna about one order of magnitude lower ($115 \, m^{-2}$, Hessler and Jumars, 1974), perhaps 10^7 species may be a better estimate but this is still a conservative estimate since accumulation of species across depth contours is much faster. However, the reliability of this estimate depends on the hypothesis that rare species are different in different areas and May (1992) rightly remarks that since half of the species found in the Grassle and Maciolek (1992) study were known to science, the implication that the total number of species is double the number of species described is also providing some kind of (minimum) estimate that is two orders of magnitude lower. Again its accuracy depends on whether the rare species are the same ones in different localities on Earth.

Conclusion

Although the marine soft-bottom benthos (as the marine plankton) occupies more than half the Earth's surface, its diversity has been rarely studied and mechanisms that maintain this diversity or its function have not been discussed extensively in the literature. Several studies of meio-and macrofauna show that local diversity may be extremely high (more than one hundred species of nematodes on a $10 \, cm^2$ and of macrofauna on a $0.1 \, m^2$ surface). The existence of such high local diversity is ascribed to temporal and spatial patchiness in food input and biological processes and structures in the sediments. Whether global biodiversity is also high is still controversial but it is not unreasonable to assume that the number of benthic invertebrate species will number in the millions instead of the about 60,000 described to date.

Acknowledgement

Publication No. 2016 of the Netherlands Institute of Ecology, Centre for Estuarine and Coastal Research, Yerseke, The Netherlands.

References

Clarke, A. (1992) Is there a latitudinal diversity cline in the sea? *Tree* 7, 286–287.
Connell, J.H. (1983) On the prevalence and relative importance of interspecific

competition. Evidence from field experiments. *American Naturalist* 122, 661–696.

Gage, J.D. (1979) Macrobenthic community structure in the Rockall Trough. *Ambio Special Report* 6, 43–46.

Grassle, J.F. and Maciolek, N.J. (1992) Deep-sea species richness: regional and local diversity estimates from quantitative bottom samples. *American Naturalist* 139, 313–341.

Grassle, J.F., Lasserre, P., McIntyre, A.D. and Ray, G.C. (1991) Marine biodiversity and ecosystem function. *Biology International* Special Issue 23.

Heip, C., Basford, D., Craeymeersch, J.A., Dewarumez, J.-M., Dörjes, J. de Wilde, P., Duineveld, G., Eleftheriou, A., Herman, P.M.J., Niermann, U., Kingston, P., Knitzer, A., Rachor, E., Rumohr, H., Soetaert K. and Soltwedel, T. (1992) Trends in, biomass, density and diversity of North Sea macrofauna. *ICES Journal of Marine Science* 49, 13–22.

Hessler, R.R. and Jumars, P.A. (1974) Abyssal community analysis from replicate box cores in the central North Pacific. *Deep-Sea Research* 21, 185–209.

Hutchinson, G.E. (1959) Homage to Santa Rosalia, or why are there so many kinds of animals? *American Naturalist* 93, 145–159.

Jumars, P.A. (1976) Deep-sea species diversity: does it have a characteristic scale? *Journal of Marine Research* 34, 217–246.

Lack, D. (1944) Ecological aspects of species-formation in passerine birds. *Ibis* 1944, 260–286.

Lasserre, P. (1992) The role of biodiversity in marine ecosystems. In: Solbrig, O.T., van Emden, H. and van Oordt P.G.W.J. (eds) *Biodiversity and Global Change*. IUBS Press, Paris, pp. 105–130.

May, R.M. (1992) Bottoms up for the oceans. *Nature* 357, 278–279.

McArthur, R.H. (1958) Population ecology of some warblers of northeastern coniferous forests. *Ecology* 39, 599–619.

McArthur, R.H. and Levins, R. (1967) The limiting similarity, convergence and divergence of coexisting species. *American Naturalist* 101, 377–385.

McGowan, J.A. and Walker, P.W. (1993) Pelagic diversity patterns. in: Ricklefs, R.E. and Schluter, D. (eds) *Species Diversity in Ecological Communities. Historical and Geographical Perspectives.* University of Chicago Press, Chicago, pp. 230–240.

Reise, K. (1985) Predator control in marine tidal sediments. In: Gibbs, P.E. (ed.) *Proceedings of the 19th European Marine Biology Symposium*. Cambridge, University Press, Cambridge, pp. 311–321.

Rex, M.A., Stuart, C.T., Hessler, R.R., Allen, J.A., Sanders, H.L. and Wilson, G.D.F. (1993) Global-scale latitudinal patterns of species diversity in the deep-sea benthos. *Nature* 365, 636–639.

Ricklefs, R.E. and Schluter, D. (1993) *Species Diversity in Ecological Communities. Historical and Geographical Perspectives.* University of Chicago Press, Chicago.

Rowe, G.T., Polloni, P.T. and Haedrich, R.L. (1982) The deep-sea macrobenthos on the continental margin of the northwest Atlantic Ocean. *Deep-Sea Research* 29, 257–278.

Sanders, H.A. (1968) Marine benthic diversity: a comparative study. *American Naturalist* 102, 243–282.

Schluter, D. and Ricklefs, R.E. (1993) Species diversity: an introduction to the problem. In: Ricklefs, R.E. and Schluter, D. (eds) *Species Diversity in Ecological Communities. Historical and Geographical Perspectives.* University of Chicago Press, Chicago, pp. 1–10.

Soetaert, K., Heip, C. and Vincx, M. (1991) Diversity of nematode assemblages along a Mediteranean deep-sea transect. *Marine Ecology Progr. Ser.* 75, 275–282.

Soetaert, K., Vincx, M. and Heip, C. (1995) Nematode community structure along a Mediterranean shelf-slope gradient. *PSZNI Marine Ecology* (in press).

Stehli, F.G., McAlester, A.L. and Helsley, C.E. (1967) Taxonomic diversity of recent bivalves and some implications for geology. *Bulletin of the Geological Society of America* 78, 455–466.

Thorson, G. (1957) Bottom communities (sublitoral or shallow shelf) In: *Treatise on Marine Ecology and Palaeoecology*. Geological Society of America, pp. 461–534.

Underwood A.J. and Petraitis, P.S. (1993) Structure of interdal assemblages in different localities: how can local processes be compared? In: *Species Diversity in Ecological Communities. Historical and Geographical Perspectives*. The University of Chicago Press, Chicago, pp. 39–51.

Linkage between Ecological Complexity and Biodiversity

Hiroya Kawanabe
Centre for Ecological Research, Kyoto University, 4-1-23 Shimosakamoto, Otsu 520-01, Japan

Importance of Relations' Diversity

Under the name of biodiversity, what kinds of things should be considered? At present, it is well known, even by the general public, that there are many levels of diversity such as genetic, species, higher taxa, ecosystem or landscape. All of these levels are included in biodiversity, but it is difficult to say that they are actually part of it (i.e. part of the diversity of living materials or 'creatures' themselves).

I would like to point out, however, that there is another kind of biodiversity: the complex sophisticated interactions among life forms, such as their sociality, competition, cooperation or symbiosis, predation, parasitism, and all of their interrelations. Such diversity of ecological relations represents, not less, but more biodiversity than that which is visible among living materials or 'creatures' as mentioned above.

I would like to highlight two very simple examples. The first is the competitive cooperation or exploitative mutualism among fishes, first discovered in cichlids of Lake Tanganyika at the end of the 1970s. For instance, there are several species of scale-eaters, which pick up scales of other fish individuals swimming around them. In most cases, the success ratio of feeding for each scale-eater is rather low – less than 10% on average. If individuals belonging to different species come together for attacking, however, the success ratio for any individual increases up to 25% or more on average. Victim individuals might be much more vulnerable in terms of their capacity to escape two kinds of different attacking behaviour simultaneously. This kind of mutualistic situation in fishes was also detected among benthic animal feeders, piscivores, algal feeders, etc. Furthermore, in the case of all scale-eaters, their mouths are never located at the terminal, but are twisted to the left or right side. This is genetically determined; it depends on simple

Mendelian inheritance. These dextral and sinistral mouths in the same species also make mutualism for scale-eating species.

The second example refers to the 'cry substance' of plants. A plant leaf attacked by herbivorous mites, for instance, releases particular chemical substances, not as a direct defence against the mites, but to attract carnivorous mites which attack the herbivorous ones. Particular chemicals are released from particular species (i.e. different substances are released for calling different carnivores).

These two simple cases demonstrate with ease the complex sophisticated interactions which exist among various life forms.

From Ecosystems to Genes

From Genes to Ecosystems: A Research Agenda for Biodiversity was published by IUBS, SCOPE and UNESCO (Solbrig, 1991). It is an exellent book, well organised and well written. However, I am inclined to add one thing to this book: a viewpoint 'from ecosystems to genes'.

Accumulated evidence strongly suggests that ecological complexity, the complex sophisticated interactions among various forms of life, together with intricate heterogeneous habitat structure, play a key role in promoting biodiversity in nature. A clear example for external inputs promoting biodiversity can be found in the immunological responses of an organism against non-self material invading the body. Diversity in the antibodies and restructured DNA sequences produced in immunological cells are based on the necessity to counter the diversity in potential antigens present in the environment.

Plant and animal interactions provide another rich source of analogous examples. Plants develop defence systems against attacks by a variety of herbivores. Their hard structure made of cellulose and other cell-wall substances partly serves as a basic physical defence. Secondary substances which plants produce and store in cytoplasm are effective chemical defences. Animals that attack plants also develop means for coping with these defences. This coevolutionary process may lead to the creation of amazing products and an ability to produce them on both sides. This is an example for the case in which species interactions, through evolutionary processes, generate novel biological traits, thus enhancing biodiversity.

Recent advances in community ecology extend this pairwise scheme of species interactions to that of more complex interactions involving a third organism. 'Cry substances' released by plants, as mentioned above, is an excellent example of this. An attacked plant may affect its uninfested neighbouring plants by initiating their defence reactions. Even insects feeding at different times or on different parts of a plant may have a substantial effect on the quality of resources available to one another (other insects?). Such indirect effects mediated by the host plant are more common than previously thought.

Therefore, it is clear that if such interactions had not always been occurring in evolutionary processes, and were not continuing to occur at present, such genetic mutations would not have taken place and would not have been maintained up to now. These interactions are also beginning to show clearly that variability and flexibility of species interactions and indirect effects among community components have also created biodiversity at genetic and species levels.

Ecological complexity enhances biodiversity through evolutionary and biogeographical processes. On the contrary, its degradation should quickly cause the diversity in phenotypic and genotypic traits to decline.

For the conservation of biodiversity, it is not sufficient to preserve living organisms or their gametes alone, because keeping animals and plants in zoological and botanical gardens or their gametes in freezers cannot conserve the full range of biodiversity they exhibit in nature, due to the loss of the ecological complexity they enjoy in their original habitats. Moreover, for promoting richer biodiversity in the future, we should encourage more complexity in biological communities with their environments at present.

Ecological Complexity for Promoting Biodiversity

We proposed a project, 'Symbiosphere: Ecological Complexity for Promoting Biodiversity', as a part of the first topic area, 'The ecosystem function of biodiversity' in the IUBS-SCOPE-UNESCO programme, Diversitas (Kawanabe et al., 1993). This project might also be greatly related to the second topic area, 'Origin, maintenance and loss of biodiversity'.

Importance of Intensive Studies in Research Belts

The Diversitas programme should be carried out, ideally, all over the world. We need to concentrate our efforts, however, in a few study areas. So, initially, around five intensive research areas were proposed. In the case of extremely intensive investigations (the first degree of intensive studies), I agree that research should be concentrated in a few areas only. For the second degree of studies, however, we need to investigate in a few belts from the North pole across the Equator to the South pole.

All over the world, deserts, or grasslands at least, can be found in the middle of every belt except one. It is situated in the Western Pacific region and Asia in a broad sense. Forests spread from Far East Russia, through Japan, Korea and China, then the Philippines, Indochina, Thailand, Malaysia, Indonesia and New Guinea, to the eastern end of Australia and New Zealand. Some other reasons initiating and maintaining higher biodiversity can also be explained in this belt.

Under these circumstances, 'International Network of Diversitas in West Pacific and Asia', (DIWPA), was proposed in 1993, officially launched in 1994 and the first workshop will be held in 1995 (Kawanabe and Wada, 1995).

References

Kawanabe, H., Ohgushi, T. and Higashi, M. (eds) (1993) SymBiosphere: ecological complexity for promoting biodiversity. *Biology International*, Special Issue no. 29.

Kawanabe, H. and Wada, E. (1995) International network of DIVERSITAS in Western Pacific and Asia. (unpublished paper presented at the XXVth General Assembly of IUBS, 5-9 September 1994, Paris).

Solbrig, O. (ed.) (1991) *From Genes to Ecosystems: A Research Agenda for Biodiversity*. IUBS, SCOPE and UNESCO, Paris.

Biotic Interactions and the Ecosystem Function of Biodiversity

12

H.A. Mooney
Department of Biological Sciences, Stanford University, Stanford, CA 94305, USA

Introduction

We are continuing to see depletions of the Earth's biotic diversity, as well as an increasing global homogenization of the biota. Here I focus on some of the possible ecosystem consequences of these trends. To do this I draw on two programmes that have addressed various dimensions of the problem. One is the SCOPE (Scientific Committee on Problems of the Environment) programme on biological invasions and the other, a SCOPE programme on the ecosystem function of biodiversity. The invasions example (species additions) is of interest since invasions can directly influence species diversity as well as providing a surrogate for understanding the possible ecosystem consequences of species removals. I focus here primarily on a single dimension of the ecosystem functioning of biodiversity issue – that of the consequences of disruptions of biotic interactions. The general topic is treated more fully in Schulze and Mooney (1993).

Impact of Species Additions

The extensive findings of the SCOPE invasions programme were summarized in Drake *et al.* (1989). In relation to the ecosystem impact of invaders it was concluded that these can be considerable and include impacts on all ecosystem processes (Table 12.1). These invasions can change the whole character of ecosystems, moving them to a new structural state with a very different diversity structure. For example, D'Antonio and Vitousek (1992) have noted the dramatic impact of invasive grasses, particularly in tropical regions, on ecosystem structure and function. Land use change often promotes the invasions of these grasses that in turn alter the fire cycle, community

Table 12.1. Processes impacted by invaders with an example.

Ecosystem property	Example
Community structure	Beaver (Naiman et al., 1988)
Biogeochemistry	Myrica (Vitousek and Walker, 1989)
Fire regimes	Tropical grasses (D'Antonio and Vitousek, 1992)
Erosion	Feral pigs (Vitousek, 1986)
Geomorphology	Spartina (Macdonald et al., 1989)
Hydrological cycles	Tamarisk (Vitousek, 1986)

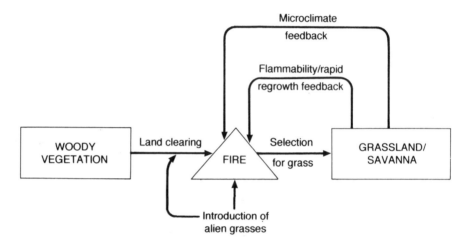

Fig. 12.1. An example of a keystone invader. The invasion of tropical grasses into tropical areas, aided by land use change, alters the successional pattern, as well as major ecosystem processes relating to biogeochemistry (from D'Antonio and Vitousek, 1992).

succession, and biogeochemical cycles. Once established, the woody, diverse systems are short circuited in the successional cycle (Fig. 12.1).

The SCOPE invasions programme focused on terrestrial systems and phenomena as shown in Table 12.1. Impacts on aquatic systems are equally impressive as shown by the following examples.

Aquatic invaders

Some of the most recent dramatic examples of the impact of invaders have come from aquatic systems. In 1982, ballast water from a ship travelling from the coast of the Americas was evidently released in a Black Sea port. These waters contained a jellyfish-like ctenophore that took hold and by the mid 1990s constituted as much as 95% of the wet weight biomass of the Black Sea and resulted in the closing down of fisheries in the connected Azov Seas (Travis, 1993).

These exchanges are not only one way. In either 1985 or 1986 ballast water from a ship from Europe was introduced into the Great Lakes of North America and with it the larvae of the bivalve, *Dreissena polymorpha*, the zebra mussel, a native of southern Russia. This mussel took hold and quickly spread through the waterways of the US and Canada. It is already having a large impact on the economy due to biofouling and impacts on navigation, boating and sport fishing. The economic impact could reach 5 billion dollars by the year 2000. These animals have enormous filtering capacities and at expected densities, could filter the whole of Lake St Clair and the western Basin of Lake Erie in as little as 10 days (Ludyanskiy *et al.*, 1993). Obviously then they can have a large impact on the biotic structure of these water bodies. Mackie (1991) gives the scenario where the mussel very effectively removes the seston through filtration, and biodeposits most of the nutrients in the water to the lake to the floor, thereby decreasing primary and zooplankton production, and dependent fisheries, but increasing the development of the benthic community

Norse (1993) recently noted the large number of cases of major impacts of introductions on marine systems including the introduction of the European periwinkle (*Littorina littorea*) that caused major changes in the structure of the intertidal communities all along the North American Atlantic seaboard, the Asian green alga (*Codium fragile tomentosoides*) that impacted the mollusc fisheries on the Atlantic coast of the US, and the Asian eelgrass, *Zostera japonica* that became established along the northwestern coast of North America, altering the nature and diversity of mudflat communities. Movement through the Suez Canal of the Red Sea jellyfish (*Rhopilema nomadica*) resulted in their establishment in the Mediterranean Sea where they have grown to very large populations that are adversely impacting fisheries, power plants and tourism in Israel.

It is clear then that the addition of a single species can have considerable impact on ecosystem structure and function. We know that not all introductions become established and not all of those that do have a major ecosystem impact. We do know, however, that those that have had such impacts have been introduced both purposefully as well as accidentally. Invaders may not only affect ecosystems directly, but also indirectly by causing extinctions of native species.

Invading species cause extinctions

Part of the cause of the loss of species on a global scale is due to the impact of invading species at a local scale. For example, 40 of the total 613 endangered and threatened species in the US are listed in this category because of the impact of an invading species on them (US Congress, 1993). One of the most dramatic instances of an invading species causing extinctions is the case of the invading brown tree snake *Boiga irregularis* into Guam sometime in the 1940s or 1950s. By the early 1980s the snake had occupied the entire island and reduced the ten species of native woodland birds to extinction or rarity (Fig. 12.2) (Savidge, 1987). It is estimated that half of all the insular extinctions

156   H.A. Mooney

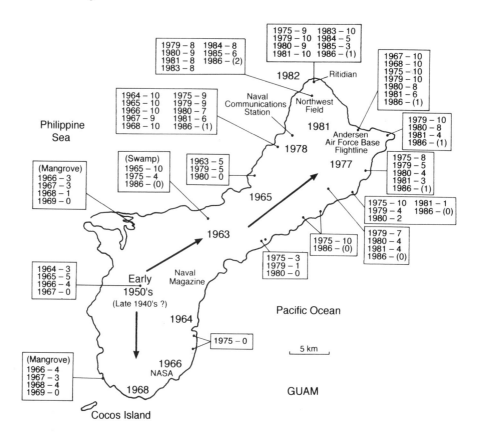

Fig. 12.2. Path of invasion into Guam of the brown snake (*Boiga irregularis*). Numbers in boxes refer to the census number of the 10 species of woodland birds found on the island through time and indicate the most likely impact of the snake on the avifauna (from Savidge, 1987).

of bird species may have been caused by invading species, particularly rats (Diamond, 1984)

Lizard species in Guam have also either had their populations severely reduced or have become extinct due to the activities of the brown snake (Rodda and Fritts, 1992). Unravelling the cause and effects of changes in populations on such heavily impacted islands is difficult since, for example, at the same time the brown snake was building up to extraordinary population sizes (100 snake ha^{-1}) a kingfisher became extinct that was a lizard predator, which would have increased the population of lizards. Complicating matters further, introductions of carnivorous lizards and a shrew during this period may have also contributed to the loss of the other non-carnivorous lizards (Table 12.2). It is clear that these systems are now considerably out of balance due to population invasions and extinctions.

Table 12.2. Probable consequences of change in lizard predator populations on two species of prey lizards on Guam (Rodda and Fritts, 1992)

Predator species	Impact on these native prey species of the predators	
	Skink (*Emoia caeruleocauda*)	Gecko (*Gehyra oceanica*)
Introduced brown snake (*Boiga irregularis*) irruption 1960-85	Negative	Negative-predominate effect
Kingfisher (*Halcyon cinnamomina*) extirpation 1960-85	Positive	Neutral
Introduced carnivorous skink (*Carlia fusca*) irruption 1965-	Negative-predominate effect	Neutral
Introduced shrew (*Suncus murinus*) irruption 1955-1965	Negative	Neutral

Locally extinct species become invaders

One way that we have been able to test the degree of influence that a species deletion may have on the function of an ecosystem is to view those cases where a species has been deleted and then has returned. Even though the effects of the deletion may not be documented the effects of the return have been. One such case is that of the beaver (*Castor canadensis*) which was nearly extirpated from the United States. In recent years due to the protection of this animal and the loss of its predators, it has increased greatly in abundance. In certain areas there has been a dramatic change in whole landscape processes through the renewed activities of the beaver, with of course a large increase in wetland areas with concomitant increases in community types as well as major alterations to carbon, nitrogen and water cycles.

Impact of Deletions

Although our view of the ecosystem impact of invading species is fragmentary we certainly have enough information to see that a single species addition can totally alter ecosystem function. I now examine briefly some information on the alternate case of species removals.

The paucity of information

What direct evidence do we have on the impact of species deletions on ecosystem function? Not much since surprisingly few controlled experimental studies have been done on this topic. However, we have enough to see trends.

For example, humans, through their history, have been selectively removing large mammals from ecosystems. Many of these large mammals have been driven to extinction locally, such as the buffalo and wolf. In more recent times there have been large decimations of mammals of the tropical forests. Dirzo and Miranda (1991) have made comparisons of tropical rain forests in Mexico where there have been major defaunations through hunting and those where the fauna has not been significantly altered. They found a major reduction in the diversity in the understorey plants in the mammal-poor forests evidently as a result of the release from grazing pressure of the understorey. The many examples of harvesting by humans of a particular organism provide a rich base in which to examine the significance of a given species in the functioning of an ecosystem. However, there have been few studies of the total system consequence of these deletions.

Brown and Heske (1990) have provided direct experimental evidence of the consequences of the impact of deletions, in this case a functional group. They experimentally removed a guild of three kangaroo rats in a desert ecosystem which resulted in the conversion of desert scrubland into desert grassland. This massive keystone effect was due to the selective predation of the rodents on large seeds as well as their major disturbance to germination microsites. The keystone effect of the rats was even larger than that caused by grazing cattle alone (Brown and Heske, 1990). This study concentrated on the community responses of deletions and did not provide an analysis of the ecosystem impacts of the changes, although they are no doubt considerable given the major shift in dominant life forms.

Not all deletions will have equal community and ecosystem impacts. There is now considerable evidence that some species play disproportionately large roles in an ecosystem and that their loss results in a major restructuring and functional operation of the system as demonstrated many years ago (Paine, 1966). An example of a likely keystone species, which has an important ecosystem role, is the red land crab found on Christmas Island in the Indian Ocean. This crab (*Gecarcoidea natilis*) has been shown to be a major controller of vegetation structure through predation effects on seeds (O'Dowd and Lake, 1991) and seedlings (O'Dowd and Lake, 1990). They also greatly influence the nutrient balance of the habitat by concentrating leaf litter at their burrows (O'Dowd and Lake, 1989). Finally, they are responsible for making rain forest habitats of the island resistant to invasion by the giant African land snail (Lake and O'Dowd, 1991). Obviously this species controls major ecosystem properties and its loss would have a cascading effect on the functioning of the rain forests of Christmas Island.

The example of the red crab indicates a single species playing many roles in the ecosystem. This is not unusual for island species where there are not many members of any functional guild. This makes islands particularly good places to study species function and has led to the prediction that following a given species deletion, substitution by functionally equivalent species will be less likely on islands than on the comparable mainland because there should be fewer species per functional group on islands (Cushman, 1995).

Landscape processes

One of the important conclusions of the SCOPE biodiversity programme is that the consequences of losses in landscape diversity and species diversity do not scale one to the other. In other words there are landscape level phenomena that are independent of species richness. Landscape level phenomena can, however, feed back on species properties and interactions. An example of this comes from a recent study of landscape heterogeneity caused by agriculture where it was shown that lack of habitat connectivity released insects from predator control. Fragmentation of habitats in the agricultural landscape is a major threat to biological diversity, which is greatly determined by insects. Isolation of habitat fragments resulted in decreased numbers of species as well as reduced effects of natural enemies (Kruess and Tscharntke, 1994).

Some Predictions on Deletions

The SCOPE biodiversity programme concluded that the greatest impact on species deletions will be in those systems where there are limited numbers of representatives of a particular functional group, especially if the species is a large dominant. To this conclusion can be added the distinction that the 'loss of species which have qualitatively different effects on ecosystem processes (i.e. effects on inputs/outputs or disturbance) will have greater impact on ecosystem processes than will loss of species which differ in size or RGR and therefore, their effects on rates or stocks in ecosystems'. (Chapin and Reynolds, unpublished) Thus putative species 'redundancy' is a buffer against external forcing factors such as climatic extremes, or in the words of Chapin and Reynolds, 'The more species there are in a functional group, the less likely that any extinction event or series of such events will have serious ecosystem consequences'.

Conclusions

The following concluding points can be made:

1. Invasives species (additions) impact all ecosystem functional properties.
2. Invasives can cause extinctions.
3. Extinctions (deletions) and invasives can totally alter biological interactions.
4. Systems with little apparent species functional redundancy are the most susceptible to dramatic effects of species losses.
5. One cannot always scale directly from ecosystem to landscape sensitivity of diversity loss since different processes come into play.

References

Brown, J.H. and Heske, E.J. (1990) Control of desert-grassland transition by a keystone rodent guild. *Science* 250, 1705-1707.

Cushman, J.H. (1995) Ecosystem-level consequences of species additions and deletions on islands. In Vitousek, P.M., Loope, L.L. and Adersen, H. (eds) *Islands: Biodiversity and Ecosystem Function*. Springer Verlag, Berlin.

D'Antonio, C.M. and Vitousek, P.M. (1992) Biological invasions of exotic grasses, the grass/fire cycle and global change. *Annual Review of Ecology and Systematics* 23, 63-87.

Diamond, J.M. (1984) Historic extinctions: a rosetta stone for understanding prehistoric extinctions. In: Martin, P.S. and Klein, R.G. (eds) *Quaternary Extinctions*. University of Arizona Press, Tuscon, Arizona, pp. 824-862.

Dirzo, R. and Miranda, A. (1991) Altered patterns of herbivory and diversity in the forest understory: a case study of the possible consequences of contemporary defaunation. In: Price, P.W., Lewinsohn, T.M., Fernandes, W. and Benson, W.W. (eds) *Plant-Animal Interactions*. Wiley Interscience, New York, pp. 273-287.

Drake, J.A., Mooney, H.A., di Castri, F., Groves, R.H., Kruger, F.J., Rejmánek, M. and Williamson, M. (eds) (1989) *Biological Invasions. A Global Perspective*. John Wiley, Chichester.

Kruess, A. and Tscharntke, T. (1994) Habitat fragmentation, species loss, and biological control. *Science* 264, 1581-1584.

Lake, P.S. and O'Dowd, D.J. (1991) Red crabs in rain forest, Christmas Island: biotic resistance to invasion by an exotic snail. *Oikos* 61, 25-29.

Ludyanskiy, M.L., McDonald, D. and MacNeill, D. (1993) Impact of the zebra mussel, a bivalve invader. *BioScience* 43, 533-544.

Macdonald, I.A.W, Loope, L.L., Usher, M.B. and O. Hamann. (1989) Wildlife conservation and the invasion of nature reserves by introduced species: a global perspective. In: Drake, J.A., Mooney, H.A., di Castri, F., Groves, R.H., Kruger, F.J., Rejmánek, M. and Williamson, M. (eds) *Biological Invasions. A Global Perspective*. John Wiley, Chichester, pp. 215-255.

Mackie, G.L. (1991) Biology of exotic zebra mussel, *Dreissena polymorpha*, in relations to native bivalves and its potential impact in Lake St. Clair. *Hydrobiologia* 219, 251-268.

Naiman, R.J., Johnston, C.A. and Kelley, J.C. (1988). Alteration of North American streams by beaver. *BioScience* 38, 753-762.

Norse, E.A. (eds) (1993) *Global Marine Biological Diversity: A Strategy for Building Conservation for Marine Conservation*. Island Press, Washington, DC.

O'Dowd, D.J. and Lake, P.S. (1989) Red crabs in rain forest, Christmas Island: removal and relocation of leaf fall. *Journal of Tropical Ecology* 5, 337-348.

O'Dowd, D.J. and Lake, P.S. (1990) Red crabs in rainforest, Christmas Island: differential herbivory of seedlings. *Oikos* 58, 289-292.

O'Dowd, D.J. and Lake, P.S. (1991) Red crabs in rain forest, Christmas Island: removal and fate of fruits and seeds. *Journal of Tropical Ecology* 7, 113-122.

Paine, R.T. (1966) Food web complexity and species diversity. *American Naturalist* 100, 65-75.

Rodda, G.H. and Fritts, T.H. (1992) The impact of the introduction of the colubrid snake *Bioga irregularis* on Guam's lizards. *Journal of Herpetology* 26, 166-174.

Savidge, J.A. (1987) Extinction of an island forest avifauna by an introduced snake. *Ecology* 68, 660–668.

Schulze, E.-D. and Mooney, H.A. (eds) (1993) *Ecosystem Function of Biodiversity*. Springer Verlag, Berlin.

Travis, J. (1993) Invader threatens Black, Azov Seas. *Science* 262, 1366–1367.

US Congress, O.T.A. (1993) *Harmful Non-Indigenous Species in the United States*. US Government Printing Office, Washington, DC.

Vitousek, P.M. (1986) Biological invasions and ecosystem properties: can species make a difference? In: Mooney, H.A. and Drake, J.A. (eds) *Ecology of Biological Invasions of North America and Hawaii*. Springer Verlag, New York, pp. 163–176.

Vitousek, P.M. and Walker, L.R. (1989) Biological invasion by *Myrica faya* in Hawaii: plant demography nitrogen fixation and ecosystem effects. *Ecological Monographs* 59, 247–265.

13 Relations between Biodiversity and Ecosystem Fluxes of Water Vapour and Carbon Dioxide

E.-D. Schulze
Lehrstuhl Pflanzenökologie, Universität Bayreuth, 95400 Bayreuth, Germany

Fluxes of water and carbon dioxide create the interface between atmosphere and terrestrial ecosystems. In a mutual feedback the vegetation cover is determined by the climatic conditions of the atmosphere, but at the same time, the plant cover has strong effects on the partitioning of energy into latent and sensible heat and thus modifies the vapour and temperature conditions in the lower troposphere. It is an important question to what extent biodiversity interacts in this linkage.

Based on the theory of gaseous transport through the boundary layer of the canopy, Jarvis and McNaughton (1986) showed that it is mainly the vegetation structure which determines the coupling between vegetation and atmosphere. Tall structures of trees are closely coupled to the vapour pressure deficit of the atmosphere whereas low plant structures remain uncoupled and evaporate at a rate which is determined by net radiation. Following this analysis, structure and not species composition is important in the dissipation of sensible and latent heat.

Kelliher *et al.* (1993) compared not only conductance but also evaporation of differently structured canopies, taking coniferous forest and grassland as extremes. In this analysis maximum conductance was quite independent of plant height because aerodynamic conductance was generally small. Also responses to light and vapour pressure deficit were remarkably similar, except for deviations in the light response due to effects of sunlight at low angle being of greater significance for trees than for grasslands. However, it also become clear from this study that species composition was not important for regulation of transpiration. The biophysical basis for this observation was clarified in a study by Hollinger *et al.* (1994) who showed, that the conceptual model of diffusive transport does not hold in plant canopies. Turbulent transport overrides any boundary layer effect, and species-specific responses.

Studies at the canopy level appear to integrate to an extent, that

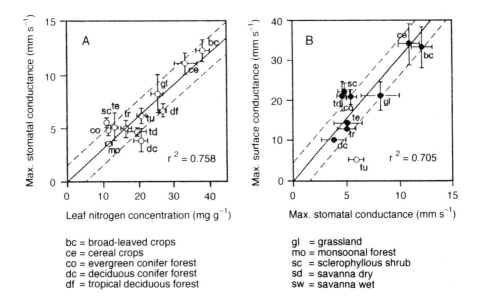

bc = broad-leaved crops
ce = cereal crops
co = evergreen conifer forest
dc = deciduous conifer forest
df = tropical deciduous forest

gl = grassland
mo = monsoonal forest
sc = sclerophyllous shrub
sd = savanna dry
sw = savanna wet

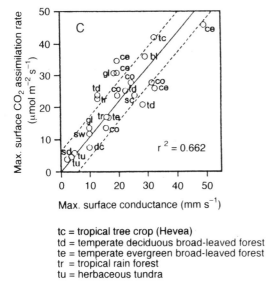

tc = tropical tree crop (Hevea)
td = temperate deciduous broad-leaved forest
te = temperate evergreen broad-leaved forest
tr = tropical rain forest
tu = herbaceous tundra

Fig. 13.1 (A) Relationships between maximum stomatal conductance and nitrogen concentration ($y = 0.3012\ x$, SE of y:1.358, r^2:0.758; Tab. 1). (B) Maximum surface conductance and stomatal conductance ($y = 2.996\ x$, SE of y:4.495, r^2:0.705), and (C) maximum surface assimilation and maximum surface conductance ($y = 1.051\ x$, SE of y:6.806, r^2:0.620), (from Schulze et al., 1994).

species-specific effects are out-averaged. Körner (1994) investigated if species specific differences exist in leaf rather than in canopy conductance. In this study also species-specific effects could not be detected. Natural vegetation types exhibited rather similar maximum leaf conductances. In fact, the mean maximum stomatal leaf conductance was $218 + 24$ mmol m^{-2} s^{-1} for 151 investigated species. This would suggest, that even at the leaf level, species are quite similar with respect to maximum conductance for water vapour.

The link between canopy and leaf processes was established by Schulze *et al.* (1994). The variation in leaf conductance (see Körner, 1994) was related to nitrogen nutrition. Differences in leaf structure result in differences in nitrogen contents (Fig. 13.1). A linear relation exists between leaf nitrogen content and leaf conductance, which contains specific leaf weight as an additional variable. If leaf conductance is known, maximum canopy conductance is closely related to maximum leaf conductance, because in most vegetation types evaporation reaches a maximum at a leaf area index of 3 (Kelliher *et al.*, 1995). Following the theory of turbulent transport (Hollinger *et al.*, 1994), maximum CO_2 assimilation is in turn linearly related to canopy conductance for water vapour. This relation between vapour and CO_2 flux is not a simple function of woody versus herbaceous species, but again driven by nutrition.

We can conclude that canopy fluxes of water vapour and carbon dioxide are driven by climatic factors, mainly net radiation and vapour pressure deficit. Plant structure determines the coupling of the fluxes to the atmosphere and whether fluxes are more related to net radiation or vapour deficit. Leaf structure determines maximum conductance via effects of specific leaf weight, but this effect is a response to leaf nitrogen content. Thus plant nutrition determines leaf conductance, canopy conductance and eventually carbon assimilation. All these relations are quite insensitive to direct effects of species. The effect of species on ecosystem fluxes of water vapour and CO_2 is an indirect one. Site fertility is greatly dependent on the vegetation cover (Schulze and Chapin, 1987; Schulze, 1994), but this interaction has not yet been included in modelling canopy processes.

References

Hollinger, D.Y., Kelliher, F.M., Schulze, E.-D. and Köstner, B.M.M. (1994) Coupling of tree transpiration to atmospheric turbulence. *Nature* 371, 60–62.
Jarvis, P.G. and McNaughton, K.G. (1986) Stomatal control of transpiration. *Advances in Ecological Research* 15, 1–49.
Kelliher, F.M., Leuning, R. and Schulze, E.-D. (1993) Evaporation and canopy characteristics of coniferous forests and grasslands. *Oecologia* 95, 153–163.
Kelliher, K.M., Leuning, R., Raupach, M. and Schulze, E.-D. (1995) Maximum conductances for evaporation of global vegetation types. *Agriculture and Forest Methods* 73, 1–16.

Körner, Ch. (1994) Leaf diffusive conductance in the major vegetation types of the globe. *Ecological Studies* 100, 463–490.

Schulze, E.-D. (ed.) (1994) *Flux Control in Biological Systems*. Academic Press, New York, 594 pp.

Schulze, E.-D. and Chapin, III F.S. (1987) Plant specialization to environments of different resource availability. *Ecological Studies* 61, 120–148.

Schulze, E.-D., Kelliher, F.M., Körner, Ch., Lloyd, J. and Leuning, R. (1994) Relationships between plant nitrogen nutrition, carbon assimilation rate, and maximum stomatal and ecosystem surface conductances for evaporation: A global ecology scaling exercise. *Annual Review of Ecological Systematics* 25, 629–660.

14 Inventorying and Monitoring Biodiversity

P.B. Tinker
*Plant Sciences Department, University of Oxford,
South Parks Road, Oxford OX1 3RB, UK*

The Problem

It is by now a truism that biodiversity is a vital global concern. From initial warnings by scientists it has grown to a world issue, linked with sustainability and development. Its importance was finally established in the Biodiversity Convention.

This importance rests on several arguments. The first is ethical: it states that as we are part of life ourselves, there is a responsibility to ensure that the diversity of life is preserved, to the best of our ability. Molecular biology and biochemistry have emphasized how close the basic relationships are between all species, including ourselves. The second argument concerns sustainability, and demands that we should preserve biodiversity for the benefit of our own and succeeding generations. We benefit from the use of many species now, and it is likely that many more may prove useful in the future, even though it may be difficult to predict which these will be. In the debate about sustainability and how it should be measured, one of the criteria of environmental damage is its reversibility. The most serious and unacceptable damage is the least reversible, and there are few things as irreversible as a biological extinction.

With these imperatives it is astonishing that at present we know so little about global biodiversity (May, 1992). Those species that have been named and classified are about 1.5 million, but this is only a fraction of the total number – the 'Grail' number that we do not know, but that is probably between 5 and 30 million. Whereas some groups, such as the mammals, birds and vascular plants are fairly well covered, most of the rest are not. There are also whole areas, such as marine biodiversity and soil biodiversity, which are particularly poorly studied, and in which very large numbers of new species may be discovered. However, the reasons why these areas lag behind

© 1996 CAB INTERNATIONAL. *Biodiversity, Science and Development: Towards a New Partnership* (eds F. di Castri and T. Younès)

are the practical difficulties of access, and any major initiative in these will be highly expensive. The total work of inventorying the biodiversity of the world is therefore a massive task, even if it stops at the species level. Biodiversity is defined as the genetic diversity both between and within species (Holdgate, 1994), but our immediate task clearly relates to the first part. Within-species diversity is a most important issue, for example in regard to plant and animal diseases and the resistances to them, and the many and varied characteristics that determine interactions with the environment, but for the present, attention to these questions is likely to focus on cultivated species.

It is not sufficient to observe, identify and classify organisms, however. It is essential also to record the physical and chemical properties of the environment in which they are found, so as to help define ecosystem relationships. Also, the world at present is in a state of pervasive change, with changes in habitats, pollution, biological introductions, and probably in climate. The need for regular monitoring of sensitive areas and species is possibly even greater than the immediate need for full identification of all species, because it gives warning of changes that may lead to shifts in populations, extinctions or population explosions.

The Resources

If this remaining task is enormous, much has been done in the past. One of our greatest resources in this work lies in the great curated collections of the world, which contain a massive store of taxonomic information in their physical collections and their databases, as is described Chapter 15.

The requirement for monitoring introduces a further set of practical problems. The regular and systematic quantitative sampling that is necessary makes monitoring each species a major task. New questions such as where to monitor, what to monitor, how frequently to monitor, and what environmental variables should be monitored at the same time, appear in any monitoring initiative, such as that contained in the *DIVERSITAS* programme (di Castri and Younès, 1994). It is quite evident that even if we could identify every species on earth, we cannot possibly monitor even a small fraction of the total in a systematic and detailed way. The subject of this discussion is therefore a highly practical one, based upon the availability of resources, and it is essential that more resources are found for these tasks (Cracraft, personal communication).

Both monitoring and inventorying demand highly trained manpower, mainly in systematics. It is probable that any one systematist cannot identify and classify more than a few hundred new species in a working lifetime, which emphasizes the magnitude of the task ahead. Quick or less rigorous methods may leave behind them many problems in classification and phenology, that have to be cleared up later. It seems impossible simply to proceed at much the same rate as now, and accept that it will take centuries to complete the inventorying of biodiversity, because the current rate of extinctions is removing species at such a frightening rate. It is not possible to say how many species are becoming extinct, because we do not know how many species there are, but some estimates suggest that up to 50% of terrestrial species will be lost by the middle of next century, at the projected rate of loss of habitat.

It is therefore evident that inventorying must be speeded up, and by a very considerable factor. Gamez (Chapter 16) describes how the INBio programme in Costa Rica is aiming to inventory, to a considerable degree of detail, all of Costa Rica's conservation areas, as well as much of the agricultural area. This is a large and valuable programme, which uses innovative methods to cover the ground rapidly, but it is not certain how easy it would be to replicate the programme elsewhere. There are a great number of detailed questions to be decided about inventorying, in particular those dealing with scale, which may require multiple study sites. Long-term studies are also needed, which links with the problems of monitoring.

The Need for Priorities

Faced with this situation, and a demand for more resources, politicians and decision-makers will almost certainly reply: 'You must prioritize'. If everything cannot possibly be done now, what is most important? There is an attempt to prioritize in the Biodiversity Convention itself (Biodiversity Convention, 1992). It identifies, for priority attention, species that are threatened with extinction; wild relatives of domesticated species; species of economic value; species of social, scientific or cultural value; and species of importance for conservation research, e.g. indicator species. This is a very formidable list, and there is no attempt to prioritize between these categories.

This highlights a major problem: many interest groups are involved in this debate, and each will have its own agenda and its own priority list. It is difficult to see how a generally accepted view of priorities can be agreed, but this is certainly desirable. The scientific community should have the most detached and also the best informed approach to this issue, and it carries a heavy responsibility to do all it can to agree and advocate a set of priorities, or at the very least a set of options from which priorities can be derived. An outline of such a proposal for an international approach was made by di Castri *et al.* (1992). If the scientific community can give united support to a single proposal or set of linked projects it will carry great weight.

The immediate practical reason for protecting biodiversity of plants lies in their potential use in plant breeding for agriculture. Hodgkin (Chapter 31) gives a good insight into one area where work is moving rapidly ahead. As is to be expected, it is the economic imperative which has proved most compelling.

Sometimes there is an implied suggestion that inventorying and monitoring must be completed before the priorities of conservation can be fully settled. Although this may be true in a strict sense, in practice this would cause an extremely damaging delay. Conservation may be targeted at areas known to have high biodiversity, even if most of this diversity is uncatalogued and unclassified. This is the global strategy behind the 'Centres of Plant Diversity' programme (Heywood and Davis, 1995), in which priority is given on the basis of high diversity, levels of endemism, number of habitat types, number of plants adapted to special conditions, and level of threat of destruction of the habitat. This process has targeted well over 200 sites, but of these only 15% could be regarded as having adequate protection. It is clearly necessary that this type of priority protection, to save the largest part possible of the worlds plant biodiversity, should also be provided with programmes for inventorying and monitoring what is there. The *DIVERSITAS* programme theme on 'Genetic resources of wild species' addresses this topic, and *in situ* conservation will clearly need sampling, inventorying and monitoring of populations. These activities are therefore integral with that of conservation.

Inventorying, Monitoring, and Global Change

It was mentioned above that biodiversity is linked to sustainability and development. The preservation of biodiversity (or a very large part of it) may be a precondition for a world that one could regard as sustainable, but development is the main danger to biodiversity. Global change is defined by the International Geosphere–Biosphere Programme as change in atmospheric composition, change in climate and change in land use. It is the last of these, with the huge shifts of land from forest into other uses, which is the main cause of loss of habitat and of biodiversity at present. There is also likely to be a major impact on plants and on the organisms that live on plants, by the progressive increase in atmospheric carbon dioxide. This will have a differential effect on the growth and water use of different species, and will change the chemical composition of plant tissue in ways that will alter its food value. The expected change in climate has not yet been detected with certainty, but this also will carry a major threat to biodiversity when it arrives. The information which has been obtained about past climatic changes shows very clearly how large shifts in geographical locations and in plant associations have occurred. Indeed, one of the integrating measures of global change could be taken as the effect on biodiversity. The effects of global change on populations and ecosystems is the main focus of the IGBP core project 'Global Change and Terrestrial Ecosystems', which therefore has strong interests in biodiversity, in particular its importance for ecosystem function. In this role there is close collaboration with *DIVERSITAS*.

The present rate of change in world systems, and the expectation that this change will accelerate and grow more complex, has focused attention on the need to set up integrated multidisciplinary monitoring schemes. These should include, but do not restrict themselves to, various measures of biodiversity such as the monitoring of selected species. These schemes are a very hopeful sign for the future. Professor Franklin (personal communication) explained the activities of the US long-term ecological research sites. This network was intended for research, but it is now adding a monitoring function, and developing into an international network. In Britain there is the Environmental Change Network, which has the specific task of monitoring change, and includes observations on a series of species in this. There are other examples of national initiatives, for example in China. The Global Terrestrial Observing System is planned to be much larger, as the title implies. It is in the planning stage at present, under the sponsorship of a series of bodies including the International Council of Scientific Unions (ICSU).

These ideas on long-term monitoring raise many difficult practical and theoretical questions. Sites must be well defined and representative of ecosystems or biomes; they must be under close management control, and have defined degrees of protection. There is also the difficult question of whether to site them where rapid change is expected, for example on ecotones and boundaries, or to aim to test the stability of populations in well-defined ecosystems. The UNESCO-MAB (United Nations Educational, Scientific and Cultural Organization – Man and Biosphere Programme) Biosphere Reserves seem very well fitted to play a major part in such activities, as they all have a conservation interest, and most will contain different intensities of land use.

There are thus a number of different initiatives in progress at present, all approaching the need to inventory and monitor biodiversity from different points of view. However, these are not sufficient. A collective effort to determine the priorities, and to concentrate the available resources on these is surely an essential precondition for the massive injection of new resources that are needed.

References

Biodiversity Convention (1992) Annex 1. UN Conference on Environment and Development, Rio, 1992.

di Castri, F. and Younés, T. (1994) DIVERSITAS: Yesterday, today and a path towards the future. *Biology International* 29, 3–23.

di Castri, F., Vernhes, J.R. and Younés, T. (1992) Inventorying and monitoring biodiversity. *Biology International*, Special issue 27.

Heywood, V.H. and Davis, S.D. (1995) Introduction. Centres of Plant Diversity . A Strategy for their Conservation. IUCN/WWF (in press).

Holdgate, M.W. (1994) Ecology, development and global policy. *Journal of Applied Ecology* 31, 201–211.

May, R.M. (1992) How many species inhabit the earth? *Scientific American* 267, 42–48.

Monitoring and Inventorying Biodiversity: Collections, Data and Training

N.R. Chalmers
The Natural History Museum, Cromwell Road, London SW7 5BD, UK

Introduction

Those nations that have ratified the Convention on Biological Diversity are required to inventory and monitor their own biodiversity. This is an onerous task, given that only a small fraction of the organisms living within the boundaries of most countries have so far been discovered, identified, scientifically named and classified (Groombridge, 1992).

A country seeking to fulfil its obligations under the Convention may well set up an action programme to inventory and monitor its biodiversity. This will necessarily be a large and expensive undertaking. However, the burden can be considerably lightened because such a programme need not start from scratch given that there already exists a substantial body of scientific information about the biodiversity of many countries of the world. Much of this information is held in scientific collections of animals, plants and microorganisms, which are housed in museums, herbaria and specialist research institutes around the world. It is therefore well worth a government considering how much it should invest in the development of information infrastructure in order to gain access to this information.

The purpose of this chapter is to describe the kinds of information held in such collections, and to show how this information can boost countries' efforts to understand, conserve and use their own biological diversity sustainably. Initially, the information flow will be mainly in one direction from nations with major scientific collections to nations in which they are poorly developed. However, this phase will rapidly be replaced by two-way traffic, for as countries carry out their own biodiversity action plans, many of them will build up their own scientific collections of organisms, plus large databases, and from these, information can flow back into an international network, to the benefit of the worldwide community.

Biological Collections

The number of biological specimens housed in scientific collections around the world is not known exactly but is immense. Howie (1993) estimates that there are at least 6500 museums and other institutions in the world holding such collections. From a survey of 100 of them, he estimates that the total number of biological objects held in scientific collections worldwide is of the order of 2.5–3.0 billion, excluding microorganisms.

The bulk of these collections are in European, North American and Australian institutions. In the United Kingdom, for example, The Natural History Museum, and The Royal Botanic Gardens of Kew and Edinburgh house over 75 million objects in their collections; while more than 70 major zoos and aquaria house over 64,000 living vertebrates; and a network of ten institutions houses some 68,000 strains of microorganisms (Stork, 1994).

The items in these collections are at their most useful when they are well curated; that is when they are prepared so as to display to greatest effect their most distinctive characteristics, when they are protected against decay and damage, when they are stored in such a way that they can readily be located and examined, and when they are accompanied by documentary information. This information will ideally include at least the scientific name of each object, the name of the collector, and the date and location at which the object was collected.

Biological collections have been amassed by the older museums and botanic gardens for over 200 years, and the majority of the documentation is still held in the ledger books and record cards on which the information was originally inscribed. Nowadays, however, this documentary information and the images of the objects are being converted in many institutions into electronic form. In addition, collections continue to expand, and as new objects are added to them, both visual and documentary information is captured electronically from the outset. The value of this information is greatly enhanced when it is backed up by detailed bibliographic information from the scientific literature, and when it is accompanied by vigorous research and information dissemination programmes within the institutions concerned.

How can this mass of information on objects, most of which are in centres that are remote from areas rich in biodiversity, help countries blessed with this richness to conserve their biodiversity and to use it sustainably?

The Utility of Information from Collections

Information from biological collections may be useful for at least three reasons. First, it may show which organisms are present in, or absent from, a defined geographical area, and this in turn may be used to guide decisions on conservation action, on the maintenance of environmental quality or on human or veterinary health policy. Second, it may be used to assess changes in distributions of organisms over time or space, and relate these to factors such as the impact of human activities. Third, it comes in a structured form which allows one to explore with economy of effort the properties, whether harmful or beneficial, of related organisms.

Geographically related information

A nation should ideally be able to call up by computer a list of all of the species collected within its boundaries and now held in the world's scientific collections. We are still a long way from achieving this goal, but nevertheless, museums, botanic gardens and related institutions are able to provide impressive lists of several major groups of organisms within their collections within specified geographical regions. Clearly, the more they are able to pool the information from their separate collections the more comprehensive are the lists they are able to provide.

A few examples illustrate this, the first of which relates to the Order Coleoptera, the beetles. This group of insects contains some 300,000–400,000 described species, which is the highest number of species known within any order of animals or plants (Hammond, 1992). The beetle collections in The Natural History Museum in London are the most species-rich of any in the world, with more than 200,000 named species represented. As a 'kickstart' to the inventorying process required under the biodiversity convention, a simple inventorying of the collections has been carried out, with a view to making available computer lists of the species in the collection and the countries from which they originate. A database has been created which contains information about specimens in every 100th drawer of the 12,000 drawer collection. For each selected drawer, we record the species that are present, number of specimens of the species, the presence or absence of particularly important specimens known as 'types' (these are the specimens that were used as the basis of the first published scientific description of the species), the country of origin, and body size. This draconian selection has provided a better understanding of the number of species in the named collection and the number of species for each country. For example, as a result of this exercise, we now estimate that the collection holds some 15,000–20,000 species of named Brazilian beetles, and although this list of names does not, of course, cover all of the Brazilian beetles, it nonetheless provides a substantial contribution towards inventorying one of the world's most important group of organisms in one of the most biodiversity-rich countries of the world.

A second example is the *Flora Mesoamericana* project, the goal of which is to name and describe all of the vascular plants of Mesoamerica, from southern Mexico to Panama. Three principal collaborating institutions provide several staff and reference collections. These are the Universidad Nacional Autónoma de México, the Missouri Botanical Garden and the Natural History Museum. In addition, several hundred botanists from other institutions are contributing to this project, and they in turn draw upon information from dozens of other scientific collections. The project will describe and name all of the 18,000 or so plants in one of the world's richest areas of biological diversity, and will provide finer-grained information on their geographical distribution. The first hard copy volumes of the project are now being published (Davidse et al., 1994) and these will be followed by electronic copies that will be available through Internet.

The above two examples relate to terrestrial environments. A third, by way of contrast, relates to one of the most abundant groups of animals found in aquatic environments. These are the copepods, which are diminutive relatives of the crabs and lobsters. They have colonized virtually every habitat from 10,000 m down in the deep oceans to 5000 m up in the Himalayas. They are by far the dominant group in the marine plankton, on which the oceanic food chains depend, and they often dominate in freshwater plankton as well. They inhabit marine sediments, and although many species are free-living, others have become parasites of almost every major animal group from sponges and corals to fish and mammals. Copepod skin and gill lice are serious pests of commercial importance in both marine and freshwater fish farms.

No overview of this diversity exists, and problems abound with the identification of copepods. Relying heavily on the collections and the Museum's library resources, scientists at The Natural History Museum are approaching completion of a massive synthesis of all 200 or more families and over 1600 genera of copepods.

These three examples illustrate the wealth of information already available in scientific collections. As this information becomes transferred onto databases, and as these become accessible through international computer networks, so it will become possible for countries to assemble inventories of their documented biodiversity.

The biodiversity present within a geographical area is of course rarely static, and the Convention of Biological Diversity rightly focuses on the need to monitor changes in biodiversity over time. Scientific collections not only provide documentary evidence of such changes, but also help to put them in perspective. An example comes from the seashore habitats, which are of great importance since the majority of people in the world live within 100 km of the seashore, and directly or indirectly have an impact upon it.

A unique perspective of community membership and function in such near shore habitats can be obtained from the palaeontological record. Over 100 years' worth of data in the form of many thousands of specimens of microfossils lodged in curated collections in natural history museums are invaluable and an irreplaceable resource for such studies. These data indicate that the species composition of near shore marine habitats was highly variable over the past 55 million years and that each identified segment of geological time, including the modern world, has its own unique and significantly large complement of species susceptible to local or total extinction by virtue of their rarity and restricted geographic distribution. Destruction or disturbance of these shallow marine habitats through, for example, pollution or via introduction of foreign species from ship's ballast water has the potential to affect severely marine biodiversity in extensive areas of the world's shallow seas (Buzas and Culver, 1994).

Inventories are themselves valuable since they summarize the total diversity of species present in a specified geographical area. However, geographically related information on individual species within a total can also be useful. Thus, many species have the potential to be used as assays of the quality of the air, water or soil found in a particular area. Lichens, for example, have been used as indicators of air quality in Great Britain (e.g. Purvis, 1992), and of forest health in Thailand (Wolseley *et al.*, 1994). To give a specific example, records in the Natural History Museum show that the lichen *Lobaria pulmonaria* was collected in parts of Britain in the middle of the 19th century where it is no longer found today. From Museum records we know that the geographical range of this organism has decreased substantially during the century, and that this is associated closely with the increase of air pollutants such as sulphur dioxide. More recently, since United Kingdom legislation has reduced the concentrations of air pollutants, the geographical range of *Lobaria pulmonaria*, and other sensitive lichens is increasing again.

Information on the identity and geographical distribution of organisms harmful to human beings, and to livestock, or on pests of crops is also highly important. Thus the spread of the New World screwworm fly (*Cochliomyia hominivorax*) into Libya posed a serious medical and veterinary threat, and its successful eradication by 1992 depended on the accurate identification of the fly based on information held in scientific collections (Hall, 1992).

Finally, information about the geographical location of species can be used to help decisions to be made about conservation priorities. The best known examples relate to charismatic vertebrates, such as giant pandas and tigers, but sophisticated decision-making procedures, notably WORLDMAP, have now been developed (e.g. Pressey *et al.*, 1993; Williams *et al.*, 1993) to provide indices of the importance of the biodiversity found in particular areas, judged by predefined criteria. One criterion might be, for example, the total number of species found in a particular area; another might be the extent to which an area contributes to the worldwide conservation of a particular species or group of species. Yet another will take into account the uniqueness of the group in question, judged from a scientific point of view. Again, these procedures depend crucially on the accurate identification of specimens together with information on their geographical distribution.

Monitoring human impact on biological diversity

Human activities can affect biodiversity both dramatically and in more subtle ways. In the latter case the effects may only be revealed by careful monitoring of the distribution and abundance either of individual species or of clusters of species within an area. An example comes from a study of schistosomiasis in the Senegal River Basin (Vercruysse *et al.*, 1995). Schistosomiasis is a major parasitic disease, infecting an estimated 200 million people in 74 countries. Serious new outbreaks of schistosomiasis are now occurring in the Senegal River Basin and high prevalences and intensity of infections are now relatively commonplace in areas previously free of schistosomiasis. A team led by Dr

Vaughan Southgate, together with Dr David Rollinson, Dr Anne Kaukas, Mr Mike Anderson and Mr David Brown (MRC) in collaboration with groups from the University of Gent, Belgium and the University of Leiden, The Netherlands, supported by the European Community, is currently monitoring the situation in the Senegal River Basin. The team is determining the causes of the new outbreaks and is characterizing the schistosomes of human beings, domestic stock and their intermediate hosts.

The Senegal River Basin is a typical Sahelian area with less than 300 mm of rainfall per year. At the end of each dry season, salt water from the Atlantic Ocean used to intrude approximately 200 km upstream. The relative high salinity of the river water prevented it being used for irrigation to develop rice agriculture. To overcome this problem a dam was constructed in 1985 at Diama near the mouth of the river. It is clear from recent data and comparison with that prior to the construction of the dam that physical and chemical changes have taken place in the intervening period. In particular, reduced salinity favours the increase and spread of snail populations. Once the intermediate hosts are established in new habitats it is easy to envisage how new transmission foci are created, especially when considering the mobility of infected people. One of the puzzles is how many strains of *Schistosoma haematobium* exist in the Senegal River Basin. The application of molecular biological techniques to separate isolates of the parasite which are transmitted by different species of intermediate host is giving clues to the existing levels of heterogeneity. It is vital that this heterogeneity is understood if the disease is going to be eventually controlled. An added complication is the reported possible emergence of drug-resistant strains of schistosome in the region. Detailed monitoring of the spread of both the intestinal and urinary forms of schistosomiasis in the Senegal River Basin and the underlying causes will hopefully prevent a repetition of such consequences occurring elsewhere in the developing world.

Again, it is important to emphasize that the ability to carry out such an investigation depends upon the existence of reference collections of hosts and parasites, together with access to the sophisticated molecular techniques that are required to identify otherwise indistinguishable hosts and parasites.

A particularly important area of monitoring involves the assessment of different regimes of forestry management on the biological diversity within the forests concerned. Thus the effects of logging on plant and insect diversity is currently being analysed in Belize and Malaysia by scientists from The Natural History Museum, and comparison of the ant fauna of leaf litter in Ghana (Belshaw and Bolton, 1993) shows that, against theoretical expectation, the leaf litter of primary forest, secondary forest and monoculture cocoa plantations contained the same ant species in about the same proportions. Belshaw and Bolton (1993) discuss possible explanations for this counterintuitive finding. The agricultural and secondary forest sites that they studied had been cleared several years prior to their study, and so it is possible that, as time passed, ant species from the leaf litter of primary forest nearby reinvaded these sites. Certainly their work shows the need for long-term detailed studies of species composition and richness in disturbed areas if the dynamics of recovery from disturbance are to be understood.

These kinds of investigations can help scientists to advise decision-makers on the effects of different land-use regimes upon biodiversity. They rely totally on the accurate identification of the species concerned, and this in turn relies upon the presence of trained taxonomists and of reference collections that the taxonomists can use as a standard against which to compare organisms under investigation.

Structured information

The science of taxonomy not only encompasses the naming and identifying of organisms, it also provides a rigorous and logically consistent method for classifying them in a way that reveals how closely or distantly related they are to one another. For example, all butterflies are more closely related to one another, than they are to beetles. Beetles and butterflies in turn, as insects, are jointly more closely related to each other than they are to crabs and lobsters. All of these together, as a group known as arthropods, are more closely related to each other than they are to human beings, lions, fish, or any other animals with backbones that are classified in a group together as the vertebrates.

This hierarchical kind of classification has immense practical value in biology, for it provides a simple strategy for exploring effectively and efficiently the possible benefits and harmful consequences of biological diversity. If an organism exists which is beneficial to human beings, then it is not at all efficient to search at random among all known organisms for similar benefits, but it is extremely efficient to look at its close relatives to see if they confer similar, or even greater benefits. By the same argument, if we know of an organism that is harmful, then the best strategy is to look also at its close relatives to see if they too are harmful. The world's biological diversity is so great, and its beneficial and harmful properties so imperfectly known, that an exploration of 'closest known relatives' rather than a random search is by far the more economical and effective strategy.

A striking example of this concerns the natural product taxol. *Systematics Agenda 2000* (1994) sets out very clearly the importance and power of the science of taxonomy as follows:

> The natural product taxol – shown to be a powerful drug agent against ovarian and breast cancer – is derived from the bark of the Pacific Yew (*Taxus brevifolia*). Unfortunately, it takes the bark of three trees to provide sufficient taxol for a single cancer patient, and the trees are killed in the process. A random search for plants containing similar products might have taken many years, but an understanding of the evolutionary relationships of the Pacific Yew quickly led researchers to examine its close relatives. From this work, it was discovered that a small quantity of leaves from the European Yew (*Taxus baccata*) can also be used to produce taxol, eventually at less cost, and with no harm to the European Yew itself.

This is but one of many examples that could have been quoted. The additional problem-solving power that is generated by information organized in hierarchies of relationship rather than in simple lists is quite exceptional.

Training

The previous sections of this chapter have emphasized the scientific and technological capabilities that need to be developed if countries are to fulfil their obligations under the Convention on Biological Diversity.

Clearly each country that is to undertake a programme of sustainable use of its biodiversity must have a sufficient number of skilled and trained personnel. The number of such scientists and information technologists does not at present match the needs of individual nations (Gaston and May, 1992), for, generally speaking, countries with the greatest biodiversity do not have the greatest concentration of taxonomists or specialists in information technology. In addition, those groups of organisms that are the most diverse, in terms of the numbers of species that they contain, do not have the greatest number of scientists directed towards their study. Mammals, birds and flowering plants are well served, but other groups by and large receive relatively little attention.

Training programmes can solve this shortfall, provided that care is taken to ensure that the training programmes are totally relevant to the concerns and biodiversity of the countries who are providing the students. The Darwin Initiative, funded by the British Government, has initiated several such training programmes.

There is, however, another need, and this is not so much for training in the specific scientific or technological skills relating to biodiversity, but in the abilities that are needed to manage biodiversity institutions. This is not a trivial task, and its importance is increasingly being recognized by international funding agencies. It takes considerable management expertise to create and run successfully a national institute that is able to inventory and monitor the biodiversity of a country, that is able to curate reference collections, and that is able to provide information relevant to the country's development and conservation programmes. This is not simply an issue of scientific or technological expertise; it is one of managerial and political skill. There are some outstanding models of how this can be achieved in both biodiversity-rich tropical countries, and biodiversity-poorer temperate countries, but it is vital if the conservation of biodiversity is to be truly compatible with development, for these scarce skills to be developed.

Conclusion

The inventorying and monitoring of a nation's biodiversity might, at first sight seem to be a mundane, and possibly unimportant matter. I hope that I have provided evidence to the contrary. First, if it is to be carried out successfully, it requires great scientific skill. Second, if the benefits are to be fully realized, it needs considerable technological, managerial and diplomatic experience. These qualities are worth developing. The future of our planet depends upon it.

References

Belshaw, R. and Bolton, B. (1993) The effect of forest disturbance on the leaf litter ant fauna in Ghana. *Biodiversity and Conservation* 2, 656–666.

Buzas, M.A. and Culver, S.J. (1994) Species pool and dynamics of marine palaeocommunities. *Science* 264, 1439–1441.

Davidse, G., Sousa, M. and Chater, A.O. (1994) *Flora Mesoamericana Vol. 6. Alismataceae and Cyperaceae*. Universidad Nacional Autónoma de México, the Missouri Botanical Garden, the Natural History Museum, London.

Gaston, K.J. and May, R.M. (1992) Taxonomy of taxonomists. *Nature* 356, 281–282.

Groombridge, B. (ed.) (1992) *Global Biodiversity: Status of the Earth's Living Resources*. Chapman & Hall, London.

Hall, M.J.R. (1992) Identification of flies in the genus *Cochliomyia* (Diptera: Calliphoridae). Unpublished report prepared for FAO.

Hammond, P.M. (1992) Species inventory. In: Groombridge, B. (ed.) *Global Biodiversity: Status of the Earth's Living Resources*. Chapman & Hall, London, pp. 17–39.

Howie, F.M. (1993) Natural science collections: extent and scope of preservation problems. In: Rose, C.L., Williams, S.L. and Gisbert J. (eds) *International Symposium on the Preservation and Conservation of Natural History Collections*, Vol. 3. Consejeriá de Educación y Cultura, Madrid, pp. 97–110.

Pressey, R.L., Humphries, C.J., Margules, C.R., Vane-Wright, R.I. and Williams, P.H. (1993) Beyond opportunism: key principles for systematic reserve selection. *Trends in Ecology and Evolution* 8, 124–128.

Purvis, O.W. (1992) Establishment of a lichen monitoring programme in Wales. *Part 1: Skomer*. Commissioned report. The Natural History Museum, London.

Stork, N. (1994) Conservation outside natural habitats. In: *Biodiversity: The UK Action Plan*. HMSO, London, pp. 83–90.

Systematics Agenda 2000: Charting the Biosphere (1994) Published by the Committees of Systematics Agenda 2000, London.

Vercruysse, J., Southgate, V., Rollinson, D., De Clercq, D., Sacko, M., De Bont, J. and Mungomba, L. (1995) Studies on transmission and schistosome interactions in Senegal, Mali and Zambia. *Tropical and Geographical Medicine* (in press).

Williams, P.H., Vane-Wright, R.I. and Humphries, C.J. (1993) Measuring biodiversity for choosing conservation areas. In: LaSalle, J. and Gauld, I.D. (eds) *Hymenoptera and Biodiversity*. CAB International, Wallingford, pp. 309–328.

Wolseley, P.A., Moncrieff, C. and Aguirre-Hudson, B. (1994) Lichens as indicators of environmental stability and change in tropical forests of Thailand. *Global Ecology and Biogeography Letters* 4, 116–123.

16 Inventories: Preparing Biodiversity for Non-damaging Use

R. Gámez
Instituto Nacional de Biodiversidad, Apt. 22-3100 Santo Domingo de Heredia, Costa Rica

The central point of this chapter is that we must use wild land biodiversity if we are to save it, and furthermore, that biodiversity inventories make biodiversity available for intellectual and economic uses by all sectors of society, including science.

This assertion constitutes the core of Costa Rica's biodiversity conservation programme, and is the outcome of over two decades of conservation efforts and the evolution of ideas. Such processes lead to the realization that sustainable utilization of natural resources for social and economic development, is the only way to guarantee that today's tropical biodiversity is conserved and will still be present in our country one hundred years from now.

In Costa Rica our Government aims to set aside a quarter of the country to the conservation into perpetuity of wild land biodiversity. This portion of the territory is expected to save 70–90% of the extant biodiversity. These estimates derive from applying the crude overall knowledge that is already available in the country's conservation, taxonomic, and ecological communities to the choice of the protected areas.

These state-owned wild lands, seven large patches designated as 'conservation areas', constitute another type of land use, the 'biodiversity crop'. In other words, wild land biodiversity is expected to provide different types of goods and services to all sectors of society: from scientists to school children; from biodiversity prospectors of pharmaceutical products to ecotourists; from watershed managers to foresters.

The government is currently conducting an administrative reorganization process that will eventually place the custody (the authority and responsibility), of the wild biodiversity in the conservation areas in the hands of local communities. These communities are expected to become the direct economic and intellectual beneficiaries of the existence of the conservation areas, and its main allies and defenders.

© 1996 CAB INTERNATIONAL. *Biodiversity, Science and Development: Towards a New Partnership* (eds F. di Castri and T. Younès)

Therefore, to facilitate non-destructive use of biodiversity, it is then necessary to know on the taxonomic level what biodiversity exists in the protected wild lands, from genes to ecosystems, in order to make this biodiversity available to intellectual, spiritual or economic users. In our view, an inventory is responsible for determining: (i) what biodiversity is present (the taxonomy/systematics); (ii) where it is (the microgeography); (iii) how it can be accessed; and (iv) how to begin to know its natural history and its most basic traits. It is the preparation of a biocatalogue of the wild lands for multiple types of users.

It is our priority to conduct the inventory as described above, intensively and with urgency to those areas conserved for biodiversity, as a prerequisite for its non-damaging use. I would like to stress here that there is plenty of room and indeed a pressing need for the participation and collaboration of both the national and the international taxonomic community in this effort. The international collaborator stands in an ideal position to collaborate both in the identification of the organisms and in the development of local taxonomic literacy.

The difficulty and cost of conducting a biodiversity inventory depends on several factors, including how much is already known, how complex species richness is, which users cover the costs, and who are the anticipated users.

Who conducts the inventory is also a major consideration in our programme. Inventories have traditionally been conducted by the professional taxonomists, collectors, and naturalists, including indigenous groups and communities. Our current vision for our national biodiversity inventory includes the additional attractive element of moving beyond a classical scientific exercise. It also constitutes an excellent educational experience for the local communities who are also the custodians of the wild land biodiversity, through their direct involvement and participation in the inventorying initiative.

The experience over nearly 5 years at INBio with the parataxonomist programme has clearly demonstrated the scientific and social benefits of involving rural communities in the cataloguing of the wild land biodiversity.

In addition, the inventory offers the equally important opportunity mentioned previously for on-the-job training for biodiversity biologists from many diverse levels of formal education through the tutorial guidance of the international taxonomic collaborator.

In our view, an All Taxa Biodiversity Inventory (ATBI) aimed at cataloguing a few biodiverse areas in the world is a major tool for involving all sectors of society in the gathering, management, and use of biodiversity information. Costa Rica is presently conducting the planning process for its ATBI in the Guanacaste Conservation Area. The Costa Rican ATBI is a pilot project in its own right on a global scale. It will answer questions of a scientific and biological nature, as well as other questions regarding the practical implications of a major effort of this sort. Above all, it will address the social and economic issues required to promote the non-damaging use of wild lands.

The real art in promoting non-destructive sustainable uses of tropical biodiversity and ecosystems is maximizing those uses that involve information extraction and ecosystem services. The proper valuation and equitable distribution of the benefits coming from the utilization of wild land biodiversity, highlighted by the Convention for Biological Diversity, will be fundamental to the country and its new biodiversity conservation programme.

The sharing of biodiversity information also constitutes an aspect of singular importance. Inventory information must be put into computerized formats so that it can be continuously managed, added to, and rearranged according to diverse users' needs. The connection to global electronic information management and distribution systems appears to be the logical next step.

At present the process commonly comes across technical, social, and economical problems associated with the management of information and the power it confers to its depositories. We expect that in the next few years all of these obstacles will be removed and rapid and efficient information management and dissemination will facilitate the non-destructive use of tropical biodiversity. One major obstacle encountered is finding people capable of translating and transmitting that information to the users, and subsequently returning their feedback to the system.

These complex processes may require the creation of new organizations or the reorganization of others. National Biodiversity Institutes or their equivalent, like INBio in Costa Rica, CONABio in Mexico, The National Museums of Kenya, and Lembaga Ilmu Pengetahuan Indonesia (LIPI) in Indonesia, are examples of institutions that bear the primary responsibility for conducting national inventories and feeding the results into the national as well as the global user community.

To conclude, I must address the issue of assigning priorities in biodiversity inventorying and monitoring activities. My comments are based on our experience in Costa Rica, and probably may come to represent a view common to many tropical countries.

The conceptual statements about user-oriented biodiversity inventories made previously, explicitly indicate that inventories should be focused on those places and taxa where an inventory is necessary as determined by scientific, social, and economic reasons. The same applies to the question of monitoring. When a society decides which species and processes it wants to monitor for some particular reason, then science can figure out how to monitor those species and processes. Which species to monitor is then the outcome, rather than the question.

With know-how and time on their hands, scientists can easily figure out methods for monitoring and inventorying. Monitoring programmes should not be started blindly. The same philosophy applies to inventories. There is little reason to start an inventory process somewhere, until there is a user process (an INBio-like organization) in place.

I realize that this perspective may conflict with some of the dearest traditions to scientists and academics. Many of us believe that many of those traditions must be significantly altered. The main concern now is the future of biodiversity, and how to link biodiversity-science and development and how to establish new partnerships with other sectors of society.

The organization and philosophy of biodiversity inventorying and monitoring, information management and priority setting for Costa Rica will allow us, if we succeed, to enjoy most of our biodiversity centuries and millennia from now.

References

Gámez, R. (1991) Biodiversity conservation through facilitation of its sustainable use: Costa Rica's National Biodiversity Institute. *Trends in Ecology and Evolution* 6, 377–378.

Janzen, D.H. (1992) A south–north perspective on science in the management, use, and economic development of biodiversity. In: Sandlund, O.T. and Schei, P.J. (eds) *Proceedings in the Norway/UNEP Expert Conference on Biodiversity*, Trondheim, Norway. NINA, Trondheim, Norway, pp. 100–113.

Janzen, D.H. (1994) Priorities in tropical biology. *Trends in Ecology and Evolution* 9, 365–367.

Lovejoy, A. and Gámez, R. (1995) INBio as a pilot project: a new approach to the management of biodiversity. In: *Proceedings of the International Conference on Biosphere Reserves*. Seville, Spain (in press).

17 Inventorying and Monitoring Flora in Asia

K. Iwatsuki
*Botanical Gardens, Faculty of Science, University of Tokyo, 3-7-1 Hakusan, Bunkyo-ku, Tokyo 112, Japan**

Introduction

Fundamental knowledge of biodiversity is insufficiently available and urgent collaboration is needed to elucidate it. In this short summary of a case study in East and Southeast Asia, this vast problem will be discussed under six topics.

Basic Information on Biodiversity

Kingdom Plantae in a narrower sense including Bryophyta, Pteridophyta, and Spermatophyta includes some 230,000 recognized species at the moment, and actual numbers are estimated between 300,000 and 500,000 according to the lumpers and splitters, respectively. As compared with the other kingdoms, Plantae is a well studied group as regards its basic species diversity. Species structure is, however, traced only in some particular species, and detailed biosystematic studies are badly needed even in most of the well-described species.

Global synthesis of plant information is expected, and is actually surveyed, although insufficient information about flora of various plants on the Earth prevents such a trial being promoted. The International Organization for Plant Information (IOPI) is expecting to edit a checklist of vascular plants and it will be developed into the Species Plantarum Project to compile the flora of the whole Earth. In order to undertake such an ambitious trial, we need more complete information on local flora.

---

* Present address: Faculty of Science, Rikkyo University, 3-34-1 Nishi-ikebukuro, Toshima-ku, Tokyo 171, Japan

© 1996 CAB INTERNATIONAL. *Biodiversity, Science and Development: Towards a New Partnership* (eds F. di Castri and T. Younès)

Floristic Research in East and Southeast Asia

Regional florae have been observed and compiled in various regions in East and Southeast Asia; according to states, districts, and even geographical areas, both in national and international projects. Some of the representative floristic surveys in East and Southeast Asia are as follows.

Japan

'Flora of Japan' published in English beginning in 1993 and will be completed within six years in six volumes.

China

'Florae Republicae Popularis Sinicae' has been issued in Chinese since the end of 1950s. An estimated 85 volumes and more than 120 issues are still in the process of being compiled by Chinese botanists in an extensive national project. A new publication 'Flora of China' in English is in preparation in the context of a joint Chinese–American project which aims to publish in English a revised and concise edition of Flora. The first instalment of an expected 25 volumes was published in 1994, and all volumes are expected to be completed before the year 2000. Various regional florae have been published recently in China. Among them 'Flora of Taiwan' is a most prominent example. Six volumes were completed in the 1960s, and now a revised edition is under preparation, its first instalment was completed early in 1994. It is unfortunate to note that no projects are currently under way on the flora of Korea.

Indochinese Peninsula

Flora of Vietnam, Laos and Cambodia, written in French, covers three countries and is now being promoted by the Museum National d'Histoire Naturelle in Paris, and contributions are made internationally. Flora of Thailand is under survey in an international project. I myself have contributed by recording more than 600 species of pteridophytes for this Flora project, but only 2000 species among 5000 known species have been compiled, and some 15,000 species are thought to be present in this country. When I started my survey on the pteridophytes in 1965, there were only some 250 species recorded in Thailand, and among the approximate 620 species enumerated in the Flora, there are many which are tentatively and collectively recognized pending further biosystematic research on them.

Malesia

Flora of the Malay Peninsula began to be revised in the 1950s, but only two volumes have actually been published, and our collective references are still rather incomplete (i.e. Ridley's 'Flora of Malay Peninsula' or Hooker's 'Flora of British India' in the last century). However, the Malay Peninsula is included in the regions under survey by 'Flora Malesiana', which was initiated in the 1950s and is still actively surveyed in an international collaboration. The area covered by this project is very wide, and it is one of the richest areas of flora on the Earth. It may take some time to complete this project, but this is one of the most important and also one of the most successful floristic projects in the world. In the Flora Malesiana area, the flora of the Philippines is under survey separately. It is most unfortunate to note here that a most active promoter of this project, Dr Benjamin Stone, passed away rather suddenly, and I am afraid for the future of this project. And, again, it is regrettable to note that Burma, or Myama has hardly been studied recently, and it is not possible to compile a modern flora there, as the rich and interesting flora is as yet little studied.

A basic floristic survey in East and Southeast Asia is actively being carried out at the moment, although the extremely high species diversity there requires more man power to overcome difficulties in field surveys as well as herbarium studies.

Biological Analysis of Biodiversity

Biodiversity, including species diversity and genetic diversity within each taxon, is one of the most important targets of biological sciences at the moment. We often speculate the number of species rather easily, without any accurate biological concept of species. Modern biological techniques can contribute to elucidate the biological structure of species as well as the phylogenetic relationships among various taxa. Without promoting such basic biological research, we cannot expect to analyse biodiversity in an ideal way. In this respect, we should have a closer union between basic floristic surveys and advanced biosystematic analysis of biodiversity. Unfortunately, however, these two areas of research are often promoted in completely different directions.

It is not overly ambitious to elucidate a genetic basis for creating taxonomic characters in our biological techniques, and it would mean that the origin of adaptive characters could be drawn by detecting origins of genetic information. And, we should note that creation of taxonomic characters is based on the evolution of biodiversity in general, as no organism can live alone.

Thus, analysis of the biological nature of biodiversity is an attractive subject in biology, and I understand that young scientists with good biological backgrounds are interested in this field. As lack of people contributing to elucidate biodiversity is a vital problem we are faced with, it is necessary to have more talented young scientists working in this field. It is important to have more scientific impact of biodiversity in order to raise the interest of young students in this field.

Threatened Plant Species

In 1989, we edited the *Plant Red-Data Book in Japan*. Many local naturalists throughout Japan contributed immensely in compiling such information.

Following such a national project, we have at the moment a network of amateur and professional plant naturalists to observe the dynamics of flora of Japan. Detailed information on threatened species is in databases, and we expect to utilize such data to work towards better conservation strategy.

Although up to the 1960s there was a variety of pollution brought about by industrial activities, we have learned significantly from a number of environmental crises and we have succeeded to a fairly good degree in overcoming them. Still, increased human activity poses a serious threat to biodiversity maintenance of beautiful nature in Japan.

Furthermore, as far as the endangered species problem in Asia is concerned, we also need more people to deal with this aspect. There are tentative lists of endangered species in various countries and areas, but most of them are extracted only from the experiences of individual botanists. An accurate observation of flora dynamics over a long-term period is necessary to give a scientific basis to the Red-Data Book, but it can not be expected to have such a scientific basis in most of East and Southeast Asia. Even within Japan, a small area, we have more than a hundred well-trained plant naturalists to contribute: for all of Asia, how many contributors would we need? Basic training courses are indispensable to promote such a contribution. At the moment, only a small number of botanists are collaborating in putting such a course together, but we have our own research work and educational duties.

Biodiversity loss is an urgent crisis to overcome, and basic information is badly needed here.

Conservation Biology

There are several biologists in Japan, although not very many are interested in promoting biological conservation studies. There is no room here to discuss this issue in detail, except to say that in order to have sound foundations for nature conservation, we need more accurate scientific grounding for conservation activities. Even in Japan, however, we have no particular institute or laboratory to accommodate projects on conservation biology, and all the research projects are at the moment carried out by taxonomists and/or ecologists who have their own scientific subjects to analyse. Promotion of conservation biology should be made to invite more scientists to take part in it.

I have to apologize again that we have hardly any contributions in this field in most of the other areas in East and Southeast Asia. Even as regards basic work on floristics, there are a lot of problems to be dealt with; particularly the urgent need to train talented scientists, and here I am afraid to note that we need more scientists who can give this training to young people.

Reintroduction Project

When biodiversity is threatened, it is expected to recover its natural habitat. Reintroduction of seriously endangered species is expected in this sense. We have some fairly successful stories of the reintroduction of endangered plant species of Ogasawara Islands, reproduced in our Botanical Gardens. Such a project is also promoting a concept of nature conservation among the general public, and it is a good subject that can be introduced through the mass media. This is again closely related to conservation biology, and our biological studies on endangered plant species were greatly advanced in connection with the reintroduction project. In relation to this, we launched a campaign not to disturb any genetic equilibrium within populations introducing individuals from other local populations: this is also a social education on genetic variation within any taxa.

Various trials have been made to reintroduce endangered species into their native localities in various countries. Most of them are just trials and no biologically sound basis is expected from it. We need to have a guideline for reintroduction in general, and this has been under preparation for a long time. General guidelines are rather difficult to develop, because cultivation and reintroduction of wild plant species are different according to the species concerned; there is a diversity in growth habit of plant species, and, the most important point is to maintain natural equilibrium, not disturbing biodiversity in nature by any artificial activities. More contributions are expected from specialists of conservation biology, and once again, I have to note that we are suffering from a shortage of personnel, or excellent biologists who can assist in conservation strategies, and this depends on supporting a system for such scientists.

Monitoring Biodiversity at Global Scales

A.E. Lugo
International Institute of Tropical Forestry, USDA Forest Service, PO Box 25000, Rio Piedras, Puerto Rico 00928–5000

Introduction

A fundamental challenge that we face as a civilization is to conduct our activities without damaging global biodiversity. At the outset of the human species, other species and natural conditions on the planet dominated its activities. Human influence over the rest of the biota increased with the harnessing of new energy sources. Rapid use of non-renewable resources, often at the expense of other species, assured human success. Today, powered by fossil fuels, people have the capacity to change dramatically the world's biodiversity. However, our survival systems are still dependent on the products and services of natural ecosystems. Without them, we cannot sustain current levels of development. Moreover, we know that important sectors of the world's biodiversity have eroded to dangerously low levels.

It is now necessary to manage the whole Earth in order to reverse damage to its biodiversity, improve stewardship of remaining resources, and assure the survival of civilization even as fossil fuel supplies run down. Necessary activities, among the many required to manage the Earth, are inventorying and monitoring of its biodiversity. This chapter focuses on the monitoring of biodiversity at global scales.

Definitions and Concepts

Invariably, biodiversity is defined as species richness. However, biodiversity is much more than species richness. The term 'biodiversity' refers to all manifestations of life on Earth. As such, biodiversity encompasses multiple scales of space and time from those that are as small as a virus to the Earth itself (Fig. 18.1). These systems operate at time steps ranging from seconds to

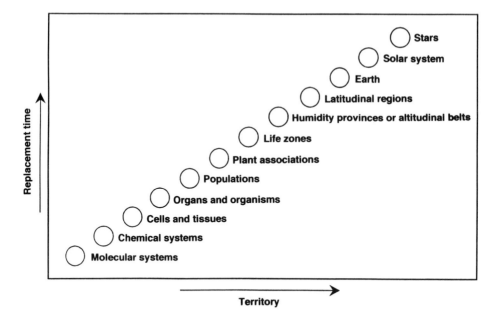

Fig. 18.1. A hierarchy of biotic organization in scales of space and time. The size of the circles and their location along the time and space scales are arbitrary for illustrative purposes. Ecosystem classification, study, and management should be organized around these biotic hierarchies in order to target ecosystem management actions to the most realistic units of function. Care is needed when using hierarchies that are not based on biotic units of organization (from Lugo, 1988).

eons. Biodiversity is the hierarchy of living systems that together constitute and support all life on our planet.

Scales of space (measured in units of area) are not the same as scales of biotic organization. Biotic scales represent the hierarchy of life that ranges from cells, to organisms, to populations, to communities and to life zones (Holdridge, 1967). Spatial scales refer to geographic area whereas biotic scales refer to biotic complexity. Biotic complexity and spatial scale are not equivalent. For example, size of organisms affects the amount of space that they occupy. But only sometimes does the spatial scale coincide with the biotic scale, i.e. the Earth is the largest spatial scale available to us and it is also the largest unit of biotic organization. Political units of space usually do not coincide with biotic units, although in islands they nearly coincide. In this chapter I differentiate between spatial and biotic scales.

The term 'ecosystem' is not a spatial term. The term refers to the functional attributes of biotic systems above the population level of organization. Ecosystem functions occur at all spatial scales. Ecosystems occur in a drop of water, a gram of soil, or the Earth as a whole. The spatial boundaries of ecosystems are arbitrary. However, classification and mapping of ecosystems is possible if the system of classification is hierarchical, and sensitive to ecosystem forcing functions and ecological constraints.

Global Scales and the Monitoring of Biodiversity

Until recently, it was impossible for scientists to approach, in any practical way, the complexity implied in the concept of biodiversity. Biodiversity catalogues focused on individual species, a job that is perhaps 20% complete for all life forms on Earth after several hundred years of scientific activity. A fundamental issue being addressed in this International Forum, is the strategy to follow in the cataloguing of the world's biota. A committee of scientists recommended that between 1980 and 1985 the United States Government should double, in constant dollars, the funds dedicated to biological inventories in the tropics and increase by 50% the number of professional systematists engaged in studies of tropical organisms (National Research Council, 1980). They also recommended studies at the ecosystem level as a second priority. Since the publication of these recommendations, scientific activity in the tropics has increased considerably.

No one questions the need for a greater taxonomic effort at the species level in the tropics. However, such an effort cannot occur at the expense of other sectors of the biodiversity hierarchy, particularly those at scales beyond organisms and populations. Inventories of ecosystems and monitoring of their areas and biotic states are just as critical to society as monitoring of the individual species within these systems. I argue that the priority for scientific resources is for monitoring biodiversity at national, international, and global scales as opposed to monitoring only the species at local scales. My proposal includes all biodiversity scales and does not play one scale against the other. There are four reasons for his proposal.

- Human activity now affects biodiversity at all scales. The effect is most dramatic at the larger geographic scales where it has long-term consequences for the functioning of individuals and ecosystems at all spatial scales. Global change not only changes the atmosphere and scars the Earth's surface, but it also affects species and organisms at local geographic scales. Paradoxically, solutions to many species problems at local geographic scales occur at the larger scales of biotic organization and geographic distribution. Yet, our knowledge at these large scales as well as long-term aspects of biodiversity are inadequate.
- Ecosystem-level and long-term study of the biota are recent scientific endeavours that occur at a time when human activity transcends local geographic scales. We know that to arrive at solutions to ecosystem management problems we must address the same, a smaller, and a larger level of biotic organization as that of the problems being addressed. We also know that it is dangerous to extrapolate solutions from lower to larger scales of biotic organization. As systems become larger and more complex, new properties emerge and these require consideration. At a time of decreasing resources and increasing urgency to solve problems, science must focus its attention on areas where discovery will be most effective in solving problems.
- New technologies are available to allow scientists to address questions that

a few decades ago were impossible to address effectively. For example, the availability of satellite images of low and high resolution, of fast computers, and Geographic Information Systems (GIS) greatly facilitate the analysis and monitoring of biodiversity at biotic scales beyond the species. These technologies are tools that allow us to quickly assess effects of human activity. They give us time to adjust and modify management strategies and to put our activities in compliance with guidelines that lead us towards sustainable relationships with global biodiversity.
- The focus on large scales of biodiversity analysis and monitoring does not preclude the inference of the status of individual species on the landscape. Modelling based on physical requirements such as light, temperature, water, and nutrients can be used to assess individual species composition on a global basis and provide the linkage between taxonomists working at small geographic scales and other scientists working at larger scales. These linkages also require interpretation of ecological information to enrich model output. This approach assures that the focus on large geographic scales and more complex levels of biotic organization will not exclude smaller scale research. The analysis must be hierarchical and coordinated across geographic and biotic scales.

To be effective, however, we need unprecedented levels of international cooperation and rapid access to information. Such international cooperation increases and concentrates the resources available to solve problems while avoiding duplication of effort, incompatible approaches, and neglect of important sectors of the biosphere. Fortunately, technological improvements in communications and transportation facilitate international cooperation and mitigate the disadvantages of distance and isolation. Moreover, the development of technologies and methodologies for large-scale monitoring of biodiversity is reaching such a level of maturity and standardization that the time is ripe for creating global networks of governments, national and international non-governmental organizations, and academia to accelerate collective understanding of global biodiversity.

Global Monitoring of Biodiversity

There is sufficient knowledge and technology to attempt a 'wall-to-wall' ecosystem map of the world. Such a map should be the product of international cooperation, using GIS technology and consistent protocols and definitions. These protocols and definitions should be of such nature as to be suitable for satellite monitoring and ground checking. To accomplish this, an ecosystem classification system is required. The selection of the ecosystem classification system will be a challenge. Available systems of classification may not be suitable because they were developed for other purposes before the availability of new technology that facilitates the handling of large databases. They may not be suitable for evaluating, monitoring, and managing biodiversity at all spatial and temporal scales. Worse yet, they may not be flexible enough to avoid conceptual pitfalls (such as confusing spatial and biotic scales of organization) or to take full advantage of available technology (such as being consistent with satellite images suitable for monitoring from space).

An ecosystem classification system for the evaluation, monitoring, and the eventual management of the global biodiversity must be (Lugo, 1994):

- based completely on quantitative data;
- as objective as possible;
- reflective as closely as possible of the forcing functions of ecosystems;
- hierarchical;
- convenient for expanding or contracting spatial scales;
- useful for anticipating global climate change;
- applicable to the whole world;
- demonstrably ecologically sound;
- consistent with the principles of climatic classification and vegetation function; and
- amenable to accepting new databases as a means of sharpening the analysis.

After the development of such a classification system and 'wall-to-wall' mapping of the systems, it will be possible to conduct long-term monitoring of the status of all ecosystems, including forests, wetlands, estuaries, etc. Brown *et al.* (1993) outlined the logistics of implementing such an approach to tropical ecosystem monitoring and its applications to forest biomass, fragmentation, hydrologic cycle, and biodiversity.

Coupling the Monitoring of Species Richness with Global Scale Assessments of Biodiversity

The cataloguing of species and monitoring of species richness can benefit from the assessment of global resources from space and vice versa if these activities are combined. The following is a four phase proposal from Brown *et al.* (1993) for achieving this coupling in tropical forests.

First, is the continuing effort to synthesize quantitative patterns of species richness in tropical forests. Such a synthesis would take advantage of current ecological understanding of how species numbers change in time and space in different ecological zones. This analysis should build on the few generalizations accepted by most ecologists who deal with the problem of explaining the patterns of species richness (Table 18.1). Of particular value in this initial phase would be the degree to which the variation in species numbers is explained by climatic factors, ecological zones, area considerations, ecosystem maturity, and ecosystem condition (disturbed, degraded, fragmented, etc.). Correlations such as these would allow extrapolation of data to larger spatial scales.

A second phase, also in progress in the scientific community, would be

Table 18.1. Some generally accepted statements about patterns of species diversity based on scientific observations.

- The diversity of most groups of plants and animals increases toward the tropical latitudes, and within tropical latitudes species diversity increases towards the equator.
- The number of species (plants and animals together) per unit area of land is usually greater in the tropical rain forests and lower in high latitudinal ecosystems.
- Tropical forests have a higher diversity than high-latitudinal forests in all scales of measure: within habitat, between habitats, and landscapes.
- Within a given altitudinal belt in the tropics, the number of tree species increases with increasing rainfall.
- Across altitudinal belts, the number of tree species and bird species decreases with increasing altitude.
- In islands, the number of animal species increases with the size of the island and with proximity to a continent. If the topography of the island is rugged, the number of species increases.
- Tropical mountainous areas (particularly high mountains) support a larger number of species (plant and animal) than equivalent temperate mountains because environments range from lowland tropical at the base to alpine conditions at the summit.
- Ecosystems with severe limiting factors such as high salinity, flooding, or freezing temperatures have reduced species richness.
- As rain forests recover from disturbance by acute events such as hurricanes or fire, species diversity increases in the disturbed area until a peak is reached after several decades.
- High-yield systems under intensive human management are species-poor, and increase in species richness to values observed in less disturbed and more mature systems when management pressure for high yield is reduced.
- The landscape is not uniform in terms of its species diversity. Often pockets with a large number of species can be interspersed within areas of low species diversity.

the testing of suggested correlations of species richness with specific field situations. This requires a network of plots where species counts of any plant or animal group can take place. To be useful for national inventories of biodiversity, these plots should be established for all major land uses and forest types as opposed to only monitoring mature forest stands. A broader inventory strategy provides better information on how different successional states and changes in forest cover contribute to regional biodiversity.

A third phase involves the incorporation of biodiversity parameters into routine field inventories. Forest inventory data are useful for making inferences regarding species richness of trees in tropical forested landscapes. Many inventories (both past and present) made a substantial effort to identify trees to the level possible (e.g. to the family or genus level if not to the species level), and to include all trees above some minimum diameter limit. These inventories have tended to be ignored by the scientific community based on the belief that they have not been conducted by professional botanists, which is not always true. Although such data are typically summarized in final reports, sometimes interim reports and/or raw data are available which can be used to draw some inferences about the richness and evenness in the distribution of tree species. Research in this area is ongoing, but there is a need for greater effort in future assessments.

Table 18.2. Databases required, in GIS format, to meet the forest classification needs at national to global scales.

Data needs	Scale		
	Global	Regional	National
Environmental factors			
Annual rainfall/seasonality			
Temperature			
Growing season length			
Elevation			
Slope classes			
Soils (texture and quality)			
Potential vegetation			
Ecofloristic zones			
Human factors			
Actual vegetation cover			
General forest types			
Forest use (protection, timber, minor forest products, etc.)			
Fragmentation			
Demography			
Resolution	Low		High
Aggregation capability	High		Low

At the ground level, slight modification in traditional inventory methods would yield large returns in our ability to assess tree species richness. Experienced local botanists should be included in the inventory team, to train field crews in proper species identification, as well as to identify rare or unknown trees from samples returned by field crews. All tree species should be included in the inventory. Other vascular and non-vascular plants need to be inventoried using smaller subplots. Raw data should be made accessible to appropriate scientists and other users. Complete tree counts and identification must be part of the normal tropical forest timber inventories.

Equally important during field work is the measurement of environmental factors at the same scale as the measurement of biotic ones. Traditionally, field inventory work has not included environmental measurements, and this greatly limits the ability of scientists to interpret biological data. Table 18.2 contains suggestions of the kinds of variables that should be added to inventory work. Ultimately, it would be desirable to develop models of expected plant species richness given environmental and human factors as well as geography. This would allow for the use of GIS databases to refine sampling for monitoring tree species richness by increasing sampling in areas of high diversity and high expected change.

The fourth phase of the proposal is the development and application of modelling techniques for the estimation of biodiversity. Fairly accurate predictions of species richness can be attained if the databases from the previous three phases are diverse in terms of types of forests and land uses, groups of organisms covered, and environments described. When properly calibrated, such models should become the predominant tool for extrapolating results from small-scale work to larger scales and to predict potential biodiversity changes associated with human use of the landscape.

Acknowledgements

I thank Sandra Brown and Andrew Gillespie for their contributions to this chapter. I also thank F.N. Scatena, J. Francis, and W. Edwards for reviewing the manuscript. The work is in cooperation with the University of Puerto Rico.

References

Brown, S., Lugo, A.E. and Gillespie, A.J.R. (1993) Strategies for environmental monitoring and assessment of tropical forest ecosystems. In: Report of the UNEP/FAO expert consultation on environmental parameters in future global forest assessments. *GEMS Report Series no 17*, pp. 115–140.

Holdridge, L.R. (1967) *Life Zone Ecology*. Tropical Science Center, San José, Costa Rica.

Lugo, A.E. (1988) Diversity of tropical species: questions that elude answers. *Biology International*. Special Issue 19, 37 pp.

Lugo, A.E. (1994) *Ecosystem Management in the USDA Forest Service*. Manuscript available through the International Institute of Tropical Forestry, USDA Forest Service, Rio Piedras, Puerto Rico. 65 pp.

National Research Council. (1980) *Research Priorities in Tropical Biology*. National Academy of Sciences, Washington, DC.

Some Problems of Inventorying and Monitoring Biodiversity in Russia

V.E. Sokolov and B.R. Striganova
Institute of Animal Evolutionary Morphology and Ecology, Russian Academy of Sciences, 33, Leninsky Prospect, Moscow 117071, Russia

Summary

The national strategy of biodiversity conservation in Russia is based on a regional approach and a combination of different tactics regarding sustainable use and restoration of biodiversity in degraded territories. Some problems particular to Russia concerning biodiversity are discussed considering both intra- and interpopulation, species, and community levels. We underline the need for specialized studies of recent speciation and extinction rates in different groups and the separation of natural and anthropogenic factors determining these processes.

Introduction

The territory of Russia occupying the northern part of Asia and Northeastern Europe is characterized by a complex climate, varying geomorphic forms, extensive lakes, a large river network, protruding coastlines on the Arctic Ocean and seas of the Pacific basin.

Complicated ecological conditions produce a high diversity of natural landscapes and ecosystems, some of which are unique on Earth (Lake Baikal, forest ecosystems of the maritime territories of Far-East Russia retaining tertiary features, forest-steppe in European Russia etc.).

Russia's extensive territory is characterized by the combination of regions with high density of industrial infrastructures, high population density and high level of the technogenic stress on natural ecosystems (Moscow, St Petersburg, Toljatti-Samara regions, Ural etc.), regions of intensive agricultural land use (Stavropol, Krasnodar regions) with extensive cultivated territories, and less-developed regions (tundra, taiga) with low population density and undisturbed natural communities.

Importance of Russia's Biodiversity within a Global Context

Water and forest ecosystems of Russia represent valuable resources of biosphere functioning, maintaining water balance and atmosphere gas regime. The territory covered by forests is approximately 7,500,000 km². Forests of Russia average 23% of the world forest resources. The carbon storage in the living forest phytomass reaches 41.10^9 tons.

Bog areas in Russia contain even more carbon storage, varying between 115 and 160.10^9 tons. Yearly fixation of atmospheric carbon dioxide by bogs compensates 10–12% of its yearly global output from burned fuel.

The total area of fresh water bodies (lakes and artificial reservoirs) in our country reaches 270,000 km² and the length of the river net averages 460,000 linear km.

The mountain systems in Southern and Eastern regions of the country are characterized by a high level of species diversity and endemism of angiosperm plants and animals. A number of recent centres of active speciation were recorded in mountain landscapes (North-East Caucasus, Altay, and mountains of the Far East), their impact on global taxonomic diversity being of importance.

For example, the fauna of myriapods of the order Diplopoda on the Russian Plain includes 70 species/subspecies, belonging to 32 genera (Lokshina and Golovatch, 1979). The diplopod fauna of the Great Caucasus exceeds 200 species, only three species being found in both these regions. The similarity coefficient (after Jaccard) of diplopod faunas between these regions does not exceed 2.5% (Golovatch, 1986). Mountain flora and fauna communities are known as refugia and centres of dispersion, maintaining the biodiversity of plain territories.

The Far East of Russia represents a strongly pronounced centre of the so-called 'neo-endemism' of some groups of land molluscs, namely the Bradybaenidae family. This family is represented in the Far East by the single genus *Bradybaena* comprising 16 species. The variations of shells and some anatomic features in representatives of the same species were found to be so significant, that in other groups they might correspond to the generic rank (A.A. Shileyko, Moscow, 1994, personal communication). For example, the size of the largest species of the family, *Bradybaena maacki*, varied from 20 to 36 mm in diameter and from 17 to 32 mm in height. A similar situation was recorded in *B. middendorffi*, *B. ussuriensis* etc. The areas of the species mentioned were restricted by separate island territories. Additional observation showed that these species manifested continuous variability series. This suggests a high level of plasticity connected with the diversity of their microhabitats. The island species, *B. weyrichii* (Sakhalin, Kuril Islands) form separate subspecies on different islands (Shileyko, 1978).

Such recent centres of speciation representing an area of active micro-evolutionary processes are of special importance as sources of both the present and future biodiversity on Earth.

Because of this situation, the national biodiversity conservation strategy in Russia requires a regional approach and the combination of different tactics for the sustainable use of biodiversity or its restoration in devastated territories. This chapter considers some problems of biodiversity inventorying and monitoring specific to Russia.

Inventorying Problems: Multilevel Approach

In 1994, the National Programme 'Biodiversity of Russia' was launched. Studying different levels of biodiversity represents a special section of the National Research Programme. One of the aims of this programme is to inventory taxonomic diversity of all groups of organisms inhabiting the territory of our country. This task has to be solved on both subspecies, species, and superspecies levels.

Inventorying of species diversity

An unequal degree of studies concerning separate groups of organisms, namely protozoans, some groups of invertebrates (mainly insects), micromycetes, etc., has required the intensification of flora and fauna studies, especially in remote and poorly accessible regions of Russia. Expert evaluation of both known and proposed sizes of some groups shows that species diversity is really much higher than currently thought. Evaluations were given on the basis of numbers of new species described in the current literature. For example, the number of known Coleoptera and Lepidoptera species in Russia and adjacent territories average 90–92% from the supposed ones, but that of Hymenoptera and Diptera only 62 and 47%, respectively (I.S. Kerzhner, St Petersburg, 1993, personal communication). Discovery of new taxa takes place at present even among large forms of mammals, whose taxonomy seems to be sufficiently studied. For example, in 1992 a new species and new genus of the antelope *Pseudooryx*, from Vietnam, was described.

The study of taxonomic diversity of some groups and the reconsideration of the general system of living organisms created by traditional methods of phenosystematics are developed on the basis of the wide application of genosystematic methods.

Genosystematics, developed in Russia by the scientific school of Academician A.N. Belozersky, has been introduced into taxonomic studies of different groups of organisms and helps to solve different problems of sibling and polytypic species determination.

Inventorying of subspecies diversity

Perennial complex research in the Far-East region has completed genetic diversity characteristics of a number of spawning populations of economically valuable fish species – *Oncorhynchus keta*, *O. nerka* and *O. gorbuscha* from the salmonid group. Using methods of enzyme electrophoresis for marking corresponding structural genes, comparative studies involved about 30 populations of *O. keta*, 20 of *O. gorbuscha* and seven of *O. nerka*. Populations inhabiting the Sakhalin-Kuril region, coasts of the Okhotsk Sea, the Kamchatka Peninsula, and river basins of the Amur and Anadyr received special detailed study. The methods used allowed quantitative evaluation of intra- and interpopulation diversity and their dynamics in time and space to be traced in areas of active fishery.

Interpopulation level
Gene pool variations were found in some populations of *O. keta*. If the enzyme characterictics are presented using circle graphics, and radii express frequencies of separate hereditary protein types, then each population is characterized by its specific 'genetic profile'. Using cluster analysis it is possible to classify the genetic similarity of populations under study. Differences in populations obtained are related to geographic distances. Genetic features allow individuals of separate populations in mixed aggregations to be identified, which is of importance for the regulation of marine fishery in order to maintain natural diversity.

Intrapopulation level
The Far-East population of *O. nerka* studied in the Kamachatka river was characterized by the presence of two forms of males. The first is big, migrates from the sea to the river for spawning, and reaches maturation at 5–7 years. The other is dwarf (small-sized), lives in a river for their entire life-cycle, matures after 3 years and ceases growth after maturation. These males are characterized by a lower spawning capacity.

Both these forms differ in their heterozygous level. Big-sized males are preferred by females during spawning and gain an advantage, but in dry years they cannot come up-stream, in such cases dwarf males maintain the population recruitment. Genetic studies revealed differences of protein loci between both big and dwarf males of *O. nerka* which are constantly present in spawning populations.

River and marine harvesting of *O. nerka* was traditionally carried out at the beginning of spawning migration of fishes, big males being caught in the first turn. Therefore, smaller males artificially received a reproductive advantage, resulting in the decrease of both size characteristics of a population and its total productivity. Harvesting tactics destroyed the evolutionary established mechanisms which maintained intrapopulation diversity.

These examples show that the study of gene pool diversity of resource species and its monitoring have to be developed at population and subpopulation levels, taking into consideration adaptive tactics used by separate populations in different types of natural habitats. Biodiversity inventorying at species and subspecies levels requires intensification of taxonomic studies of all biotic groups using both traditional morphological methods and genosystematic analysis as well.

At the same time, problems of the general system of large taxa of organisms belonging to different kingdoms are also of importance for aims of inventorying current biodiversity. Some groups require detailed analysis. For example, approximately 4000 species are described in the protozoan system (Vickerman, 1992). However, genetic distances between different species within separate traditional groups (for example in Amoebae) are greater than those between plants and vertebrates (Schlegel, 1991).

Biodiversity inventorying of biotic communities and ecosystems

Using the ecosystem level for biodiversity inventorying seems to be of particular importance for evaluating biosphere functions of biodiversity and predicting its dynamics in time and space. The territory of Russia represents a favourable area for analysing regularities of biodiversity change in biotic communities along three main ecological gradients: latitudinal, climatic (from west to east), and altitudinal.

The National Programme 'Biodiversity of Russia' proposes to give priority to developing ecosystem and community studies, considering them as fundamental elements of inventorying. These studies in Russia will be based on the existing network of state reserves and other types of protected territories, providing a multitude of undisturbed natural terrestrial ecosystems in different soil–climatic zones.

The principal methodological approach in these ecosystem studies is to analyse structural-functional patterns. Assessment of integral indices of ecosystem functioning (e.g. carbon dioxide flow, respiration rates, primary production, relations between accumulation and mineralization of organic matter, relation between living biomass and storage of dead organic remains, etc.) has to be combined with a consideration of structural and dynamic patterns of microbial, plant and animal communities. Taxonomic, hierarchy and trophic structures, differentiation in the degree of ecological niches, competition and coexistence phenomena etc. have to be taken into account. Analysis of taxonomic composition should include species and higher taxonomic levels and consideration of endemic, rare, endangered and other groups of species needing special conservation efforts.

The necessity to develop a differential approach has to be underlined: indices of total taxonomic richness of separate communities, abundance of separate groups of organisms, assessment of the matter and energy allocation between components of an ecosystem are not sufficient without analysis of species composition.

For example, comparative soil–zoological studies were carried out in forest ecosystems in the southern region of Far-East Russian, where areas of tertiary mixed forests, including some endemic tree species, have been preserved in State Reserves. The soil macrofauna was investigated in plots of primary cedar–broad-leaved and broad-leaved (oak–birch) stands in comparison with secondary oak forests. The total abundance of soil animals in primary forests was three times higher than in the oak-stand because of low population densities of mesophyllous groups (Oligochaeta, Myriapoda, Scarabaeidae beetle larvae etc.). But the most essential differences were observed in the taxonomic composition of the soil fauna: the predominant oligochaete in primary forests was the Far-eastern endemic species *Dravida ghilarovi* (from the East-Asiatic family Moniligastridae). This species was absent in the secondary forests being replaced there by the widely distributed *Eisenia nordenskioldi* (Lumbricidae family). Only one soil species of Hirudinea, *Orobdella whitmani*, was found under cedar stands. The endemic species of Diplopoda, *Levizonus thaumasius*, is characteristic of this region. These diplopods were numerous in primary forests whereas in secondary oak-stands only single individuals were found. The total taxonomic richness reached 22 and 21 species in primary forests and only 14 in the secondary one (Striganova, 1980).

In recreation suburban forests in Moscow we observed an increase in phytomass values of grass cover on one of the last stages of the recreation digression. This increase was observed during the destruction of a forest stand, when separate trees were intermittent with open plots. Analysis of grass and herb composition showed that typical forest forms have been replaced by species characteristic of dry grass communities and ruderal plots. This stage preceded the final destruction of plant communities (Karpachevsky, 1981).

These examples demonstrate that the consideration of integral characteristics of communities and ecosystems under study seems to be insufficient to assess their real biodiversity status. The complex approach to biodiversity inventorying determines the necessity of both taxonomic and biogeographical studies and analyses of adaptive strategies, life-cycles and population dynamics of key-species. This allows biotic mechanisms to be revealed, which maintain homeostatic multispecies communities necessary for inventorying biodiversity. The combination of taxonomic, areal, field, and experimental ecological investigations requires the elaboration of a system of methods, which is proposed to be carried out in the framework of the National Biodiversity Programme (Sokolov *et al.*, 1994).

Institutions Involved in the Biodiversity Inventorying and Monitoring

Methods of biodiversity assessment at all levels of life organization are being developed in various institutes of the Russian Academy of Sciences: the Zoological and Botanical Institutes (St Petersburg) – 'alma mater' of the native zoology and botany; the Institute of Evolutionary Animal Morphology and Ecology (Moscow), which actually represents the head institution in the field of fundamental ecology in Russia; the Institute of General Genetics (Moscow) which is developing problems of fundamental and applied genetics; the Institute of Paleontology (Moscow), developing problems of historical dynamics of the biodiversity. These are the principal institutes of the Department of General Biology of the Russian Academy of Science. Inventorying and monitoring biodiversity on the regional level are carried out by institutes of the Academy's regional branches – the Karelian Institute of Biology (Petrozavodsk), the Marine Biological Institute (Murmansk), the Institutes of Biology (Ufa, Kazan, Novosibirsk), the Institute of Plant and Animal Ecology (Ekaterinbourg), the Institute of Forest Research (Krasnojarsk), the Soil-Biological Institute and the Institute of Marine Biology (Vladivostok), the Institute of Biological Problems of the North (Magadan). This is a non-exhaustive list of the scientific centres studying problems of biodiversity in Russia. Specialists working in these institutes, together with university professionals in biology departments of Russian Universities, represent the scientific community capable of tracing the current state and dynamics of biodiversity in all regions of Russia.

Biodiversity Monitoring

Biodiversity monitoring has to be developed in different directions to solve problems such as: (i) dynamics of biodiversity of natural ecosystems; (ii) dynamics of the structure and functioning of biotic communities; (iii) dynamics of biodiversity of separate taxonomic groups of organisms. The IUBS-SCOPE-UNESCO proposals for an international network (di Castri *et al.*, 1992) underlined the necessity for combining 'extensive' and 'intensive' monitoring methods. Intensive monitoring proposes controlling taxonomic diversity of focal groups in selected sites. Extensive monitoring deals with all types of biomes and controlling the diversity of the main structural–functional components of ecosystems.

Extensive monitoring on the ecosystem level will be carried out in Russia on the basis of the biosphere reserve system. We now have 17 biosphere reserves in different regions and in nearly all natural zones of our country.

On a regional scale, diversity of undisturbed territories and human-made infrastructures, frequency of patch types, gradients, etc. will be monitored. The landscape approach requires use of distant methods of the monitoring and measuring of integral indices of the functional structure of ecosystems considering the balance of production–destruction rates.

Intensive monitoring will be carried out in special experimental stations considering dynamics of both the taxonomic composition of key groups of organisms and quantitative structure parameters. For example, in the Moscow region at the experimental biogeocenotic station 'Malinki', observations of soil dynamics, forest stands, plant communities, macro- and micromycetes and soil invertebrate animals were carried out for 30 years. This allows us to describe recent trends of changes of different ecosystem components.

We cannot agree with the position in the proposals that intensive monitoring can be restricted by qualitative control only. Early features of unfavourable changes of populations of key-species can be traced namely on the quantitative level: restriction of topic preference, occurrence, decrease of the population densities etc.

Monitoring of taxonomic diversity of separate groups of organisms has to consider both current and historical dynamics of speciation and extinction. There are different underlying reasons for a decrease in population density and disappearance of separate species.

One of them is natural extinction, which is the part of the evolutionary process. Paleontological data show that in the course of evolution, mass extinction of different groups of animals was repeatedly recorded. The last took place in the late Pleistocene and embraced many species of large-sized mammals. Reasons for this extinction are still being discussed. Climatic changes and human activity are considered to be of importance. In addition, competition (e.g. *Castoroides ohioensis* and *Castor canadensis*; *Arctodus simus* and *Ursus arctos*, etc.) or an excessive specialization which prevented the development of fitness to new environmental conditions (mammoths).

Human activity seemed to play the essential role in the extinction of some mammal species during the Holocene. The total annihilation of some forms has taken place, for example, *Hydrodamalis gigas* ('marine cow'). This representative of the Syrenia order was discovered in the 1740s near the Komandor Islands. These big, hardly mobile valuable animals were exterminated by trade-trappers over the course of the following 27 years. The extinction of large-sized *Bos primigenius* was also related to human activity, coinciding in time with large-scale forest clearing in Central Europe (Sokolov, 1986).

During the last 2000 years, 106 species/subspecies of mammals have disappeared (Harper, 1945). Thirty-three forms went extinct between the beginning of this period and 1800, 33 forms between 1800 and 1900 and 40 between 1901 and 1944. A number of risk factors were present as far as nature conservation was concerned: habitat degradation, overharvest, influence of introduced species, decline of available resources, annihilation of species with aims to protect plant, livestock and human populations.

Analysis of natural and anthropogenic factors involved in recent extinction processes has to consider natural extinction rates related to the evolutionary history of separate groups. An overview of paleontological data concerning the geological dynamics of mammal genera suggests that replacement rates of species and genera, e.g. an extinction–speciation balance, differed significantly in separate groups. Table 19.1 shows calculations of the beta-diversity (after Whittaker) of genera composition of some orders of mammals from the moment of their origin up to the end of the Pleistocene. Paleontological data summarized by N. Kalandadze and A. Rautian (in Sokolov, 1986) were used (see Table 19.1).

Beta-diversity indices of the different time periods varied widely from 1.6 to 5.8. The orders with maximum indices exceeding 5.0 can be outlined. It includes primates, Marsupialia and Insectivora which are characterized by especially high rates of both speciation and extinction processes. It resulted in a cardinal replacement of the generic composition in these groups in the course of their historical development and promoted their evolutionary progress.

Current shifts in population dynamics of key-species and community structures found in different natural landscapes are sometimes hardly inter-

Table 19.1. Biodiversity dynamics of genera in some orders of mammals in the geological time-scale.

Order	Geological periods	Total no. of fossil genera	No. of current genera	Beta-diversity (after Whittaker)
Triconodonta	Triassic-late Cretaceous	19	-	5.1
Symmetrodonta	Triassic-late Cretaceous	12	-	4.4
Marsupialia	Early Cretaceous-present	251	85	5.8
Insectivora	Early Cretaceous-present	287	66	5.7
Primates	Late Cretaceous-present	247	66	5.8
Cetacea	Eocene-present	214	38	4.2
Tubulidentata	Paleocene-present	5		2.5
Lagomorpha	Paleocene-present	59	12	3.4
Rodentia	Paleocene-present	914	372	4.0
Artiodactyla	Eocene-present	630	88	3.2
Proboscidea	Eocene-present	38	2	2.4
Sirenia	Eocene-present	21	2	1.6

preted. Many of them are considered as consequences of the climatic trend or anthropogenic impacts. Natural cyclic phenomena and shifts caused by outside factors are sometimes hardly distinguishable, especially if we deal with perennial cycles. That is why, in addition to the well-known patterns of season dynamics, it is necessary to study and monitor the long-term dynamics of ecosystems determined by sun activity cycles (11 and 30 years) and climatic fluctuations over the century.

Over 250 years of observations of forest dynamics in central Russia have demonstrated significant changes of forest communities, whose cyclic nature can be recorded on the basis of observations during the life span of some human generations (Skrjabin, 1959).

These observations were carried out in the Voronezh region in extensive Usman forest massif (now called the State Voronezh Reserve). Pines from this forest massif were used as a source of ship timber. Therefore, both the delivery of timber and restoration of valuable pine stands were carefully monitored. The first recordings date back to the end of the 17th century (Savelov, 1697, cited by Skrjabin, 1959). The main tree species in this region used for building were *Pinus sylvestris* and *Quercus robur*. Areas occupied by these forest stands varied during the period of observations, following climatic fluctuations. The climatic trend showed the following consequences.

- Late 17th century: drought conditions, overdrying of bogs, predominance of oak areas, displacement of pine stands to sand hills.
- Early 18th century: increase of the climatic humidity, restoration of pine stands on felled plots, restriction of oak areas.
- Mid-18th century: new wave of humidity increase, broadening of pine stands.
- Late 18th-early 19th century: high precipitation, increase of the temperature

amplitude with very cold winters (features of the continental climatic trend). A broad distribution and active restoration of pine stands.
- Mid-19th century: decrease in precipitation, overdrying of bogs, expansion of oak.
- 1870–1880s: shift to climatic humidity, active restoration of pine on felled plots.
- Late 19th–early 20th century: decrease in precipitation.
- 1910–1940s: decrease in winter temperatures, expansion of pine during comparatively humid periods and the competitive success of oak in dry periods.

Determining limits of cyclic perennial fluctuations of natural ecosystems in different time scales allows natural and anthropogenic changes of ecosystem biodiversity to be distinguished and both the dynamics of biodiversity and thresholds of the stable equilibrium of ecosystems related to biodiversity to be predicted.

Conclusion

The aims of inventorying, assessment of current biodiversity and sustainable use of biological resources all require the elaboration of special scientific strategies to carry out this programme. The combination of intensive and extensive approaches and a multilevel conception of biodiversity on genetic, taxonomic, and synecological levels, must involve many specialists. Biodiversity represents an interdisciplinary problem at the junction of different scientific branches.

Another feature determining the complexity of biodiversity monitoring is the assessment of biodiversity changes in time and space using multiscale approaches. Spatial scales vary from intrahabitat structures to the global level. Time scales vary from seasonal fluctuations to geological life history.

All of these problems can be solved through collaboration between different national programmes and broad international cooperation in specific scientific fields. Ratification of the Convention on Biodiversity and publication of international documents have to be accompanied by the initiation of private international scientific programmes on the study, monitoring and conservation of separate components of global biodiversity involving specialists from different countries.

References

di Castri, F., Vernhes, J.R. and Younès, T. (1992) Inventorying and monitoring biodiversity. *Biology International*, Special Issue 27, 1–28.

Golovatch, S.I. (1986) The distribution and faunogenesis of millipede of the USSR European Part. In: Chernov, I.Ju. (ed.) *Phylogenesis and Phylocenogenesis*, Nauka Publ., Moscow, pp. 92–138. (in Russian)

Harper, F. (1945) *Extinct and Vanishing Mammals of the Old World.* Special publication, 12. American Committee for Intern. Wild Life Protection. New York Zoological Park, NY.

Karpachevsky, L.O. (1981) *Forest and Forest Soils.* Forest Industry Publ., Moscow.

Lokshina, I.E. and Golovatch, S.I. (1979) Diplopoda of the USSR fauna. *Pedobiologia* 19, 381-389.

McArthur, R.H. and McArthur, J.W. (1961) On bird species diversity. *Ecology* 42, 594-598.

Schlegel, M. (1991) Protist evolution and phylogeny as described from small subunit ribosomal RNA sequence comparisons. *European Journal of Protistology* 27, 207-219.

Shileyko, A.A. (1978) Terrestrial molluscs of the super-family Selicoidea. *USSR Fauna* New series, 117. Mollusca. Vol. 3, 6. Nauka Publ., Leningrad.

Simpson, E.H. (1949) Measurement of diversity. *Nature* 163, 688.

Skrjabin, M.P. (1959) Surveys of the history of Usman Forest Massif. *Proceedings of Voronezh State Reserve* 8, 3-118.

Sokolov, V.E. (1986) *Rare and Vanishing Animals. Mammals.* High School Publications, Moscow.

Sokolov, V.E., Striganova, B.R., Reshetnikov, Ju.S., Chernov, Ju.I. and Shatunovsky, M.I. (1994) Russian Biodiversity Conservation Programme. *Biology International* 28, 29-32.

Striganova, B.R. (1980) Veraenderung der trophischen Struktur der Tiergemeinschaften in Waldboden bei der Storung der Pflanzendecke. In: Schubert, R. and Schuh, J. (eds) *Bioindikation auf der Ebene der Population un Biogeozonosen* 5, Halle/Saale, pp. 3-9.

Vickerman, K. (1992) The diversity and ecological significance of Protozoa. *Biodiversity and Conservation* 1, 334-341.

Whittaker, R.H. (1975) *Communities and Ecosystems.* Macmillan, New York.

Wilson, E.O. (1988) The current state of biological diversity. In: Wilson, E.O. (ed.) *Biodiversity.* National Academy Press, Washington, DC, pp. 3-18.

20 Inventory and Monitoring for What and for Whom?

P.B. Bridgewater
Australian Nature Conservation Agency, Canberra, Australia

Inventory and monitoring are both classic information-rich activities. Both have research elements which underlie them, but too often when these activities are discussed it is in terms of the research needed to carry them out, instead of the operational rules necessary to provide results. In other words, information is not an end product; the application of information is.

We ignore at our peril the warning of Prendergast and colleagues in *Nature* last September who state:

> where conservationists are under pressure to make rapid decisions on reserve placement, we suggest that a strategy based solely on the two most popular criteria, diversity (species richness) and rarity, and on only one or a limited number of taxa, may fail to provide adequate protection for many other organisms.

I will add a further warning – we forget also at our peril that ecology, which underpins these studies, is a four-dimensional subject, involving the coordinates of land, sea, air and time.

To decide on conservation priorities decision-makers need access to data and information about biophysical processes, ecosystems and communities, species populations, and genetic characteristics within species populations, for the region under consideration, be that a catchment, a state or a continent. Once the requisite data and information are available an analytical procedure is needed to identify key candidate areas which could be managed for their identified values. And then we need to plan a monitoring programme, with performance indicators to determine the level of our success.

It would take enormous resources and many years to complete anything resembling a comprehensive inventory of any nation's biodiversity at ecosystem/community, species, and genetic level. Accelerating rates of land use change, and increasing human use of marine resources, mean that measures to conserve biodiversity must be put in place now. We should therefore question the views of some systematists that we need to know what is there before we can manage it. Such sentiments are basically anticonservation in focus, and unhelpful to managers of biological resources. This is not to say we do not need inventory – of course we do. But it can equally well proceed alongside monitoring as being seen as necessary information we need before we act. Of course, the use of new information technologies, especially worldwide networking, is tailor-made to enhance inventory effort and use. But we must beware of the political imperative to release information widely in a conservative way, and we must take full note of the rising tide of concern for intellectual property.

An example of the application of inventory information is the Australian attempt to evaluate our protected area system, and set targets for conservation effort. Although we have uneven data for land and sea we have approached the problem in the following way.

- Recognize the need to include both the species and the land(sea)scape scale of biological diversity. Genetic diversity is probably best catered for using other methods.
- Utilize some taxa for which there are sufficient data (gathered under the Australian Biological Resources Study) to give an accurate and reliable picture of diversity and endemism to guide us to species-based 'hotspots'. The key taxa were *Eucalyptus* species, birds and butterflies.
- Utilize some land(sea)scape measures to give us an indication of the conservation status of landscapes.
- Using an index of threatening process to develop an index of threat to land(sea)scapes/key taxa.
- Combine the species and land(sea)scape levels to produce a filter of need for maximum focus on potential *in situ* site identification.

These solutions are not the best or only solutions. Rather, they represent a baseline of what can be done with existing data sets and relatively simple Boolean overlay and weighting of attributes. This is a mechanism for identifying areas of importance for biodiversity conservation, which are under threat. These indicative areas therefore provide a basis for further detailed conservation assessments, possibly as collaborative projects between nature conservation agencies of the different jurisdictions. The procedure presented here can be readily repeated should new data become available or should modifications be required to the weightings which were used to derive the maps. We are attempting to move this exercise into the marine environment.

In the discussion which follows, the definition of monitoring is taken as 'a system of continued observation, measurement and evaluation for defined purposes'. Ecologically sustainable development requires proper management. Proper management requires the understanding of patterns and processes in biotic systems and the development of assessment and evaluation procedures which assure healthy patterns and processes in biotic systems.

In an era where the activities of people are the dominant force influencing biological systems, assessment and evaluation procedures that assure protection of biological resources must include long-term monitoring. Long-term monitoring of the environment is essential to distinguish patterns of change from noise and to characterize the patterns. To formulate a plan to monitor changes to the environment, consideration must be given to what is meant by long-term. If the aim of ecologically sustainable development is accepted, a monitoring system which assesses sustainability of the environment must be in place for as long as we intend to pursue that aim.

Information is not an end product; the application of information is. Monitoring can support three key functions.

1. Environmental policy and management decisions by government, including definitions of programme objectives and priorities and selection of specific regulatory or enforcement actions.

2. Identification and definition of environmental problems which are not now recognized or which may emerge in the future.

3. Evaluation of impacts on environmental quality resulting from specific governmental policies, programmes or actions.

Long-term monitoring will require site tenure and sites should be selected on the basis of landscape representativeness if the gathered information is to be meaningful at a fine scale, but interpretable on a continental scale. Legally protected areas which form a national nature conservation system, based on landscape representativeness, can provide site tenure security. Here, in global terms, the importance of the Biosphere Reserves network, or the natural side of the World Heritage ledger cannot be ignored.

A minimum data set must be developed which will reveal at least two basic facts: whether there is change and whether the change or lack of change is due to human activity or other natural processes.

Any national long-term programme to monitor changes to the environment should include the following points.

1. Biological, chemical and physical factors must be integrated to comprise the data gathered for assessment and evaluation.

2. Terrestrial, marine, freshwater and atmospheric systems, and the boundaries between them, should be monitored.

3. For security of tenure and management purposes, long-term monitoring sites should be legally protected areas, and where possible, within a national (or international) nature conservation system based on landscape representativeness.

4. Those data gathered should be directly related to the formulation of the objectives and priorities of environmental policies and the evaluation of the effectiveness of policies.

5. Specific regulatory and/or enforcement actions should flow from the environmental management goals derived from the evaluation of those data gathered.

6. The assessment of those data gathered should be used not only to determine historical events, but to predict future or heretofore unrecognized problems.

7. A national long-term environmental monitoring programme is sufficiently important to warrant statutory protection. At the international level, the Convention on Biological Diversity should be providing that protection.

Finally, who will do all the monitoring we really need? We know we do not have enough scientists, so if we do not design monitoring regimes which emphasize local ownership, control and participation in this international effort, we are doomed to fail from the start. Again, the Biosphere Reserves network, enhanced and energized, offers a positive way forward.

21 | Biodiversity Conservation and Protected Areas in Tropical Countries

G. Halffter

Instituto de Ecologia, A.C., Apartado Postal 63, 91000 Xalapa, Veracruz, Mexico

Introduction

The most frequent response to the growing threats to the survival of biological diversity has been the creation of new protected areas. If these areas provide some measure of protection (which is not always true), then they are useful in themselves. However, it is important to reflect upon the question of whether or not the creation of protected areas, by itself, can assure the conservation of a substantial portion of present biodiversity under the conditions that prevail in tropical countries.

The politics and customs of temperate industrialized nations cannot be extended automatically to the tropical regions of the world. Approaches to conservation in tropical countries must be based on real world, practical methods that guarantee immediate and medium term protection. Those approaches that do not take into account rural poverty in 'developing' nations and that do not take into account disintegrating traditional socioeconomic structures and simultaneous migration and population increase are not feasible. Good intentions not withstanding, these factors have a deciding effect on the success of efforts to preserve biodiversity.

For good intentions (political or biological) to extend beyond mere rhetoric, actions must be based on real information gathered *in situ* on the ecological, economic, social, cultural (perception of nature), and political conditions of each region in question. In this context I will comment upon the place of rustic activities in discussions of biodiversity. I use 'rustic' to distinguish certain ways of exploiting natural resources that are different from those employed by 'intensive' exploitation. Rustic approaches include traditional techniques of extractive resource usage as well as many forms of farming, ranching, forestry, and fishing used now but imperiled by the imposition of 'intensive' exploitation (efficient or not), uncontrolled urbanization, extensive pollution and a host of activities lacking environmental considerations promoted by outside forces (including problems stemming from population increases).

As Alcorn (1991) has pointed out, modern society did not invent biodiversity conservation. We are moving from an 'archaic' approach associated with production to a modern one based upon protected areas. But are the same possibilities and modalities associated with this change in temperate countries achievable in the tropics? In the United States, with one of the most extensive and efficient systems of protected areas, modern analysts express doubt that protected areas alone can effectively conserve present animal and plant species richness and their evolutionary processes (see Hudson (1991) for a thought-provoking analysis).

Analysis of the possible ways that protected areas could achieve biodiversity conservation in the tropics can be based on two scenarios. In the first, protected areas are surrounded by areas of varying degrees of rustic usage located between the protected areas and zones of intensive exploitation; landscape ecology would govern the interface between protected areas and adjacent areas of rural usage. In the second, intensive usage is the norm and its negative effects are compensated for by the lack of any exploitation (except tourism) of the protected area. My preference for the first coincides with the conclusion that biodiversity conservation cannot be separated from consideration of natural resource utilization.

Biodiversity Conservation and the Needs of Local Populations

Rustic usage

Rustic usage is considered here because I believe that its preservation and improvement is a fundamental part of an intelligent forward-thinking political approach to biodiversity conservation. My interest in rustic exploitation of resources by no means stems from an opposition to either intensive exploitation or the complete elimination of protected areas. Where there is sufficient land and water, in addition to favourable environmental and economic conditions, intensive exploitation is both justified and necessary. Where it is possible to conserve biodiversity in totally protected areas for at least the intermediate term without creating social conflicts, such areas are absolutely desirable. But in the tropics, the majority of the landscape is neither under sustained protection nor intensive usage. It is in these areas that well managed rustic usage can substitute for preventable environmental degradation provoked by poorly planned development. As di Castri (1992) pointed out, the principal cause of environmental degradation has been the failure to unite development and environmental concerns in a single political approach. The environment, economics and society comprise a complex, integrated system in which there are different legitimate alternatives to solving a given problem. In the political context of biodiversity, we must accept the fact that different approaches exist to solve the problems, and that they include many acceptable methods and results. The current environmental enigma lies at the interface of three systems: the ecological, the economic, and the social, all of which must be considered to arrive at useful plans of action.

The most prevalent modern approach to conserving biodiversity is the establishment of as many protected areas as possible. In recent times, as a result of ideas advanced by the MAB Program of the United Nations Educational Scientific and Cultural Organization (UNESCO), the exclusion of local activities from protected areas has not been imposed as an indispensable requirement in designated biosphere reserves. Nevertheless, in general, rustic exploitation of natural resources is not contemplated.

It is interesting to consider to what degree the insistence on exclusion of traditional use of resources from protected tropical areas is the result of a true ecological analysis and whether or not social, economic and political problems have been considered. Or, more simply, whether such an exclusive position is a poorly conceived attempt to extrapolate experience from highly industrialized temperate countries to tropical situations. The next section discusses whether or not protected areas by themselves are sufficient to assure the conservation of tropical diversity. I believe that without rustic areas it is impossible to conserve a significant proportion of the current biological diversity.[1]

I consider rustic usage as diverse forms of utilization of biotic resources that are distinct from intensive usage. The separation is not always clear, and between the two there are many intermediates. Efficiency, *per se*, is not a distinguishing criterion since rustic and intensive usage can be efficient or inefficient. Rustic usage is associated with a heterogeneous approach to utilizing resources. Different plants are cultivated; farming is associated with livestock and use of forest resources. Agricultural chemicals are only rarely used, and then sparingly, as are farming and ranching machinery. The agricultural enterprise is labour intensive, even at the cost of economic efficiency. Family businesses, communal efforts and cooperatives dominate the economic structure. Outside financing is uncommon and crops are sold in local and regional markets (a few highly valued products may be exported). The primary strategy is based more upon long-term, stable usage than on maximizing production.

Intensive use seeks to maximize short-term production by strong reliance upon machinery, fuel, fertilizers, pesticides, and credit. Such exploitation favours large monocultures (not only of a single species, but also of single varieties) covering large, homogeneous areas. Crops are destined for national and global markets.

Toledo (1994) provides an in depth examination of the characteristics of rustic producers, whom he refers to as 'peasants', as opposed to intensive producers, whom he calls 'farmers'. In 1993, Toledo referred to 'premodern peasant production' in the same sense that I employ 'rustic use' here.

In general terms, 'traditional use' is equivalent to 'rustic use' as defined here.[2] There are, however, exceptions. Some 'traditional uses' (e.g. whaling) are indeed intensive. Moreover, some 'rustic usage' may employ modern means (e.g. limited use of machinery), but always with an emphasis upon stability, diversity and employment, not upon maximum production.

Rustic usage has not prevented the survival of a world rich in biotic diversity or, in Alcorn's (1991) terms 'archaic conservation'. Tropical forests and other tropical ecosystems are not devoid of human population. Millions of people inhabit tropical ecosystems (World Conservation Monitoring Centre, 1992, Chapter 27). Many recent publications document how traditional exploitation of natural resources has permitted the survival of biological diversity (Mexico: Toledo et al., 1985; Rojas, 1990; Toledo, 1990; Boegue and Barrera, 1993; Gomez Pompa et al., 1993; Leff and Carabias, 1993; Amazonas: Posey, 1983; American Tropics: Oldfield and Alcorn, 1991; general: Altieri and Hecht, 1990).

One kind of activity commonly considered as traditional usage is extraction. At first glance, such an association seems contradictory because the purpose of many extraction processes is to supply merchandise for external markets. Extraction activities demonstrate the complex relationship between socioeconomic structure and the sustainability of natural resources. As May (1992) points out, local populations have no control over the commercialization of their products, and they are the first to resent fluctuations in external demand. When demand is great, price increases do not benefit local populations, but it does result in pressure to produce more from their local resources.

A question that has so far escaped serious scrutiny is the ability of traditional production to absorb new technology without its affecting overall strategy. New technology can increase returns and increase the efficiency of rustic exploitation without changing its principal conceptual bases.

Conservation and local populations

A change from a nearly stable situation to one of rapid degradation of biodiversity can be a result of an increase in local populations. Nevertheless, a clear understanding of the realities of tropical countries must extend beyond the issue of population increase to include economic, political, and social elements, including those that stimulate immigration to those ecosystems we wish to protect.

The vision of the role and responsibilities of local populations is beginning to change. Clay (1991: 272) observes the following:

> The most severe threats to the world's fragile environments come from population pressure (usually from the dominant society and only rarely from indigenous people), legal restrictions on land rights in general and communal rights in particular, international debt, and the imposition of imported technologies or models of resource use that are inappropriate in fragile areas and create financial dependencies. Excluding populations from their traditional areas has disastrous consequences for the future of that environment, because those peoples are an integral part of the environmental dynamics.

According to McNeely (1988: xi), 'Biological resources are often under threat because the responsibility for their management has been removed from the people who live closest to them, and instead has been transferred to government agencies located in distant capitals'. These references represent a complete change in attitude towards the relationship between conservation and local populations.

The generalization can be made that in Latin America the demographic pressures threatening tropical forests do not come from indigenous populations but rather from recent immigrants, including those participating in official colonization programmes, where dire need forces them to overexploit their new surroundings. Immigration to forest regions is a demographic escape valve for hopeless social conflicts in other regions. The colonization of the Selva Lacandona in the Mexican state of Chiapas is a prime example of the threats posed by immigration to the tropical forest ecosystem.

The tropical world abounds with examples of the unsustainable use of biotic resources for immediate economic reasons. 'The fundamental constraint is that some people earn immediate benefits from exploiting biological resources without paying the full social and economic cost of resource depletion, instead, these costs (to be paid either now or in the future) are transferred to society as a whole' (McNeely, 1988: vii–iii).

The survival of 'archaic' conservation associated with rustic or traditional use of natural resources is affected by (i) outsiders who seek immediate and intensive exploitation and (ii) changes precipitated by contact with a consumer society that results in the abandonment of traditional forms of exploitation. Increased demand for consumer goods (often at inflated prices) increases pressure on the supply of natural resources, including land underexploitation.

Land ownership determines land usage and usage of natural resources. Traditional groups – and their proprietary systems (often collectives or communals) – are poorly equipped to resist colonizing groups as well as large, expanding land holdings. Indigenous groups under this kind of pressure are pushed into new forest areas, where they still are available.

In Mexico, the transformation of the countryside is reflected by the decrease in the agricultural component of the GNP from 16% in 1970, to 7% in 1990. During this period, the country changed from net exporter to net importer of principal basic food products, including corn and milk (Gordillo, 1993; Mestries, 1991).

It is possible that the most important factor affecting success in protected areas in tropical America is the active participation of local populations – participation as shareholders as well as workers. For active participation to occur, conservation cannot be seen as being imposed from the outside. Reserves that continue the traditional uses and culture of local populations are very helpful. This is a difficult challenge since traditional uses can be distorted by contact with a consumer society.

Large-scale cattle ranching

Of all the changes suffered by the Latin American tropics, the most brutal has resulted from large-scale cattle ranching operations.[3] Even though ranching does not yield high economic returns per unit of area utilized, it is an intensive activity that occasions profound changes in the landscape. Few forms of production are more strikingly different from traditional exploitation than large-scale ranching. The expansion of ranching activities has been analysed for various locations. In the Mexican state of Veracruz, which has supported a livestock industry since the 16th century, induced pasturage rose from 21.6% of surface area in 1940 to 50% in 1993. The expansion has come at the expense of forests and agricultural land once devoted to traditional uses (see Rodriguez and Boegue, 1992; Barrera and Rodriguez, 1993, for Amazonia see Hecht, 1992).

Large-scale ranching not only drastically reduces forests, leaving in its wake relict islands of the native habitat, but it also displaces the majority of the rural population from its land, its resources and its traditional modes of life. As several authors note, large-scale ranching exists to satisfy the demand for inexpensive meat in industrialized societies (principally the United States and the large cities in developing countries). Its beneficiaries within local populations comprise a small minority. The livestock industry in general provides few jobs (in Amazonia, one job per 1000 hectares according to Hecht, 1992). Fewer jobs results in rural emigration to the cities and also provokes local hostilities.

Given good soil and adequate slope, induced pasturage can stabilize. More frequently, however, forest removal increases not only erosion but also adversely changes the structure and porosity of the soil. Soil becomes compacted and loses its capacity both to support biota and for aqueous and gaseous interchange, with a resultant loss in fertility. The strong political and financial impetus to expand large-scale ranching has not considered the ecological consequences of expansion into fragile lands condemned to degradation once the native flora is removed.

Frequently, deforestation is not related to the requirements of cattle raising. Conversion of forests into pastures has a narrow connection with various unproductive, but highly profitable activities: cheap credit, tax incentives and land speculation (McNeely, 1988). Clearing land is a form of insurance in inflationary societies where land investment is a less risky alternative to other economic options, especially in agriculture. An indirect indicator of the importance of credit is the fact that, since 1990, the rate of forest destruction in Amazonia has decreased not because of any political decision, but because of a reduction in available credit to expand land holdings.

Protected Areas and Biodiversity Conservation

Biodiversity conservation is unquestionably benefited by national systems of protected areas. What is open to discussion is their efficacy and viability in the medium term as these systems have been implemented in the greater part of tropical America.[4] In this context, I will focus my observations on two points: (i) are presently constituted systems of protected areas sufficient to preserve a significant portion of their biodiversity? and (ii) what strategies are being used to implement these protected areas?

The latter point serves to indicate the limitations of conventional strategies. In my opinion, these limitations are determined by (i) the lack of a relationship between protected areas and the development programmes that determine the use of natural resources, (ii) economic and social factors intrinsic to countries in the American tropics, and (iii) ecological factors.

Regarding (i) above, one must keep in mind that the use of natural resources is determined by competing pressures. The governments of many tropical countries have created by decree protected areas that merely add to the already long list of 'paper parks and reserves'. On one hand, the national scientific community, public opinion, and the political desire for a 'good international image' push for the protection of biodiversity. On the other hand, national and foreign business interests which stand to profit from the 'mining' of natural resources produce competing pressures – as do displaced rural populations lacking land. Other pressures come from a complex group of economic interests and local or regional politicians complaining about the opening of new lands to cultivation and, ultimately, cattle raising. The situation becomes complicated (and this is an important point) because the governmental agencies that make the decisions about protection are not the same as those charged with forestry, agricultural and ranching development – indeed, there may be no programmed connection between them. Under circumstances that make it difficult to satisfy everyone, one way out of the problem is to declare certain areas protected and abandon the rest to unrestricted intensive use. 'The unwritten philosophy of nature preserves is that nature can be preserved on one side of the fence so that exploitation can continue unabated on the other' (Cooperrider, 1991: 146).

When protected areas are planned and created without consulting the affected local populations, the possibilities of their providing real protection are very limited. They cannot resist human pressures. Fortunately, the foregoing negative scenario does not always happen, but the exceptions make it the rule.

As mentioned above, conventional strategies are limited also by social and economic factors prevalent in the tropical Latin American countries. These factors, which include population growth, the lack of opportunities for young people, the exigencies of international markets, the inertia of local interests, and an uneven distribution of wealth, all combine to create a hostile frontier attitude toward natural areas which have not yet been totally transformed (see especially Toledo, 1989; Tudela, 1990), even while knowing that the new use is not the most appropriate nor productive in the long term.[5] In the face of wholesale colonization, of opening up new lands to ranching, monocultural agriculture, or to haphazard colonization (which can lead to unbridled immigration), protected areas can accomplish little by themselves. We cannot lose sight of the fact that protected areas are submitted to the same pressures as the rest of the territory.

Frequently we forget that the greater part of the large protected areas in the American tropics support local populations. Likewise, large extensions of 'virgin' rainforests that we desire to preserve also have native inhabitants. Population densities may be low, but not always. Inhabitants may be aboriginal or they may include outsiders in search of homesteads. Removal of resources from local control has led to rebellions and revolutions – confrontations that seldom end easily and rapidly and that never have a positive effect on biodiversity conservation. Social instability accentuates the worst predatory pressures. If we wish to plan for effective medium-term conservation, we must avoid situations where the creation of protected areas becomes the centre of social unrest. That is not to say that protected areas should not be created. On the contrary, an intelligent prioritization of the use of natural resources has not only ecological value, but also social value. What is necessary is that the interests, customs and culture of local populations be taken into account, that local populations be included in the decision-making process, and that they share in the benefits of conserving biodiversity (among recent references, see McNeely, 1988; Alcorn, 1991; Parks, 1994).

Besides economic and social reasons, there are ecological reasons for believing that conservation of tropical diversity cannot be restricted to protected areas (likewise, perhaps for non-tropical biodiversity, see Hudson, 1991). Tropical ecosystems are spatially highly heterogeneous and vary microgeographically in significant ways, especially in mountainous areas. In the best case, protected areas contain only a part of the existing biodiversity. Moreover, they interrupt not only migratory movements, but also population and interpopulational replacements in animals. By impeding natural movements, there is a reduction in genetic diversity, especially in large vertebrates and associated flora. Much evidence exists that protected areas become islands of genetically impoverished floras and faunas.

Only a few years ago, such arguments would have surfaced only occasionally since ecological 'insularity' on continents was the exception in the tropics. In Costa Rica we are approaching a system of 'islands' represented by protected areas. The situation could become a reality in other Latin American and tropical countries without the positive element that good management and protection, that exist in Costa Rica, represent. Speaking of the United States, Chadwick (1991) says, 'The plain fact is that most of our existing preserves have the same problems as fragments of habitat elsewhere across the nation. They are too small and isolated to guarantee the long-term survival of many of their wild residents'. Chadwick also notes the local extinction of 42 kinds of native mammals from 14 US parks.

The foregoing leads me to believe that in large part the success of conservation strategy depends on what happens in large extensions of land not intensively exploited by man. In these areas large proportions of plant and animal species can survive and maintain adequate genetic flow and geographic range. This is the situation outside of reserves and protected areas, without which a system of protected areas is insufficient to promote biodiversity conservation.

Summary

The efficacy of a system of protected areas in biodiversity conservation is enhanced by including in its plan provisions for certain types and levels of human activity, as envisioned by the UNESCO-MAB (United Nations Educational, Scientific and Cultural Organization – Man and Biosphere) Programme. Human activity can favour survival of biodiversity and, above all, can be a long-term guarantee against invasion and inappropriate uses of natural resources. The same is true in those reserves where some extraction of natural resources is permitted only when no changes in the ecosystem result.

Interposed between protected areas and those intensively manipulated exists a large expanse of territory where rational and well-planned rustic use of biological resources can coexist with an effort to conserve biodiversity. The key to a successful relationship lies in planning land use according to its ability to withstand exploitation. In planning, it is important to bear in mind that protected areas alone are insufficient. As Alcorn (1991) points out, the key points of conservation political strategy for the 21st century will be the search for mechanisms to strengthen conservation in rustic areas as well as ways to assure that the transition from traditional subsistence economics to global capitalism includes the ethic and mechanisms for conservation both in the protected areas and in those dedicated to rustic usage. Presently, there is little effort and minimal funding to support conservation in regions supporting rustic usage.

Overall success in biodiversity conservation resides in a clear and explicit plan for each region in each country based on consultation with local populations and accepted by all affected government representatives and decision-makers. The plan must address what resources require protection in the national interest and how biodiversity and protected areas contribute to a sustained economic development and to the welfare of each citizen.

It is clear that successful planning of this sort requires great scientific and technological research within each area, not simply an extrapolation of the results of research elsewhere. Strong differences in ecology, economic status and social and political conditions demand national plans based on the work of a given nation's own scientific and technological community. Careful and intelligent planning is the great challenge we face as the 20th century comes to a close.

Acknowledgments

In various ways I have benefited from the cooperation of Narciso Barrera-Bassols, a geographer at the Instituto de Ecologìa. I much appreciate his help.

Notes

1. For example, less than 10% of the alleles of the principal crops (and even less for related native plants) occur inside protected areas (World Conservation Monitoring Centre, 1992, Chapter 34).
2. Anthropologists (and many ecologists) have paid close attention to the rustic use of biotic resources by indigenous populations (referred to as 'aborigines', 'traditionals', 'trivial populations', etc.; see the excellent reviews by Toledo (1994), Oldfield and Alcorn (1991) and Redford and Padoch (1992)). A commonly overlooked fact is that in many industrialized countries, whole regions continue farming, livestock raising and forestry in ways that conserve much rustic usage. This occurs because geographic and edaphic factors are not conducive to intensive, mechanized exploitation, because product quality would be compromised by massive production, or because of an inherent cultural resistance to changing traditional patterns of family life based on self-sufficiency and long-term stability.
3. Intensive cattle ranching is an economic and cultural phenomenon with its own characteristics that is ever-expanding in the American tropics. N. Barrera-Bassols (1994, Xalapa, personal communication) says that the explosive increase in cattle raising using primitive production methods (extensive pasturage covered by introduced grasses; emphasis on quantity; few technological components; absentee landlords; and low return per hectare) has resulted in increasing competition between cattle and rural populations for productive space and food.
4. Kux (1991: 297) points out that

   > Since 1972, the number of protected areas throughout the world has grown by 47% – with most of this increase in the Third World (Miller 1984). Nine of the ten countries that have set aside over 10% of their land as protected areas are developing countries ... While these efforts are impressive from an international conservation perspective, creating protected areas is only a first step in solving the problem. Whether these areas do, in practice, conserve biological diversity (even though in most cases this was not a criterion used to establish protected areas), and whether they are properly managed with adequate staff and budget are questions that may well be answered negatively in most developing countries (Machlis and Technell, 1985).

5. In Amazonia, 'The forest destruction now at hand might have been tolerable if the replacement land uses – agriculture and pasture – were sustainable. As it stands now, agriculture and pasture are but short moments of production in a larger process of degradation.' (Hecht, 1992: 381).

References

Alcorn, J.B. (1991) Ethics, economies and conservation. In: Oldfield, M.L. and Alcorn, J.B. (eds) *Biodiversity: Culture, Conservation and Ecodevelopment*, Westview Press, Boulder, Colorado, pp. 311–346.

Altieri, M. and Hecht, S. (eds) (1990) *Agroecology and Small Farm Development*. Westview Press, Boulder, Colorado.

Barrera, N. and Rodriguez, H. (eds) (1993) *Desarrollo y Medio Ambiente en Veracruz. Impactos Económicos, Ecológicos y Culturales de la Ganaderìa en Veracruz*. Fundaciôn Friedrich Ebert/CIESAS-Golfo/Instituto de Ecologìa, Mèxico.

Boegue, E. and Barrera, N. (1993) Producción y recursos naturales en los territorios étnicos: Una reflexión metodológica. In: Warman, A. and Argueta, A. (eds) *Nuevos enfoques para el estudio de las etnias indìgenas de México*. CIIH-UNAM/Miguel Angel Porrúa Editorial, México, pp. 91–118.

di Castri, F. (1992) Scientific Program UNCED'92. Conference June 1, 1992 (unpublished). Universidade Federal do Rio-UNESCO. Rio de Janeiro, Brasil.

Clay, J. (1991) Cultural survival and conservation: lessons from the past twenty years. In: Oldfield, M.L. and Alcorn, J.B. (eds) *Biodiversity: Culture, Conservation and Ecodevelopment*. Westview Press, Boulder, Colorado, pp. 248–273.

Cooperrider, A. (1991) Introduction: Part III, Reintegrating humans and nature. In: Hudson, W.E. (ed.) *Landscape Linkages and Biodiversity*. Island Press, Washington, DC.

Chadwick, D.H. (1991) Introduction. In: Hudson, W.E. (ed.) *Landscape Linkages and Biodiversity*. Island Press, Washington, DC, pp. xv–xxvi.

Gomez-Pompa, A., Kaus, A., Jiménez-Osornio, J., Bainbridge, D. and Rorive, V.M. (1993) México. In: *Sustainable Agriculture and the Environment in the Humid Tropics*. National Academy Press, Washington, DC, pp. 483–548.

Gordillo, G. (1993) La problemática del campo en la modernizaciún. In: Blanco, J.J. and Woldenberg, J. (eds) *México a Fines de Siglo*. Vol. 2. CONACULTA and Fondo de Cultura Económica, México, pp. 309–341.

Hecht, S.B. (1992) Valuing land uses in Amazonia: colonist agriculture, cattle, and petty extraction in comparative perspective. In: Redford, K.H. and Padoch, C. (eds) *Conservation of Neotropical Forests: Working from Traditional Resource Use*. Columbia University Press, New York, pp. 379–399.

Hudson, W.E. (ed.) (1991) *Landscape Linkage and Biodiversity*. Island Press, Washington, DC.

Kux, M.B. (1991) Linking rural development with biological conservation: a development perspective. In: Oldfield, M.L. and Alcorn, J.B. (eds) *Biodiversity, Culture, Conservation and Ecodevelopment*. Westview Press, Boulder, pp. 295–316.

Leff, E. and Carabias, J. (eds) (1993) *Cultura y Manejo Sustentable de los Recursos Naturales*. 2 vols. CIIH-UNAM/Miguel Angel Porrua Editores, Mèxico.

May, P.H. (1992) Common property resources in the Neotropics: theory, management progress, and an action agenda. In: Redford, K.H. and Padoch, C. (eds) *Conservation of Neotropical Forests: Working from Traditional Resource Use*. Columbia University Press, New York, pp. 359–378.

McNeely, J.A. (1988) *Economics and Biological Diversity: Developing and Using Economic Incentives to Conserve Biological Resources*. IUCH, Gland, Switzerland.

Mestries, F. (1991) La crisis ganadera: La modernizaciûn en la encrucijada. *Cuadernos Agrarios* 1, 97–111.

Oldfield, M.L. and Alcorn, J.B. (eds) (1991) *Biodiversity: Culture, Conservation and Ecodevelopment*. Westview Press, Boulder, Colorado.

Parks, vol. 4, no. 1. February 1994. *Building Community Support in Protected Areas*.

Posey, D.A. (1983) Indigenous ecological knowledge and development of the Amazonia. In: Moran, E. (ed.) *The Dilemma of Amazonian Development*. Westview Press, Boulder, Colorado, pp. 225–255.

Redford, K.H. and Padoch, C. (eds) (1992) *Conservation of Neotropical Forests: Working from Traditional Resource Use*. Columbia University Press, New York.

Rodriguez, H. and Boegue, E. (eds) (1992) *Medio Ambiente y Desarrollo en Veracruz*. CIESAS-Golfo/Instituto de Ecologia/Fundación Friedrich Ebert, México.

Rojas, T. (ed.) (1990) *Agricultura Indigena: Presente y Pasado*. Ediciones de la Casa Chata, CIESAS, Mèxico.

Toledo, V.M. (1989) *Naturaleza, Producción, Cultura: Ensayos de Ecologia Politica*. Universidad Veracruzana, Xalapa, México.

Toledo, V.M. (1990) The lesson of Pàtzcuaro: nature, production and culture in a indigenous region of Mexico. In: Oldfield, M. and Alcorn, J. (eds) *Culture and Biodiversity Conservation and Development of Biological Resources under Traditional Management*. Westview Press, Boulder.

Toledo, V.M. (1993) La conservación indigena de la biodiversidad en Mèxico: evidencias empìricas (Conference). *Symposium: Biodiversidad en Iberoamérica: Ecosistemas, Evolución, Procesos Sociales*, 19–24 October 1993, Mérida, Venezuela.

Toledo, V.M. (1994) La apropiación campesina de la naturaleza: Un anàlisis etnoecológico. Unpublished PhD Thesis, Universidad Nacional Autónoma de México.

Toledo, V.M., Carabias, J., Mapes, C. and Toledo, C. (1985) *Ecología y Autosuficiencia Alimentaria*. Siglo XXI Editores, México.

Tudela, F. (ed.) (1990) *Desarrollo y Medio Ambiente en América Latina y el Caribe: Una Visión Evolutiva*. Ministerio de Obras Públicas y Urbanismo, Madrid.

World Conservation Monitoring Centre (1992) *Global Biodiversity: Status of the Earth's Living Resources*. Chapman & Hall, London.

22 Conservation of Coastal-Marine Biological Diversity

G. Carleton Ray
Department of Environmental Sciences. University of Virginia, Charlottesville, VA 22903, USA

> When you assemble a number of men, to have advantage of their joint wisdom, you inevitably assemble with those men all their prejudices, their passions, their errors of opinion, their local interests and their selfish views. From such an assembly, can a perfect production be expected?
> Benjamin Franklin, addressing the American Constitutional Convention
> (Quoted from Time Magazine, October 1990)

The Present Context

Marine conservation is like a raindrop in a hurricane – looking for a place to land, but swept aside by more formidable forces and neglect. Society seems locked onto a growth mentality, while conservation is usually expressed as concern for 'flagship' species and habitats, such as whales and coral reefs. Relatively little attention is paid to ominous signs of environmental pollution and an epidemic of coastal algal blooms. Fisheries seem out of control, at the same time that demand for sea-food products rapidly increases to feed ourselves and our pets and to fertilize our fields. Coastal-marine systems remain misunderstood and mismanaged, despite the fact of their huge productivity, their harbouring of the bulk of humanity, and their content of the greatest concentration of Earth's biological diversity. And governments and non-governmental conservation organizations alike seem not able to contrive a coastal-marine conservation strategy that is environmentally or socially consistent or credible.

Solutions to problems of coastal-marine biodiversity conservation must be centred in what may be termed an 'ecocultural revolution', based on new ways for humans to conceive their planet and to manage their affairs. This requires that the special interests of management agencies, scientists, and conservation organizations be set aside in favour of a new union. Foremost, the conservation of coastal-marine systems must be understood in the context of their scale, the land- and seascape diversity of their ecosystems, and the interactions and controls of the various ecosystem components. This requires, as a first priority, the acquisition of the scientific and social knowledge on which conservation can act. The challenge is no less than to harmonize two chaotic, non-linear, geographically divergent, non-equilibrium systems that also operate on very different time-space scales – society and environments. As Costanza *et al.* (1993) point out, there is an 'inherent mismatch between the characteristics of the ecological system on the one hand and of the human institutions developed to manage it on the other.' The role of science – both natural and social – is to seek solutions to this mismatch.

The accountability for this ecocultural revolution may lie in the scientists themselves, as they are the agents of environmental knowledge, and with programmes such as *Diversitas* for facilitating the development of the information base on which conservation may act.

The diversity of coastal-marine systems

It has been popular and persuasive to place biological diversity in the context of species diversity and richness. Thus, those interested in tropical forests and coral reefs alike have been quick to point out that because these environments contain more species than elsewhere (we think!), and because there is an extinction 'crisis' (it is believed!), therefore these areas deserve number one priority. This establishment of priorities based on species deserves further consideration.

Hutchinson (1959) was probably correct in saying: 'The marine fauna, although it has at its disposal a much greater area than the terrestrial, lacks this astonishing diversity. If the insects are excluded, it would seem to be more diverse . . .' The global species richness of oceanic–pelagic systems is, indeed, generally low, although there are areas as high as terrestrial systems (Angel, 1993). The greatest richness tends to occur in regions of low productivity at boundaries between different water masses where different faunas are mixed, but the locations of these boundaries are unstable and may be altered seasonally over hundreds of kilometres. Further, the majority of open-ocean species are rare, but it would be a mistake to conclude that they play a minor role in community dynamics or ecosystem function (McGowan, 1990).

Thus, if high species richness becomes a criterion for conservation action, not only will the open ocean receive low priority, but regions important in ecological function may be overlooked. Furthermore, conclusions about which areas are more or less species-rich must take account of dramatic new discoveries, which seem to occur almost daily – Grassle *et al.* (1990) have found that the deep sea is 'not the impoverished desert that was envisioned by most biologists'.

Clearly, there are more dimensions to diversity than a narrow focus on species. Phylum diversity is expressive of more basic genetic, form, and function differences than seemingly endless repetitions on the themes of beetles or worms, and in this sense, marine systems exceed the land by twice (May, 1988). High phylum diversity may reflect the age and extent of the oceans, which not only twice exceed the land in surface area, but are also about two orders of magnitude greater in biological volume, implying that biogeography is yet another way to look at diversity. For example, Hayden *et al.* (1984) described 21 types of oceanic and coastal-margin realms and 45 coastal provinces, compared to the eight realms and 193 biogeographic provinces identified by Udvardy (1975) for the land. The inclusion of mid-water, deep-water, and off-shelf benthic areas of the total ocean could amplify marine biogeographic provinces to at least 300 (Ray *et al.*, 1992). Province-level diversity is a representation of biological diversity at the most basic, global scale.

In short, it would appear that by any measure except total described species, marine systems are globally about twice as diverse as the land, a not surprising conclusion from paleohistorical and evolutionary points of view. This is not said to establish a one priority over another, but to require, only, that we take notice.

The 'coastal zone' and the 'Marine Revolution'

The conservation of marine biological diversity faces its greatest challenges within the coastal zone, where the greatest proportion of humanity has, at least since the Agricultural Revolution, always lived. This zone includes the coastal forests and wetlands of the coastal plains, as well as estuaries, bays, lagoons, and continental shelves (Fig. 22.1). Ketchum (1972) called it: '... a natural entity, but with flexible boundaries ... the broad interface between land and sea where production, consumption, and exchange processes occur at high rates of intensity'. The coastal zone includes only approximately 8% of Earth, but may account for about as much biological productivity as the remainder of the sea or the rest of the land (Holligan, 1990; Holligan and de Boois, 1993). One reason for this exceptional productivity is that the coastal zone is uniquely the place on Earth where atmosphere, land, and sea meet.

The 'Marine Revolution' represents a massive entry by humanity into ocean space and, though concentrated in the coastal zone, affects all the seas and oceans. It follows the Agricultural and Industrial Revolutions as a significant change in human relationships with the global environment (Ray, 1970). Its effect is to accelerate, once more, human capacity to exploit this planet and to increase its carrying capacity for civilization. In developmental terms, the effects on marine systems are about at the point where most terrestrial environments were in about the mid-1800s when the Industrial Revolution was in full swing. The big marine species have been, or are being, massively depleted and, if trends are not reversed, some that are still abundant may suffer similar fates as waterfowl or, some of them, the passenger pigeon.

As the Marine Revolution has crept up on us, we have begun to recognize the global impact of humans on the coastal zones and oceans. These impacts may already have proceeded to the extent that we will never know of past conditions. Systems may have 'flipped' to other steady or unsteady states,

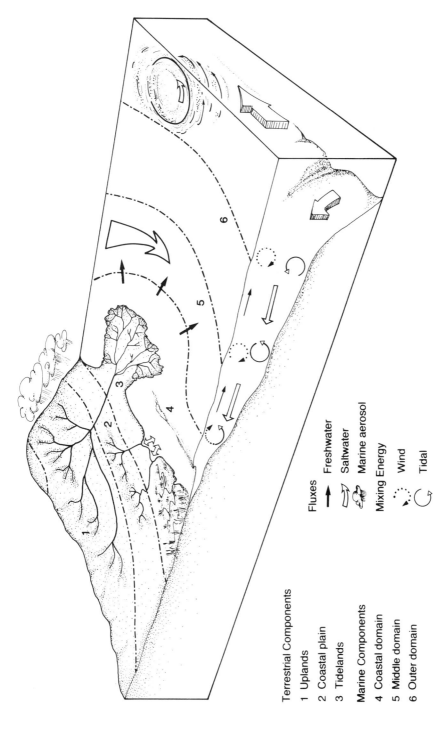

Fig. 22.1. The coastal zone. This zone consists of both terrestrial and marine components, tied together functionally by the processes shown (from Ray and McCormick-Ray, 1989).

Terrestrial Components
1 Uplands
2 Coastal plain
3 Tidelands

Marine Components
4 Coastal domain
5 Middle domain
6 Outer domain

Fluxes
Freshwater
Saltwater
Marine aerosol

Mixing Energy
Wind
Tidal

making both preservation and restoration problematic. Signs of malady abound. There is a global epidemic of coastal algal blooms, many of them highly toxic, that appear to be caused by factors of human perturbation (Anderson, 1994). The states of estuaries are critical. The previous condition of one of the world's richest estuaries, Chesapeake Bay on the US east coast, can only be interpreted from the writings of early explorers and colonizers, such as Captain John Smith, and from uncertain estimates of fishery yields. The Bay's most significant functional elements, oyster reefs and beds of submerged aquatic vegetation, have almost disappeared and could take decades to centuries to restore under even the most favourable of conditions.

Historical 'revolutions' have separated us from our environmental roots. It seems ironic that the loss of coastal habitats is exacerbated by coastal peoples, when many of these people are in the best position to notice how these changes affect them directly. Widespread extinction appears not to have occurred, except near shore (Carlton, 1993), but there exist more subtle, more systemic, and more difficult changes to address – many of them too subtle for conservation to care to address.

The hierarchical structure of coastal-marine systems

Any attempt to match environments and institutions requires, first and foremost, some understanding of the structure and function of ecosystems. Enough information is on hand on this subject to make major strides in conservation application. In brief, both individual species' distributions and the composition of the community of organisms are functions of the suitability of the environment as habitat, and also of the hierarchical relationships among regional attributes, habitats, and the organism's or community's ability to influence environmental conditions though feedbacks (Ray et al., 1995). This approach requires that coastal-marine systems be considered primarily as mosaics of habitats, ecosystems, or what may be termed 'land–seascapes' (Ray, 1991; Costanza et al., 1993). Hierarchy theory (O'Neill et al., 1986) posits that large-scale features act as 'controls' on smaller-scale features. Detailed species inventories are not required for this approach to biological diversity, as the mosaic of the land–seascape is the surrogate.

Fig. 22.2 provides an example of ecological controls on biodiversity and their feedbacks from the United States east coast. Biogeographic provinces are defined by strong spatial and temporal gradients under the control of large-scale, even global, water-body patterns, currents, atmosphere, and climate. A recently discovered example of a global input, which under certain circumstances could act as a control at the level of the biogeographic province, is the work of Swap et al. (1992) who estimated that the annual input of 13 Mtons of Saharan dust enters the central Amazon basin in bursts that accompany major, wet-season systems. They suggest that: 'Any strategy designed to preserve the Amazon ... should equally concern itself with the interrelationships between the rain forest, global climate and arid zones well removed from Amazonia.' Similarly, M. Garstang (University Virginia, 1994, personal

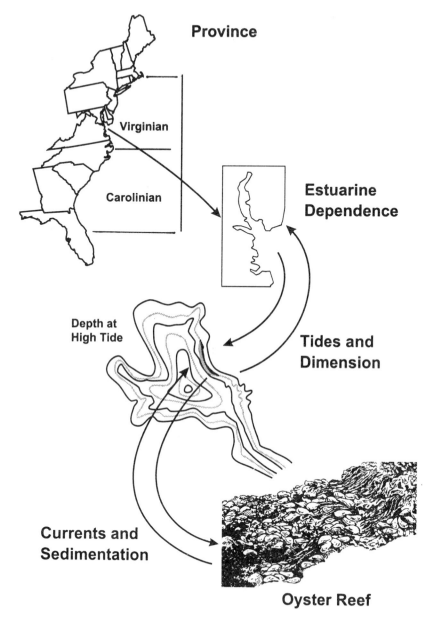

Fig. 22.2. The levels of habitat description for a US East Coast estuary, the Chesapeake Bay. Atmospheric and ocean climate control the ecotonal gradients that determine the boundary conditions of biogeographic provinces. Within these provinces, a variety of estuaries occur and certain distinctive habitat features occur with each of these, such as oyster reefs. These reefs are the only hard, benthic structures of these estuaries, and act as habitats for a multitude of species (Fig. 22.3). Should oyster reefs be deleteriously affected by either human activities or stochastic events, their fauna would be affected, also affecting the ecological functions of the estuary. Such feedbacks would predictably also affect faunal interactions of the adjacent fauna of the shelf. See text for further explanation (from Ray et al., 1995).

communication) estimates that massive amounts of aerosols of terrestrial origin, containing trace nutrients, also enter coastal and oceanic waters and that the productivity of some of those areas may depend to a large extent on these inputs.

Next smallest in the hierarchy are land-seasheds, estuaries, lagoons, bays, water masses, and other within-province physical entities. For the US east coast, watersheds are defining characteristics for biodiversity. Ray and Hayden (1992) have presented a means whereby watersheds may be characterized according to variable associations of their five subdivisions: uplands, coastal plain, tidelands, nearshore entrainment volume, and offshore entrainment volume. Various combinations of these result in differing physical structures and habitat arrangements, with profound effects on the biota.

The interacting mosaic of habitats and their ecological processes within each watershed, for example of hard and soft bottoms, wetlands, coastal forests, etc., must be examined at a still smaller scale. This is the scale at which species presence–absence is determined and which relates habitat state and dynamics to species life history. An example for coastal systems is the control exerted by overwash disturbance on barrier islands, which produces marked changes in number and types of plants (Fahrig et al., 1993).

This hierarchical approach, therefore, conceives a series of 'filters' that act to define species' presence in time and space, and that also can act as controls on behaviour. That is, the 'potential' for species presence is dependent on a set of habitat requirements defined at scales relevant to the natural history of the organism. This model incorporates natural history, ecological functions, and environmental conditions and provides the basis for a classification of environments on which conservation of ecosystems may be based.

An example of feedbacks: the oyster reef

I have presented a top-down, hierarchical system. However, bottom-up ecological 'feedbacks' may also occur. A notable example is that of reefs formed by the eastern oyster, *Crassostrea virginica*. These reefs play host to a wide variety of organisms (Fig. 22.3). Oyster reefs are formed by a lattice of oysters that are cemented together, providing a substrate for other living forms to attach, hide, and feed. They predominate at the interface of land and water in the mesohaline portion of estuaries (Galtsoff, 1964; Bahr and Lanier, 1981). The oyster reef is important as the only hard, biogenic structure of east coast estuaries in an otherwise mobile sand-and-mud environment. It provides the basis for a community with a high metabolism, a landscape feature essential for many attributes of estuarine dynamics, and a station for transient, estuarine-dependent species. A few of these species are endemic to oyster reefs, but most will also adopt other hard substrates, such as jetties and even discarded cans and bottles, as habitat.

The ecological role of the oyster reef is analogous to that of tropical coral reefs. Its structure introduces benthic roughness and exerts a fundamental control on the benthic boundary layer. This, in turn, affects channel circulation

Fig. 22.3. The biological diversity of attached fauna on a single cluster of oysters. The number of phyla involved may exceed the total found in all terrestrial habitats (from Winslow, 1882).

and current flow, which in turn affects other habitats, larval transport and settlement of many organisms, and phytoplankton encounters important to the oysters themselves and to other benthic suspension feeders. The oyster reef also has a marked influence on water quality through the filtering activities of feeding oysters, and supports a benthic community by deposition of faeces and pseudofaeces (Newell, 1988). Thus, this species and the structures it creates play essential roles in estuaries and their biological communities.

The oyster reef presents an excellent example of ecological feedbacks that can cascade up through larger and larger scales to the coastal system as a whole. As such, it presents a model for the understanding of multiscale effects, the role of perturbations, and conservation needs. In theory at least, the effects of perturbations at the local scale of the reef itself may offer at least partial explanations for the 'functional diversity' of coastal systems (Steele, 1991). That is, if the oyster reef is perturbed, so will be the estuary itself, to some degree, and as estuaries are major feeding and breeding areas for a host of shelf species, the fauna of the shelf will also be affected to some degree. Thus, the functional relationships among the biota and physical conditions will ultimately be altered. That this has already occurred is probable. The ecological importance of oyster reefs is being lost as disease, over-harvest, and changed estuarine ecosystems all interact further to destroy these structures.

The need for research on land- and seascape ecology

Odum (1992) listed as one of his 20 great ideas in ecology: 'An expanded approach to biodiversity should include genetic and landscape diversity, not just species diversity.' His intent was to incorporate species-level questions with those that are systems-oriented. Early in the development of the IUBS–SCOPE–UNESCO programme on 'the ecosystem function of biodiversity' called *DIVERSITAS*, Simpson (1988) similarly concluded that biological diversity should be placed in the context of ecosystem structure and function, and should address questions dealing with (i) diversity at the level of the landscape, (ii) species diversity within biotic communities, and (iii) connections between changes in landscape diversity and species diversity within mosaics.

The need for the 'landscape' approach is dictated by the principles listed in Fig. 22.4, all of which are based on our current knowledge. First and foremost, there is not a place on Earth that is not critical for some form of life. In addition, the vast majority of species require a mosaic of habitats at different scales during various stages of their life histories. These two facts alone have two consequences. First, they dictate an hierarchical approach to management. Second, the number of species in any area becomes a poor criterion for establishing priorities for conservation, as this would require very difficult value judgements to be made. The other principles relate to the facts that we know little of species and their life histories and interrelationships, and that a concentration on ecological pattern and process is mandated.

> No environments are devoid of species dependency
>
> Every species requires a mosaic of habitats for its continued existence
>
> Most species are naturally rare
>
> Species distribution and life histories are largely unknown
>
> Cause–effect relationships of human-caused vs. "natural" threats remain highly uncertain
>
> Coastal and marine ecosystems are especially dynamic and non-linear

Fig. 22.4. Summary of principles relating species diversity, habitats, conservation and research.

All of these factors argue strongly against a 'hot-spot' mentality in determining priorities for conservation action.

Maintenance of biological diversity depends on a hierarchy of controls at the level of the landscape (Urban et al., 1987; Swanson et al., 1988; Bridgewater, 1993; Franklin, 1993). As Barrett and Bohlen (1991) suggest: '... landscape ecology considers the development and dynamics of spatial heterogeneity, the spatial and temporal interactions and exchanges across the landscape, the influences of spatial heterogeneity on biotic and abiotic processes, and the management of this spatial heterogeneity for society's benefit and survival.' With respect to coastal-marine issues, the term 'landscape' deserves expansion into 'land- and seascape', as the same general principles apply to both terrestrial and aquatic systems (Ray, 1991).

This is the context in which questions about maintenance of biodiversity may best be addressed. Many examples come to mind. For example, biogenic structures, patchy food supply, and small-scale disturbances may be contributing factors for the high diversity of deep-sea communities (Grassle et al., 1990; Grassle, 1991). The high species diversity of coral reefs may be maintained by (i) physical disturbances, such as tides and storms and (ii) the reproductive strategies of the corals themselves (Loya, 1990). As sea levels rise, the relative proportions of marshlands to barrier islands to lagoon areas will be altered (Hayden et al., 1991), with probable widespread effects on coastal communities. On a global scale, widespread extinctions are not likely to result from the magnitudes of global environmental change we are now experiencing, but widespread alterations in species distributions and community composition are probable, as experiences with the El Niño have demonstrated (Glynn, 1988; Ray et al., 1992). The land–seascape approach may also be applicable to pelagic–oceanic communities, but is more difficult as the mechanisms for such mobile systems are quite different from those for sessile ones such as forests and reefs (McGowan, 1990).

The land–seascape approach is also relevant to the species and community levels of interaction. For example, human perturbations such as chronic oil pollution require small-scale species and community-level research. Even low levels of oil pollution may affect the reproductive and/or physiological capacities of corals (Loya, 1990). Similar effects of chronic pollution may be widespread, e.g. the reduced phagocytic responses of molluscs (McCormick-Ray, 1987) or recent experiments that seem to have confirmed that pollutants, accumulating in higher trophic levels, can suppress the immune response of marine mammals (Raloff, 1994). However, these responses can not be translated into conservation until and unless they are placed within a multi-level matrix of interactions. (NB It is ironic that the experimental studies required for determining the effects of pollution on marine mammals may not presently be possible in many countries, including the US, due to concern for 'animal welfare'.)

Socio-economics and Institutional Constraints

If ecological models, such as the one presented above, seem sound, then why are they not the focus for conservation and management practice? My purpose here is not necessarily to defend the hierarchical model *per se*, but to identify some constraints that pervade our strategic and decision-making processes and that affect our conservation goals.

'Misplaced concreteness'

A large part of our modern conservation predicament lies in Alfred North Whitehead's notion of the 'fallacy of misplaced concreteness', which 'flourishes because the disciplinary organization of knowledge requires a high level of abstraction . . .' (Daly and Cobb, 1989). To 'abstract', in this sense, is to leave out many of the qualities of a complex subject, sometimes to the point of *reductio ad absurdum*. In the case of economics, the abstraction is to treat nature as a commodity. For fisheries management, it is to concentrate on the output side, e.g. 'yield', rather than on input, e.g. ecosystem dynamics, and natural history (Odum, 1989). In the case of conservation, it is to 'market' crises and endangered, high-profile, 'charismatic' species, which carries with it much problematic legal (Rohlf, 1990) and ecological (Karieva, 1994) baggage.

Placing too high a priority on species may actually contribute to long-term conservation problems, particularly for poorly known coastal-marine environments. Raven and Wilson (1992) suggest a 'fifty-year plan for biodiversity surveys, concentrating on well-known groups', on the premise that 'hot spots', of endemism, species richness, and endangered species would be revealed. There are a host of problems with this approach, especially for coastal-marine systems. First, the species composition of any area is highly variable, the more so as spatial and temporal scales are reduced to the level of most aquatic sampling regimes (Jackson, 1991). Second, the high mobility of many marine species, and 'the variety of responses to environmental change, especially the diverse space and time scales with which organisms react to each other and to the environment' (Steele, 1991) results in a host of 'moving targets'. Third, the huge costs, long time, and great uncertainty that are involved in species inventories and surveys usually result in community descriptors, not in models for ecosystem-management regimes.

'Social traps', and 'worrisome gaps'

A social trap is defined by Odum (1989; see also Platt, 1973; Cross and Guyer, 1980; Costanza, 1987) as: 'A situation where a short-term gain is followed in the long term by a costly or deleterious situation not in the best interest of either the individual or society.' Odum also draws attention to worrisome gaps: 'A good way to assess the predicament of humankind is to consider the gaps that must be narrowed if humans and the environment . . . are to be brought into more harmonious relationships.'

The leading global conservation organization, the World Conservation Union (IUCN) has institutionalized the gap between species and ecosystems in its Species Survival Commission and its Commission on National Parks and Protected Areas. The latter commission has recognized, that individual protected areas exist as patches within larger systems, are highly vulnerable, and may not be sustainable (IUCN, 1993), but its protected-area strategy is not yet clear about solutions to the problems that: 'biodiversity is not a "set-aside" issue that can be physically isolated in few, or even many, reserves'. On the contrary, the entire landscape matrix is involved (Franklin, 1993). For coastal and marine areas, this matrix involves entire biogeographic regions, or larger, for which integrated national and international networks of protected areas would be required to maintain biodiversity (Ray and McCormick-Ray, 1992).

The fact is that most coastal-marine protected areas have been established *ad hoc*, without reference to one another, and are primarily 'people-managed', with little reference to scales of ecological interactions. Indeed, very little science has gone into protected-area selection and establishment, demonstrating, rather, the 'art of the possible' by using the selection criteria of natural beauty, valued species or communities (e.g. coral reefs), or recreation as justification. This is not entirely to be criticized, as the increased coverage of protected areas, worldwide, has been remarkable during the past four decades. But the time has come to make their selection, establishment, and management ecologically sensible as well.

This illustrates a further gap, that between ecosystem science and conservation or management practice. Closing this gap is a two-way street. Both 'sides' seem ambivalent about the other. A mood that many 'environmentalists' perpetuate is that we already have more knowledge than we can apply, an attitude that displays a profound misunderstanding of how knowledge is gained, how it can be applied, and how the results can be monitored. On the other hand, science is an opera with its own share of prima donnas. As Ehrlich and Daily (1993) state: 'The dominant social paradigm of ecologists is still too much focused on high-powered, environmental testing of trivial hypotheses rather than on finding ways to improve understanding of (and generating action to solve) important problems.'

The 'tyranny of small decisions'

These three constraints – misplaced concreteness, social traps, and worrisome gaps – have combined with deadly efficiency to impede coastal-marine conservation efforts, which remain dominated by what has been termed the 'tyranny of small decisions' (Kahn, 1966; Odum, 1982). As Odum has observed: 'No one purposely planned to destroy almost 50% of the existing marshland along the coasts of Connecticut and Massachusetts . . . or to reduce the annual surface flow of water into the Everglades, to intensify the effects of droughts, or to encourage unnaturally destructive fires.' Yet these effects are exactly what has happened.

Fisheries is a prime example of this. It is by far the dominant biologically based human exploitation activity of the coasts and seas. Fisheries activities have major, as yet unquantified, effects on biological diversity, analogous to forest clear-cutting (Hammer et al., 1993). It is simply inconceivable that the massive removal of fishes from their systems is ecologically benign, as deterministic fisheries models do assume. As Munro et al. (1987) have said: '. . . anything that changes the biomass of one part of the community will have some effect on all other parts of the community'. An exemplary case of this once again concerns the eastern oyster. Kennedy and Breisch (1993) describe attempts to manage this valuable species from about 1820 onwards. Despite the accumulation of scientific insights and clear warnings of overexploitation, a vociferous oyster industry, combined with a short-sighted management community and ill-devised models, has ruled over a catastrophic oyster decline, ignoring the ecological role of the oyster reef, as well as the near ecological collapse of the Chesapeake Bay.

For fisheries, all of these constraints are evident. Misplaced concreteness is reflected by 'yield' rather than ecological models. A social trap is that fisheries are largely managed as common property resources. Serious worrisome gaps exist among fisheries, ecology, and oceanography. The resulting tyranny is that management, thus conservation, remains so dominated by sectorized abstractions, as well as by just plain greed, that we will probably see even more massive depletions and economic disruptions before solutions are found.

Taking up the Challenge

Clearly, we are at a strategic juncture, beyond the point of: (i) proving that conservation of the coasts and seas is urgent, (ii) issuing vague recommendations, (iii) publishing yet more synthesis volumes, or (iv) justifying the need for conservation to be science-based. A strategy is required, defined by Morison (1958) as: '... the art of defeating the enemy in the most economical and expeditious manner'. Thus, strategy falls between goals and objectives and between policy and tactics. Most conservation strategies are not clear about this relationship, mostly because they do not offer a clear, operationally applicable 'game plan' consistent with a strong scientific base.

This problem is illustrated by the *Global Biodiversity Strategy* (WRI, 1992), which lists seven major components for catalysing biodiversity conservation. 'Basic and applied research on biodiversity conservation' gets attention only under the seventh category, that of 'Expanding human capacity to conserve biodiversity' and the research category is not operationally related to the other categories. The marine counterpart to this strategy (Norse, 1993) presents eleven categories of activity, of which 'Strengthening the knowledge base' is eighth, again not clearly related to the others from an operational point of view. Both of these documents accomplish the important purpose of heightening awareness of their subjects. Nevertheless, they are deficient in illustrating how science and conservation may be more closely linked and what the outcomes may be. Angel's (1994) review of the latter is pertinent: '... the conservation movement seems too divorced from the scientific base from which must come the knowledge it seeks, the technology it requires for monitoring and the ability it needs to understand change and to predict ecosystem response.'

A role for DIVERSITAS

From the beginning of *DIVERSITAS*, the marine portion has been singled out (with the microbiological portion) for special attention (di Castri and Younès, 1990). A Marine Working Group has convened several meetings during the past two years and has suggested that an International Marine Biodiversity Program (IMBP) consist of the following components (Grassle *et al.*, 1991; Lasserre *et al.*, 1994).

1. *Networks of sites for research and monitoring.* Coastal marine laboratories and field stations, together with their associated logistical capabilities – personnel, ships, remote-sensing capabilities, collections, etc. – provide entry points into the coastal-marine environment. Marine laboratories are proposed for inclusion in a global research network that could be representative of biogeographic regions and interconnected by communication links.

2. *Long-term measurements.* Major developments in observing and modelling coastal-marine biodiversity will require information on several spatial and temporal scales, from local to regional to global, and from subannual to

decadal and longer. Biological, physical, and chemical properties of nearshore environments can now be measured continuously. The data would be assembled into regional to global information systems.

3. *Theory and experiments.* Major advances in understanding marine ecological processes have come from combinations of descriptive observations and theory. These advances need to be integrated into the larger framework of coastal-ocean models and the natural patterns of physical and chemical variation.

4. *Education and training.* These dual objectives are fundamental to research in general, but especially towards research on conservation and management. Fortunately, the bulk of marine laboratories are, by virtue of their association with universities and museums, *de facto* educational and training institutions. In addition, by means of modern communications technology, the marine environment can also be brought into the classroom through interactive television and real-time transfer of data.

I maintain that *DIVERSITAS* is, inherently, an incipient strategy for conservation. Initially, research is emphasized, but the logic of this suggested marine programme is clear. For example, the network could evolve into, or derive from, protected areas, which together comprise the least deleteriously affected areas. The measurements, theory, and experiments could provide essential tools for conservation action. The educational and training functions could provide for public understanding and could supply the people necessary to do the job.

Conclusion

The thrust of this chapter is that coastal-marine conservation is fragmented, with an ill-defined scientific base and a poorly defined strategy for conservation. Coastal-marine systems suffer from relatively little attention and much less understanding than is the case for terrestrial systems. Thus, key questions concern how our knowledge of coastal-marine systems can be amplified, made apparent to policy makers and the public, and applied to conservation imperatives. It is equally essential to determine how systems can become sustainable, given human perturbations and environmental change.

Biodiversity, sustainability, and environmental change have become inextricably bound objectives for a global Sustainable Biosphere Initiative (Lubchenco *et al.*, 1991; Huntley *et al.*, 1992). It is neither clear to what extent this Initiative will include coastal-marine systems, nor how its results may influence conservation. A paper by Ludwig *et al.* (1993), influenced heavily by the characteristics of fisheries, challenged the very principle of sustainability, not because it is impossible, but largely because it is improbable, given human behaviour and difficulties in scientific understanding. This paper inspired a Forum of the Ecological Society of America, in which Levin (1993) pinpointed the problem: 'What sustainability does require ... is attention to the management of systems under uncertainty, to the linkages among physical, biological, and socioeconomic systems, and to the interface between science and policy.' In that same Forum, Holling (1993) pointed out that systems present moving targets, and that: 'It is this science of integration and synthesis that has been ill served by funding agencies and universities.' These statements also apply to the traps and gaps within and among our own scientific endeavours and to conservation of marine environments.

The evidence points strongly to the conclusion that we now live in an unsustainable world. But whether human 'revolutions' have increased the carrying capacity of the Earth or not is debatable. It may equally be true that changes in human behaviour have been driven by population increase, rather than the reverse. Nevertheless, there is irrefutable evidence for the need for the ecosystem perspective and there are some signs that this perspective is gaining in resource management. The 'large marine ecosystem' (LME) concept places emphasis on ecosystem processes (cf. Sherman et al., 1993 and reference to prior LME volumes). About 95% of annual yields as biomass of marine fisheries come from 49 LMEs, but virtually every one of them is overfished, polluted, and has suffered widespread habitat destruction. Attention is also being drawn to the important effects of climate change on fisheries (Kennedy, 1990). Finally, oceanography, climatology, and fisheries may be drawing closer together; the US GLOBEC (Global Ocean Ecosystems Dynamics) research programme promises to conduct research on links between climate variables, plankton production, and fisheries production.

I suggest that the application of the techniques of land- and seascape ecology could become the integrating factor for the conservation of coastal-marine biological diversity. This approach requires that conservation of biological diversity be increasingly directed away from individual species and more towards ecosystems in quantifiable, holistic ways. Addressing the concreteness of extinction should receive high priority only for cases that are solvable biologically and ecologically in the long term, and which result in strong sociological feedbacks for ecosystem perspectives.

Who will be accountable?

Present coastal-marine conservation theory and practices are not only woefully incomplete, but also lack a sufficient scientific base and cohesion. Under these conditions, how can conservation and sustainability become possible? Even if these goals are attainable, how will we ever know? And who will be accountable? Perhaps it is us – the scientific community.

A return to 'tradition' or to the wisdom of local peoples to solve problems does not seem to offer much hope. McNeely and Pitt (1985) have examined 'the human dimension in environmental planning' predicated on the reasonable thesis that:

> We must provide the means for local people, who maintain ecologically
> sound practices, to play a primary role in all stages of development in the
> area they identify with, so that they can participate and benefit directly, in a
> manner which is consistent with their values, time frames, and decision-
> making processes.

But what are the 'means', what is 'ecologically sound', and whose 'values', are to be promulgated? And, how can 'local peoples' know of regional problems beyond their experience or scopes? Klee (1985) encapsulates the problem: 'Pacific cultures were once highly effective, and, if supported or adapted to modern conditions, could continue to be so.' But how those practices are to be adapted is critical. Polunin (1985) goes even farther by admitting that too little is known of the status of tradition to be able to assess its effectiveness.

The fact is that 'tradition' is difficult to define. One example with which I am intimately familiar, is the Eskimo tradition of hunting marine mammals. The tradition is partly cultural and partly material, but sustainability does not lie in traditional methods, and probably never did. Nor does it lie in the material impositions of a market economy, which drive overuse. This example leads me to conclude that traditional knowledge is indispensable as knowledge, but knowledge and its application into conservation practice are two very different things. A fisherman may possess exquisite knowledge of the natural history of his quarry, but this may as easily lead to overexploitation as to abstinence. This is to say that there is no question that some societies have apparently existed for centuries in a sustainable manner (Johannes, 1978), but we may never know the real effects of long-term human use, as it may be virtually impossible to trace the course of ecosystem sustainment under human use for periods much farther back than the late 1800s when detailed data on environmental use began to accumulate.

Surely, a portion of the solution to our conservation dilemma lies in ethics. Beatley (1994) provides a particularly illuminating way of looking into the complex relationships of humans and Nature. He draws attention to two intersecting axes – one describing the extreme from utilitarian to duty-based approaches, and the other from anthropocentric to ecocentric points of view. Four sectors emerge. Within the utilitarian–anthropocentric sector are the prevalent practices of consumption and growth of most market-oriented cultures. Within the duty-based-biocentric sector are some traditional cultures, animal rights groups, and many ecologists. With respect to the coasts and oceans, it is not clear where the science of conservation fits. Perhaps this is a good thing, for it may mean that we still have time – but not much time – better to define ourselves for the joint purposes of coastal-marine research and its application to conservation.

One thing is certain. Marine and coastal conservation require not only a systems-oriented approach in carrying out the themes of *Diversitas*, but also the forwarding of the information gained to conservation organizations and management agencies, including devices for the inclusion of humanity and human ethics into the global, coast-and-ocean matrix. Perhaps, if we are clever, we will be able to initiate an 'ecocultural revolution', wherein the information gained from our research will be so compelling that conservation will be unavoidable. Perhaps it behoves the scientists themselves to form the new union that is indispensable for global, systems-level conservation by both land and sea. And, hopefully, *Diversitas* provides some fertile ground in which these changes can be nurtured.

Acknowledgements

I am indebted to many colleagues for the thoughts expressed in this chapter, among whom are: Peter Bridgewater, Chief Director, Australian Nature Conservation Agency; J. Frederick Grassle, Director, Institute of Marine and Coastal Sciences, Rutgers University; Pierre Lasserre, Director, Division of Ecological Sciences, UNESCO; Alasdair McIntyre, University of Aberdeen; John C. Ogden, Director, Florida Institute of Oceanography; Frank H. Talbot, James Cook University, Townsville, Australia; and Michael Garstang, Bruce P. Hayden, M. Geraldine McCormick-Ray, and Thomas M. Smith, all of the Department of Environmental Sciences, University of Virginia.

The writing of this chapter was supported by the Global Biodiversity Fund of the University of Virginia, through grants of the Sara Shallenberger Brown of Louisville, Kentucky, and the Munson and Henry Foundations of Chicago, Illinois. I also wish to thank my colleagues on the US National Committee of the IUBS and Talal Younès and Colleen Adam of the IUBS Secretariat in Paris for their support and encouragement.

References

Anderson, D.M. (1994) Red tides. *Scientific American* August, 62–68.
Angel, M. (1993) Biodiversity of the pelagic ocean. *Conservation Biology* 7, 760–772.
Angel, M. (1994) Review of *Global Marine Biological Diversity: A Strategy for Building Conservation into Decision Making*, Norse, E.A. (ed.) 1993. Island Press, Washington, DC, *Nature* 367, 126–127.
Bahr, L.M. and Lanier, W.P. (1981) *The Ecology of Intertidal Oyster Reefs of the South Atlantic Coast: a Community Profile*. Office of Biological Service, Fish and Wildlife Service, US Dept Interior, Washington, DC.
Barrett, G.W. and Bohlen, P.J. (1991) Landscape ecology. In: Hudson, W.E. (ed.) *Landscape Linkages and Biodiversity*. Island Press, Washington, DC, pp. 149–161.
Beatley, T. (1994) *Ethical Land Use*. The Johns Hopkins University Press, Baltimore, MD.
Bridgewater, P.B. (1993) Conservation strategy and research in Australia – how to arrive at the 21st century in good shape. In: Moritz, C., Kikkawa, J. and Doley D. (eds) *Conservation Biology in Australia and Oceania*. Surrey Beatty, Chipping Norton, pp. 17–25.
Carlton, J.T. (1993) Neoextinctions of marine invertebrates. *American Zoologist* 33, 499–509.
Costanza, R. (1987) Social traps and environmental policy. *BioScience* 37, 407–412.

Costanza, R., Kemp, W.M. and Boynton, W.R. (1993) Predictability, scale, and biodiversity in coastal and estuarine ecosystems; implications for management. *Ambio* 22, 88–96.

Cross, J.G. and Guyer, M.J. (1980) *Social Traps*. University of Michigan Press, Ann Arbor.

Daly, H.E. and Cobb Jr, J.B. (1989) *For the Common Good: Redirecting the Economy Toward Community, the Environment, and as Sustainable Future*. Beacon Press, Boston.

di Castri, F. and Younès, T. (1990) Ecosystem function of biological diversity. *Biology International* Special Issue No. 22, IUBS, Paris.

Ehrlich, P.R. and Daily, G.C. (1993) Science and management of natural resources. *Ecological Applications* 3, 558–560.

Fahrig, L., Hayden, N. and Dolan, R. (1993) Distribution of barrier island plants in relation to overwash disturbance: a test of life history theory. *Journal of Coastal Research* 9, 403–412.

Franklin, J.F. (1993) Preserving biodiversity: species, ecosystems, or landscapes? *Ecological Applications* 3, 202–205.

Galtsoff, P.S. (1964) The American oyster *Crassostrea virginica* Gmelin. *Fishery Bulletin of the US Fish and Wildlife Service* 64, 1–480.

Glynn, P.W. (1988) El Niño-Southern Oscillation 1982–1983: nearshore population, community, and ecosystem responses. *Annual Review of Ecology and Systematics* 19, 309–345.

Grassle, J.F. (1991) Deep-sea benthic biodiversity. *BioScience* 41, 464–469.

Grassle, J.F., Maciolek, N.J. and Blake, J.A. (1990) Are deep-sea communities resilient? In: Woodwell, G.M. (ed.) *The Earth in Transition: Patterns and Processes of Biotic Impoverishment*. Cambridge Univeristy Press, Cambridge, pp. 353–393.

Grassle, J.F., Lasserre, P., McIntyre, A.D. and Ray, G.C. (1991) Marine biodiversity and ecosystem function. *Biology International* Special Issue 23. IUBS, Paris.

Hammer, M., Jansson, A. and Jansson, B. (1993) Diversity change and sustainability: implications for fisheries. *Ambio* 22, 97–105.

Hayden, B.P., Ray, G.C. and Dolan, R. (1984) Classification of coastal and marine environments. *Environmental Conservation* 11, 199–207.

Hayden, B.P., Dueser, R.D., Callahan, J.T. and Shugart, H.H. (1991) Long-term research at the Virginia Coast Reserve. *BioScience* 41, 310–318.

Holligan, P. (ed.) (1990) *Coastal Ocean Fluxes and Resources*. IGBP Global Change Rept no. 14. International Geosphere Biosphere Programme, Stockholm.

Holligan, P. and de Boois, H. (1993) *Land–Ocean Interactions in the Coastal Zone (LOICZ): Science Plan*. IGBP Global Change Rept No. 25 International Geosphere Biosphere Programme, Stockholm.

Holling, C.S. (1993) Investing in research for sustainability. *Ecological Applications* 3, 552–555.

Huntley, B.J., Ezcurra, E., Fuentes, E.R., Fujii, K., Grubb, P.J., Haber, W., Harger, J.R.E., Holland, M.M., Levin, S.A., Lubchenco, J., Mooney, H.A., Neronov, V., Noble, I., Pulliam, H.R., Ramakrishnan, P.S., Risser, P.G., Sala, O., Sarukhan, J. and Sombroek, W.G. (1992) A sustainable biosphere: the global imperative the international sustainable biosphere initiative. *Bulletin of the Ecological Society of America* 73, 7–14.

Hutchinson, G.E. (1959) Homage to Santa Rosalia or why are there so many kinds of animals? *American Naturalist* 93, 145–159.

IUCN (1993) *Parks for Life: Report of the IVth World Congress on National Parks and Protected Areas*. IUCN, Gland, Switzerland.

Jackson, J.B.C. (1991) Adaptation and diversity of reef corals. *BioScience* 41, 475-482.

Johannes, R.E. (1978) Traditional marine conservation methods in Oceania and their demise. *Annual Review of Ecology and Systematics* 9, 349-364.

Kahn, A.E. (1966) The tyranny of small decisions: market failures, imperfections and the limit of economics. *Kyklos* 19, 23-47.

Karieva, P. (1994) Ecological theory and endangered species. *Ecology* 75, 583.

Kennedy, V.S. (1990) Anticipated effects of climate change in estuarine and coastal fisheries. *Fisheries* 15, 16-24.

Kennedy, V.S. and Breisch, L.L. (1993) Sixteen decades of political management of the oyster fishery in Maryland's Chesapeake Bay. *Journal of the Environment* 16, 153-171.

Ketchum, B.K. (ed.) (1972) *The Water's Edge: Critical Problems of the Coastal Zone*. Massachusetts Institute of Technology Press, Cambridge, MA.

Klee, G.A. (1985) Traditional marine resource management in the Pacific. In: McNeely, J.A. and Pitt, D. (eds), *Culture and Conservation: the Human Dimension in Environmental Planning*, Croom Helm, London, pp. 193-202.

Lasserre, P., McIntyre, A.D., Ogden, J.C., Ray, G.C. and Grassle, J.F. (1994) Marine laboratory networks for the study of biodiversity, function, and management of marine ecosystems. *Biology International*, Special Issue no. 31, IUBS, Paris.

Levin, S.A. (1993) Science and sustainability. *Ecological Applications* 3, 545-546.

Loya, Y. (1990) Changes in the Red Sea coral community structure: a long-term case history study. In: Woodwell, G.M. (ed.) *The Earth in Transition: Patterns and Processes of Biotic Impoverishment*. Cambridge University Press, Cambridge, pp. 369-384.

Lubchenco, J., Olson, A.M., Brubaker, L.B., Carpenter, S.R., Holland, M.M., Hubbell, S.P., Levin, S.A., MacMahon, J.A., Matson, P.A., Melillo, J.M., Mooney, H.A., Peterson, C.H., Pulliam, H.R., Real, S.A., Regal, P.J., and Risser, P.G. (1991) The sustainable biosphere initiative: an ecological research agenda. A report from the Ecological Society of America. *Ecology* 72, 371-412.

Ludwig, D., Hilborn, R. and Walters, C. (1993) Uncertainty, resource exploitation, and conservation: lessons from history. *Science* 260, 17-36.

May, R.M. (1988) How many species are there on Earth? *Science* 241, 1441-1449.

McCormick-Ray, M.G. (1987) Hemocytes of *Mytilus edulis* affected by Prudhoe Bay crude oil emulsion. *Marine Environment Research* 22, 107-122.

McGowan, J.A. (1990) Species dominance-diversity patterns in oceanic communities. In: Woodwell, G.M. (ed.) *The Earth in Transition: Patterns and Processes of Biotic Impoverishment*. Cambridge University Press, Cambridge, pp. 395-421.

McNeely, J.A. and Pitt, D. (eds) (1985) *Culture and Conservation: the Human Dimension in Environmental Planning*. Croom Helm, London.

Morison, S.E. (1958) *Strategy and Compromise*. Little, Brown, Boston and Toronto.

Munro, J.L., Parrish, J.D. and Talbot, F.H. (1987) The biological effects of intensive fishing upon coral reef communities. In: Salvat, B. (ed.) *Human Impacts on Coral Reefs: Facts and Recommendations*. Muséem Naturelle D'Histoire et Ecole Pratique des Hautes Études Antenne de Tahiti Mus., E.P.H.E. Moorea, Polynésie française, pp. 41-49.

Newell, R.I.E. (1988) Ecological changes in Chesapeake bay: Are they the result of overharvesting the American oyster, *Crassostrea virginica*? In: *Understanding the*

Estuary: Advances in Chesapeake Bay Research. Chesapeake Research Consortium Publication 129. CBP/TRS 24/88, pp. 536–546.

Norse, E.A. (ed.) (1993) *Global Marine Biological Diversity: A Strategy for Building Conservation into Decision Making.* Island Press, Washington, DC.

Odum, E.P. (1989) *Ecology and our Endangered Life-Support Systems.* Sinauer, Sunderland, MA.

Odum, E.P. (1992) Great ideas in ecology for the 1990s. *BioScience* 42, 542–545.

Odum, W.E. (1982) Environmental degradation and the tyranny of small decisions. *BioScience* 32, 728–729.

O'Neill, R.V., DeAngelis, D.L., Waide, J.B. and Allen, T.F.H. (1986) *A Hierarchical Concept of Ecosystems.* Princeton University Press, Princeton, NJ.

Platt, J. (1973) Social traps. *American Psychologist* 28, 641–651.

Polunin, N.V.C. (1985) 'Traditional marine practices in Indonesia and their bearing on conservation. In: McNeely, J.A. and Pitt, D, (eds) *Culture and Conservation: the Human Dimension in Environmental Planning.* Croom Helm, London, pp. 155–179.

Raloff, J. (1994) Something's fishy: marine epidemics may signal environmental threats to the immune system. *Science News* 146, 8–9.

Raven, P.H. and Wilson, E.O. (1992) A fifty-year plan for biodiversity surveys. *Science* 258, 1099–1100.

Ray, C. (1970) Ecology, law, and the 'Marine Revolution'. *Biological Conservation* 3, 7–17.

Ray, G.C. (1991) Coastal-zone biodiversity patterns. *BioScience* 41, 490–498.

Ray, G.C. and Hayden, B.P. (1992) Coastal zone ecotones. In: Hansen, A.J. and di Castri, F. (eds) *Landscape Boundaries: Consequences for Biotic Diversity and Ecological Flows.* Springer-Verlag, Berlin, pp. 403–420.

Ray, G.C. and McCormick-Ray, M.G. (1989) Coastal and marine biosphere reserves. In: Gregg Jr, W.P., Krugman, S.L. and Woods Jr, J.D. (eds) *Proceedings of Symposium on Biosphere Reserves,* Fourth World Wilderness Conference US National Park Service and US MAB, Washington, DC, pp. 68–78.

Ray, G.C. and McCormick-Ray, M.G. (1992) *Marine and Estuarine Protected Areas: A Strategy for a National Representative System within Australian Coastal and Marine Environments.* Report for Australian National Parks and Wildlife Service, Canberra, Australia.

Ray, G.C., Hayden, B.P., Bulger Jr, A.J. and McCormick-Ray, M.G. (1992) Effects of global warming on the biodiversity of coastal-marine zones. In: Peters, R.L. and Lovejoy T.E. (eds) *Global Warming and Biological Diversity.* Yale University Press, New Haven, pp. 91–104.

Ray, G.C., Hayden, B.P., McCormick-Ray, M.G. and Smith, T.M. (1995) Landscape diversity of the US east coast coastal zone, with particular reference to estuaries. In: Ormond, R. and Gage, J. (eds) *Marine Biodiversity: Patterns in the Sea.* Cambridge University Press, Cambridge (in press).

Rohlf, D.J. (1990) Six biological reasons why the Endangered Species Act doesn't work – and what to do about it. *Conservation Biology* 5, 273–282.

Sherman, K., Alexander, L.M. and Gold, B.D. (1993) *Large Marine Ecosystems: Stress, Mitigation, and Sustainability.* AAAS Press, Washington, DC.

Simpson, B.B. (1988) Biological diversity in the context of ecosystem structure and function. *Biology International* no. 17, 15–17.

Steele, J.H. (1991) Marine functional diversity. *BioScience* 41, 470–474.

Swanson, F.J., Kratz, T.K., Caine, N. and Woodmansee, R.G. (1988) Landform effects on ecosystem patterns and processes. *BioScience* 38, 92–98.

Swap, R., Garstang, M. and Greco, S. (1992) Saharan dust in the Amazon Basin. *Tellus* 44B, 133–149.

Udvardy, M.D.F. (1975) *A Classification of the Biogeographical Provinces of the World*. IUCN Occasional paper no. 18, IUCN, Gland, Switzerland.

Urban, D.L., O'Neill, R.V. and Shugart Jr, H.H. (1987) Landscape ecology. *BioScience* 37, 119–127.

Winslow, F. (1882) *Methods and Results. Report of the Oyster Beds of the James River, Va., and of Tangier and Pocomoke Sounds, Maryland and Virginia*. US Coast and Geodetic Survey, Government Printing Office, Washington, DC.

WRI (1992) *Global Biodiversity Strategy: Guidelines for Action to Save, Study, and Use Earth's Biotic Wealth Sustainably and Equitably*. World Resources Institute, Washington, DC.

23 Biological Conservation in a High Beta-diversity Country

J. Sarukhán[1], J. Soberón[2]* and J. Larson-Guerra[2]

[1]Universidad Nacional Autónoma de México, Torre de Rectoría 6° Piso, Cd. Universitaria, DF 04510, México; [2]Comisión Nacional para el Conocimiento y Uso de la Biodiversidad, Fernández Leal 43, Barrio de la Concepción, Coyoacán, DF 04020, México

Introduction

The conservation of biodiversity, as well as its destruction, are problems that transcend economic, political, religious and cultural frontiers as it is clearly expressed in the Convention on Biological Diversity (UNEP, 1992). In this sense, it may seem difficult to justify the existence of a national, in our case Mexican, view of the problems involved with the conservation of biodiversity. However, neither human needs, culture and capabilities nor nature's wealth are uniformly distributed throughout the planet. As a result, different views on the issues of biodiversity conservation have been put forward and they will, in many cases, be the framework for the development of national and international policies and strategies.

In this chapter we will argue in favour of a particular view of the biodiversity challenge that is based on, first, an ecological and biogeographical fact: Mexico's high beta-diversity; and second, on some national social realities: the highly disaggregated distribution pattern of rural population and the extent of its cultural variety. These facts lead us to emphasize that knowledge, to increase our understanding of biodiversity (identity, distribution and processes) and a multiple strategy of biological resource use (to promote rational development), are the key to sound policy design and actions that favour biological conservation.

---

*Corresponding author.

Table 23.1. Mexico's species richness and endemism in selected taxa.

Taxa	World	Mexico (% of world)	Endemic to Mexico (%)
Flowering plants	250,000	22,000(c. 10)	52
Gymnosperms	600	71(12)	No data
Pteridophytes	12,000	1,000(8)	c. 20
Amphibians	4,000	284(7)	60
Reptiles	6,550	717(11)	51
Birds	9,672	961(10)	9
Mammals	4,327	439(10)	31
Freshwater fishes	8,411	347(4)	23
Papilionoidea	20,000	2,237(11)	9

A High Beta-diversity Developing Country

Megadiversity as a result of high beta-diversity

Mexico has an area of 1,953,162 km² including 0.31% of insular territory (INEGI, 1993). It is located north of the Equator and about half of the territory lies south of the Tropic of Cancer. There are eleven morphotectonic provinces (Ferrusquía-Villafranca, 1993) and the country is topographically complex, one-third of Mexico has slopes greater than 25° (Tamayo, 1949 in Flores-Díaz, 1974). There are 20 biotic provinces (Alvarez and de Lachica, 1974). Toledo and Ordóñez (1993) identified six ecological zones: humid tropic (11.3% of terrestrial surface), subhumid tropic (20.5%), humid temperate (0.5%), subhumid temperate (16.9%), arid and semiarid (50.7%) and alpine (less than 0.1%).

Mexican territory roughly contains 1.5% of the World's total emerged lands (INEGI, 1991), but it contains an overall 10-12% of the known species (Toledo and Ordóñez, 1993) and species endemism percentages within groups range from 9% to 60% depending on the group (Table 23.1). In general, the explanation for this high species richness lies in a complex geologic history, varied climate and the fact that the Neartic and Neotropical realms meet in Mexico (Ramamoorthy et al., 1993). However, measuring diversity is more than merely counting species present in a country. Specific distribution patterns and local diversity have important consequences for the conservation of biodiversity and the use of biological resources. Species diversity measures have been classified in alpha, beta and gamma types in order to explicitly recognize the different spatial scales of some of the processes that determine species richness (Whittaker, 1972; Wilson and Shmida, 1984). Gamma (or regional) diversity is the one that makes Mexico a megadiversity country (Mittermeier, 1988; Toledo, 1988) and the main suggested determinants at this level are historical factors (Shmida and Wilson, 1985). Alpha (or local) diversity measures for different taxa in different types of ecosystems do not qualify specific communities in Mexico as particularly rich compared to other similar

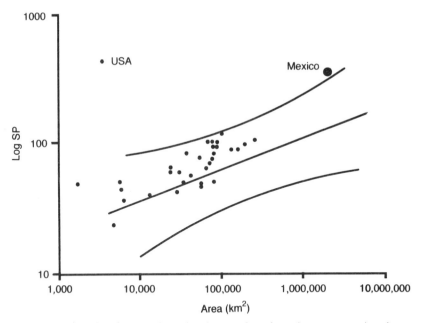

Fig. 23.1. Beta-diversity of mammal species. Log number of species vs. state size. Average number of species of mammals (including Chiroptera) against the log of size of states. Using the regression line, the predicted value for the whole country is far below the observed value. This means that México has many more species than expected for its size, (from Arita, 1993).

communities throughout the world (e.g. tropical rain forest trees in Meave, 1989; butterflies in Raguso and Llorente, 1991).

In order to explain this contrasting pattern between high gamma (regional) and an average (relative to similar ecosystems) alpha (within habitat) diversity, it is useful to look at among-communities diversity. Beta-diversity measures the turnover in species when one moves along a gradient or through different habitats. This species replacement is a good measure of community turnover and habitat heterogeneity (Wilson and Shmida, 1984). Habitat heterogeneity and historical geologic and biotic factors combine to produce high beta-diversity. As of yet, there are no standard means to measure and compare beta-diversity, however there are excellent examples that clearly show its meaning. In Fig. 23.1 the states of Mexico are plotted according to size and mammal species richness. The predicted species richness for the total area of Mexico considered as a whole is much lower than the observed value, which means that states do not share many species: that is, the size of the lists may be similar but their species composition is different.

In addition to species level richness, Mexico is also an important centre of origin and diversification of cultivated plants (Fowler and Mooney, 1990) and as such, houses the wild and managed genetic resources of many useful species and their relatives (i.e. maize, agave, beans, peppers and *Amaranthus* spp. the 'happiness seed') (Hernández-Xolocotzi, 1993). This kind of genetic diversity is closely related to culture and it is not a mere gift of nature. Domestication is a long evolutionary and historical process whose legate belongs to humanity, however, the *in situ* conservation, use and further maintenance of the evolutionary viability of these resources is a daily activity of indigenous and peasant people in Mexico and elsewhere (Leff and Carabias, 1993 and references therein).

Population distribution

The long-term demographic history of Mexico is marked by the rapid decline in population size that began with the Spanish conquest in 1521. Several authors note that the pre-Columbian population was highly dispersed (Broda, 1979). Estimations about population size before Spanish arrival vary wildly. Considering Central Mexico, from the Tehuantepec Isthmus to the Lerma and Panuco Rivers, they range from 4.5 to 25 million people. Postconquest population size estimations have smaller errors and show a dramatic decline, mainly due to disease, from 6 million in 1548, to 2.5 million in 1568 and only 1 million in 1605. In 60 years there was a 75% or 97% – depending on the estimate we use – decrease in indigenous population size. The consolidation of the political and socioeconomic colonial structure during the 17th and 18th centuries favoured the recovery of rural population (INEGI, 1990 and references therein). By the middle of the 20th century the country had 6 million inhabitants. Until the first half of the 20th century Mexico was still a predominantly rural society (Fig. 23.2). It is only around 1960, for the first time in Mexico's history, that half the population was living in localities with more than 2500 citizens. This decade marks the beginning of Mexico's transition to an urban society.

It is important to note that although urbanization is the main demographic trend in Mexico, the rural population has more than doubled during this century. Rural population growth is coupled with a highly dispersed distribution pattern. Information from the 1990 census (Fig. 23.2) comes from 156,602 localities, 98% of which have less than 2500 inhabitants and 95% have less than 1000 (Fig. 23.3). Most localities are concentrated in central Mexico (see inset in Fig. 23.3) which is also a major endemism zone for taxa like birds (Benitez *et al.*, 1994). This atomization of the rural population reflects an extensive use of the territory and is illustrative of the difficulties involved in the organization of commercialization of products and the introduction of basic services like electricity, drinking water, drainage, education and health. The dispersion pattern of rural populations that prevailed during pre-Columbian times is again seen 500 years after the conquest. It is an aspect that is closely related to the problems and possible solutions related to conservation and the use of biodiversity and biological resources in a high beta-diversity country.

Cultural diversity

The situation of indigenous languages reflects cultural diversity because language facilitates cultural inheritance of information (Hernández-Xolocotzi, 1993). At the time of the conquest 120 languages were spoken and 54 were

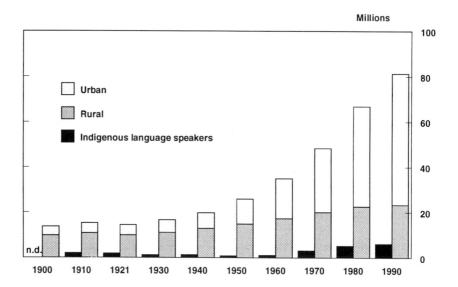

Fig. 23.2. Population growth in Mexico 1900-1990. Urban population represents those living in localities bigger than 2500 inhabitants, except 1910 (bigger than 4000) and 1921 (bigger than 2000). Indigenous language speakers are only those of 5 or more years of age (from INEGI, 1990 and INEGI, 1991).

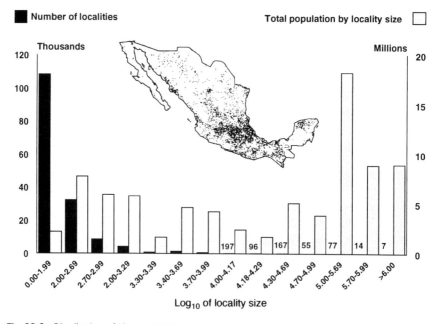

Fig. 23.3. Distribution of the population by locality size in 1990. Data from 156,602 localities; 98% of which have less than 2500 inhabitants and 95% have less than 1000. Inset map represents localities of more than 1000 inhabitants which represent less than 5% of total localities (from INEGI 1991).

in use in 1986 (Martínez, 1986 in Hernández-Xolocotzi, 1993). Figure 23.2 shows the growth of population (five years or more of age) speaking native languages. It can be clearly seen that since the 1950s, indigenous language speakers are a permanently growing population reaching approximately 6,000,000 in 1990. Official census data used for Fig. 23.2 show that 23 languages account for 98% of the indigenous language speakers.

The multiple processes involved in the relation between cultural and biological diversity still remain unclear (Bye, 1993). Nevertheless, coevolution of many plants and human culture has been well documented (Fowler and Mooney, 1990). Economic development (mainly in the primary sector) and growth of human populations are often a direct threat to biological diversity through direct habitat transformation and overexploitation. However, there are many ways to transform habitats and to use resources, and since some have proved to be much more sustainable than others (in terms of productivity and/or diversity), therein might lie clues to the future of biological resource use and biodiversity conservation. The fact that middle America is considered a major centre of plant domestication is evidence of both biological wealth and of cultural development and diversity. Many species have been domesticated in the area and this reveals a close relation between man and plants. As many as 3352 species (more than 10% of the Mexican flora) have ethnobotanical references of medicinal use and almost a quarter of the Mexican vascular flora are used by humans (Bye, 1993). Evidence of anthropogenic activities in sites once thought of as pristine habitats have been accumulating, particularly in the Mayan region (Gómez-Pompa and Kaus, 1992). For a long time, anthropologists and biologists have documented the variety of practices and the depth of knowledge that allows many Mexican indigenous groups or peasant communities to make a sustained use of their rich resource base (e.g. Alcorn, 1984; Browner, 1985; Caballero and Mapes, 1985).

Conservation Biology

The cultural diversity and population dispersal of rural Mexico shows that conservation biology cannot be attained without considering the well-being and the incorporation of the know-how of peasant and indigenous populations. Cultural diversity relates to a local scale where people directly interact with their biological resources. This is the very basis of the general process in which a nation uses biological resources. We think that this local scale, on which actors use, understand, conserve or destroy biodiversity and biological resources, is the fundamental reference to a sound strategy on biological conservation. However, when addressing conservation issues we have to articulate local, regional, national and global scales. On each scale there are different sets of actors, information and decisions. These different sets involve biology (mainly systematics, genetics and ecology), culture (rural and urban, traditional or 'occidental'), society, economics, trade, and institutions.

In the first two sections we have concentrated on beta-diversity, demography and culture. In the following sections, we will outline issues on different scales and themes, such as the achievements of protected areas in Mexico, some examples of the viability of conservation outside parks through the application of traditional and/or innovative approaches to resource use, and a commented list of the issues that need to be addressed within a national strategy for biodiversity use for conservation.

Natural protected areas in Mexico

The main lesson we can learn from the fact that Mexico is a high beta-diversity country, is that to preserve, by means of protected areas, a significant proportion of biodiversity, we would need to cover a large number of particular places and sites. Otherwise, high species turnover would lead to the exclusion of many biological communities, species, populations and genes from protected areas. For example, a workshop on the identification of priority areas for conservation in Mexico, from which a yet unpublished document is being compiled by WWF (Williams-Linera et al., 1992), mentions 198 areas that are biologically important. Excluding so much land from human use is an economic, social, and political impossibility. This implies that a strategy of biological conservation based only on natural reserves and parks in a territory with high beta-diversity is unlikely to succeed in the conservation of a big fraction of biodiversity. In a homogeneous region, regarding both topographic and biotic factors, or in a complex but unpopulated area, a strategy of preserving through protected areas may seem like a sound strategy (setting aside considerations of the demographic and genetic viability of the populations), because a high percentage of the species will be found in the reserves, but this is not so in a case like Mexico.

Nevertheless reserves are important and necessary. Mexico has made significant efforts to set aside land for sustainable use, conservation and research on biodiversity. The first national parks were created in Mexico in 1876 and 1898. During the late 1930s and beginning of the 1940s, there was a major increase in the number of protected areas, The concept then prevailing was that of a national park (shown as recreation areas in Fig. 23.4) and these areas were usually small (around a thousand hectares or less) and of very little practical use in terms of biodiversity conservation. In 1979, the first area dedicated specifically to conservation in a modern sense (a Biosphere Reserve) was decreed and their number has continued to grow. The important difference is that these areas, even if less numerous than parks, account for ten times more surface. Mexico made an important contribution to the Biosphere Reserve concept when scientific research centres and local communities were incorporated into their management scheme (Halffter, 1988). The examples of the Sierra de Manantlán Biosphere Reserve (related to the Universidad de Guadalajara) and the Mapimi and La Michilia Biosphere Reserves (Instituto de Ecología AC) are representative of the success of this approach (Williams-Linera et al., 1992).

During the 1980s and the first three years of the 1990s there has been a considerable growth in the number and surface of protected areas dedicated

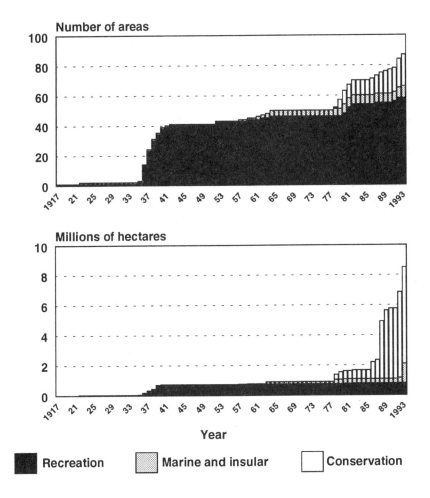

Fig. 23.4. Growth in number and surface of protected areas in Mexico from 1917 to 1993. Recreation areas represent national monuments and parks. Marine and insular groups parks, biosphere reserves and areas for the protection of wild flora and fauna. Conservation areas group continental biosphere reserves and areas for the protection of wild flora and fauna (from CONABIO's database).

entirely to conservation (Fig. 23.4). In December 1993, Mexico had 21 Biosphere reserves (or equivalent categories) accounting for 6,441,451 hectares, i.e. 3.26% of Mexico's land territory, plus 1,283,171 hectares in insular or marine protected areas, and 0.04% dedicated to recreation (national parks and monuments).

Conservation outside parks

Of course, setting aside reserves and protecting new areas will always be required because there are many cases in which the ecology of a species or groups of species justify total, or almost total isolation from human activities. However, widespread use of the countryside will continue in Mexico in the foreseeable future (as opposed to some highly industrialized countries where rural areas are being 'liberated' for conservation and restoration (Green, 1985), either by a peasant population or by producer associations and private companies.[1]

There is no doubt that large-scale transformation from one type of habitat (i.e. tropical rain forest) to a very different, simplified one (i.e. cattle pasture) will have disastrous consequences for most species. Environmental damage and loss of species is intrinsic, not incidental, to the maintenance of heavily productive ecosystems (Green, 1985). However, primary forests are often used for cocoa or coffee farming, for extractionist purposes or for selective tropical logging that do not modify the structure of the forest as drastically as some 'modern' practices do. The hard data, as opposed to island biogeography theoretical estimates, of the relation between forest fragmentation and extinction is not conclusive at all. There are many examples of extensive habitat transformation that still preserve a substantial number of species. For example, Lugo (1988) discusses the case of Puerto Rico, where 99% of primary forests has been lost to coffee (shaded) plantations and secondary forest has taken place. However, the total loss of bird species is seven out of 60 original, pre-Columbian species, instead of the more than 50% predicted by island biogeography theory, and the actual number of bird species is 97, due to exotics that have enlarged the species pool.

Brown and Brown (1992) discuss the case of the Brazilian Atlantic forest which has been reduced by 88% to a very fragmented forest, subject to a variety of human uses. According to this study there are 171 endangered species of vertebrates and invertebrates, however only six are listed as possibly extinct. Many suitable habitats for them, however, are still unexplored and directed exploration in the last 20 years has proven that several bird and butterfly species thought to be extinct were still present in previously unexplored sites. In Los Tuxtlas Rain Forest, Dirzo and García (1992) reported a forest cover reduction of 84% coupled with a clear pattern of insularization (fragmentation) of the remaining forest. Even if most large mammals have been locally lost (Dirzo and Miranda, 1991), the remnant fragments still preserve almost the complete butterfly fauna (Raguso and Llorente, 1991).

Analysis of many examples of indian or peasant economies and practices have led a number of authors to suggest not only that there are many examples of diversified use of ecosystems, but that there is no such a thing as a pristine habitat in much of the tropical area (Gómez-Pompa and Kaus, 1992 and references therein) and that many cultures have not only not degraded the diversity of their resources but may have even increased it. In a remarkable example, Nabham *et al.* (1991) show that protection of natural areas does not necessarily lead to higher species richness, as shown by the decreased diversity over a 25-year period in a protected area in the USA relative to a similar one where the O'odham people maintained their traditional ways. Robinson and Redford (1991) reviewed the evidence on sustainable use of fauna, and, although the matters are far from settled, there is no doubt of the feasibility of maintaining human populations while keeping a fairly large number of species. In Kenya, non-park areas have as rich and diverse fauna as parks (Western, 1989).

However suggestive, the data quoted above do not address the extent of modification in the ecological properties of natural communities, the genetic erosion of populations or the disappearance of small or inconspicuous fauna or flora. There are indeed particular species, or groups of them, that are very sensitive to management or environmental modifications. An example is presented by Thiollay (1992) who shows that selective logging in a Guianan primary rain forest, that affects only three trees per hectare, provoked the disappearance of most of the understory birds from the logged places. These species had not returned ten years later. In other cases, the social or ethnical details of use may be incompatible with an open, capitalistic approach to biological resource exploitation. For example, Peluso (1992) describes the failure, in Indonesia, of 'modern' methods of exploitation of valuable ironwood (*Eusideroxylon zwageri*) versus the complex and subtle native ways of exploiting it both for local use and commercial sale outside the villages.

Despite the valid counterarguments, the examples presented above support the idea that there are forms of large-scale human use of land that are compatible with the persistence of high percentages of at least the commonly monitored species. The argument is that these types of use should complement the systems of protected areas where human activities are heavily restricted.

Biological Resource Use Strategies for Conservation

Biodiversity is an umbrella concept that includes genes, species, communities, ecosystems and global ecology. This is why it is difficult to organize in a clear and organized scheme the issues related to its conservation, the causes of its destruction, the actors related to them, and the actions that need to be undertaken. When biodiversity is valued, sold, or used it becomes a biological resource. For many of the issues mentioned there are no defined positions and many of them are nowadays the subject of intense debate. The biological, cultural and social heterogeneity of a country can be seen as an obstacle to productivity. We believe that, on the contrary, this heterogeneity can be the basis for an ecologically and economically sustainable landscape and society. To do this, we need to consider many issues and the following are only some of them. Those related to valuing and using biodiversity and biological resources are emphasized, because we think that this is one of the main areas of neglect, debate and in urgent need of attention.

Biodiversity legislation

The nation and its states represent a scale on which the articulation of a national strategy can be carried out. Environmental and conservation policy in Mexico has had a complex development. Until the 1970, forestry and hunting laws were the most important legal instruments related to biological resource use and conservation. The environment was introduced during the 1970s in a public health law and it was not until the 1980s that a general law on ecological equilibrium and environmental protection was issued. This gradual growth of the legal framework on environmental policy was paralleled by the growth (in size and expertise) of the public institutions dedicated to their enforcement. The result is a nested array of legislation and technical regulations, concentrated on the ecological ordinance of territory, pollution, natural resource management, protected areas and environmental education, and involving federal, state and municipal jurisdictions (Carabias, 1990a). However, not only environmental and ecological legislation and administration are related to biodiversity issues. A strategy needs to consider fundamental issues (e.g. natural accounts, intellectual property of germplasm and knowledge, natural product commercialization and certification) that are related to legal frameworks like the North American Free Trade Agreement, the GATT and national Forestry and Agriculture Laws.

Monetary value of biodiversity

One of the points we have to explore is not only whether alternative uses of land are economically competitive in increasingly competitive and global markets, but to question whether there are viable strategies to keep this value as a concrete income for the local people who conserve biodiversity and preserve and improve genetic resources with their daily agricultural practices. There are many examples of the very significant economic value of biological resources (Table 23.3 and also Myers, 1988). Almost always, however, locals are paid only a small fraction of the final value of their products or of the germplasm they conserve. A fundamental issue of the strategy is to include in trade policy the 'natural' products that are available to those living in areas that are conservation priorities. The actual trade framework is orientated to highly manufactured products. 'Softer' products (e.g. organic coffee, 'chicle', ornamental plants) also need to be supported by the commercialization infrastructure of the nation. Many of the issues of commercialization in global markets can be approached through quality control and scientific certification systems oriented to satisfy certain markets.

National accounts and environmental services

Some recent attempts to recalculate national accounts, including wasteful use of 'natural capital', show the dramatic ecological cost of economical growth (Repeto, 1989). In the case of Mexico, water has always been a limiting resource for agricultural development (Carrasco, 1979) and soil degradation is a particularly costly process given that only about 15% of the land is suitable for agriculture and estimates of severe to total erosion range from 26% to 47% of the national surface (Toledo et al., 1989). Vegetation cover and agriculture, cattle and forestry practices are tightly related to the two

Table 23.2. Selected examples of the value of biodiversity and biological resources.

Source of value	Country or region	Benefits or damage (USD)	Reference
Extractive sustainable use of latex, nuts and fruits	Peruvian primary forest	$6640 ha^{-1}	Peters et al. (1989)
'Best' modern use (monoespecific forestry)		$4184 ha^{-1}	
Clearing of four species of medicinal plants Discount rate=5% Rotation=50 years	1/4 ha of 50 years old secondary forest in Belize	$3327	Balick and Mendelsohn (1992)
'Best' modern use (monoespecific forestry) in neighbouring Guatemala		$3184 ha^{-1}	
Export quality butterflies, very rare species are excluded	1 ha of secondary forest in Chiapas, México	$2976 ha^{-1} year^{-1}	de la Maza and Soberón (unpublished data)
Genetic contributions of cultivars of maize to crop yields	USA	– 71% of yield gains in single cross hybrids from 1930 to 1980 – Gains to N. Dakota of $2.3 million year^{-1} 1985-1989	World Conservation Monitoring Center (1992)
Genes from tomatoes collected in Perú	USA	$8 million in profit 1962	Iltis (1988)
Rattan exports	Indonesia	90 million per year in 1980s	Cornelius (1984) in Myers (1988)
Non-wood products in all, totalling 80,000 tons (1982)	Indonesia	28 million 1973 200 million 1982	Gillis (1986) in Myers (1988)

processes mentioned. Mexico is making attempts to include a group of variables called non-produced capital goods (these include soil, water, air and some types of wild vegetation) into its system of national accounts (Jarque, 1994).

Agriculture, ranching and forestry policy

Governments have seldom promoted rural production practices that preserve biodiversity or use it in a sustainable way. In Mexico, until very recently, the practice has been to promote 'productivity' by means of agrochemicals and use of improved varieties that very often are associated with drastic reductions of biodiversity (Toledo et al., 1989).

One of the reasons for this is that markets of biodiversity are often illegal or paralegal, and that the conventional economical analyses do not include the 'natural capital' depreciation due to destructive uses of nature. The North American Free Trade Agreement (NAFTA) and General Agreement of Tariffs and Trade (GATT) frameworks, as well as legislative changes concerning rural property, all involve (directly or indirectly) modification of practices that have been going on (right or wrong) for many years. Potential changes have to be carefully considered, many of them can be positive but some of them are likely to threaten biodiversity even more. Natural resource management in a wide sense (e.g. ecosystems, soils, land use patterns) have to incorporate the insight of the biodiversity approach.

Diversity and commercialization

Worldwide, the dominant trade perspective favours productivity and specialization (Bhagwati, 1993). However, ecologically sustainable productive systems need to be diverse. This implies less 'productivity' of a more diverse array of products. There is no doubt that there are many marketable resources in diverse ecosystems. Peters et al. (1989) found that a hectare of Peruvian Amazon forest subject to extractive, probably sustainable use, yielded a higher yearly income than the same hectare subjected to a conventional, 'developed' use that was destructive and unsustainable. Balick and Mendelsohn (1992) complemented the analysis by adding medicinal plants used in local markets in Belize, and they reached the same conclusion. A similar, unpublished study, by de la Maza and Soberón, using butterflies in Chajul, Chiapas also supported the result. Salafsky et al. (1993) make an interesting evaluation of two extractive tropical reserves that considers the phenology of the harvested species, the 'phenology' of markets demand and prices and the problems involved in commercialization of their products.

Perhaps we should think of the possibilities of different production and commercialization approaches for the 21st century. If we think of wild and managed territories as sources of grass for cattle, we will not go far either in conservation or in the sustainable production of food and other goods. Alternative markets and products are being developed, and many that we don't yet imagine will appear that will make different approaches to tropical forest cropping and agricultural practices very appealing in commercial terms (Prance et al., 1987). Examples of these alternatives are butterfly farms in Papua New Guinea (National Research Council, 1983), emerging international markets for soft medicine (Bonati, 1991) and plants in general, and growing groups of organic product consumers (e.g. coffee, cacao). Advancing national policies that encourage these practices and developing national standards for their production and certification will result in a sound strategy in the long-run. However, we cannot only promote national and international trade of natural products. The regional and local dimension of markets is often neglected and they should be given particular attention since they are especially important for food production, nutrition and also for germplasm conservation through the maintenance of markets for landraces (Dewey, 1978; Carabias 1990b; Prescott-Allen and Prescott-Allen, 1990).

Genetic resources

The value of known and unknown genetic resources, particularly for medicine and food, has been recognized for many years by the scientific community. Conservation of traditional agroecosystems is the most reasonable strategy for *in situ* preservation of the germplasm of several crops (Wilken, 1977; Clawson, 1985; Altieri et al., 1987; Fowler and Mooney, 1990). Many crops' genetic resources, including wild relatives, are disappearing at rates that we do not even imagine and they will be fundamental if the promises that biotechnology is making to humanity are to be fulfilled. This issue is the subject of debate since it involves intellectual property rights, strong economic interests, the ethics of humanities legacy and the limits to the commercialization of life (Altieri, 1990; Brooks and Murphy, 1989; Dahlberg, 1987; Kloppenburg and Rodriguez, 1992). Even if consensus is far from being reached, genetic resources have to be explicitly considered within a national strategy on biological resources and biodiversity.

Conclusions

We have presented a panorama composed of three dominant factors: (i) a very high spatial turnover of species over a complex landscape; (ii) a numerous, heterogeneous, and highly dispersed rural population that has a rich tradition of sustainable use of the biota and is also in urgent need of incorporation into national development, and (iii) a fairly well-developed institutional capacity to manage biodiversity and biological resources coupled with a complex policy and strategy challenge. The issues suggested as priorities within a strategy for the use of biological resources in a way that contributes to the creation of a sustainable landscape have consequences on local, regional, national, and global scales. Many of them are considered within the general conceptual framework of the Biodiversity Convention. The issues we have outlined are merely a fraction, nevertheless significant, of a multiple use strategy for biodiversity conservation. Perhaps many of the issues are related to the legislative and operative implementation of the recommendations and obligations signed by Mexico, and more than a hundred other countries in Rio in 1992.[2] A national view of biodiversity starts from biological facts and processes, and in Mexico, the main pattern is heterogeneity. It has to build upon those aspects directly related to the resources. Its success, however, is also related to national and international administration and policies. A strategy for biodiversity use and conservation has to be as umbrella-like as the concept of biodiversity itself.

Notes

1. Article 27 of the Mexican Constitution establishes the faculty of the nation to impose modalities to property and to regulate the use of natural resources and oversee their conservation. Modifications introduced in 1992 will facilitate the formation of bigger productive associations (thus managing larger areas under agroindustrial schemes) than were previously possible (Téllez, 1993).
2. Mexico ratified the Convention on 11 March 1993.

References

Alcorn, J.B. (1984) Development policy, forests and peasant farms: reflections on huastec-managed forests contributions to commercial production and resource conservation. *Economic Botany* 38, 389–405.

Altieri, M.A. (1990) How common is our common future? *Conservation Biology* 4(1), 102–103.

Altieri, M.A., Anderson, M.K. and Merrick, L.C. (1987) Peasant agriculture and the conservation of crop and wild plant resources. *Conservation Biology* 1(1), 49–58.

Alvarez, T. and de Lachica, F. (1974) Zoogeografia de los vertebrados de México. In: *México: Panorama Histórico y Cultural*. Vol. II. *El Escetiario Geográfico*. SEP/INAH. México, pp. 219–295.

Arita, H. (1993) Riqueza de especies de la mastofauna de México. In: Medellin, R. and Ceballos, G. (eds) *Avances en el Estudio de los Mamiferos de México*. Publicaciones Especiales, Vol. 1, Asociación Mexicana de Mastozoologia, A.C. México, pp. 109–128.

Balick, M.J. and Mendelsohn, R. (1992) Assessing the economic value of traditional medicines from tropical rain forests. *Conservation Biology* 6(1), 128–130.

Benitez, H., Villalón, R.M. and Navarro, A. (1994) Avifauna de las montañas de Michoacán: comentarios biogeográficos sobre las aves del Eje Neovolcánico. Unpublished manuscript. México.

Bhagwati, J. (1993) The case for free trade. *Scientific American* November, 18–23.

Bonati, A. (1991) Industry and the conservation of medicinal plants. In: Olayiwola, A., Heywood, V. and Synge, H. (eds) *The Conservation of Medicinal Plants*. Cambridge University Press, Cambridge, pp. 141–145.

Broda, J. (1979) Las comunidades indigenas y las formas de extracción del excedente: época prehispánica y colonial. In: Florescano, E. (ed.), *Ensayos sobre el desarrollo económico de México y América Latina 1500-1975*. Fondo de Cultura Económica. México, pp. 54-92.

Brooks, H.J. and Murphy, C.F. (1989) Ownership of plant genetic material. In: Knutson, L. and Stoner, A.K. (eds) *Biotic Diversity and Germplasm Preservation, Global Imperatives*. Kluwer Academic Publishers Dordrecht, pp. 493-497.

Brown, K. and Brown, G.G. (1992) Habitat alteration and species loss in Brazilian forests. In: Whitmore, T.C. and Sayer, J.A. (eds) *Tropical Deforestation and Species Extinction*. Chapman & Hall, IUCN, London, pp. 119-142.

Browner, C.H. (1985) Plants used for reproductive health in Oaxaca. *Economic Botany* 39, 482-504.

Bye, R. (1993) The role of humans in the diversification of plant in Mexico. In: Ramamoorthy, T.P., Bye, R., Lot, A. and Fa, J. (eds) *Biological Diversity of Mexico: Origins and Distribution*. Oxford University Press, Oxford, pp. 707-731.

Caballero, J. and Mapes, C. (1985) Gathering and subsistence patterns among P'urhepecha Indians of México. *Journal of Ethnobiology* 5, 31-47.

Carabias, J. (1990a) La política ecológica de la SEDUE. In: *Estancamiento Económico y Crisis Social en México 1985-1988*. Vol. II. UAM. Mexico, pp. 315-371.

Carabias, J. (1990b) Las politicas de producción agrícola, la cuestión alimentaria y el medio ambiente. In: Leff, E. (ed.) *Medio Ambiente y Desarrollo en México*. CIIH/UNAM Mexico, pp. 329-362.

Carrasco, P. (1979) La economia prehispánica de México. In: Florescano, E. (ed.) *Ensayos Sobre el Desarrollo Económico de México y América Latina 1500-1975*. Fondo de Cultura Económica, Mexico, pp. 15-53.

Clawson, D.L. (1985) Harvest security and intraespecific diversity in traditional tropical agriculture. *Economic Botany* 39, 56-67.

Dahlberg, K.A. (1987) Redefining development priorities: genetic diversity and agroecodevelopment. *Conservation Biology* 1(4), 311-322.

Dewey, K.G. (1978) Nutritional consequences of the transformation from subsistence to commercial agriculture in Tabasco, Mexico. *Human Ecology* 6, 55-69.

Dirzo, R. and García, M. (1992) Rates of deforestation in Los Tuxtlas, a neotropical area in Southeast Mexico. *Conservation Biology* 6(1), 84-90.

Dirzo, R. and Miranda, A. (1991) Altered patterns of herbivory and diversity in the forest understory: a case study of the possible consequences of contemporary defaunation. In: Price, P.W., Lewinsohn, G., Fernandes, W. and Benson, W.W. (eds) *Plant-Animal Interactions: Evolutionary Ecology in Tropical and Temperate Regions*. John Wiley, Chicheater, pp. 273-287.

Ferrusquia-Villafranca, I. (1993) Geology of Mexico: a synopsis. In: Ramamoorthy, T.P, Bye, R., Lot, A. and Fa, J. (eds) *Biological Diversity of Mexico: Origins and Distribution*. Oxford University Press, Oxford, pp. 3-107.

Flores-Díaz, A. (1974) Los suelos de la República Mexicana. In: *México: Panorama Histórico y Cultural*. Vol. II. *El escenario geográfico*. SEP/INAH, Mexico, pp. 7-108.

Fowler, C. and Mooney, P. (1990) *Shattering: Food, Politics, and the Loss of Genetic Diversity*. University of Arizona Press, Tucson.

Gomez-Pompa, A. and Kaus, A. (1992) Taming the wilderness myth. *BioScience* 42(4), 271-279.

Green, B.H. (1985) Conservation in cultural landscapes. In: Western, D. and Pearl,

M. (eds) *Conservation for the Twenty-first Century*. Oxford University Press, Oxford, pp. 182-198.

Halffter, G. (1988) El concepto de Reserva de la Biosfera. Reprinted in: *Memorias del Seminario sobre Conservación de la Diversidad Biológica de México*. Vol. 1. UNAM/WWF, Mexico, 1991.

Hernández-Xolocotzi, E. (1993) Aspects of plant domestication in Mexico: a personal view. In: Ramamoorthy, T.P., Bye, R., Lot, A. and Fa, J. (eds) *Biological Diversity of Mexico: Origins and Distribution*. Oxford University Press, Oxford, pp. 733-753.

Iltis, H. (1988) Serendipity in the exploration of biodiversity: what good are weedy tomatoes? In: Wilson, E.O. (ed.) *Biodiversity*. National Academy of Sciences, Washington, DC, pp. 98-105.

INEGI (1990) *Estadísticas Históricas de México*. Tomo I. INEGI, Mexico.

INEGI (1991) *Anuario Estadístico de los Estados Unidos Mexicanos*. INEGI, Mexico.

INEGI (1993) *El Sector Alimentario Mexicano*. INEGI, Mexico.

Jarque, C. (1994) Cuentas nacionales y medio ambiente. In: Glender, A. and Lichtinger, V. (eds) *La Diplomacia Ambiental: México y la Conferencia de las Naciones Unidas sobre Medio Ambiente y Desarrollo*. FCE/SRE, Mexico.

Kloppenburg, J. and Rodriguez, S. (1992) Conservationists or corsairs? *Seedling*, June/July, 12-17.

Leff, E. and Carabias, J. (eds) (1993) *Cultura y Manejo Sustentable de los Recurses Naturales*. CICH/UNAM, Mexico.

Lugo, A. (1988) Estimating reductions in the diversity of tropical forest species. In: Wilson, E.O, (ed.) *Biodiversity*. National Academic Press, Washington, DC, pp. 145-154.

Meave, J.A. (1989) Estructura y composición de la selva alta perennifolia en los alrededores de Bonampak, Chiapas. Unpublished Thesis, UNAM, Mexico.

Mittermeier, R.A. (1988) Primate diversity and the tropical forest: case studies of Brazil and Madagascar and the importance of megadiversity countries. In: Wilson, E.O. (ed.) *Biodiversity*. National Academic Press, Washington DC, pp. 145-154.

Myers, N. (1988) Tropical forests: much more than stocks of wood. *Journal of Tropical Ecology* 4, 209-221.

Nabham, G., House, D. Suzan, H., Hodgson, W., Hernández, L. and Malda, G. (1991) Conservation and use of rare plants by traditional cultures of the U.S./México Borderlands. In: Oldfield, M.L. and Alcorn, J.B. (eds) *Biodiversity: Culture, Conservation and Ecodevelopment*. Westview Press, Boulder, Colorado, pp. 127-146.

National Research Council (1983) *Butterfly Farming in Papua New Guinea*. Managing Tropical Animal Resources Series. National Academy Press, Washington, DC.

Peluso, N.L. (1992) The ironwood problem: (mis) management and development of an extractive rainforest product. *Conservation Biology* 6(2), 210-219.

Peters, C.P., Gentry, A.H. and Mendelsohn, R.O. (1989) Valuation of an Amazonian rain forest. *Nature* 339, 655-656.

Prance, G.T., Balée, B.M,, Boom, B.M. and Carneiro, R.L. (1987) Quantitative ethnobotany and the case for conservation in Amazonia. *Conservation Biology* 1(4), 296-310.

Prescott-Allen, R. and Prescott-Allen, C. (1990). How many plants feed the world. *Conservation Biology* 4(4), 363-374.

Raguso, R.A. and Llorente, J. (1991) The butterflies (Lepidoptera) of the Tuxtlas Mountains, Veracruz, México revisited: species richness and habitat disturbance. *Journal of Research on the Lepidoptera* 29, 105-133.

Ramamoorthy, T.P., Bye, R., Lot, A. and Fa, J. (1993) *Biological Diversity of Mexico: Origins and Distribution*. Oxford University Press, Oxford.

Repeto, R. (1989) *Wasting Assets: Natural Resources in the National Income Accounts*. World Resources Institute, New York.

Robinson, J.G. and Redford, K.H. (eds) (1991) *Neotropical Wildlife Use and Conservation*. The University of Chicago Press, Chicago.

Salafsky, N., Dugelby, B.L. and Terborgh, J.W. (1993) Can extractive reserves save the rain forest? An ecological and socioeconomic comparison of nontimber forest product extraction systems in Petén, Guatemala and West Kalimantan, Indonesia. *Conservation Biology* 7, 39–52.

Shmida, A. and Wilson, M.V. (1985) Biological determinants of species diversity. *Journal of Biogeography* 12, 1–20.

Téllez, L. (ed.) (1993) *Nueva legislación de tierras, bosques y aguas*. FCE, Mexico.

Thiollay, J.M. (1992) Influence of selective logging on bird species diversity in a Guianan Rain Forest. *Conservation Biology* 6(1), 49–63.

Toledo, V.M. (1988) La diversidad biológica de México. *Ciencia y Desarrollo* 81, 17–30.

Toledo, V.M., Carabias, J., Toledo, C. and González-Pacheco, C. (1989) *La Producción Rural en México: Alternativas Ecológicas*. UNAM/Fundación Universo Veintiuno, Mexico.

Toledo, V.M. and Ordóñez, M.J. (1993) The biodiversity scenario of Mexico: a review of terrestrial habitats. In: Ramamoorthy, T.P., Bye, R., Lot, A. and Fa, J. (eds) *Biological Diversity of Mexico: Origins and Distribution*. Oxford University Press, Oxford, pp. 757–777.

UNEP (1992) *Convention on Biological Diversity*. UNEP/CBD/94/1. Interim Secretariat for the Convention on Biological Diversity, Geneva.

Western, D. (1989) Conservation without parks: wildlife in the rural landscape. In: Western, D. and Pearl, M. (eds) *Conservation for the Twenty-first Century*. Oxford University Press, Oxford, pp. 158–165.

Whittaker, R.H. (1972) Evolution and the measurement of species diversity. *Taxon* 21, 213–251.

Wilken, G.C. (1977) Integrating forest and small-scale farm systems in middle America. *Agroecosystems* 3, 291–302.

Williams-Linera, G., Halffter, G. and Ezcurra, E. (1992) Estado de la biodiversidad en México. *Acta Zoológica Mexicana*. No. Especial. La Diversidad Biológica de Iberoamérica.

Wilson, M.V. and Shmida, A. (1984) Measuring beta-diversity with presence–absence data. *Journal of Ecology* 72, 1055–1064.

World Conservation Monitoring Center (1992) *Global Biodiversity: Status of the Earth's Living Resources*. Chapman & Hall, London.

24 Conserving Biodiversity: The Key Political, Economic and Social Measures

J.A. McNeely

IUCN, The World Conservation Union, Rue Mauverney 28, 1196 Gland, Switzerland

Introduction

Society is reaching a crisis point in its relationship with living resources. It is responding by showing greater concern about biological diversity, which offers an entry point into addressing some of the most important political, economic, and social issues of our time. These issues present alternative, sometimes contradictory, approaches to seeking an appropriate relationship between people and resources. An examination of these issues can help modern society to choose the wisest course of development. While these issues are all interrelated in a highly complex way, I will conclude with some practical suggestions for implementing relatively simple measures to reverse the loss of biodiversity.

Political, Economic, and Social Issues in Conserving Biodiversity

Politics and biodiversity

Conservationists have been campaigning for many years to have their issues put on the political agenda. A number of indications suggest that under the banner of 'conservation of biological diversity', a measure of success may finally be at hand. The Global Environment Facility (operated by the World Bank in cooperation with the United Nations Development Programme (UNDP) and the United Nations Environment Programme (UNEP)) has provided $300 million for biodiversity since 1991. A new Convention on Biological Diversity was signed by 157 nations at the United Nations Conference on Environment and Development ('the Earth Summit', held in Rio de Janeiro in June 1992) and entered into force at the end of 1993. Governments,

© 1996 CAB INTERNATIONAL. *Biodiversity, Science and Development: Towards a New Partnership* (eds F. di Castri and T. Younès)

industry, academics, and conservation organizations everywhere are addressing the biodiversity issue through meetings, publications, programmes and on-the-ground action.

The attention given to biodiversity comes not a moment too soon, at least according to the biologists who are warning that a crisis of epic proportions is just around the corner (e.g. Diamond, 1987; Myers, 1987; Wilson and Peter, 1988; Raven, 1993). But action to convert this welcome flurry of interest into significantly changed behaviour will need to overcome a number of formidable political obstacles lurking in the shadows.

It is ironic that as the scientists arguing on behalf of conserving biodiversity become more successful, they may lose relative influence. When biodiversity was seen as a rather narrow issue of a few endangered species or national parks, politicians were content to leave the issue with scientists and specialized agencies, but as measures to conserve biodiversity start to affect the way people consume resources – as advocated by *Agenda 21*, the *Global Biodiversity Strategy* (WRI, IUCN, UNEP, 1992), and various other environmental master plans – politicians are asserting more control over biodiversity concerns.

A fact of modern life is that the important decisions affecting biodiversity are taken by politicians, through a political process which is only partly informed by science. All biodiversity problems reflect a conflict of interests between alternative uses of resources, and politicians survive by balancing the conflicting interests of their various constituencies; scientists and conservationists are not the only ones with a say on what is to happen in a particular habitat or to a species. For example, scientists have found that the annual runs of adult salmon in the Colombia River Basin in the USA have declined by an estimated 75.85%. But groups with an interest in the policy response to this observed decline include electric utilities environmental advocates, the barge industry, recreational boaters, agricultural irrigators, logging and mining interests, the aluminium industry, government agencies, and commercial, sports, and tribal fishing interests (Hyman and Wernstedt, 1991). Political forces will ultimately decide what to do about the salmon, but diverse values and viewpoints need to influence the decision-making process through rational application of biological and economic analysis. Although science cannot resolve the inevitable political disputes, it can help to predict the outcome of any particular management action or policy. Policy-makers cannot possibly pick their way through the options, or even see what the options are, without good science.

Far better information is required for decision-making about biodiversity, and science is what generates this information. Current knowledge about the structure and function of ecosystems is inadequate, as an increasingly knowledgeable and sceptical public asks questions that cannot be answered with a suitable degree of certainty (Binkley, 1992). Interest groups fill the information vacuum with politically convenient interpretations of 'truth'. Since the latitude for choice is wide and the objective capacity to screen alternatives is limited, the policy environment becomes volatile, presenting considerable difficulties for both politicians and conservationists alike.

Policy-makers would prefer to have certainty about the effects of their policies on biodiversity, but science cannot provide this certainty, so the policy-makers must work, as scientists do, by assessing the risks and probabilities. Political decisions may well take science into consideration, but the political appointees and bureaucrats who make policy are responsive above all to the groups that provided support for their elected leaders. And since political aspirations do not always coincide with conservation priorities, a natural tension characterizes the relationship between politics and science. As a result, despite seeing themselves the gatekeepers to observable evidence, scientists may be little more than concerned bystanders when policies are formulated for the problems of conserving biodiversity.

Conservationists face two formidable problems in trying to keep biodiversity on the public agenda (Tobin, 1990). First, no easily-identified opponent is available against which conservation forces can be rallied; unlike headline-makers such as the hole in the ozone layer, the Exxon Valdez, and Chernobyl, no newsworthy disasters have yet involved humans and the loss of biodiversity. Second, the loss of biodiversity has no immediate impact on life-styles; if we are losing dozens or hundreds of species per day, as many experts assert, then we are already living with the consequences of extinction without any discernible effects on our daily lives (at least not yet).

One important problem may be the perception of reality by the general public. Whereas most people are concerned about the loss of species, few see any particular relationship between their own life-style and the loss of biodiversity. As Daniel (1990) has pointed out, people in the industrialized countries appreciate the natural world, but what they appreciate is not their interconnection with it. Instead, they give a sentimental or aesthetic value to it, as something beautiful or peaceful or magnificent, or as a reminder of their frontier experience, as the formidable foe against which they struggled and prevailed. Rural people, on the other hand, tend to see biodiversity in strictly practical terms of resources, health, and welfare.

This leads to a disparity between the principle of conserving biodiversity and the practice of conserving individual species. In many cases, preserving species such as rhinos, elephants, or tigers provides primarily abstract benefits to individual members of the general – largely urban – public, while the people who are expected to make economic sacrifices by restricting their activities in the habitat of these species tend to be large ranchers, forestry interests, or developers who are very effective in conveying their concerns to politicians; or small farmers over whom conservation agencies have little influence. So whereas the general public may support biodiversity conservation in the abstract, the support by rural people for specific action tends to be much weaker because they pay more of the costs and perceive fewer of the benefits.

This is the political reason why the conservation of biodiversity remains more talk than action. Since politicians want to deliver benefits, not constraints, those who advocate policy changes need to become much more politically astute if they wish to have the impact they desire. They will need to build on science to demonstrate the real benefits of conserving biodiversity to farmers, ranchers, and foresters of conserving biodiversity, balance the attention given to loss of biodiversity with concern for sustainable use of harvestable species, and build a broader constituency among business, the public, and academics. As the following sections suggest, economic and social factors will be crucial in this endeavour.

Economics and biodiversity

It is clear that the economic and cultural power of expanding capital is now reaching all parts of our globe, bringing unequal exchange as the order of the day. Economics is an especially important field because it dominates political debates. However, certain elements of its method and theory may, when inappropriately applied, contribute to serious social, political, and ecological disruption. For example, economists commonly use cost–benefit analyses and projections of comparative cost-effectiveness in their identification and evaluation of problems and solutions. They argue that the production of a good is by definition economic only when the total benefits (and their distribution) exceed the total costs (and their distribution). In principle, this must include the costs of dealing with loss of biodiversity, pollution, atmospheric warming, and so forth; but in practice such factors are commonly ignored, or 'externalized', by economists. Further, the production of a good may well be profitable in a commercial sense even if it is not economic; financial returns can rise as a forest or fishery is destroyed and profits are made because the policy, regulatory, or inspection system is unable to force the full costs to be paid. Externalizing environmental costs may in fact be the usual situation, and one of the main reasons biodiversity is being lost. To an economist, accustomed to weighing things like savings, investments, and growth, ecological concerns are secondary considerations to be factored into the larger econometric model. Ecologists studying the complex relationships of living things to their environment, however, know from experience that treating nature like a limitless resource leads to inevitable collapse.

Economists argue that environmental degradation is bearable so long as the gains from the activities causing the degradation (such as clearing a forest for agriculture) are greater than the benefits of preserving the areas in their natural form. The idea that there is some optimum stock of natural assets based on this comparison of costs and benefits is central to modern economic thinking. But it assumes that the full benefits of preserving the areas in their original form can be assessed accurately, and that the gains from the activities are also accurately assessed as the balance of gains over costs. Some economists, however, question how well life-support functions, such as the contributions of biodiversity to geochemical cycles or ecosystem functioning, can be captured by cost–benefit analysis, suggesting instead that in the face of uncertainty and irreversibility, conserving what remains could be a sound risk-averse strategy (Pearce *et al.*, 1990).

Assigning monetary values to biodiversity to put decision-making on a more economically defendable footing implies that biodiversity can be reduced to the simple metric of money. But plants, animals, and societies are exceedingly complex, require a wide variety of materials derived from various sources to remain healthy. Giving a cash value to biodiversity or its components, however, 'forces the great range of unique and distinct materials and processes that together sustain or even constitute life into an arbitrary and specious equivalence' (Rappaport, 1993). Further, not all problems affecting biodiversity can be adequately characterized or described in quantitative, let alone monetary, terms. As Morowitz (1991) put it, 'we are often left trying to balance the "good" of ethics with the "goods" of economics'. It is difficult to assign economic values to biodiversity because species extinction cannot be reversed no matter how much money is spent, the preferences of future generations are impossible to predict, it is difficult to balance present benefits and future costs, and commodity value and moral value can be totally different. It is often necessary to compare what is economically beneficial to individuals against what is beneficial to society as a whole; and of course the latter judgement is ultimately a political one.

Rappaport (1993) suggests that the relationship of economics to ecology is a relationship of the instrumental to the fundamental, and, at the same time, a relationship of the conventional (the rules of which are relatively easy to modify) to the natural, the laws of which are virtually inflexible. Thus society has given higher valuation to economic growth than to biodiversity. 'The instrumental, in claiming the place of the fundamental', Rappaport points out, 'degrades fundamental value to the status of mystifying ideology at the same time that it generates social injustice, urban turbulence, savings and loans disasters, and environmental degradation and, further, reduces the capacity of social systems to deal with troubles as they emerge'. Thus the narrow and special interests of the commercial subsystem of our society have been elevated to the status of society's basic values, and consumption is overwhelming conservation.

On the other hand, it can be argued that biodiversity is decreasing at least partly because so few genetic traits, species, or ecosystems have market prices, the negative feedback signals that equilibrate market economies. Many economists contend that prices – provided they accurately reflected environmental costs – could act as part of a negative feedback system to keep use of biological resources in a closer equilibrium with their sustainable availability. To the extent that the value of biological diversity could be included in the market system, economists argue, markets could help assist in the conservation of biodiversity.

Social issues and biodiversity

Throughout the world, resources historically have been managed by diverse human societies which typically gave social and symbolic value to land and resources beyond their immediate extractive value. These symbolic relationships were based on ecological principles that supported a system of social, ethical, and economic rules that was highly adaptive in the on-going struggle to maintain a viable equilibrium between natural resources, individual desires, and the demands of society (see, for example, Maybury-Lewis, 1992; Suzuki and Knudtson, 1992). The traditional symbolic values helped enable societies to avoid over-exploitation, maintain linkages with nature, and live within the limits imposed by the availability of resources and technology.

Over the past several generations, the World's highly diverse and often localized cultural adaptations to local enviromental conditions have been profoundly disrupted in most places by a global culture increasingly characterized by very high levels of material consumption, at least for a privileged minority (Durning, 1992; Douthwaite, 1992). Economic growth based on the use of fossil fuels as an energy source, greatly expanded international trade, and improved public health measures has spurred such a rapid expansion of human numbers and consumption of resources that new approaches to resource exploitation have been required. These approaches, often involving powerful machinery, sophisticated technology, and arcane economic instruments, have overwhelmed the conservation measures that local communities had developed from their long experience of surviving in an uncertain world.

Living within the limits of their local ecosystems forces people to develop detailed knowledge about resource management. Their behaviour directly affects their own survival and they and their descendants may live or die depending on their success as resource managers. But cultural mechanisms that have been developed as adaptations to the environment over tens or hundreds of generations are quickly cast aside when trade frees people from traditional ecological constraints, changing them from what Dasmann (1975) calls 'ecosystem people' who are adapted to their local ecosystem into 'biosphere people' who can draw from the resources of the entire world. 'Ecosystem people' live with close feedback between their welfare and the way they treat their environment (they have a 'tight feedback loop'), whereas the implications of the behaviour of 'biosphere people' toward their environment is well insulated from their welfare (they have a very 'loose feedback loop').

A major effect of economic development has been to convert as many of the world's peoples as possible from ecosystem people into biosphere people, making them part of the global consumer culture. The loss of cultural diversity typically means that people are less well adapted to specific local conditions, though they may be able to contribute more to the global economy; in economic terms, they are more productive (at least in the short term). People in even the most remote areas now are expected to produce more for distant markets, and to open up their territories for resource exploitation and tourism. Many of them welcome the material goods which result, but the cost is often loss of control of their own resources and a loss of at least certain elements of their own culture. Development breeds dependence, and the institutional control of resources by local communities tends to be strongest when they are the most independent. Once they become integrated into larger systems, the social and economic centre of gravity shifts away from the community and local institutions become increasingly marginalized politically (Murphree, 1993).

Many powerful influences in the world take the view that the loss of community autonomy is not a significant problem, seeing cultural homogeneity and the ecological interdependence among nations as basically desirable. Indeed, the World Commission on Environment and Development has called for greatly expanded interdependence through enhanced flows of energy, trade, and finance (WCED, 1987). On the other hand, some have suggested that such interdependence – making the world a single global system – is the ultimate source of the global depletion of resources. When we are all part of one system connected by powerful economic forces, it becomes very easy to overexploit any part of the global system because other parts will soon compensate for such overexploitation (Dasmann, 1975). The damage may not even be noticed until it is too late to do anything to avoid permanent degradation. Perhaps worse, global interdependence enhances the domination of the economically powerful, yet requires a support structure which in itself can be very fragile – the impacts of changes in oil prices, the stock market, exchange rates, interest rates, and local wars demonstrate the point.

Agenda 21, the manifesto issued by the Earth Summit, stated the principle that 'An open trading system, which leads to the distribution of global production in accordance to comparative advantage, is of benefit to all trading partners'. The system of trade which has enabled the entire globe to be exploited has led to great prosperity for those who have been able to benefit from the expanded productivity, but it has often led to devastation of local ecosystems and the loss of traditional values. Free traders maintain that liberalized markets will solve environmental problems by promoting more efficient use of natural resources, and that increased revenues will lead to decreased environmental damage. Others contend that global markets actually undermine efforts to protect the environment, and that increasing revenues are precisely the problem, leading to overconsumption of biological resources. And in a world characterized by the international mobility of capital, labour, resources, and knowledge in pursuit of absolute advantage, comparative advantage has little validity as an argument for free trade. An obvious conflict exists between the ideals of an international policy of free trade and a national policy which ensures that prices accurately reflect costs, including costs of dealing with impacts of resource exploitation on the environment. A country which reflects environmental costs in its prices will be at a disadvantage in free trade against those who do not do so. Therefore, national protection of a basic policy of internalization of environmental costs constitutes a clear justification for tariffs on imports from a country that does not internalize its environmental costs (Daly, 1992). This has been dramatized by the American insistence that tuna caught by Mexican fishermen be harvested in a way that does not lead to the incidental loss of dolphins, and the subsequent Mexican contention that this was contrary to the General Agreement on Tariffs and Trade (GATT).

In fact, the GATT agreement – warmly welcomed by most governments – has significant negative implications for biodiversity and could itself undermine the work initiated by the Earth Summit (Prudencio, 1993). It prevents countries from imposing export or import bans which they may want to use to protect their own forests, dolphins, or elephants; and permits the export of products such as pesticides or toxic wastes that are prohibited in the producer country but sold to other countries (usually in the developing world). It discourages the use of trade measures to influence environmental policy outside a country's territory, even though such issues as oceans, atmospheric pollution, wildlife trade, and biodiversity are issues that concern all countries and where trade may be wielded as an effective policy tool. GATT therefore undermines international environmental agreements, such as the Convention on Biological Diversity, through its prohibition of trade measures that could enable countries that play by the rules of an international agreement to penalize others that do not.

After all, a global market is not a community or a political system. Whereas the rules of democracy, with voters making decisions about the fundamental choices of society, apply within national boundaries, in the global market, companies can often elude decisions they do not like and look across national boundaries to find lower wages, more permissive environmental laws, or laxer financial regulations (Dionne, 1993). Although trade will allow some countries to live beyond the ecological carrying capacity of their own territory, all countries cannot possibly do this; no matter how much world trade may expand, all countries cannot be net importers of raw materials and natural services. Free trade allows the ecological burden to be spread more evenly across the globe, thereby buying time before facing up to the limits, but at the cost of eventually having to face the problem simultaneously and globally rather than sequentially and nationally (or even locally) (Daly, 1992).

This is how cultural diversity – often the result of local adaptations to local systems of resources – is now threatened by the new global consumer culture, which is spreading through television, trade, government control, subsidies, and other means. The dominant economic forces in the world are using a vast government machinery to facilitate foreign exchange earnings through international trade. This economic expansion has the implicit (and sometimes explicit) goal of promoting more complete exploitation of biological resources. As an inevitable result, traditional management systems that were effective for thousands of years become obsolete in a few years, replaced by systems of exploitation which bring short-term profits for relatively few and long-term costs for many. This leads to the loss of both biological diversity and cultural diversity.

Conserving Biodiversity

Chaos or order?

Considerable uncertainty surrounds conservation of biodiversity. On the biological side, knowledge will always be insufficient about the characteristics and functions of ecosystems, populations, and organisms. Demographic parameters and population sizes are notoriously difficult to estimate, especially for those species threatened with extinction. Even parameters that are relatively easy to measure may exhibit pronounced and unpredictable fluctuations from year to year. Biological and economic uncertainty sometimes serves simply to weaken arguments and thus can become the basis for delaying action. New approaches to complexity theory and chaos (Prigogine and Stengers, 1984) may help shed light on this problem of uncertainty.

Entropy, the tendency toward disorder and chaos, is an existential anxiety that contributes to religion and philosophy, and is confirmed by everyday, common-sense observation of the biological cycles of growth and decline. However, this idea of increasing disorder – entropy – is usually countered by asserting, often at a religious level, that society has purpose and order. The current belief in 'sustainable development' can be seen as the latest manifestation of our faith that development is not leading us down the path to entropy. Other cultures recognize a continuous cycle of ritual creation, destruction and re-creation, as an important mechanism of cultural and biological survival, calling for the reaffirmation of links with past and future generations, together with the expression of concern about the future well-being of society (Reichel-Dolmatoff, 1976).

The Nobel Prize-winning astrophysicist Peter Kafka (1991) suggests that chaos is inevitable. 'In a very sophisticated way', he says, 'the entropy law seems to have conquered the Earth, an open dissipated system in which we thought it wouldn't be valid. While everybody was still worrying and quarrelling about the resources, we have been filling up and blocking the sinks.' He concludes that the onset of global instability is probably unavoidable.

> Growth of evolutionary speed itself seems to be an evolutionary success as long as the errors can be pushed to the borders, i.e. until the global scale has been reached. This acceleration must certainly take place when evolution on a planet reaches the level of mental structures. The discovery of the laws of nature will start technological progress because this provides more power. When deadly consequences of this 'progress' are felt on the critical time scale (the own life-time) by a majority, insight into the logical pre-conditions of creation may become dominant in the global society of minds.

When we perceive cultures as part of systems which include the environment, it is clear that the dynamics of cultural change begin as a relatively slow deviation in a part of the system and then develops into major modifications, closely following models of complexity and chaos. New mathematical models of chaos demonstrate that even small events can have big consequences. In one experiment at the Thomas Watson Research Center in New York, scientists dropped 35,000 grains of sand on to a pile one by one; as the sides grew too steep – in some cases by only a single grain of sand – avalanches would make the pile collapse. Then it would start growing steeper again until it was time for the next avalanche. Similarly, a species can survive for millions of years in virtually the same form, and then abruptly change to another form, or become extinct altogether. It appears, then, that biological evolution proceeds at the boundary between order and chaos, or rather cycling between the two.

The majority of real world species (for example, most insects, plants, and vertebrates) consist of multiple populations weakly coupled by migration or other sources of gene flow, and in this circumstance chaos can actually reduce the probability of extinction. Allen *et al.* (1993) show that although low densities lead to more frequent extinction at the local level, the decorrelating effect of chaotic oscillations in populations reduces the degree of synchrony among populations and thus the likelihood that all are simultaneously extinguished. Similarly, when the planet consisted of thousands of loosely connected cultures of different sizes, local cultures could ebb and flow freely, helping our species adapt to constantly changing conditions.

When there is too much order, the system becomes frozen and cannot change; pressure is built up until there is a chaotic change, whereupon the system 'forgets' its previous order and takes on a different form. A major problem for the world is what has been termed 'hyper-coherence', produced by centralization and policies from higher-order systems which cannot possibly know and understand the local environments as well as the individual rural villagers do. When these highly centralized systems become overconnected, they can be very unstable as they become dependent on central hierarchical control so that change in one part affects all the others too directly and rapidly (Flannery, 1972).

Chaos is of practical importance for both biodiversity and human welfare. The Project on Enviromental Change and Acute Conflict, sponsored by the American Academy of Arts and Sciences and the University of Toronto, has found considerable evidence that scarcities of renewable resources are already contributing to dislocations and violent conflicts in many parts of the world. These conflicts may foreshadow more violence in coming decades, particularly in poor countries where shortages of water, forests, and fertile land are already producing considerable hardship. Unlike unrenewable resources, such as fossil fuels and iron ore, renewable resources are linked in highly complex systems, and the overuse of water, soil, or forests can lead to many unforeseen, simultaneous environmental crises. Scarcities often produce insidious and cumulative social effects, such as large migrations and economic disruption that in turn lead to ethnic strife, civil war, and insurgency. We ignore the implications of entropy at our peril.

The issue of tenure

Modern forms of development have often removed the responsibility for managing biological resources from the people who live closest to them, and instead have transferred this responsibility to government agencies located in capitals which are distant from the resources and remote from the realities of rural life. Given the power of the global trading system, and its accountability to commercial interests and higher levels of government rather than to local institutions, we should not be surprised that forests, grasslands, marine habitats, and the species they support have been grossly overexploited. The government agencies have seldom developed the resource-management capacity and political clout which is sufficient to counteract the new-found technological capacity to exploit, especially in times of shrinking government expenditures for resource management and calls for expanding consumption of resources.

Tenure over resources is the critical issue in conserving biodiversity. Without secure tenure, rural communities can only afford to consider their own short-term interests. They are compelled to exploit resources for maximum immediate gain, regardless of future consequences for themselves, their children, the resource base, or biodiversity.

On the other hand, security of tenure offers opportunities for communities to gain benefits from their resources, so governments should consider returning at least some nationalized resource systems, such as forests and wildlife, in at least some places to community-based tenure systems. Such systems can often be more cost effective than government-sponsored forms of management, and putting resource management back in the hands of local communities or land-owners helps governments divest themselves of responsibilities for functions they have proven incapable of providing adequately.

The full implications of such 'privatization schemes' need to be considered. Transferring the control of access rights from a national to a local authority puts power into those making the local decisions and living with the consequences of these decisions. As Murphree (1993) points out, the way that natural resources are used in any particular place and time is the result of conflicting interests between groups of people having different objectives. Seldom does any one group dominate, and resources can be used in a number of different ways at the same time and place. So the variation in resource management is part of an on-going process involving different interests and struggles of the various actors; some local actors are likely to benefit more than others, thereby creating new tensions in the community. This requires a dynamic and adaptive response rather than a static one.

In many situations, it may be most appropriate for government management authorities to focus on playing a regulatory, facilitating, and coordinating role in managing biological resources. Thus rural communities or land-owners which share the land with wildlife might be given the opportunity to manage and utilize wildlife resources in return for foregoing other land-use options as well as protecting the resources from illegal harvesting. It may well be that well-organized community institutions offer the most cost-effective option for conserving biodiversity when based on appropriate incentives and local social control mechanisms (such as public opinion).

Further, involving local communities in efforts to conserve nature simply makes good sense. Although they often are the last to be consulted about proposed developments, local communities are usually the first to know about changes in the environment. The central government institutions which are setting policy cannot possibly know and understand the local environments as well as the local people do, and activities planned and undertaken by communities guided by local knowledge are certain to be more culturally sensitive and less disruptive than centralized programmes which tend to operate in terms of highly aggregated and simplified information (Rappaport, 1993). Finally, action at the local level strengthens rather than undermines local institutions and is thus empowering. So instead of attempting to correct disorders as understood by outsiders through programmes imposed on local systems, conservation action should be designed to strengthen the capacity of local systems to correct disorders the local people themselves experience. Therefore, ways need to be found to support local institutions or processes that can lead to culturally appropriate as well as substantively effective conservation programmes.

The use of incentives, charges, and other market instruments

Regulations have been the mainstay of environmental policies and resource protection in virtually all countries, often being the preferred method of control by governments (Kröller, 1992). For politicians, regulations offer a way to hide the true costs of environmental protection and thus avoid distributional conflicts which would arise by imposing new taxes and charges; for bureaucracies, command and control rules are a source of power and influence; some interest groups, such as trade unions, may prefer regulations to economic instruments because an increase of costs due to a suppression of subsidies is politically less acceptable in view of a potential loss of jobs and social benefits; and environmental pressure groups, non-governmental organizations (NGOs), and even the public at large often plead in favour of a regulatory framework, rather than market instruments, as a more certain and predictable way of ensuring protection of biodiversity. Regulations have proven their value, though often at a high cost in terms of litigation and bureaucratic interventions; but biodiversity is too complex to be managed by a regulatory framework such as that used for endangered species (Dudley, 1992). So to make further progress, the regulatory framework now needs to be complemented by market instruments such as economic incentives; and some people contend that in any case economic incentives are likely to be the most effective means of influencing human behaviour as it affects biodiversity (McNeely, 1988).

Most economic incentives to date have been used to promote resource exploitation; indeed, these might more accurately be termed 'perverse incentives'. Globally, about $1 trillion per year is spent to subsidize resource consumption, thereby providing a significant obstacle to conservation (Panayotou, 1994). Although across most of the developed world agriculture accounts for only 3–5% of GDP, in 1991 the OECD nations spent a total of US$322 billion on agricultural support and subsidies (Butler, 1992); much of this has had significant negative impacts on biodiversity, both at home and abroad.

Agricultural subsidies are unquestionably important to modern societies, but it is possible to continue farm income support in ways that are not coupled to commodity production. For example, income support could be provided in terms of incentives for soil, water and wetland conservation, support for diverse crops and livestock breeds, and other measures that would conserve biodiversity.

Incentives, such as subsidies, tax differentials, and other fiscal mechanisms, can be used effectively to divert land, capital, and labour towards conservation. They can ensure more equitable distribution of the costs and benefits of conserving biological resources, compensate local people for losses suffered through regulations controlling exploitation, and reward the local people who make sacrifices for the benefit of the larger public. Incentives are clearly worthwhile when they help conserve biological resources, at a lower economic cost than that of the economic benefits received.

To function effectively, incentives require some degree of regulation, enforcement, and monitoring. They must be used with considerable sensitivity if they are to attain their objectives, and must be able to adapt to changing conditions. Each particular setting will have its own challenges, calling for a site-specific design of the package of incentives and disincentives. The role of government in providing the policy framework is crucial, because government is able to mobilize the necessary funds and to take the political decisions that affect the distribution of costs and benefits.

Establishing a system for conserving biodiversity will require a combination of incentives and disincentives, economic benefits and law enforcement, education and awareness, enhanced land tenure and control of new immigration (especially if the buffer zones around protected areas are targeted for special development assistance). The key is to find the balance among the competing demands, and this will usually require a site-specific solution.

Various other market instruments could have a significant positive impact on biodiversity. Repetto *et al.* (1992), for example, analysed a wide range of environmental charges, including effluent charges on toxic substances and vehicle emissions, recreation fees for use of the national forests and other public lands, product charges on ozone-depleting substances and agricultural chemicals, and the reduction of subsidies for mineral extraction and other commodities produced on public lands. This sample of potential environmental charges would reduce a wide range of damaging activities while raising over $40 billion in revenues. Recreation fees in national forests, for example, could yield US$5 billion per year. These findings refute the argument that environmental quality can be obtained only at the cost of lost jobs and income. Instead, providing a better framework of market incentives by restructuring revenue systems can simultaneously improve environmental quality and make economies more competitive. Environmental charges would tax people on their use of energy, amenities, and the amount of waste they generate rather than on their salaries, property, and profits. Environmental charges would give people an attractive incentive for savings. At present, the only way most people can reduce their tax bill is to work less and earn less income, or cheat. Environmental charges would give them the option of reducing their tax bills by acting on their principles – by saving energy, recycling, or bicycling to work. 'Virtue is its own, but not necessarily its only, reward,' say Repetto and his colleagues.

Systemic solutions or small victories

The kind of public policies which are most popular are those which call for modest changes in current practices to address immediate, proximate causes rather than imposing comprehensive changes in deeply imbedded social behaviour. As Tobin (1990) has pointed out, popular policies coincide with prevailing public opinion and do not require law-abiding people to change their lifestyles or cause them great inconvenience; they distribute material benefits to a majority or to a politically significant and effectively organized minority; they provide more benefits than costs, giving advantages to policies with easily monetized values, such as goods traded in the marketplace or development that provides jobs; and provide concentrated benefits today while deferring and diffusing costs (the popularity of this approach is indicated by the US federal budget deficit).

Policies to conserve biodiversity tend to have the opposite characteristics. They call for fundamental changes in the way people relate to the environment; for example, IUCN, WWF, and UNEP (1991), Ophuls and Boyan (1992), Haila and Levens (1992), and Piel (1992) have all recently called for limits on rates of resource use, following up the calls of Schumacher (1974) and Daly and Cobb (1989) for a minimal frugal steady state as the appropriate form of a post-industrial society – a fundamental transformation of world view. Prescriptions for a sustainable future tend to restrict access to resources, expect people to forego material benefits, assign values to resources that are elusive or difficult to measure, and pay today for abstract future benefits.

And indeed, as Rappaport (1993) says, any adequate theory for correcting the ills of modern society must be systemic, 'for we are facing such a multiplicity of quandaries, dilemmas, crises, inequities, iniquities, dangers, and stresses ranging from substance abuse, homelessness, teenage pregnancy, and prevalence of stress disease among minorities to global warming to ozone depletion that they cannot all be named, much less studied'. Breaking the system into its component parts might facilitate 'problem-solving', but this has the danger of causing more problems than it solves by isolating complex systems from their contexts. The underlying principle of complexity theory is that traditional reductionism – that is, seeking to understand systems by breaking them down into components and analysing the interactions between them – cannot provide an adequate understanding of complex systems. Rather than ever-more-detailed analyses of fine internal structure, conserving biodiversity requires moving toward a more comprehensive view which synthesizes contributions from many sectors (WRI, IUCN, UNEP, 1992).

But complexity can sometimes go too far. As with bureaucracies, some of our approaches to conservation seem to be too tightly bound, or 'hyper-coherent', to use the technical term of social scientists (ecologists have a similar concept: 'over-connectedness', Holling (1986); while foresters use terms such as 'over-mature'). Conservationists call for such measures as more research, more science and technology, more government regulation, improved standards of professional ethics, better forest regulations, better enforcement of penalties for infractions, stronger judicial systems so that offenders are more likely to be brought to justice, more inclusive planning measures, environmental impact assessments, and many others. Although most of us will be convinced that such measures are useful and important, they almost invariably lead to increased numbers of bureaucrats in government and increased numbers of office staff in the regulated industries to handle the increased paper work. An extended process of planning and approval for new projects gives additional burdens both to regulators and the regulated – the system is becoming increasingly inefficient, with more and more effort spent maintaining the system rather than promoting new production. This leads to more costs, higher taxes, and so forth, ultimately resulting in total paralysis, revolutionary change, or even more overexploitation.

On the other hand, defining manageable portions of the problem to be tackled seems to be an important strategic device for politicians. The failure of the California Environmental Protection Act of 1990, which contained 30 different environmental provisions, suggests that comprehensive solutions are unlikely to work because they stimulate opposition from a multitude of economic interests, each disagreeing with a relatively small part of the whole. And it is simply too overwhelming to think concurrently of whole litanies of problems; the response is to sink into passive despair. Instead, building a series of 'small wins' creates a sense of control, reduces frustration and anxiety, and fosters continued enthusiasm (Heinen and Low, 1992). But if these 'small wins' are to be real victories, they must contribute to the overall strategy for conserving biodiversity:

- giving management responsibility and tenure rights to the people most directly involved;
- ensuring that prices fully reflect environmental costs;
- providing economic incentives to encourage individual behaviour which is in the long-term benefit of the larger society;
- providing the best available science to support political decisions; and
- seeking a diversity of local solutions to local problems.

Conclusion

Having made the very important conceptual advance to treat biodiversity as a necessary asset to support a sustainable way of life, we now need to find ways to avoid wasting this asset. Rather than pretend that we can create wealth by stripping assets, we need to employ all factors of production, biodiversity included, in the most efficient way possible so that human society can derive the optimal sustainable benefit from it.

Throughout history, local societies have ebbed and flowed as their wisdom was tested against the criterion of adaptability. Those societies that were able to develop the wisdom, technology, and knowledge to live within the limits of their environments and adapt to changes ground them were able to survive. Other societies overexploited their resources, sacrificing sustainability and adaptability for the perceived benefits of immediate wealth. Some people believe that modern industrial civilization will somehow avoid the fate of earlier civilizations, using the free market to overcome its abuses of the environment (Simon and Kahn, 1984). But the future is uncertain, the global consumer culture has not passed the test of time, and many indicators give cause for serious concern.

While Fukuyama (1992) contends that we have reached 'the end of history', and the liberal-democratic consumer society has seemingly become the ultimate to which humanity can aspire, most people hope that history will in fact continue. But if the consumer society proves to be not sustainable, the transition to whatever comes after it will be extremely messy, for the world will have no coherent conception of what it is one might be working towards (Ekins, 1993). On the other hand, perhaps the thousands of local cultures adapted to locally available resources collectively can offer a coherent alternative to the dead-end of consumerism; the diversity of cultures can help maintain the diversity of nature.

Vaclàv Havel, then Prime Minister of Czechoslovakia, speaking at the World Economic Forum in Davos, Switzerland, on 4 February 1992, put diversity into a historical perspective:

> The fall of communism can be regarded as a sign that modern thought – based on the premise that the world is objectively knowable, and that the knowledge so obtained can be absolutely generalized – has come to a final crisis. This era has created the first global, or planetary, technical civilization, but it has reached the limit of its potential, the point beyond which the abyss begins. The end of communism is a serious warning to all mankind. It is a signal that the era of arrogant, absolutist reason is drawing to a close and that it is high time to draw conclusions from that fact. Communism was not defeated by military forces, but by life, by the human spirit, by non-science, by the resistance of Being and man to manipulation. It was defeated by a revolt of colour, authenticity, history in all its variety and human individuality against imprisonment within a uniform idealogy. . . . Things must once more be given a chance to present themselves as they are, to be perceived in their individuality. We must see the pluralism of the world, and not bind it by seeking common denominators or reducing everything to a single common equation.

Havel is a poet turned politician, and his message reaches out to all of us. We need both biological diversity and cultural diversity, and we need diversity in our approaches to conservation. Our capacity to adapt to change will be based on the decisions we make today. If we continue to abuse our life support systems and the biological diversity upon which these systems are built, we and our children will pay the price. If, on the other hand, we decide that it is time to transform our societies into sustainable ones that live within the limits of the productivity of nature, then our descendants will sing our praises.

References

Allen, J.C., Schaffer, W.N. and Rosko, D. (1993) Chaos reduces species extinction by amplifying local population noise. *Nature* 364, 229-232.
Binkley, C.S. (1992) Forestry after the end of nature. *Journal of Forestry* 90(10), 33-37.
Butler, D. (1992) Why exorbitant farming subsidies have to end. *World Link* 5(4), 6-10.
Daly, H. (1992) Free trade, sustainable development and growth: Some serious contradictions. *Ecodecision* June, 10-13.
Daly, H.E. and Cobb, J.B. Jr (1989) *For the Common Good: Redirecting the Economy Towards Community, the Environment and a Sustainable Future.* Beacon Press, Boston.
Daniel, J. (1990) Remembering the sacred family. *Orion* (Summer), 9-14.
Dasmann, R. (1975) National parks, nature conservation, and 'future primitive'. *Ecologist* 65(5), 164-167.
Diamond, J.M. (1987) Extant unless proven extinct? Or, extinct unless proven extant? *Conservation Biology* 1, 72-76.
Dionne, E.J. (1993) Free trade is on a collision course with democracy. *International Herald Tribune*, 1 April 1993, p. 4.
Douthwaite, R. (1992) *The Growth Illusion: How Economic Growth has Enriched the Few, Impoverished the Many, and Endangered the Planet.* Green Books, Devon.
Dudley, J.P. (1992) Rejoinder to Rohlf and O'Connell: biodiversity as a regulatory criterion. *Conservation Biology* 6, 587-589.
Durning, A.T. (1992) *How Much is Enough? The Consumer Society and the Future of the Earth.* W.W. Norton, New York.
Ekins, P. (1993) The sustainability question: are there limits to economic growth? *Traces* 36-39.
Flannery, K. (1972) The cultural evolution of civilizations. *Annual Review of Ecology and Systematics* 3, 399-426.
Fukuyama, F. (1992) *The End of History and the Last Man.* Free Press, New York.
Haila, Y. and Levens, R. (1992) *Humanity and Nature: Ecology, Science and Society.* Kluto Press, New York.
Heinen, J.T. and Low, R.S. (1992) Human behavioural ecology and environmental conservation. *Environmental Conservation* 19, 105-116.
Holling, C.S. (1986) Resilience of ecosystems: local surprise and global change. In: Clark, W.C., and Munn, R.E. (eds) *Sustainable Development of the Biosphere.* Cambridge University Press, Cambridge, pp. 292-317.
Hyman, J.B. and Wernstedt, K. (1991) The role of biological and economical analyses in the listing of endangered species. *Resources* (Summer 1991), 5-9.
IUCN, WWF, and UNEP (1991) *Caring for the Earth: A Strategy for Sustainable Living.* IUCN, WWF, UNEP, Gland.
Kafka, P. (1991) Intrinsic limit to the speed of innovation and its relevance for the question 'Where are they?' In: Heidmann, J. and Cline, M.J. (eds) *BioAstronomy: The Search for Extraterrestrial Life*, Bezland, Heidelburg, pp. 340-343.
Kröller, E. (1992) Report on the Development Cooperation Directorate/Development Centre workshop on the use of economic instruments for environmental management in developing countries. OECD, Paris.

Maybury-Lewis, D. (1992) *Millennium: Tribal Wisdom and the Modern World*. Viking Penguin, New York.

McNeely, J.A. (1988) *Economics and Biological Diversity: Developing and Using Economic Incentives to Conserve Biological Diversity*. IUCN, Gland.

Morowitz, H.J. (1991) Balancing species preservation and economic considerations. *Science* 253, 752–754.

Murphree, M.W. (1993) *The Role of Institutions*. Paper prepared for the Liz Claiborn-Art Ortenburg Foundation Workshop on Community-based Conservation, Airlie, Virgina, October.

Myers, N. (1987) Tackling mass extinction of species: a great creative callenge. *The Horace M. Albright Lecture in Conservation*. University of California, Berkeley.

Ophuls, W. and Boyan, A.S. Jr (1992) *Ecology and the Politics of Scarcity Revisited: The Unravelling of the American Dream*. W.H. Freeman, San Francisco.

Panayotou, T. (1994) Financing mechanisms for Agenda 21 (or How to pay for sustainable development). Paper presented at meeting on Financial Issues of Agenda 21, 2–4 February, Kuala Lumpur, Malaysia.

Pearce, D., Barbier, E. and Markandya, A. (1990) *Sustainable Development: Economics and Environment in the Third World*. Edward Elgar, London.

Piel, G. (1992) *Only One World: Our Own To Make and Keep*. W.H. Freeman, San Francisco.

Prigogine, I. and Stengers, I. (1984) *Order out of Chaos: Man's New Dialogue with Nature*. Bantam Books, New York.

Prudencio, R.J. (1993) Why UNCED failed on trade and environment. *Journal of Environment and Development* 2(2), 103–109.

Rappaport, R.A. (1993) Distinguished lecture in general anthropology: The anthropology of trouble. *American Anthropologist* 95, 295–303.

Raven, P. (1993) Biological extinction: Its scope and meaning for us. Paper presented to the Annual Meeting of the Council for Advancement of Science Writing's 31st Annual 'new horizons in science' briefing, St Louis, Missouri, 2 November 1993.

Reichel-Dolmatoff, G. (1976) Cosmology as ecological analysis: a view from the rainforest. *Man* 11, 307–318.

Repetto, R., Dower, R.C., Jenkins, R. and Geoghegan, J. (1992) *Green Fees: How a Tax Shift can Work for the Environment and the Economy*. World Resources Institute, Washington, DC.

Schumacher, E.F. (1974) *Small is Beautiful: A Study of Economics as if People Mattered*. Blond and Briggs; London.

Simon, J.L. and Kahn, H. (1984) *The Resourceful Earth: A Response to Global 2000*. Basil Blackwell, London.

Suzuki, D. and Knudtson, P. (1992) *Wisdom of the Elders: Sacred Native Stories of Nature*. Bantam Books, New York.

Tobin, R. (1990) *The Expendable Future: US Politics and the Protection of Biological Diversity*. Duke University Press, Durham, Michigan.

WCED (World Commission on Environment and Development) (1987) *Our Common Future*. Oxford University Press, Oxford.

Wilson, E.O. and Peter, F.M. (1988) *Biodiversity*. National Academy Press, Washington, DC.

WRI, IUCN, UNEP (1992) *Global Biodiversity Strategy: Guidelines for Action to Save, Study, and Use Earth's Biotic Wealth Sustainably and Equitably*. WRI, IUCN, UNEP. Washington, DC.

25 Biodiversity Conservation in the New South Africa

B.J. Huntley

National Botanical Institute, Private Bag X7, 7735 Claremont, Cape Town, South Africa

Introduction

South Africa, at the southern tip of Africa, occupies 4% of the continent's and 0.8% of the world's total land area. Its geographic position, flanked by the warm Agulhas Current on the east and the cold, upwelling Benguela Current on the west, accounts for a diverse range of climatic conditions. A narrow coastal plain, steep escarpment, high mountains and extensive interior plateaux result in sharp topographic gradients. Tropical to temperate climates, rain forest to desert habitats, an extended period of geological stability and the absence of Pleistocene glaciations, account for the country's extremely rich biodiversity. South Africa possesses 8% of the world's vascular plants, 7% of the world's bird species and 6% of the world's mammal species. Its flora is especially rich, with the Cape Floristic Region including over 8600 species, of which 68% are endemic to this 'littlest Floral Kingdom' sensu Takhtajan (1986).

For over two decades, South Africa was excluded from UNESCO and other United Nations environmental initiatives, and the international academic boycott severely constrained communication between South African scientists and their colleagues throughout Africa and abroad. Following the historic transition to a democratic government in April 1994, the opportunity to share South African expertise and experience with conservation biologists throughout the world has become an exciting reality. This chapter reviews recent advances in biodiversity conservation in South Africa, against the backdrop of the comprehensive synthesis undertaken five years ago (Huntley, 1989a) and in the context of the new sociopolitical order.

Table 25.1. Species richness of selected non-marine taxa reproducing in South Africa (1,221,000 km^2)

Taxon	No. of species	Taxon	No. of species
Fishes	112	Bryophyta	776
Amphibians	84	Pteridophyta	237
Reptiles	286	Gymnospermae	39
Birds	600	Monocotyledonae	4377
Mammals	227	Dicotyledonae	13,972
Invertebrates	278,000	Fungi	20,000

Sources: Animals and Fungi (Siegfried and Brooke, 1994); Bryophyta and phanerogams (Arnold and de Wet, 1993)

Patterns of Diversity: What Are the Units of Biodiversity, How Many are there, and How are they Distributed?

The popular definition of biodiversity, embracing the richness of life on Earth, from genes to species to ecosystems, sets a daunting task for anyone wishing to undertake an assessment of the biodiversity of even a small area of low diversity. Fortunately, South Africa has attracted and/or produced several hundreds of biologists over the past two centuries, has a network of more than twenty natural history museums and herbaria, and possesses an unusually strong core of conservation biologists in its universities, science councils and nature conservation agencies. As a consequence, a reasonable estimate of the country's biological richness, at least at the species level, can be made for all vertebrate, bryophyte, pteridophyte and vascular plant taxa (Table 25.1), and detailed checklists are available for a wide range of taxa for intensively studied areas such as the Kruger National Park (Table 25.2). Siegfried and Brooke (1994) point out, however, that even within the best known South African vertebrate groups, we need to recognize 10% more species than are presently accepted due to the presence of cryptic and under-recognized species among those vertebrate taxa already described in the literature. It is estimated that a further 10% await discovery or description.

While the debate attending the choice of an appropriate unit for the measurement of biodiversity continues (Bond, 1989; McIntyre et al., 1992; Noss, 1990; Vane-Wright et al., 1991; Walker, 1992), the biological species concept will probably have to serve us for some years to come (Hockey et al., 1994). Some species are clearly more critical to the maintenance of ecological processes and biodiversity than others. This is demonstrated by the dependence for pollination of 15 of the most spectacular fynbos species of red-flowered plants on a single species of butterfly (Johnson and Bond, 1994; Bond, 1994). The extinction of this single 'keystone' species would lead to the extinction of an ecosystem function and with it the extinction of 15 species of flowering plants, the roles of which in other ecosystem processes would likewise disappear.

Table 25.2. Checklist of animals and plants recorded in the Kruger National Park, South Africa. Area: 19,485 km^2

Taxon	No. of species	Taxon	No. of species
Fishes	49	Spiders	96
Amphibians	35	Solifuges	23
Reptiles	119	Scorpions	28
Birds	492	Ticks	23
Mammals	147	Centipedes	22
Molluscs	85	Millipedes	22
Crustaceans	8	Plants	1922

Source: Zambatis (personal communication)

The biogeography of South Africa's flora and fauna has been reviewed in Werger (1978) and in many subsequent papers. Although the patterns of distribution of taxonomic groups provide useful insights into their phylogeny, they fall short of guiding decisions on where conservation effort should be directed. Species richness and endemism in terrestrial mammals display opposite patterns of distribution in South Africa, Fig. 25.1 (Siegfried and Brown, 1992). The same is true of birds and dragonflies breeding in South Africa, where species richness is highest in the north-east, lowest in the south-west, where the highest concentration of endemics is found (Siegfried, 1992; Samways, 1992).

The flora of south Africa is exceptionally rich, with 18,625 species of phanerogams, of which at least 80% are endemic, a level of endemism more typical of oceanic islands than continental regions. Cowling and Hilton-Taylor (1994) provide a detailed review of the centres of plant diversity and endemism (Davis and Heywood, 1994) in southern Africa and describe eight biodiversity 'hot-spots' (sensu Myers, 1990), seven of these in South Africa (Fig. 25.2). The Cape Floristic Region (Rebelo, 1994) with 8600 species of which 6000 are endemic to the Region's 90,000 km^2, and with 1406 species listed as Red Data taxa, is described by Myers (1990) as the 'hottest' of the world's biodiversity hot-spots.

Plausible explanations for the extraordinary diversity of the Cape Floristic Region have eluded generations of botanists, but a clearer picture of the ecological and historical factors is emerging, in which soils, topography, climate and fire regime as well as inherent ecological characteristics of certain lineages probably play significant roles (Cowling and Holmes, 1992; Cowling and Hilton-Taylor, 1994). Pollination studies (Manning and Linder, 1992; Steiner and Whitehead, 1990, 1991; Steiner et al., 1994) are also providing clues to the evolutionary mechanisms involved in the rapid radiation of the South African flora.

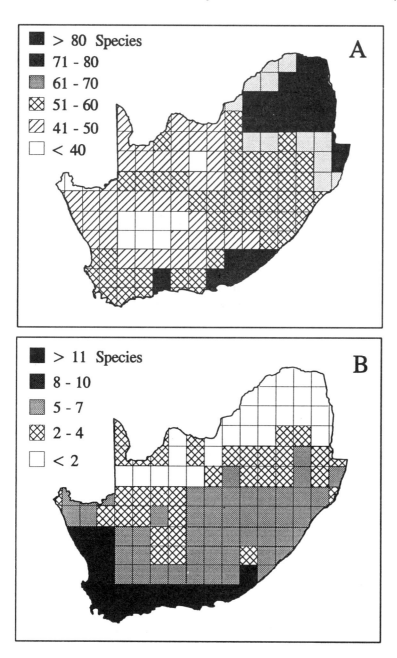

Fig. 25.1. Species richness (A) and numbers of endemic species (B), recorded within one-degree squares, for resident, terrestrial mammals of South Africa (from Siegfried and Brown, 1992).

Fig. 25.2. Floristic 'hot-spots' of South Africa (from Cowling and Hilton-Taylor, 1994).

Protected Areas: What is Protected and How Effective is the Network?

As was the case elsewhere in Africa, the location of South Africa's earliest national parks and reserves was determined by the presence of spectacular large mammals, rather than through objective biodiversity analyses. Attempts to redress the imbalance created by this approach, based on a quantitative assessment of the adequacy of protected areas in relation to vegetation types, were initiated 20 years ago (Edwards, 1974). Despite further assessments at a subcontinental (Huntley, 1978; Huntley and Ellis, 1984) and regional scale (Jarman, 1986), an optimal network is still lacking. Edwards (1974) and Huntley and Ellis (1984) used the best available vegetation mapping categories as their units of biodiversity. In the case of South Africa, this was Acocks's (1953) 'Veld Type'. The synthesis of these assessments indicated that the lowland fynbos, succulent karoo and highveld grassland ecosystems of South Africa were the least adequately protected in southern Africa, a situation that pertains today, twenty years later.

Although the problem remains essentially unchanged, research approaches, technologies and available information have advanced substantially. During the 1970s and 1980s, two generations of Red Data Books for South Africa's flora and vertebrate fauna were published (Ferrar, 1989), allowing a detailed assessment of the status and trend of individual species, while the computerized database of over 710,000 voucher specimens in the National Herbarium, Pretoria, provides an electronic information system for the entire South African flora (Gibbs-Russell and Gonsalves, 1984; Arnold and de Wet, 1993). These data sets have been used, in conjunction with additional field information and geographic information systems, in advanced gap-analysis studies (Lombard et al., 1992; Rebelo and Siegfried, 1992; Lombard, 1993; Rebelo, 1993; Rebelo and Tansley, 1993).

The evaluation of the existing network of marine reserves and the identification of additional sites for protection has been undertaken on the basis of zoogeographic and functional characteristics by Emanuel et al. (1992). Their study builds on the substantial information base developed by several decades of marine biological research in South Africa (Branch et al., 1994).

The question of protected area size, the so-called SLOSS (single large or several small) debate has been addressed in diverse ecosystems. Empirical studies of island-biogeographic influences on species composition and survival in fynbos, forest, and 'island' communities (Bond, 1989; Cowling and Bond, 1991; Dean and Bond, 1994) have provided the most convincing accounts yet of fragmentation effects from African ecosystems. The results of these studies indicate that the minimum area required for the maintenance of many of the target plant taxa is far less than the average size of existing protected areas in the ecosystems involved. This conclusion might not apply for animals within these communities.

National parks and publicly-owned nature reserves in South Africa now exceed 72,000 km^2, or 5.9% of the country's land area. Despite the shortcomings of the network with regard to the coverage of succulent karoo, lowland fynbos, and highveld grassland ecosystems described above, the protected area system is remarkably effective at the species level. Siegfried (1989) reports that the country's 576 formally protected areas include reproducing populations of 92% of the amphibian, 92% of reptilian, 97% of avian, and 93% of mammalian species recorded in South Africa. These statistics are based on incomplete data sets – over 50% of the protected areas lack checklists for any taxon – suggesting that the actual percentage of species protected might be even higher than the above estimates.

The value of the protected area network is not only reflected in terms of its inclusion of the vast majority of South Africa's biological diversity within less than 6% of the country's land area, but also in terms of the viability of its plant and animal populations. Although sophisticated Population and Habitat Viability Analysis methodologies have recently been developed, analysis for South African species have not yet been published. The successes of species protection programmes have been significant and in some cases spectacular. Elephant (Hall-Martin, 1992), black rhinoceros, white rhinoceros (Hughes, personal communication), mountain zebra, bontebok and many other species have recovered from the brink of extinction

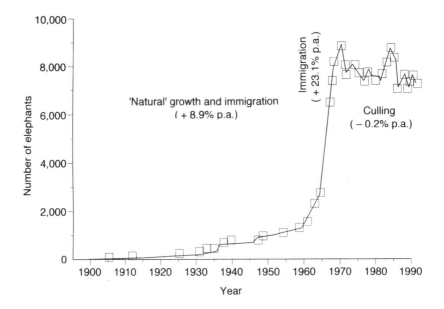

Fig. 25.3. Trend of the African elephant population in the Kruger National Park, 1900-1990 (from Hall-Martin, 1992).

at the end of the 19th century to healthy, widely dispersed populations today (Fig. 25.3).

Although conservation actions have prevented imminent extinctions in the 20th century, they had occurred in historic, and prehistoric times due to human intervention. Siegfried and Brooke (1995) record the extinction of eight species of larger mammal in South Africa in prehistoric (130,000 BP to 1500 CE (Common Era)) and of two species in historic (1500 CE to present) times. The extinction rate since the arrival of European settlers in 1652 approximates one species per 100 years, that for the prehistoric period, one extinction per 16,500 years. Siegfried and Brooke (1995) suggest that the eight extinctions between 130,000 BP and the Holocene were part of the 'overkill' phenomenon which was, however, substantially less potent in its impact in Africa than in the Americas and Australasia.

The development of South Africa's protected area network has focused on terrestrial species and ecosystems, with limited attention being directed to the marine environment. This situation is remarkable given the long tradition of marine biological research in the country and the extensive series of proposals for coastal, estuarine and marine reserves (Tinley, 1985; Heydorn, 1989, Hockey and Buxton, 1989; Shelton, 1989, Robinson and de Graaff, 1994). Hockey and Buxton (1989) conclude that if the existing coastal protected areas were extended to include the intertidal zone, at least 9% and up to 35% of each major habitat type within each biogeographical province would be protected. Even without such extensions, it is probable that the rich

Table 25.3. The number of species breeding in the terrestrial and freshwater ecosystems of South Africa and the number considered to be showing definite population decreases in the local Red Data Books

Taxonomic group	No. of species	No. decreasing	% decreasing
Fishes	107	16	14.9
Amphibians	84	9	10.7
Reptiles	286	8	2.8
Birds	600	41	6.8
Mammals	227	33	14.5
All taxa	1304	107	8.2

Sources: Macdonald (1992)

marine biodiversity of South Africa's 3000 km coastline is as effectively protected as the country's terrestrial ecosystems.

Monitoring the Status of Species: Measuring Success Within and Outside the Protected Area System

Within the 6% of South Africa included in the protected area system, the monitoring of the size, composition and health of many of the vertebrate and higher plant communities forms an integral part of management programmes (Joubert, 1983). For the rest of the country, the remaining 94%, broad-scale monitoring of key indicator species, making extensive use of volunteer groups, has been achieved to an encouraging degree.

The South African Red Data Books (RDBs) (Ferrar, 1989) were compiled by wide consultation among biologists and conservationists and have provided a coarse-grained base to monitoring. Macdonald (1992) used the Red Data Books to demonstrate decreases from 2.8 to 14.9% of the populations of species from different taxonomic groups included in the RDBs, over the approximately ten years between the first and second editions of the books (Table 25.3).

The South African Red Data Books have used the standard IUCN definitions for the various categories of threat (Wells et al., 1983). These criteria, despite their well-recognized deficiencies, have served well in a large and in some cases poorly documented fauna and flora. The recent advances in defining categories of threat (Mace and Lande, 1991; Mace and Stuart, 1994) offer greater predictive power, but have thus far only been applied in South Africa to cycads (Osborne, 1995).

Another approach to broad-based monitoring is provided through the activities of the South African Bird Ringing Unit, which over the past 21 years has accumulated data on over 1.37 million birds of 830 species banded with SAFRING rings (Underhill and Oatley, 1994). Using age structure analyses of curlew sandpiper populations at Langebaan Lagoon (a South African Ramsar site) Underhill (1987) was able to shed light on the dynamics of lemming population cycles in Siberia, 12,000 km distant!

Atlassing projects are also generating important information on the dynamics of South African bird populations (Harrison, 1992), particularly with regard to local movements and long-term trends in abundance correlated with environmental variables. More recently, a Protea Atlassing project, similar to the highly successful Banksia Atlas of Australia, has been launched to stimulate the involvement of the public in flora conservation, and to generate information of use to conservation biologists (Rebelo, 1991).

The need for active monitoring of populations of rare or threatened species is gaining recognition among both amateur and professional conservationists. Stimulated by the global network of IUCN's Species Survival Commission, 97 South Africans are now involved in developing action plans for selected species at global (Cactus and Succulent Action Plan; Oldfield, personal communication), continental (African Rhino Action Plan; Brooks, personal communication; Felid and Hyaena Action Plans; Mills, personal communication) or national levels (Cycad Action Plan; Osborne, 1990).

In the short term, the socio-political events and the associated direct and indirect effects of human activities such as urbanization, industrial and agricultural development, deforestation, pollution, invasion by alien plants and animals, etc., will exert the most significant impacts on the fauna and flora of South Africa (Huntley, 1989b). Longer-term impacts of global change, through the increase of 'greenhouse' gases and of tropospheric ozone depleting chemicals, will probably only be evident late in the 21st century. Global warming effects would be particularly important for the narrow elevational endemics, especially those of the Cape fold mountains, as has been suggested for lycaenid butterflies by Samways (1993b). The consequences of increased UV-B radiation have already been shown to reduce regeneration success and seedling survival in South African plants (Musil and Wand, 1994) particularly in areas of low soil fertility. This finding will be of considerable consequence if it is generally applicable in the low-nutrient soils of the Cape fynbos, one of the most specious ecosystems in the world.

Conservation Biology: The South African Biodiversity Knowledge Base

The development of a predictive understanding of the structure and functioning of South Africa's terrestrial, inland water and marine ecosystems has been a central goal of environmental studies in the country since the early 1970s. Following the conclusion of the International Biological Programme, a series of closely linked national Cooperative Scientific Programmes (CSPs) were established to draw together researchers and managers working in a wide diversity of ecosystems, with the primary objective of finding solutions to environmental problems best addressed by cooperative interdisciplinary effort. The CSPs ran from 1975 to 1990, producing several hundred graduates at MSc and PhD level and over 1500 research papers, theses and management guidelines.

The principal findings of the CSPs have been integrated in a series of synthesis volumes, on savanna ecology (Huntley and Walker, 1982; Scholes and Walker, 1993), fynbos ecology (Kruger *et al.*, 1983; Cowling, 1992; Davis and Richardson, 1994), inland water ecosystems (Davies and Walmsley, 1985; Allanson *et al.*, 1990; Walmsley and Davies, 1991; Davies *et al.*, 1993), sandy beaches (Brown and McLachlan, 1990), the Benguela upwelling system (Crawford *et al.*, 1988; Payne *et al.*, 1992), the Agulhas Bank (Hutchings, 1994), marine systems (Branch *et al.*, 1994), large mammal management (Owen-Smith, 1983), ecological effects of fire (Booysen and Tainton, 1984), ecology of invasive biota (Macdonald *et al.*, 1986), biodiversity (Huntley, 1989a), ecosystem function of biodiversity (Davis and Richardson, 1994), global climate change (Anon., 1990), and many more. Evaluations of two of the major CSPs (on terrestrial ecosystems and on the Benguela upwelling system), have been published by Huntley (1987) and Rothschild and Wooster (1992).

The knowledge base and core of expertise created by the CSPs provide the intellectual resources essential to the development of a modern Conservation Biology of South Africa, currently supported by the recent establishment of two postgraduate MSc degree courses in conservation biology and resource ecology at the Universities of Cape Town and the Witwatersrand, respectively. Further underpinning of the knowledge base has been provided by the publication of new or revised editions of standard works on marine fishes (Smith and Heemstra, 1986; Gon and Heemstra, 1990), mammals (Skinner and Smithers, 1990), birds (Maclean, 1993), insects (Scholtz and Holm, 1985; Samways, 1994), plants (Arnold and de Wet, 1993) and environmental management and legislation (Fuggle and Rabie, 1992). There is a need for a new monographic account of the reptile fauna of South Africa, especially of the lizards, which include many endemic taxa. Furthermore, increased support for biological surveys and taxonomic research, especially on lower plants, lower vertebrates and most invertebrate groups is needed if a comprehensive understanding of South Africa's biodiversity is to be achieved.

The primary publications on which the above syntheses are based have appeared in the open literature, particularly South African journals such as the *South African Journal of Botany, South African Journal of Science, South African Journal of Wildlife Management, South African Journal of Zoology* and the 'house' journals of research and conservation agencies such as *Bothalia, Koedoe* and *Lammergeyer*. Penetration of the international conservation literature has been limited, possibly reflecting the impacts of the academic boycott but also relating to the size of the Southern African research community. Of the 551 full-length papers published in *Conservation Biology* and in *Biological Conservation* during the years 1990 to 1993, only 21 were authored by South Africans. This statistic is not necessarily representative of the impact of South African research in biology. Pouris (1989), on the basis of an evaluation of the Science Citation Index records for the 1981–1985 period, demonstrated that South Africa ranked fourth in the world in ornithology, eighth in ecology and tenth in zoology in terms of papers published and citations listed in the SCI.

Biodiversity Management: Applying Information and Understanding

The ecological research agenda in South Africa over the past 20 years has been extremely productive in terms of scientific papers published and degrees conferred, but its real value must be assessed in the extent to which the knowledge generated is implemented in the improved management and sustainable development of the country and indeed, of the subcontinent. Such knowledge transfer is often slow and restricted in its impact. Despite these frustrations, important advances have been witnessed in the fields of invasive plant control (Macdonald, 1988a,b, 1992), fire management (Van Wilgen and Wills 1988; Van Wilgen et al., 1994), wildflower harvesting (Davis, 1992), catchment management (Le Maitre et al., 1993), river management (Davies et al., 1993), marine resources (Siegfried, 1994) and integrated environmental management (Quinlan, 1993).

Although the vast majority of ecological studies and management actions relate to the needs of vertebrates and higher plants, belated but important attention is now being given to some insect groups. The first reserve created specifically for an invertebrate species was the Ruimsig Butterfly reserve, established in 1985 to protect one of the last populations of a local endemic *Aloeides dentatis*. Priorities for further action are presented in the Red Data Book: Butterflies (Henning and Henning, 1989), and Samways (1990, 1992, 1993a, 1994) provides detailed proposals on conservation measures for montane grasshoppers, dragonflies and other insect groups.

Gradually, conservation biologists and resource managers are becoming equipped with the conceptual and technical skills necessary to make scientifically sound decisions for conservation – the ultimate goal of conservation biology training (Crowe and Siegfried, 1993). The inclusion of such knowledge within 'expert' and other decision support systems has not been successful. Despite the heuristic value of such approaches, which enjoyed much interest in the 1980s (Huntley, 1988), none of those developed in South Africa have been applied by field managers. More appropriate has been the development of broad conceptual guidelines communicated in the process of numerous workshop meetings, where researchers and managers have integrated theory and practise relating to such issues as culling seals (Butterworth, 1992), management of large mammals (Ferrar, 1983), invasive plant eradication or control (Macdonald, 1988a,b), management of riparian vegetation (Bosch et al., 1994), forest fragments in exotic timber plantations (Everard, 1993), fire frequency and season (Van Wilgen et al., 1994), and developing national strategies for biodiversity conservation (FRD, 1992).

Perhaps the most valuable biodiversity management skills that have been developed are those that are not normally reported in the open literature. The hands-on experience of wildlife managers in game capture and translocation (Novellie and Knight, 1994), culling and processing the products of hundreds of elephant and buffalo, and tens of thousands of antelope; habitat management through the controlled use of fire and water resources; collection and cultivation of indigenous plants in the *ex situ* genebanks of botanical gardens; these and many more skills are integral to achieving the ambitious goals of a comprehensive biodiversity conservation strategy.

Partnerships: Sharing Experience and Expertise

South Africa was expelled from the General Assembly of the United Nations on 12 November 1974, and until its readmission on 24 June 1994, its ability to participate in international projects and to share its experience and expertise, most especially with its fellow African nations, was severely curtailed. Happily, the situation is now reversed, with South Africans not only being free to travel to all countries of the globe, but also for colleagues from other states being welcome to engage in a wide range of partnerships with their South African colleagues.

An exciting array of opportunities for expanding old or initiating new partnerships is possible in the new South Africa. These include the following.

Expertise

As has been indicated above, the expertise in environmental sciences available in South Africa is substantial compared with many other developing countries. Walmsley and Walmsley (1994) note that in 1991 the environmental science expertise in South Africa, according to the country's Central Statistic Services, totalled over 12,000 professionals. In a more detailed survey, Walmsley and Walmsley (1993) found that out of a sample of 911 environmental professionals in universities, research councils and conservation agencies, biodiversity ranked second, after resource utilization, out of 16 concern areas within which research was being conducted.

The availability of such expertise, both in the form of human skills and electronic databases, offers significant opportunities for partnerships with other southern African countries in the preparation of the biodiversity assessments required by the Convention on Biological Diversity. The implementation of the Convention could provide mechanisms for mutually supportive projects throughout the region, possibly facilitated through the Southern African Development Community.

Education and training

Colvin (1992) notes that only 7% of the world's trained ecologists come from developing countries, and that in most cases, scientists from abroad (the 'north') have filled the gap, in what he aptly refers to as 'scientific ecotourism'. Although South Africa is relatively well supplied with ecologists born and trained in Africa (the South African Institute of Ecologists has 350 members), there is an urgent need to share education and training facilities more widely through Africa.

In 1991, 300,000 students were enrolled in South African universities, and of 58,000 degrees awarded annually during the 1988-1991 period, 12,800 were in the natural sciences. During the same period, 116,000 students were enrolled and 14,000 diplomas were awarded annually by the country's 12 technikons, 2900 of these diplomas being awarded in the natural sciences (FRD, 1993). While the provision of education and training at university and technikon level is adequate to meet national demand, increasing numbers of students from other African countries are anxious to make use of South African facilities. Furthermore, the need for a training college for wildlife managers, such as that at Mweka, Tanzania, has long been realized and following consultations with representatives of the Southern African Development Community (SADC), IUCN's Regional Office for Southern Africa, and Mweka, a new college for African Wildlife Management has been established on the border of the Kruger National Park in 1995.

Cooperative projects
Despite the academic boycott, South African conservationists were able to maintain many links with their colleagues, even if only at a personal level. These contacts were especially strong in the field of biodiversity research and conservation action, as demonstrated by the long-term collaboration in ichthyological studies in Lake Malawi (Skelton, personal communication), the Okavango basin (Skelton *et al.*, 1985), and through collaboration with Indian Ocean island states in the conservation of the coelacanth (Bruton and Stobbs, 1991) and marine turtles (Hughes, 1993; personal communication) and with many nations in over 30 years of research in the Antarctic and sub-Antarctic (Miller, 1991) and in the ecology of mediterranean-type ecosystems (Cowling, 1992).

Future cooperative projects include the development of transfrontier national parks, such as that proposed through the linking of an extensive area of Mozambique with the Kruger National Park's eastern border. Other potential transfrontier parks include linkage between South African protected areas and those in Namibia (Richtersveld); Botswana (Kalahari Gemsbok); and southern Mozambique (Ndumu, Tembe).

Reintroductions
Perhaps the most spectacular contribution that South African conservationists have made, and will continue to make, is in the re-introduction of species that have been depleted or become extinct in other African countries. The classic success story is that of the White Rhinoceros, reduced to approximately 20 survivors in Zululand by 1900, but now with a population of over 6300 in South African reserves. Since 1962, 3465 animals have been distributed, the majority within South Africa, but 1216 have been translocated to 32 other countries, six of these within their former distributional range in Africa.

Ecotourism

Giannecchini (1993) predicts that world tourism receipts will increase to nearly US$3 trillion in 1996, making travel and tourism the largest single industry in the world. Of the 55-60% of the world travel market that comprises leisure travel, 20-25% could be defined as nature tourism, or ecotourism. Southern Africa is still an underdeveloped ecotourism destination and a strategy embracing all southern African countries is needed to exploit the potential benefits of ecotourism to biodiversity conservation at a regional scale. Initiatives have already been taken to develop an environmental ethic within the tourism industry (the 'Hwange Principles') (Ferrar, 1993). Studies on the environmental impacts of tourist developments in national parks and reserves have become mandatory procedures in the National Parks Board and Natal Parks Board (Porter and Brownlea, 1990), and visitor education, promoting desirable resource utilization practices, and a 'user pays' approach that incorporates social costs, has been proposed (Preston *et al.*, 1991).

Indigenous plant use

The fastest growing biodiversity focus in southern Africa is that on the traditional use of indigenous plant resources. The growing awareness that primary health care will continue to be dependent on traditional medicines for the substantial majority of the region's population has placed new emphasis on a field of research neglected for many decades in South Africa (Cunningham, 1994). Advances in this field will be greatly facilitated by the establishment of a southern African network of plant scientists and conservationists, called for at a recent regional meeting of botanists from 14 African states (Huntley, 1994). A survey of research expertise in indigenous plant use (Anati *et al.*, 1994) provides a basis for developing such a network. At a national level, the topic has been stimulated by the establishment of an indigenous plant use programme (Cunningham *et al.*, 1992) and by a substantial injection of funding from the Foundation for Research Development.

Changing Paradigms: Biodiversity Conservation in the New South Africa

On 2 February 1990, the then President, Mr F.W. de Klerk announced the reform programme which culminated in the inauguration on 10 May 1994 of Mr Nelson Mandela as the President of a democratic South Africa. The dramatic political transition that has occurred since has had a profound influence on conservation policy and action in the country.

Until the transition process began, the environmental impacts of the Separate Development policy of the apartheid regime were poorly recognized, even among ecologists. In the first comprehensive review of the topic, Huntley *et al.* (1989) drew attention to the vast disparities in natural resource distribution between the ruling minority and the disenfranchised majority, and described four possible environmental scenarios that might develop in the presence or absence of substantial political and socioeconomic reform. The worst case scenario, 'paradise lost', would result from the maintenance of an isolationist fortress state, where natural resources would be plundered in a lawless society. The best case, 'rich heritage', would enjoy a stable economy and society where responsible environmental management could be afforded in a modern, free-market economy. Early indications suggest that South Africa has opted for the latter option.

The new government of national unity has placed the highest priority on its Reconstruction and Development Programme (RDP). The RDP introduces a broad-based, basic needs approach to redress the legacy of apartheid, not only in terms of socio-economic disparities, but also in terms of environmental upliftment. For the first time in the country's history, the root causes of environmental decay – poverty and disenfranchisement – are being addressed. The RDP provides policy outlines specific to the environment that will require new approaches to be developed by the existing conservation agencies and research institutes. Much of what is contained in the RDP chapter on the environment was foreshadowed in a series of important volumes on the socioeconomic/environment interface (Huntley *et al.*, 1989; Cock and Koch, 1991; Ramphele and McDowell, 1991).

Perhaps the most important policy change required by the RDP is for a participatory decision-making process around environmental issues, empowering communities to manage their natural environment. This is particularly pertinent where land tenure and access to natural resources are under dispute. Archer (1993) describes the participation process followed in negotiations between pastoralists and conservationists in the recently established Richtersveld National Park, in the desert of the extreme north-west Cape, whereas Cooper (1994) describes work with community-based organizations participating in the Natal Rural Forum, on the opposite side of the sub-continent. Such transparent procedures are regrettably novel in South Africa, where decisions on resource use and access were previously taken without any, or with only cursory, consultation with the affected parties. Good neighbour relations have become a fundamental facet in national park and reserve management (Fourie, 1994; Loader, 1994) and will, it is hoped, dispel the view held by many among South Africa's black majority that saving rhinos, elephants or rare plants is an indulgent pursuit of the affluent minority.

Interaction between the government and the public has been facilitated through the growth of conservation non-governmental organizations (NGOs), from a mere handful in the 1960s (the Wildlife Society and the Botanical Society accounting for over 90% of NGO membership at that time) to over 1000 NGOs and community-based organizations active in environmental matters in the 1990s (Geach, 1991).

Conclusion

The future is unknown and unknowable. What is absolutely clear, however, is that South Africa is on a new, democratic and exciting path. The combination of the commitment of the new government towards a stable and open economy and to environmental responsibility, the availability of an excellent technological and scientific infrastructure supporting a strong core of conservation biologists, and the goodwill being extended to the country by its neighbour states in southern Africa, bodes well. Through active partnerships at regional and international levels, South Africa can now look forward to playing a meaningful role in the global family of nations.

Acknowledgements

This chapter benefited from critical review by many colleagues, most particularly Ortwin Bourquin, Richard Cowling, Michael Samways and Roy Siegfried. Their advice and support is gratefully acknowledged.

References

Acocks, J.P.H. (1953) Veld Types of South Africa. *Memoirs of the Botanical Survey of South Africa* 28, 1–192.

Allanson, B.R., Hart, R.C., O'Keeffe, J.H. and Robarts, R.D. (1990) *Inland Waters of Southern Africa: an Ecological Perspective*. Kluwer, Dordrecht.

Anati, A., Walmsley, J.J. and Hansen, L.C.B. (1994) Survey of research expertise in indigenous plant use: sub-Saharan Africa. *FRD Programme Report Series* 12, 1–107. Foundation for Research Development, Pretoria.

Anon. (1990) Studies of global change in southern Africa. *South African Journal of Science* 86, 278.

Archer, F. (1993) Integrating conservation practices with pastoralist management strategies in the Richtersveld National Park. *Bulletin of the Grassland Society of Southern Africa* 4, 31–32.

Arnold, T.H. and De Wet, B.C. (1993) Plants of southern Africa: names and distribution. *Memoirs of the Botanical Survey of South Africa* 62, 1–825.

Bond, W.J. (1989) Describing and conserving biotic diversity. In: Huntley, B.J. (ed.) *Biotic Diversity in Southern Africa: Concepts and Conservation*. Oxford University Press, Cape Town, pp. 2–18.

Bond, W.J. (1994) Do mutualisms matter? Assessing the impact of pollinator/disperser disruption on plant extinctions. *Philosophical Transactions of the Royal Society* 344, 83–90.

Booysen, P. de V. and Tainton, N.M. (eds) (1984) Ecological effects of fire in South African ecosystems. *Ecological Studies*, Vol. 48. Springer-Verlag, Heidelberg.

Bosch, J., Le Maitre, D., Prinsloo, E. and Smith, R. (1994) Proposed research programme for riparian zones and corridors in forestry areas. *FRD Programme Report Series*. Foundation for Research Development, Pretoria.

Branch, G.M., Griffiths, C.L., Branch, M.L. and Beckley, L.E. (1994) *Two Oceans: a Guide to Marine Life in Southern Africa*. D. Phillips, Cape Town.

Brown, A.C. and McLachlan, A. (1990) *Ecology of Sandy Shores*. Elsevier, Amsterdam.

Bruton, M.N. and Stobbs, R.E. (1991) The ecology and conservation of the coelacanth *Latimeria chalumnae*. *Environmental Biology of Fishes* 32, 313–339.

Butterworth, D.S. (1992) Will more seals result in more fishing quotas? *South African Journal of Science* 88, 414–416.

Cock, J. and Koch, E. (eds) (1991) *Going Green: People, Politics and the Environment in South Africa*. Oxford University Press, Cape Town.

Colvin, J.G. (1992) A code of ethics for research in the Third World. *Conservation Biology* 6, 309–311.

Cooper, K.H. (1994) The environmental support programme of the Natal Rural Forum. *African Wildlife* 48, 27–29.

Cowling, R.M. (ed.) (1992) *The Ecology of Fynbos: Nutrients, Fire and Diversity*. Oxford University Press, Cape Town.

Cowling, R.M. and Bond, W.J. (1991) How small can reserves be? An empirical approach in Cape Fynbos, South Africa. *Biological Conservation* 58, 243–56.

Cowling, R.M. and Hilton-Taylor, C. (1994) Patterns of plant diversity and endemism in southern Africa: an overview. In: Huntley, B.J. (ed.) *Botanical Diversity in Southern Africa*. National Botanical Institute, Pretoria, pp. 31–52.

Cowling, R.M. and Holmes, P.M. (1992) Endemism and speciation in a lowland flora from the Cape Floristic Region. *Biological Journal of the Linnaean Society* 47, 367–383.

Crawford, R.J.M., Shannon, L.V. and Shelton, P.A. (1988) Characteristics and management of the Benguela as a large marine ecosystem. In: Sherman, K. and Alexander, L.M. (eds) *Biomass and Geography of Large Marine Ecosystems*. American Association for the Advancement of Science, pp. 171–221.

Crowe, T.M. and Siegfried, W.R. (1993) Conserving Africa's biodiversity: stagnation versus innovation. *South African Journal of Science* 89, 208–210.

Cunningham, A.B. (1994) Combining skills: participatory approaches in biodiversity conservation. In: Huntley, B.J. (ed.) *Botanical Diversity in Southern Africa*. National Botanical Institute, Pretoria, pp. 149–167.

Cunningham, A.B., De Jager, P.J. and Hansen, L.C.B. (1992) *The Indigenous Plant Use Programme*. Foundation for Research Development, Pretoria.

Davies, B.R. and Walmsley, R.D. (eds) (1985) Perspectives in Southern Hemisphere Limnology. *Hydrobiologia* 125, 1–263.

Davies, B.R., Snaddon, C.D. and O'Keeffe, J.H. (1993) *A Synthesis of the Ecological Functioning, Management and Conservation of Southern African River and Stream Ecosystems*. Water Research Commission, Pretoria.

Davis, G.W. (1992) Commercial exploitation of natural vegetation: an exploratory model for management of the wildflower industry in the Fynbos Biome of the Cape, South Africa. *Journal of Environmental Management* 35, 13–29.

Davis, G.W. and Richardson, D.M. (eds) (1994) *Mediterranean-type Ecosystems: Functions of Biodiversity*. Springer-Verlag, Heidelberg.

Davis, S.D. and Heywood, V.H. (eds) (1994) *Centres of Plant Diversity: A Guide and Strategy for their Conservation*. Oxford University Press, Oxford.

Dean, W.R.J. and Bond, W.J. (1994) Apparent avian extinctions from islands in a man-made lake, South Africa. *Ostrich*, 65, 7–13.

Emanuel, B.P., Bustamante, R.H., Branch, G.M., Eeekhout, S. and Odendaal, F.J. (1992) A zoogeographic and functional approach to the selection of marine reserves in the west coast of South Africa. *South African Journal of Marine Science* 12, 341–354.

Edwards, D. (1974) Survey to determine the adequacey of existing conserved areas in relation to vegetation types. A preliminary report. *Koedoe* 17, 2–37.

Everard, D.A. (ed.) (1993) *The relevance of Island Biogeographic Theory in Commercial Forestry. Environmental Forum Report*. Foundation for Research Development, Pretoria.

Ferrar, A.A. (1983) Guidelines for the management of large mammals in African conservation areas. *South African National Scientific Programmes Report 69*. Foundation for Research Development, Pretoria.

Ferrar, A.A. (1989) The role of Red Data Books in conserving biodiversity. In: Huntley, B.J. (ed.) *Biotic Diversity in Southern Africa: Concepts and Conservation*. Oxford University Press, Cape Town, pp. 136–147.

Ferrar, A.A. (1993) Redistributing the benefits of ecotourism. *African Wildlife* 47, 244–245.

Fourie, J. (1994) Comments on national parks and future relations with neighbouring communities. *Koedoe* 37, 123–136.

FRD (1992) *Biotic Diversity, Endangered Species and Red Data Books: A Future Approach*. Foundation for Research Development, Pretoria.

FRD (1993) *South African Science and Technology Indicators*. Foundation for Research Development, Pretoria.

Fuggle, R.F. and Rabie, M.A. (eds) (1992) *Environmental Management in South Africa*. Juta, Cape Town.

Geach, B. (1991) *The Green Pages*. Weekly Mail, Johannesburg.

Giannecchini, J. (1993) Ecotourism: new partners, new relationships. *Conservation Biology* 7, 429–432.

Gibbs-Russell, G.E. and Gonsalves, P. (1984) PRECIS – curatorial and biogeographic system. In: Allkin, R. and Bisbey, F.A. (eds) *Databases in Systematics*. Academic Press, London, pp. 137–153.

Gon, O. and Heemstra, P.C. (1990) *Fishes of the Southern Ocean*. JLB Smith Institute of Ichthyology, Grahamstown.

Hall-Martin, A.J. (1992) Distribution and status of African elephant *Loxodonta africana* in South Africa, 1952–1992. *Koedoe* 35(1), 65–88.

Harrison, J.A. (1992) The Southern African Bird Atlas Project databank: five years of growth. *South African Journal of Science* 88, 410–413.

Henning, S.F. and Henning, G.A. (1989) South African Red Data Book: Butterflies. *South African National Scientific Programmes Report 158*. CSIR Pretoria.

Heydorn, A.E.F. (1989) The conservation status of southern African estuaries. In: Huntley, B.J. (ed.) *Biotic Diversity in Southern Africa: Concepts and Conservation*. Oxford University Press, Cape Town, pp. 290–297.

Hockey, P.A.R. and Buxton, C.D. (1989) Conserving biotic diversity on southern Africa's coastline. In: Huntley, B.J. (ed.) *Biotic Diversity in Southern Africa: Concepts and Conservation*. Oxford University Press, Cape Town, pp. 298–309.

Hockey, P.A.R., Lombard, A.T. and Siegfried, W.R. (1994) South Africa's commitment to preserving biodiversity: can we see the wood for the trees? *South African Journal of Science* 90, 105–107.

Hughes, G.R. (1993) Report to the Department of Foreign Affairs on a visit to the Seychelles, 10–17 July 1993. Natal Parks Board, Pietermaritzburg.

Huntley, B.J. (1978) Ecosystem conservation in southern Africa. In: Werger, M.J.A. (ed.) *Biogeography and Ecology of Southern Africa*. Junk, The Hague, pp. 1333–1384.

Huntley, B.J. (1987) Ten years of cooperative ecological research in South Africa: a review. *South African Journal of Science* 83, 72–79.

Huntley, B.J. (1988) Conserving and monitoring biotic diversity: some African examples. In: Wilson, E.O. (ed.) *Biodiversity*. National Academy Press, Washington, DC, pp. 248–260.

Huntley, B.J. (ed.) (1989a) *Biotic Diversity in Southern Africa: Concepts and Conservation*. Oxford University Press, Cape Town.

Huntley, B.J. (1989b) Challenges to maintaining biotic diversity in a changing world. In: Huntley, B.J. (ed.) *Biotic Diversity in Southern Africa: Concepts and Conservation*. Oxford University Press, Cape Town, pp. xiii–xix.

Huntley, B.J. (1994) (ed.) *Botanical Diversity in Southern Africa*. National Botanical Institute, Pretoria.

Huntley, B.J. and Ellis, S. (1984) Conservation status of terrestrial ecosystems in southern Africa. In: *Proceedings of the Twenty-Second Working Session, Commission on National Park and Projected Areas*. IUCN, Gland, Switzerland, pp. 13–22.

Huntley, B.J. and Walker, B.H. (eds) (1982) Ecology of Tropical Savannas. *Ecological Studies* 42. Springer-Verlag, Heidelberg.

Huntley, B.J., Siegfried, W.R. and Sunter, C.L. (1989) *South African Environments into the 21st Century*. Human and Rosseau Tafelberg, Cape Town.

Hutchings, L. (1994) The Agulhas Bank: a synthesis of available information and a brief comparison with other east coast shelf regions. *South African Journal of Science* 90, 179–185.

Jarman, M.L. (ed.) (1986) Conservation priorities in lowland regions of the Fynbos biome. *South African National Scientific Programmes Report* 87, CSIR, Pretoria.

Johnson, S.D. and Bond, W.J. (1994) Red flowers and butterfly pollination in the fynbos of South Africa. In: Arianoustou, M. and Groves, R.H. (eds) *Plant Animal Interactions in Mediterranean-Type Ecosystems*. Kluwer, Dordrecht, pp. 137–148.

Joubert, S.C.J. (1983) A monitoring programme for an extensive national park. In: Owen-Smith, R.N. (ed.) *Management of Large Mammals in African Conservation Areas*. Haum Educational Publishers, Pretoria, pp. 201–212.

Kruger, F.J., Mitchell, D.T. and Jarvis, J.U.M. (eds) (1983) Mediterranean-type ecosystems: the role of nutrients. *Ecological Studies* 43. Springer-Verlag, Heidelberg.

Le Maitre, D.C., Van Wilgen, B.W. and Richardson, D.M. (1993) A computer system for catchment management: background, concepts and development. *Journal of Environmental Management* 39, 121–142.

Loader, J.A. (1994) National Parks and social involvement – an argument. *Koedoe* 37, 137–148.

Lombard, A.T. (1993) Multispecies conservation, advanced computer architecture and GIS: where are we today? *South African Journal of Science* 89, 415–418.

Lombard, A.T., August, P.V. and Siegfried, W.R. (1992) A proposed geographic information system for assessing the optimal dispersion of protected areas in South Africa. *South African Journal of Science* 88, 136–140.

Lombard, A.T., Nicholls, A.O. and August, P.V. (1995) Where should nature reserves be located in South Africa? A snake's perspective. *Conservation Biology* 9, 363–372.

Macdonald, I.A.W., Kruger, F.J. and Ferrar, A.A. (eds) (1986) *The Ecology and Management of Biological Invasions in Southern Africa*. Oxford University Press, Cape Town.

Macdonald, I.A.W. (1988a) Invasive alien plants and their control in southern African nature reserves. In: Thomas, L.K. (ed.) *Management of Exotic Species in Natural Communities*. Colorado State University, Fort Collins, pp. 63–79.

Macdonald, I.A.W. (1989b) The history, impacts and control of introduced species in the Kruger National Park, South Africa. *Transactions of the Royal Society of South Africa* 46, 251-276.

Macdonald, I.A.W. (1992) Vertebrate populations as indicators of environmental change in southern Africa. *Transactions of the Royal Society of South Africa* 48, 87-122.

Mace, G.M. and Lande, R. (1991) Assessing extinction threats: towards a re-evaluation of IUCN threatened species categories. *Conservation Biology* 5, 148-157.

Mace, G.M. and Stuart, S.N. (1994) Draft IUCN Red List Categories, Version 2.2. *Species* 21, 13-24.

Maclean, G.L. (1993) *Roberts' Birds of Southern Africa.* 6th edn. John Voelcker Bird Book Fund, Johannesburg.

Manning, J.C. and Linder, H.P. (1992) Pollinators and evolution in *Disperis* (Orchidaceae), or why are there so many species? *South African Journal of Science* 88, 38-49.

McIntyre, S., Barrett, G.W., Kitching, R.L. and Recher, H.F. (1992) Species triage – seeing beyond wounded rhinos. *Conservation Biology* 6, 604-606.

Musil, C.F. and Wand, S.J.E. (1994) Differential stimulation of an arid-environment winter ephemeral *Dimorphotheca pluvialis* (L.) Moench by ultraviolet-B radiation under nutrient limitation. *Plant Cell and Environment* 17, 245-255.

Myers, N. (1990) The biodiversity challenge expanded: hot-spots analysis. *The Environmentalist* 10, 243-255.

Noss, R.F. (1990) Indicators for monitoring biodiversity-hierarchical approach. *Conservation Biology* 4, 355-364.

Novellie, P.A. and Knight, M. (1994) Repatriation and translocation of ungulates into South African national parks: An assessment of past attempts. *Koedoe* 31, 115-119.

Miller, D.G. (ed.) (1991) Thirteenth Anniversary of Antarctic Treaty. Special Issue, *South African Journal of Antarctic Research* 21, 75-231.

Osborne, R. (1990) A conservation strategy for the South African cycads. *South African Journal of Science* 86, 220-223.

Osborne, R. (1995) The World Cycad Census and a proposed revision of the threatened species status for cycads. *Biological Conservation* 71, 1-12.

Owen-Smith, R.N. (ed.) (1983) *Management of Large Mammals in African Conservation Areas.* Haum Educational Publishers, Pretoria.

Payne, A.I.L., Brink, K.H., Mann, K.H. and Hilborn, R. (eds) (1992) Benguela trophic functioning. *South African Journal of Marine Science* 12, 1-1108.

Porter, R.N. and Brownlea, S.F. (1990) Integrated environmental management. A planning strategy for nature conservation developments. *Southern African Journal of Wildlife Research* 20, 81-86.

Pouris, A. (1989) Strengths and weaknesses of South African science. *South African Journal of Science* 85, 623-625.

Preston, G.R., Fuggle, R.F. and Siegfried, W.R. (1991) Environmental education in South African nature reserves: promoting desirable resource utilization practices. *South African Journal of Science* 87, 544-547.

Quinlan, T. (1993) Environmental impact assessment in South Africa: good in principle, poor in practice? *South African Journal of Science* 89, 106-110.

Ramphele, M. and McDowell, C. (eds) (1991) *Restoring the Land: Environment and Change in Post-Apartheid South Africa.* Panos Institute, London.

Rebelo, A.G. (1991) *Protea Atlas Manual: Instruction Booklet to the Protea Atlas Project.* National Botanical Institute, Claremont.

Rebelo, A.G. (1993) Using rare plant species to identify priority conservation areas in the Cape Floristic Region: the need to standardize for total species richness. *South African Journal of Science* 89, 156–161.

Rebelo, A.G. (1994) Cape Floristic Region. In: Davis, S.D. and Heywood, V.H. (eds) *Centres of Plant Diversity: A Guide and Strategy for their Conservation*. Oxford University Press, Oxford, pp. 218–224.

Rebelo, A.G. and Siegfried, W.R. (1992) Where should nature reserves be located in the Cape Floristic Region, South Africa? Models for spatial configuration of a reserve network aimed at maximizing the protection of floral diversity. *Conservation Biology* 6, 243–252.

Rebelo, A.G. and Tansley, S.A. (1993) Using rare plant species to identify priority conservation areas in the Cape Floristic Region: the need to standardize for total species richness. *South African Journal of Science* 89, 156–161.

Robinson, G.A. and De Graaff, G. (1994) *Marine Protected Areas of the Republic of South Africa*. Council for the Environment Pretoria. National Parks Board, Pretoria.

Rothschild, E.J. and Wooster, W.S. (1992) The evaluation of the first 10 years of the Benguela Ecology Programme 1982–1991. *South African Journal of Science* 88, 2–8.

Samways, M.J. (1990) Land forms and winter habitat refugia in the conservation of montane grasshoppers in southern Africa. *Conservation Biology* 4, 375–382.

Samways, M.J. (1992) Dragonfly conservation in South Africa: a biogeographical perspective. *Odonatologica* 21, 165–180.

Samways, M.J. (1993a) Insects in biodiversity conservation: some perspectives and directives. *Biodiversity and Conservation* 2, 258–282.

Samways, M.J. (1993b) Threatened Lycaenidae of South Africa. In: New, T.R. (ed.) *Conservation Biology of Lycaenidae*. IUCN, Gland.

Samways, M.J. (1994) *Insect Conservation Biology*. Chapman & Hall, London.

Scholes, R.J. and Walker, B.H. (1993) *An African Savanna: Synthesis of the Nylsvley Study*. Cambridge University Press.

Scholtz, C.H. and Holm, E. (eds) (1985) *Insects of Southern Africa*. Butterworth, Durban.

Shelton, P.A. (1989) The conservation status of pelagic ecosystems of southern Africa. In: Huntley, B.J. (ed.) *Biotic Diversity in Southern Africa: Concepts and Conservation*. Oxford University Press, Cape Town, pp. 310–327.

Siegfried, W.R. (1989) Preservation of species in southern African nature reserves. In: Huntley, B.J. (ed.) *Biotic Diversity in Southern Africa: Concepts and Conservation*. Oxford University Press, Cape Town, pp. 186–201.

Siegfried, W.R. (1992) Conservation status of the South African endemic avifauna. *South African Journal of Wildlife Research* 22, 61–63.

Siegfried, W.R. (ed.) (1994) *Rocky Shores: Exploitation in Chile and South Africa*. Ecological Studies 103. Springer-Verlag, Heidelberg.

Siegfried, W.R. and Brooke, R.K. (1994) Santa Rosalia's blessing: cryptic and under-recognized species in southern Africa. *South African Journal of Science* 90, 57–58.

Siegfried, W.R. and Brooke, R.K. (1995) Anthropogenic extinctions in the terrestrial biota of the afrotropical region in the last 500,000 years. *Journal of African Zoology* (in press).

Siegfried, W.R. and Brown, C.A. (1992) The distribution and protection of mammals endemic to southern Africa. *South African Journal of Wildlife Research* 22, 11–16.

Skelton, P.H., Bruton, M.N., Merron, G.S. and Van Der Waal, B.C.W. (1985) The fishes of the Okavango drainage system in Angola, South West Africa and Botswana. Taxonomy and distribution. *Ichthyological Bulletin of the J.L.B. Smith Institute* 50, 1–21.

Skinner, J.D. and Smithers, R.H.N. (1990) *The Mammals of the Southern African Subregion*. 2nd edn. University of Pretoria.

Smith, M.N. and Heemstra, P.C. (1986) *Smith's Sea Fishes*. MacMillan, Johannesburg.

Steiner, K.E. and Whitehead, V.B. (1990) Pollinator adaptation to oil-secreting flowers – *Rediviva* and *Diascia*. *Evolution* 44, 1701–1707.

Steiner, K.E. and Whitehead, V.B. (1991) Oil flowers and oil bees. Further evidence for pollinator adaptation. *Evolution* 45, 1493–1501.

Steiner, K.E., Whitehead, V.B. and Johnson, S.D. (1994) Floral and pollinator divergence in two sexually deceptive South African orchids. *American Journal of Botany* 81(2), 185–194.

Takhtajan, A. (1986) *Floristic Regions of the World*. University of California Press, Berkeley.

Tinley, K.L. (1985) Coastal dunes of South Africa. *South African National Scientific Programmes Report* 109, 1–297.

Underhill, L.G. (1987) Changes in the age structure of curlew sandpiper populations at Langebaan Lagoon, South Africa, in relation to lemming cycles in Siberia. *Transactions of the Royal Society of South Africa* 46, 209–214.

Underhill, L.G. and Oatley, T.B. (1994) The South African bird ringing unit: 21 years of service and research. *South African Journal of Science* 90, 61–64.

Vane-Wright, R.I., Humphries, C.J. and Williams, P.H. (1991) What to protect? Systematics and the agony of choice. *Biological Conservation* 55, 235–254.

Van Wilgen, B.W. and Wills, A.J. (1988) Fire behaviour prediction in savanna vegetation. *South African Journal of Wildlife Research* 18, 41–46.

Van Wilgen, B.W., Richardson, D.M. and Seydack, A. (1994) Managing fynbos for biodiversity: constraints and options in a fire-prone environment. *South African Journal of Science* 90, 322–329.

Walker, B.H. (1992) Biodiversity and ecological redundancy. *Conservation Biology* 6, 24–36.

Walmsley, R.D. and Davies, B.R. (1991) Water for environmental management: an overview. *Water S.A.* 17, 67–76.

Walmsley, J. and Walmsley, R.D. (1993) The environmental science research infrastructure in South Africa. *FRD Programme Report Series*, 7. Foundation for Research Development, Pretoria.

Walmsley, J.J. and Walmsley, R.D. (1994) Deployment of research and educational expertise within the environmental sciences. *South African Journal of Science* 90, 255–256.

Wells, S.M., Pyle, R.M. and Collins, N.M. (1983) *The IUCN Invertebrate Red Data Book*. IUCN, Gland.

Werger, M.J.A. (1978) Biogeographical division of Southern Africa. In: Werger, M.J.A. (ed.) *Biogeography and Ecology of Southern Africa*. Junk, The Hague, pp. 233–299.

26 Biodiversity in the Developing Countries

T.N. Khoshoo
Tata Energy Research Institute, India Habitat Centre, Lodi Road, New Delhi - 110 003, India

Introduction

Biodiversity is the sum total of plants, animals and microorganisms existing as an interacting system in a given habitat. It exists on earth in eight broad realms with 193 biogeographical provinces (Udvardy, 1975, 1984). Each biogeographical province is composed of ecosystems, which are constituted by communities of living species existing in an ecological region. At the global level about 1,604,000 species of plants, animals and microorganisms have been described so far. However, it is estimated that there are around 17,980,000 species, but a working figure is about 12,250,000 species (WCMC, 1992). These species exist on land, fresh water and in marine habitats, or occur as symbionts in mutualistic and in parasitic state with other organisms.

Needs of Developing Countries

A study of the distribution of biodiversity in the world reveals that it is maximum in equatorial, tropical and subtropical regions, but it decreases progressively as one goes to the polar regions. Nature has been very benevolent to the developing countries in the distribution of biodiversity on the Earth. Thus developing countries, located in subtropical/tropical belt, are far richer in biodiversity than the industrial countries in the temperate region. The former harbour at least 50–70% of the total biodiversity of the world. Furthermore, the Vavilovian Centres of Diversity of crops and domesticated animals are also located in some developing countries (Vavilov, 1951).

The genes from wild ancestors of crops endemic to developing countries have made a distinctive contribution in crop improvement with considerable gain to the growers. Such examples have been listed by Witt (1982), Khoshoo

© 1996 CAB INTERNATIONAL. *Biodiversity, Science and Development: Towards a New Partnership* (eds F. di Castri and T. Younès)

Table 26.1. Conservation of biodiversity: strengths and weaknesses.

Developing countries	Industrial countries
• Biodiversity rich	• Relatively poor
• Vavilovian Centres of Diversity	• Nil
• Backed by indigenous people, local technical knowledge and indigenous systems of medicine	• Largely non-existent
• Biodiversity supported by cultural diversity	• Largely absent
• Genetics, breeding and biotech base poor	• Rich base
• Largely *in situ* conservation	• Largely *ex situ*, but *in situ* for their own non-agricultural biodiversity
• Conservation not entirely science-based	• Largely science-based
• Largely subsistence or intensive agriculture	• Largely industrial agriculture
• Sustainable utilization of biodiversity: not possible without capacity building.	• Capacity in existence
• Research and development, education and training, and demonstration and extension need enhancement.	• Rich base
• Poverty	• Affluence
• Largely bioindustrial development	• Largely industrial development
• Biodiversity-rich/Technology poor	• Biodiversity poor/Technology rich

(1988, 1991) and FAO (1993). The contribution of such genes has been considerable, which has made a positive difference both in social and economic terms.

Biodiversity is, therefore, an important biological resource and a strength of the developing countries. However, it is an irreplaceable resource because its extinction is for ever. The conservation of this resource and its sustainable utilization has to be central to all developmental planning in most developing countries, because the economy in most of these countries is dependent on agriculture, horticulture, animal husbandry, fisheries, forestry, medicinal and bioindustrial products, and others.

The agri-biodiversity is interwoven with the cultural diversity in developing countries. The latter enables people to adapt to changing conditions as manifest by the selection of the particular plants and animals that they use for their sustenance. It is also related to different languages, religious beliefs, literature, art, music, recreation, tourism, and other aspects of social diversity. Furthermore, there are indigenous systems of medicine based on considerable experience and local technical knowledge handed down from historical times. Many of these systems are based on well-documented and respectable texts based on local technical knowledge.

There are differences in biodiversity distribution, conservation, and utilization between developing and industrial countries. The chief strengths and weaknesses are listed in Table 26.1.

Most organizations involved with biodiversity talk of conservation, collection, evaluation, monitoring, training etc., but in actual practice nothing tangible is happening on the ground particularly in the developing countries. Furthermore, these countries though rich in biodiversity are deficient in genetics, breeding, pharmaceutical sciences, and biotechnology. For sustainable utilization, it is most important for these countries to build expertise in these areas both for gene and drug prospecting. Conservation is no longer a function of building a fence round an area, but involves considerable upstream science and technology including biotechnology.

Keeping in view the importance of developing countries as major repositories of biodiversity, some steps need to be taken in these countries to save biodiversity for humanity at large. A fuller discussion of this aspect is given elsewhere (Khoshoo, 1994b).

Scientific and technical problems associated with conservation are, indeed, wide ranging. Sound biosystematic and experimental evolutionary, genetic and breeding studies are needed to work out the details of distributional patterns of life forms, regenerative capacity, population genetics, evolutionary biology, breeding systems, size and extent of gene exchange within the gene pools, etc. Besides, expertise would be needed in several cognate fields such as meteorology, land and water management, landscape and restoration ecology, forestry, agriculture, economics, etc.

Attention will also have to be paid to the inventorying of a priority list of biota, monitoring their populations and genetic evolutionary effects of shrinkage of populations together with remedial rehabilitative and restocking measures, dynamics of agroecosystems, ethnobiology and relationship between biological and cultural diversities and traditional methods for their conservation etc. It is equally important to augment expertise in biotechnology which is an knowledge-intensive area and leads to value-addition to the products from biodiversity. Thus the extent and nature of capabilities in the science and technology components of biodiversity will actually determine the capability of a country to conserve and sustainably utilize biodiversity (Khoshoo, 1994a).

The idea of the foregoing exercise is to build a cadre of protected areas network (PAN) conservators and S and T specialists. It is very important for programmes on PAN to be successful.

The cadre of conservators has to be backed by another cadre of well-trained conservation biologists for tropical and subtropical areas. They should be conversant with techniques of long-term conservation, breeding biology, population genetics, techniques for measuring in the laboratory and monitoring under field conditions, nature and extent of genetic and species diversity. As indicated earlier, for sustainable utilization, expertise needs to be built in the areas of drug and gene prospecting and also in biotechnology for product development.

Earlier transfer of genes was limited to closely related taxa, but biotechnology now enables such transfers across phylogenetic and taxonomic barriers, and with this bacteria have become potential chemical factories for human welfare: an example is manufacture of human insulin by bacteria. This has opened new vistas far beyond imagination (Khoshoo, 1994a).

Such capacity building is the most urgent need of the developing countries both nationally and for cooperation beyond the geographic boundaries of a country.

Biodiversity, Bioproductivity and Biotechnology

It is indeed ironical that developing countries have given to the world all agribiodiversity, they themselves have been the areas of low productivity and high population density. Even now many of these countries (particularly in Africa) fall within the 'hunger belt' of the world. Equally important is the fact that many of these centres are located in tribal belts and governments of the region are bound to 'uplift' them socially and economically. Once these communities adopt high-yielding varieties, the traditional genetically diverse but low-yielding varieties will be progressively replaced by high-yielding uniform varieties. The former are the result of thousands of years of selection. If these are lost, it would be an incalculable loss to humanity at large. The industrial countries, by making use of the science of genetics and breeding, have raised productivity. Fig. 26.1 shows that low biodiversity and low bioproductivity has been prevalent in harsh ecosystems. Underlying the pre-Green Revolution agriculture, was high diversity and low bioproductivity. The world agriculture then moved to high productivity accompanied by low diversity and the result was the Green Revolution. Today, we realize that the Green Revolution has paid its dividends and in many developing countries acute hunger is no longer a reality. However, it has not been ecologically entirely friendly. The Green Revolution has also helped the developing countries to feed themselves and not go round with a begging bowl for food. Thus it has also been a question of balancing immediate economic and social gain and largely manageable level of environmental degradation and pollution. Among other things, sustainability in agriculture, animal husbandry, fisheries and forestry will depend on the ability to combine high productivity with high diversity. Agriculture all over the world has to move towards such a broad goal. Therefore, there is needed a clarity of vision regarding the relationship between biodiversity, bioproductivity and biotechnology.

At the same time, the institutional structure that controls biotechnology should not overshadow those institutions that deal with conservation of biodiversity, and on no account ignore the rights and privileges of the local communities. While the former involves upstream science and technology, the conservation area is largely languishing even for simple and time-tested scientific and technological inputs.

The relationship between biodiversity and biotechnology is shown in Fig. 26.2. The countries of the world can be divided in four groups: (i) biodiversity-poor and biotechnology-poor; (ii) biodiversity-poor but biotechnology-rich, (iii) biodiversity-rich but biotechnology-poor, and (iv)

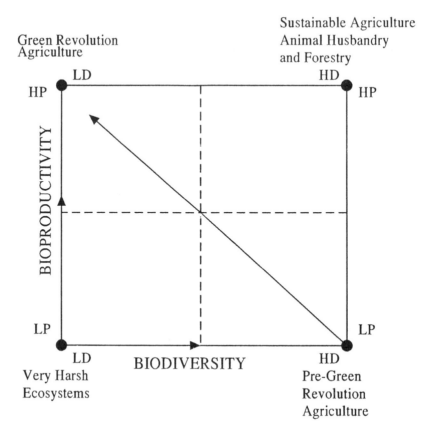

Fig. 26.1. Relationship between biodiversity and bioproductivity.

biodiversity-rich and biotechnology-rich. To the first group belong countries in the Middle East (e.g. Saudi Arabia), to the second group belong USA, Japan, Germany, France, Sweden and UK, the third group comprises southern countries like Indonesia, India, China, Malaysia, Brazil, Mexico and others in the tropical–subtropical belt, and there is no country which falls in the fourth group. At present there is a flow of biodiversity from the third group (South) to the second group (North). The extent and nature of flow of biotechnology from North to South is not commensurate with the flow of biodiversity from South to North. This is an unequal exchange and will remain so until such time as countries of the South become self-reliant in biotechnology. An important factor underlying this exchange is that whereas some countries (like India and China) do have the inherent capability to enter the fourth group (rich both in biodiversity and biotechnology), the countries of the North can not make it to the fourth group in the real sense of the word. The reason is that they do not have any worthwhile agribiodiversity growing naturally, although they do have excellent *ex situ* facilities in the form of field

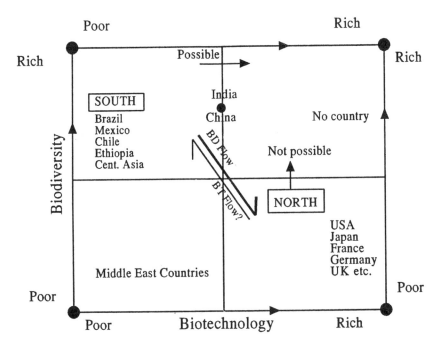

Fig. 26.2. Relationship between biodiversity and biotechnology.

gene banks, seed and other banks. The latter do not have the advantage of long-range ecological processes and organic evolution that operate under natural conditions and constantly refine and update biodiversity through mutation, recombination, and natural selection. These are the three cardinal elements of organic evolution. In essence, biodiversity stored in industrial countries is in the form of 'gene morgues', because advantages conferred by exposure to the process of natural selection are not available. In contrast to *in situ* conditions, organic evolution is virtually at a standstill or halted under *ex situ* conditions. Therefore, in all bio-banks, germplasm is preserved and not conserved in space and time. However, both *in situ* and *ex situ* conservation are necessary so as to complement and supplement each other (Khoshoo, 1994a).

The role that biodiversity plays in organic evolution and in plant and animal breeding is supported by mutation, recombination and selection. The power of recombination is indeed considerable in the evolution of biodiversity. Under organic evolution natural selection takes place. It is non-purposive and non-predirectional. The speed of evolution is dictated by the selective factors operating in the environment. If there are no perturbations, it can lead to conservation and enhancement of bio-wealth. However, under artificial selection guided by human needs, plant and animal breeding is purposive, directional and speed of evolution can be high. With the advent of biotechnology, it is possible to make gene transfers across taxonomic/

phytogenetic barriers and thus widen the germplasm base and thereby increasing the chances of sustainable development if the right choices are made, and newer and better products flow for human use.

It is most imperative for developing countries to strengthen the biotechnological base, which has become all the more important on account of the introduction of factors like intellectual property rights and patenting. If the biodiversity rich/technology poor developing countries are able to enhance this capability, then they have a chance to negotiate and bargain with strength with the industrial countries which are biodiversity poor/technology rich.

Developing countries must, therefore, come together and reach an understanding regarding various aspects, including scientific and technological, economic, social, cultural and legal issues, and collection, supply and costing of the raw material of biodiversity. Today the cost of biodiversity is the cost of collection and travel involved. If these countries remain divided (as they are today) regarding their stand on prospecting for new genes and drugs and compete among themselves, even the type of benefits that INBio has been able to get from Merck (USA), will no longer be forthcoming.

Summary

Nature has been very benevolent to the developing countries in the distribution of biodiversity on the earth. Thus the developing countries, located in subtropical/tropical belt, are far richer in biodiversity than the industrial countries in the temperate region. The Vavilovian Centres of Diversity of crops and domesticated animals are also located in developing countries. Biodiversity is thus an important living resource and strength of the developing countries. However, it is a irreplaceable resource because its extinction is for ever. The conservation of this resource and its sustainable utilization has to be central to all developmental planning in most developing countries because most of these have an agricultural economy. Incorporation of genes in crops from their wild ancestors, still available in developing countries, have, in many instances, made a colossal difference for the better in social and economic terms in agriculture.

The rich biodiversity of developing countries is also backed by indigenous people and their local technical knowledge about indigenous systems of medicine and other cognate uses. These countries are indeed treasure-troves of genes and bioactive molecules. At present, commensurate with the biodiversity wealth that these countries have, there exists no matching management, scientific and technical infrastructure. This needs to be created. In order to ensure long-term conservation and prospecting for new genes and drugs, there is an urgent need to extend to developing countries scientific and technical help in concrete terms in conservation and sustainable utilization of their biodiversity. This alone will make them self-reliant and help the whole humankind in the future.

Acknowledgements

This chapter represents contribution number 21 of a research programme in Conservation of Biodiversity and Environment jointly coordinated by TERI and the University of Massachusetts at Boston. The programme is supported in part by the MacArthur Foundation.

References

FAO (1993) *Harvesting Nature's Diversity*. World Food Day, Rome.
Khoshoo, T.N. (1988) *Environmental Concerns and Strategies*. Ashish Publishing House, New Delhi.
Khoshoo, T.N. (1991) Conservation of biodiversity in biosphere. In: Khoshoo, T.N. and Sharma, M. (eds) *Indian Geosphere-Biosphere*, National Academy of Sciences, Allahabad, pp. 178-233.
Khoshoo, T.N. (1994a) India's Biodiversity: Tasks Ahead. *Current Science* 67, 577-582.
Khoshoo, T.N. (1994b) *Conservation and Sustainable Utilization of Biodiversity in the Developing Countries: A Blue Print for Action*. Indian National Science Academy, Delhi.
Udvardy, M.D.F. (1975, 1984) *A Classification of the Biogeographical Provinces of the World*. IUCN, Gland, Switzerland.
Vavilov, N.L. (1951) The origin, variation immunity and breeding of cultivated plants. *Chronica Botanica* 13, 1-364.
WCMC (1992) *Global Biodiversity: Status of the Earth's Living Resources*. World Conservation Monitoring Centre, London.
Witt, S.C. (1982) *Genetic Engineering of Plants*. California Agricultural Land Project, San Francisco.

27 Biodiversity and Agriculture, Grasslands and Forests

M. Lefort and M. Chauvet

Bureau des ressources génétiques 57, rue Cuvier, F-75231 Paris Cedex 05, France

Introduction

For thousands of years, animal and plant species have dispersed away from their centres of origin and gradually adapted to highly diverse environments. These adaptations have led to the expression and creation of a tremendous amount of genetic diversity in the form of traditional breeds and cultivars, which have been exposed to widely differing selection pressures according to the environments and humans they encountered.

These cultivars evolved near their wild relatives, and genes constantly flowed between the two, either directly or via bridging species and spontaneous interspecific hybrids. These evolutionary dynamics, which provided the foundations of peasant farming, have long helped maintain a wide genetic diversity.

Later, the intensification of agriculture led to the homogenization of animal and plant productions on ever-larger areas and to the extensive use of farm inputs. The available genetic diversity became under-used (local breeds and cultivars were replaced by elite forms, the habitats of wild relatives were destroyed) and hence much of it was lost. However, it is now clearly accepted that the conservation of genetic resources is vital to making agriculture sustainable and to answering future food requirements. Management, conservation and utilization of genetic resources, among the major building blocks of biodiversity, should thus be reconsidered in order to support the feeding, health and industrial requirements of future generations.

The aim of this chapter is to provide a short presentation of the methods used for managing and conserving genetic diversity and of all related issues, whether scientific, technical, political, economic, or social. We shall stress the need for managing farm and forest lands dynamically and give a reminder of their ecological functions.

Managing Genetic Diversity in Production

Breeders have long realized that the very success in their field means that local varieties will disappear and be replaced by widespread modern varieties. *Ex situ* conservation of genetic resources provides at least a partial answer to this issue, but it is not enough to compensate for the effects of broadscale standardization in agriculture. Caring for diversity also implies maintaining an optimum diversity level in production systems, which combines the advantages of standardization, i.e. economies of scale and those of keeping enough available options at any time for responding to the inherent fragility of uniform systems.

Assessing the Risks

Examples of catastrophes due to the small genetic diversity of cultivated plants are common throughout history: potato blight, grapevine phylloxera, southern corn leaf blight (*Helminthosporium*). Nevertheless, the erosion of genetic resources carries on at three levels (Chauvet and Olivier, 1993).

1. Fewer species are cultivated because the plants which are best studied are those with the highest economic value and breeders are in full competition on a currently worldwide market. The relative advantage of these plants over secondary species keeps increasing, and the latter are gradually ruled out. Biotechnology is likely to take this even further because the plants which will produce new raw materials for industry are those which have been well studied and have a high yield, such as rape or maize.
2. The range of varieties within a species is also on the decrease because producers fall into line with a few main types which correspond to market constraints. The large number of cultivars is misleading insofar as they are often highly related to each other.
3. Variability within varieties also decreases because landraces are being replaced by pure lines, clones and F1 hybrids. This is in answer to a set of technical, commercial and regulation constraints.

The situation is slightly different in domestic animals. The same forces are acting, but at a different rate. Hens have followed the same changes as plants, i.e. hybrids are used. There are only a few pig populations throughout the world, Many breeds of cattle are threatened because the agrosystems which supported them have changed and because the breeding effort in these breeds has been much lower than in the mainstream breeds. Variability in the mainstream breeds remained fairly large as long as most of the selection was in the male line. The decrease in the number of selection criteria and the increasingly widespread use of artificial insemination followed by the implantation and cloning of embryos are radically changing the scene.

These changes result at least partly from decisions which economic actors chose freely. Unfortunately, they are often discovered when it is too late, once

Socioeconomic and Legal Aspects

they have become irreversible or that their negative effects have appeared. Moreover, the rate at which they occur and their magnitude are increasing, making adjustments ever harder. A risk assessment and monitoring scheme is therefore required for warning and guiding decision-makers.

Socioeconomic and Legal Aspects

Current regulations on the trade of seeds and varieties are mainly geared to the needs of large-scale trade and industry, which are in a position to afford the costs related to high agricultural and technical standards. People now realize that other opportunities exist, such as varieties for the amateur or regional produce, which require appropriate regulations. On the whole, one may even wonder whether regulations are required at all and whether they will not needlessly restrict diversity.

The natural tendency of the market is towards standardization. There are cases where competition acts on product diversity rather than on prices, but this diversity is often only apparent. For instance, a few very slight genetic changes may produce a whole range of colours from the same genetic stock. Maintaining genetic diversity of the produce requires that quality standards, certification marks and labels enable the consumer, at least in countries which are sophisticated enough to have them, to reliably identify the source and production conditions of the goods. This is what the European Union is trying to introduce by promoting an official status for local products and foodstuffs and protected designations of origin.

Just when steps are being taken in favour of the environment, agricultural policies carry on encouraging intensification and standardization in many ways. The setting of prices, grants and the terms for obtaining them and which development schemes are approved all work in much the same way. Somehow, removing incentives in favour of standardization would be far cheaper than adding new incentives in favour of biodiversity. This does require that departments, having been asked to act in one way for decades, change their attitude, which is never easy. Likewise, land-related issues, including taxation aspects, ownership and how it is handed down and schemes affecting land structures (consolidation) directly affect environmental diversity and should not be set aside.

International trade agreements (e.g. General Agreement on Tariffs and Trade, GATT) play a key role. Their basic assumption is that economic activities should ideally take place where production costs are lowest at a given time. However, agriculture and the management of natural habitats as well as human cultures are deeply rooted in a territory and cannot be transported without damage. If we really do want to introduce sustainable agriculture and development, then international agreements should be revised to fit in with the requirements of biodiversity protection.

Conservation of Genetic Resources for Future Needs: What Should we Preserve?

It is hard to predict what tomorrow's agricultural requirements will be and the environment changes fast. One should thus be cautious about stating which resources should be preserved in the long term. Although it would be desirable to keep a very broad range of resources for the medium or long term, one has to make decisions. and a practical rather than an idealistic approach is required.

Resources definitely worth maintaining include four main groups:

1. Varieties and breeds of current agricultural value which differ genetically from each other;
2. Landraces, breeds and ecotypes that are well adapted to specific regions and, although they are often less productive than the previous group, may provide genes conferring quality, adaptation and hardiness;
3. Wild parents of domesticated species, whose function as a gene pool for agriculture is now well established;
4. Genetic material obtained for research purposes (haploid, aneuploid, translocated and transgenic material, recombinant inbred lines) in order to provide research with the appropriate means of responding fast and efficiently to future agricultural needs.

As stated in the introduction, it is unrealistic to expect that we can maintain all of the above resources. Franker (1984) was the first to suggest building core collections of a limited size that would be capable of representing a broad fraction of all known genetic diversity. The advantage of core collections is that the few resources that are included are known in depth and can thus be better enhanced.

Core collections are therefore one of the main issues of long-term *ex situ* and *in situ* conservation, and sampling difficulties are crucial when designing them. Research into the following fields should be encouraged in order to make sampling strategies sounder:

- estimating diversity according to criteria which are neutral with respect to natural selection and to others which are non-neutral, and finding the best criteria for preserving those elements of diversity which are likely to answer the needs of tomorrow's agriculture;
- assessing the structure of genetic diversity in both space and time and designing a stratified sampling scheme which takes better account of how the material differentiated genetically.

Choosing what material should be preserved and how to manage it can only be improved by extending the reasoning further and by gathering more knowledge on the above points.

From Reasoned Resource Conservation

The genetic diversity of a species can be maintained in several ways, some of which let the evolutionary potential of the species express itself, and some of which do not. We shall first discuss several topics related to maintaining a status quo (true *ex situ* conservation), before tackling more dynamic forms of conservation which let the material considered carry on evolving.

True ex situ *conservation*

Purely *ex situ* conservation is associated with various types of gene banks: seed, pollen and organ banks for plants, semen and embryo banks for animals, field gene banks, especially for woody species, and zoos for animals. Although these have long been considered as an easy way of avoiding other solutions, they are essential for preserving the breeds and species currently under threat. Moreover, because genetic diversity is preserved as is, they will provide dated reference samples against which diversity changes can be measured in the coming decades.

In plants, long-term storage of seeds and cryogenic storage of organs produced *in vitro* are already widely used or under development. They now have to be applied on a large scale to a great number of species for which core collections should be made available. However, important issues have to be resolved for each method.

1. With orthodox seeds (i.e. which resist well to dehydration and hence to very low temperatures), the quality obtained after very long-term preservation (loss of vigour and viability, fragmentation of the DNA, loss of structural and functional integrity by the membranes) still has to be improved.

2. Recalcitrant seeds (i.e. which contain much water and are killed by desiccation) and cryopreservation of organs produced *in vitro*, studies on the causes of intolerance to desiccation and on the metabolic and cell disorders which appear below a certain temperature threshold (frost diseases) should be encouraged. Nevertheless, simple methods such as preservation under hypoxic conditions should also be considered.

Most of the *ex situ* conservation carried out in the field is due to arboreta (forest trees), conservation orchards (fruit trees), repositories and botanical gardens (local cultivars and wild relatives). This form of conservation is extremely cumbersome and expensive, and is now used only for a few economically valuable species. The following conditions are required.

1. Minimum stocks should be defined for each population in order to restrict genetic drift.

2. An appropriate location should be found which is devoid of contamination by parasites and remote from species with which spontaneous hybridization could occur.

3. Strict sanitary control should be provided, as well as adequate protection against fires for forest stands.

In animals, gamete storage techniques are reasonably operational in quite a few species, and should now be extended to all domestic species. The problem is how to make full use of this resource for managing the genetic diversity of the live form of the livestock of the species considered.

The cryogenic storage and transfer of embryos is far more difficult. In-depth technical and scientific studies are required for virtually each domestic species. In the rabbit for instance, which is the model species used in France, we still understand little about what mechanisms account for genetic differences in post-thawing survival (Joly et al., 1993).

Finally, one should be cautious about preserving a breed only in a frozen form, bearing in mind that the young will have learning difficulties if left on their own. However, the absence of support for preserving uneconomic breeds in developed countries in their live form inevitably leads to their being preserved solely *in vitro*. There is thus an urgent need to find other ways of making use of resources which are not directly profitable.

International information networks for zoos are now well established (ISIS, CBSG). They improve coordination between zoos and enable worldwide mating to restrict inbreeding. It would be worth increasing the awareness of zoo managers of conserving jeopardized local and national breeds in addition to the more 'exotic' species kept for cultural reasons.

There are a few final points on the organization of *ex situ* conservation which should be mentioned.

- Because conservation is very expensive, redundancy between gene banks should be as low as possible, except of course for safety duplicates. Removing redundancy is the first step towards building core collections.
- Conservation tasks should be distributed between countries which have genetic resources including support to developing countries, and the resources themselves should be able to travel freely.
- A gene bank does not have to be bound to a geographical entity. It can be organized as a network with a coordinator/contact person for the outside world and a distribution of tasks between members of the network. This form of decentralized organization is being preferred in France.

Dynamic conservation of artificial populations

The aim of this method is to keep heterogeneous gene pools as a readily usable source of variability. These pools are split up and exposed to environmental conditions which vary in space and time in order to simulate natural conditions as best as possible. This method, in addition to maintaining a wide variability in small populations, actually enables the evolutionary potential of the material to express itself, unlike the previous methods. Dynamic conservation should thus be regarded as a complement to the latter. As an example, let us now examine how bread wheat conservation is managed in France. Three bread wheat composite populations with a wide genetic basis were created. One of these was made allogamous by introducing a gene for male sterility in order to enhance hybridization within the population. Samples of these populations were propagated and experimented on during ten years over a wide network covering a broad geographical area. Results obtained after eight years clearly indicate that the populations differentiated according to their environment (David et al., 1992).

1. A North/South gradient indicated an evolution in earliness.
2. Average height of the overall population had increased because of competition between genotypes.
3. Populations differentiated with respect to: (i) resistance of adult plants to powdery mildew; and (ii) the frequency of genes conferring specific resistance to *Erysiphe graminis* f.sp. *tritici*, because of differences in powdery mildew pressure between the experimental environments.

Dynamic conservation has also been applied to barley (Allard, 1990) but unfortunately is much underused. Many questions still remain open regarding sampling (how many subpopulations should be used? in how many places and where should they be propagated?) and management (how can gene flow between them be enhanced in order to broaden their scope of response to the environment?). Answers to these questions in France will be provided by combining simulations and experiments based on the bread wheat experience.

In situ *conservation*

In situ conservation deals with keeping reproductive organisms in their natural habitat where genetic variability between and within populations is still high. It requires preliminary ecogeographical studies to determine the amount of variability present within the species and the spatial and temporal structure of that variability. Information gathered by these studies should help define:

1. which species should take priority for conservation (rarity, exposure to genetic erosion, socioeconomic value);
2. where conservation areas should be located for maintaining within species diversity and the natural diversity between species;
3. how these areas should be managed so as to avoid any loss in genetic variability.

With regard to research, little in depth knowledge is available about *in situ* conservation, much of which is actually population genetics in a natural environment. Knowledge of the following points needs to be improved promptly:

1. an estimation of the diversity of candidate species and of its ecogeographical structure;
2. sampling strategies (number of conservation areas, ecogeographical location, number of individuals per area to avoid genetic drift);
3. conditions and mechanisms for maintaining genetic variability in natural and seminatural environments.

Despite these weaknesses, programmes are being set up. One example is the European programme for the conservation of forest genetic resources (EUFORGEN) which was initiated by the 2nd European Ministerial Conference on forest protection (Strasbourg, 1990) and is managed by the International Plant Genetic Resources Institute (IPGRI) and the FAO. The aim of this project is to provide joint European management of several forest species (spruce, cork oak, poplar, and precious hardwoods) with representatives from a range of biological and genetic conditions from all over Europe.

In situ conservation is inherently run as networks. Costs can be reduced by organizing it on the scale of major world regions. Conservation areas include:

1. protected zones, in tight cooperation with the managers of these zones, whose primary task is to preserve the ecosystem as a whole, not genetic diversity;

2. specific zones managed within natural communities or within botanical gardens and conservatories;

3. agroecosystems and managed forests, with a specific organization and regular monitoring of diversity.

... To a More Integrated Management of Agricultural and Forest Land

In addition to long-term conservation discussed in the previous section, one should also contemplate using agricultural and forest land less intensively and in a more balanced way. These considerations, which in the long run should lead to a more environment-friendly use of agricultural resources, are probably one of the best ways of safeguarding a sustainable and cost-effective agriculture. They require that all parties involved cooperate tightly, from industry to the managers of rural areas, and that economists and legal experts join in.

In the plant world, incentives for expanding on-farm conservation, especially in developing countries, are one of the highlights of integrated conservation. It illustrates the efforts that farmers put into developing their own cultivars by taking advantage of natural or induced introgressions, and by selecting and propagating their own seeds. These activities belong to the dynamic management of diversity and have usually succeeded in maintaining genetic diversity within and between species and in letting it develop. Genetic diversity between species is due not only to the presence of wild parents in the neighbouring natural habitats, which enables gene flow between them and the cultivated material to occur, but also to the multiplicity of species cultivated at any one location.

This type of practice usually enabled subsistence farming to be carried out whatever the biotic or abiotic fluctuations. Unfortunately, it has been supplanted by the arrival of cultivars which are highly productive but also input-intensive and often ill-adapted and poorly responsive to environmental fluctuations. There are now plans to revert gradually to more diversified and less expensive farming methods by improving local cultivars in cooperation with local peasant communities. The planned improvements will preserve the vital components of the environment (water, soil, flora and fauna), which are the only guarantee that the system will be sustainable.

One example is an Ethiopian project for maintaining local farming practices which preserve diversity in a dozen well delimited and protected zones. This project includes educating local farmers. It highlights two difficulties which will reappear wherever such proposals are implemented.

1. Economically, reconversion has to be associated with increased acceptance of the produce, which is basically far less profitable than the produce of so-called elite varieties.

2. Socially, can two totally different farming systems coexist, one of which maximizes productivity in the short term by depleting resources, and the other which has a lower output under non-limiting conditions but which preserves biodiversity for a sustainable agriculture?

An organized network should be built around these new farming systems. This will at first be quite cumbersome because a wide variety of tasks have to be organized (collecting, storage, propagation) and because all sorts of people have to be trained initially (farmers, local and regional coordinators, agricultural technicians).

There is a strong wish in developed countries to further promote regional products as well as to preserve traditional local varieties and breeds, which are often hardier than their elite counterparts. Maintaining populations of them in an agricultural setting only makes sense if it can go beyond the collector stage, be turned into a true conservation operation and somehow be incorporated into the economic fabric. Considerations for including a list for amateurs in the official French catalogue of varieties is a first step in this direction. Subsidies by the European Union for preserving threatened local domestic breeds is another such step.

However, the real need is for restructuring agricultural systems as a whole so as to:

1. reduce production uncertainties with respect to biotic and abiotic fluctuations;

2. better integrate long-term environmental protection; And

3. improve responsiveness to the very fast changes in the agricultural markets and legislation.

Vissac (1994) has demonstrated with examples drawn from dairy cattle why it is important that we should have both alternative breed types and resources to increase the number of open options in the face of present uncertainties. He suggested launching action research initiatives involving scientists, industrial partners and country planners. Much work remains to be done in that field, which has hitherto been neglected by agronomists.

The Tarina breed in the Beaufortin district is a good example of what can be done about local breeds of small extent. This dairy breed produces a highly rated cheese (Beaufort) and its grazing contributes to maintaining skiing facilities. Because the breed is so scarce in the Beaufortin, it has to be managed in a much broader context by including individuals of the same breed introduced into other districts. Widely extended breeds are harder to assign to any single region but multibreed agricultural systems should nevertheless be supported (Vissac, 1994).

Systematic mowing of grasslands tends to make them more monospecific. Well tended grazing is far better for maintaining diversity of both plants (all species present get a chance to bear fruit) and animals (wildlife can coexist with the livestock).

In forestry, sylvicultural practices of the previous decades have also aimed for profitability: single species stands, uniform management (tillage, fertilization, tree work). These changes have decreased biological diversity and the adaptability of forests to respond to future uncertainties, particularly to major outbreaks of forest pests. One should therefore try and better integrate biodiversity conservation issues when managing timber production forests. The *Global Biodiversity Strategy* (1992) suggests some ways of doing this.

1. Key habitats should be preserved during harvesting so as to keep the heritage of natural forests.

2. Populations of the key species should be preserved in priority because they determine the structure of the community, and the presence of other species.

3. Fragmentation of natural forests should be avoided because this creates barriers for gene flow.

4. Forests should be regenerated with native species, whose yield is often higher than that of exotic species in the long run.

It is important that research should find which parameters affect variability within the populations so that they can be better integrated into the management schemes. Similarly, a better understanding of the effects of natural perturbations on forest dynamics should help simulate them in sylvicultural practice.

A final point on agroforestry: it would be beneficial if a few trees were introduced into the landscape. In addition to their aesthetic value, this would increase bird species diversity and decrease erosion risks.

Ecological Functions of Agricultural and Forest Land

Agricultural and forest land house ecosystems play various roles well beyond production, which is their main function. These matters have been seldom studied because they were so obvious in preindustrial agrosystems. The increase in environmental artificiality brought about by intensified and simplified farming reveals many difficulties related to the extinction of certain species or to the upsetting of certain functions. Agricultural and forest lands now have, in addition to their production role, an ecological and a social role in order to meet the demands for structured and enjoyable rural landscapes.

Habitat for wildlife

Many species carry out at least part of their life cycle in agricultural land. This includes weeds, birds which nest in crops, game which use it for feed and shelter and many insects. Some of these compete directly with the species being produced and are detrimental to the crop. Humankind has tried to eradicate them for thousands of years without success. But the powerful methods we now have at our disposal have drastically changed the setting. It can be stated that apart from species and ecotypes which are resistant to insecticides or herbicides, the only species that will remain in our fields are those that we have deliberately chosen. Designing more selective pesticides or slightly changing farming practices may be sufficient in some cases to offset much destruction. Nesting birds, for instance, may be protected by harvesting after the young birds have fledged or by restricting the speed of the harvesters.

Some species may be of direct use to agriculture, such as pollinating insects or predators of pests. Diversified environments should be maintained around the cultivated fields (hedgerows, grassland, woodland) so that these species may survive during periods when they cannot live off the crops. Environments like these may also play an important part as 'biological corridors', enabling species which are highly dependent on a particular type of environment to migrate from one sector to another despite a low dispersal ability, Migrating individuals may thus contribute their genes to populations that would otherwise suffer from excessive fragmentation of their habitat.

Any change in forest management inevitably affects biodiversity. It is well known that the plant species assemblage of undergrowth is strongly dependent on the major tree species in the overstorey. We know very little about how planting a restricted number of improved genotypes or even clones may affect the specific and genetic diversity of other life forms.

Conclusion

It is clear that preserving and taking advantage of biodiversity assets in agriculture and forestry is a far-reaching project which extends well outside the field of a few specialists in genetic resource centres. This matter falls within the competence of land use planning policies and is an issue for society to tackle, which in turn requires a broad democratic debate. This debate cannot occur unless the farming community joins in, because its future is directly affected and it will have to be involved and trained appropriately from an early stage. A decentralized administration will be required in which local and regional authorities are directly involved. The debate should reveal what comes under public interest and should therefore be dealt with by public organizations.

A shift in agricultural research programmes is an essential component of this policy. They will still have to include development issues but will also have to study systems which are far more complex than those considered in a pure production context.

These ambitions will only succeed if current attitude barriers are overcome. Agriculture no longer exclusively belongs to the farming community. Likewise, nature does not belong to conservationists. Our societies are becoming ever more urban and we biologists and scientists should perhaps help them make new plans for managing land in a sustainable way which takes into account the wishes and needs of various nature users whilst leaving as many options open as possible for the coming generations.

References

Allard, R.W. (1990) The genetics of host–pathogen coevolution. Implication for genetic resources conservation. *Journal of Heredity* 81, 1–6.

Chauvet, M. and Olivier, L. (1993) *La Biodiversité, Enjeu Planétaire. Conserver Notre Patrimoine Giénétique*. Paris, Sang de la Terre.

David, J.L., Savy, Y., Trottet, M. and Pichon, M. (1992) Méthode de gestion dynamique de la variabilité génétique. Example d'un réseau expérimental de populations composites de blé tendre. In: *Complexes d'Espèces, Flux de Gènes et Ressources Génétiques des Plantes*. (Proceedings of international colloquium in homage to Jean Pernès, Paris, 8–10 January 1992), BRG (Bureau des Ressources Génétiques)/Lavoisier, Paris, pp. 337–350.

Frankel, O.H. (1984) Genetic perspectives of germplasm conservation. In: Arber, W., Illmensee, K., Peacock, W.J. and Starlinger, P. (eds) *Genetic Manipulation: Impact on Man and Society*. Cambridge University Press, Cambridge, pp. 161–170.

Global Biodiversity Strategy (1992) Washington, WRI, IUCN and UNEP. French version: *Stratégie mondiale de la biodiversité* (1994). Paris, BRG (Bureau des Ressources Génétique)/Comité français pour l'IUCN.

Joly, T., Théau-Clément, M., Drouet-Viard, F. and de Rochambeau, H. (1993) Application de la cryoconservation des embryons à la protection des ressources génétiques chez le lapin. In: *Ressources Génétiques Animales et Végétales. Méthodologies d'Etude et de Gestion*. (Proceedings of the colloquium in Montpellier, 28–30 September 1993), BRG/INRA. ISARA Lyon, INRA Toulouse, INRA Tours, pp. 267–278.

Vissac, B. (1994) Populations animates et systèmes agraires. l'exemple des bovins laitiers. *INRA Productions Animales* 7(2), 97–113.

28 Managing Biodiversity in Canada's Public Forests

P.N. Duinker
Faculty of Forestry, Lakehead University, Thunder Bay, Ontario Canada P7B 5E1

Introduction

Conservation of forest biodiversity is taking centre stage on Canadian forest management and policy agendas at every turn. Citizens are demanding that forest biodiversity be maintained, even if they scarcely know what it means or how it might be achieved. Forest users and managers are putting biodiversity into their goal statements, yet few have developed firm ideas as to how to translate the concepts into concrete targets, and then to design appropriate and implementable schedules of biodiversity-conserving actions.

The purpose of this chapter is to examine some trends in Canada with respect to managing biodiversity in publicly owned forests. I begin by summarizing information about Canada's forests and their administration and management. Shifts in management philosophy, largely in response to public opinion, are described, and the biodiversity emphases in recent Canadian forest-policy documents highlighted. Then I discuss some definitions for forest biodiversity, as well as some approaches to biodiversity conservation. A few key threats to biodiversity conservation are briefly described. Conclusions follow identification of the principal research and development needs in this area.

Forests and Their Management in Canada

Canada (population 28.8 million) has a land area of some 921 million ha, of which 416 million ha (or 45% of the total) is classified as forest (Canadian Forest Service, 1994). Canada's forest area comprises about 10% of the world's forests. Some 71% of the forest land is owned by the provincial governments, 23% by the federal government, and only 6% by private concerns. This dominating public ownership of forest land extends to all provinces and territories except those in the Maritimes (New Brunswick, Nova Scotia, and Prince Edward Island), where private ownership is greater. This chapter focuses on biodiversity management on public forests because of the predominant public ownership.

Canada's forests are indeed diverse (Canadian Forest Service, 1994), and range from the Carolinian deciduous forests of southern Ontario to the tree line across the north, and from the boreal forests stretching across the country to the coastal coniferous rainforests of British Columbia. Native tree species number some 140 (Hosie, 1969). More than 500 species of terrestrial vertebrates find their habitats, for all or part of their life requirements, in Canadian forests (Bunnell and Kremsater, 1990).

Management philosophies for forests have undergone significant evolution in the past century. Sustained yield has been a guiding (though elusive) concept for a long time, and it was a cornerstone in multiple-use thinking in North America in the 1960s and 1970s. Multiple use gave way to integrated forest management in the 1980s, where managers try to balance management explicitly for an ever-widening array of forest values and benefits. In concert with the concept of sustainable development, the latest management approach is called ecosystem management (e.g., Ontario Forest Policy Panel, 1993). This tries to balance utilitarian forest values with a focus on keeping forest ecosystems in good condition, i.e. as wholesome, functioning ecosystems. Economy is thus balanced with environment. Biodiversity conservation is a key cornerstone of forest ecosystem management, and public involvement in management planning is another.

The Mandate to Manage Forests for Biodiversity Conservation

For many decades, people have had concerns for species extinctions caused by human activities, and for environmental problems like pollution. In the past 5–10 years, biodiversity has become an overarching concept that embraces a host of environmental concerns. For many people, biodiversity signifies most, if not all, of what looking after nature and the environment is all about. And, despite Canada's vast marine, freshwater, wetland and agricultural ecosystems, the public usually focuses its biodiversity attention to forests. This may be a reflection of the profound influence forests have on Canadians' ideas about nature, the fact that two-thirds of the 300,000 identified species that live in Canada are forest-dwelling (Canadian Forest Service, 1994), and the frequency with which conservation and environmental groups focus their attention, and thus the media's as well, on forest issues. Regardless of the reasons, the forest community in Canada has taken note, and pays much attention in its discussion forums and policy statements to biodiversity.

At a national level, strong language regarding forest biodiversity conservation and related issues appears in the Canada Forest Accord, the latest national forest strategy (Anon., 1992), the final report of the National Round

Table on the Environment and the Economy (Thompson and Webb, 1994), and the Draft Canadian Biodiversity Strategy (Federal-Provincial- Territorial Biodiversity Working Group, 1994). Using Ontario as an example, similar strong language appears in the 'Policy Framework for Sustainable Forests' (Ontario Ministry of Natural Resources, 1994), the Ontario Forest Industries' Association's guiding principles and code of forest practices (OFIA, 1993), and the new Crown Forest Sustainability Act (Bill 171, 3rd Session of Ontario's 35th Legislature).

Canadians clearly want their public forests to be managed in such a way that biodiversity is conserved. The forest community seems to have accepted this responsibility, in declarations at least, and is now in the process of making its declarations operational. Biodiversity conservation projects abound, and are stimulated and supported in a variety of ways. For example, Wildlife Habitat Canada, a national non-profit organization, financially supports a wide range of research and development projects related to forest biodiversity. As another example, under Canada's Green Plan, the Canadian Forest Service is sponsoring a series of 'model' forests across Canada and abroad. The model forests are to attempt new approaches to sustainable forest development, a key element of which is biodiversity conservation.

Conceptions of Forest Biodiversity

A useful entry point into defining biodiversity is to dissect the term. 'Bio' refers to life, and 'versitas' in Latin means variety (Canadian Forest Service, 1994). Thus, biodiversity means variety of life. To expand on this, a commonly used definition of biodiversity comes from the US Office of Technology Assessment (1987): biodiversity is 'the variety and variability among living organisms and the ecological complexes in which they occur'. There are many other definitions, and such a plurality is to be welcomed.

To make biodiversity an operational concept in forest management, it is necessary to be explicit about what is considered to be part of it, and how the parts or elements are to be measured. Most people agree that biodiversity has many facets. Kimmins (1992a) spoke of: (i) genetic diversity; (ii) within-ecosystem species diversity (so-called alpha species diversity); (iii) among-ecosystem species diversity (beta species diversity); (iv) within-ecosystem structural diversity; (v) among-ecosystem structural diversity; and (vi) temporal diversity. Noss (1990) presented a hierarchical characterization of biodiversity, with two axes forming a matrix. One axis is composed of composition, structure, and function, and each of these sets of ecological attributes can apply at each of the following four scales: (i) regional landscape (or forest, for our purposes); (ii) community/ecosystem (stand); (iii) population/species; and (iv) gene pools.

Thus, it seems that forest biodiversity includes all the ways we have of realizing and characterizing the variety of life in forests. We include not only composition and structure of organisms (parts, wholes, assemblages), but also the processes in which organisms are engaged, and, most significantly, the ecosystems that form the habitat for organisms and are defined in terms of both biotic and abiotic elements. Clearly, biodiversity is exceedingly complex. However, we can develop ways to simplify it sufficiently so that forest managers can understand some of the biodiversity of their forest and begin to manage in such a way that it is conserved.

Approaches to Conservation of Forest Biodiversity

Guidelines for indicator selection

There has to be a way, or several ways, of measuring biodiversity before forest managers can explicitly plan for it. Noss (1990) provided the following guidelines for choosing indicators:

> Ideally, an indicator should be (1) sufficiently sensitive to provide an early warning of change; (2) distributed over a broad geographic area, or otherwise widely applicable; (3) capable of providing a continuous assessment over a wide range of stress; (4) relatively independent of sample size; (5) easy and cost-effective to measure, collect, assay, and/or calculate; (6) able to differentiate between natural cycles or trends and those induced by anthropogenic stress; and (7) relevant to ecologically significant phenomena.

Ian Thompson of the Canadian Forest Service (1993, personal communication) suggests that indicators be readily measurable in the woods, repeatable (i.e., giving the same results if several competent investigators measure it), and predictable (i.e., amenable to reliable prediction). All these traits of indicators are important, but there is special significance in Thompson's last point. If we have no way to generate defensible forecasts (Duinker and Baskerville, 1986) for indicators of biodiversity, we have no defensible basis to plan for it. The key here is to link forest management strategies with biodiversity indicators in explicit cause–effect models.

Approaches

Guidance for forest managers on biodiversity conservation is beginning to appear on the Canadian scene (e.g. Duckworth and Fleming, 1993). However, no forest managers can yet claim to be taking a fully comprehensive approach to biodiversity conservation in forest management. The traditional way to handle concerns for wildlife habitat, a key part of the overall biodiversity picture, has been to apply stand-scale guidelines. In Ontario, for example, many sets of such guidelines (e.g. Ontario Ministry of Natural Resources, 1988) are used in forest-management planning. They are applied either in a comprehensive way (i.e. wherever timber-management operations will occur) for generalist species such as moose, or in a specific way for habitat specialists when foresters, biologists or members of the public identify discrete 'areas of concern'.

Despite its simplicity and operational appeal, the guidelines approach to biodiversity conservation is suffering criticism (e.g. Duinker, 1989a). The main drawback is that guidelines thwart much-needed learning about the relationships between species and their habitats, learning that can come only from explicitly setting goals for wildlife habitat across large forest landscapes and over long periods of future time (Baskerville, 1985; Duinker, 1987, 1989a). Adaptive management for forest biodiversity conservation, as for resource and environmental systems in general (Holling, 1978; Walters, 1986; Lee, 1993), requires preparation of explicit forecasts of biodiversity responses to planned forest interventions, and comparison of these forecasts with measurements of the actual responses during and after implementation of the interventions (see Duinker and Baskerville, 1986; Duinker, 1989b, for discussions of forecasting and monitoring in resource management and environmental assessment).

The guidelines approach to forest biodiversity conservation is gradually giving way to an adaptive management approach using explicit forecasts of the expected behaviour of elements of forest biodiversity under a management regime. Some practitioners are taking a species-by-species approach (e.g. Bonar et al., 1990), while others are starting with ecosystem diversity (e.g. the White River Forest example described below). A combination of these two approaches is appealing (Duinker, 1993), and is summarized below.

A proposed approach

Two key departure points for the proposed approach are as follows.

1. The approach must be simple (Salwasser et al., 1984). Complex approaches will be unappealing to managers and stakeholders alike. I believe strongly that a simple approach is as technically legitimate as any complex and more comprehensive approach, and indeed more powerful because of higher potential understanding and acceptance by those using it.

2. Among-stand diversity is the place to start (Miramichi Pulp and Paper Inc., 1992; Welsh, 1992; Perera, 1992; Society of American Foresters Task Force, 1991; Booth et al., 1993).

I propose the following indicators/scales to begin addressing forest biodiversity in management of public forests in Canada:

1. Among-stand diversity – composition and structure; and
2. Special species – habitat carrying capacities; population viability.

Among-stand diversity

All forest managers in Canada have some sort of database characterizing the stands that make up their forests. The stands in these databases represent reasonable forest landscape units for diversity analysis. Regardless of whether the forest stand inventory is digitally mapped, managers can use histograms to analyse both the richness (number of members of a class) and evenness (proportion of the class made up of each member) of stand types, stand development stages (or age classes), stand sizes, and even stocking levels and site classes.

Here is a simple example, using age classes. Suppose it is reasonable to class all stands into 20-year age classes. A forest has a relatively rich age-class distribution if all classes up to age 200 year are represented, and a poor one if only classes up to age 60 are represented. If the older stands represent a significant element of that forest's biodiversity, and there is reason to believe that under natural conditions the forest would have some area in the older age classes, then the manager may wish to plan for the existence of stands in each of the older age classes.

Suppose a forest has stands in four ages classes up to age 80 years. If 90% of the area is in the two classes below 40 years, and 10% in the two above, we would say the age-class structure is uneven, or unbalanced. The manager may well determine that forest biodiversity would be improved with more balance in the age-class structure. A goal here, perhaps somewhat arbitrary, might be to ensure that each class among the four age classes has at least 15% of the total forest area. This is a simple example because it does not address spatial patterns of the area in the age classes (e.g. patch sizes, locations, and connectivity). However, it is a start that can be made even when forest inventories are not digitally mapped.

I believe that histograms are better than numerical indices of diversity. They convey more information, and are intuitively easier to grasp. If the forest inventory is mapped, then the manager can examine how the richness and evenness of key inventory variables vary across the forest landscape. This might be done by dividing the entire forest into meaningful subunits, and examining the differences among histograms for each subunit. The manager can also examine spatial patterns using a wide variety of measures being developed by landscape ecologists (e.g. McGarigal and Marks, 1994). A simple measure relating to shape of stands, in relation to their size, might be the length of stand boundaries per unit area (say, $m\,ha^{-1}$, or $km\,km^{-2}$), which can be interpreted, if done properly, as an edge/area ratio. Depending on which variables are used to define the stands and their boundaries, not all boundaries are likely to represent biologically significant edge, so such measures require considerable care in their design, execution, and interpretation. Stand definitions in timber-oriented forest inventories may not yield ecologically significant patches (Plinte *et al.*, 1995).

What makes the among-stand diversity approach attractive is that forest managers are increasingly making use not only of digitally mapped inventories but also of dynamic forest inventory simulation models. Some of these are lump-sum or aggregating models, so results of among-stand diversity analyses can not be mapped. Others, though, such as HSG (Moore and Lockwood, 1990), work on a stand-by-stand basis, and permit a wide range of quantitative analyses to be performed on inventory projections.

Special species

As stated above, most of the concern under the banner of forest biodiversity focuses on native species that are most threatened by human activities. For many reasons, humans look upon living nature through 'species' eyes – moose, budworm, sugar maple, and skunks represent very different 'units' of nature. Forest managers cannot escape the desire by many members of the public to look after special species.

Which species? For forest biodiversity, Probst and Crow (1991) recommended choosing species that have: (i) specialized habitat needs; (ii) low densities; (iii) large home ranges; (iv) poor dispersal and colonizing abilities; and (v) susceptibility to local extinction. Other criteria might include the degree to which a species: (i) can be used to indicate the status of habitat for other species, or even ecosystem 'health'; (ii) is a critical element of the forest foodchain; or (iii) is socioeconomically important.

Because forest managers influence habitat but may not be in charge of population management such as hunting or trapping, I recommend that they begin figuring species into their biodiversity planning using habitat supply analysis (e.g. Duinker *et al.*, 1991; Greig *et al.*, 1991). Measures include areas in habitat goodness classes, such as those used in most so-called Habitat Suitability Indices, or, better yet, habitat carrying capacities in terms of individuals per unit area. Carrying capacities can be calculated in a 'gross' form, without recognition for population dynamics of the species of concern (e.g. Duinker *et al.*, 1993), or in a 'net' form that accounts for population dynamics (e.g. Duinker, 1986).

Species-related goals might include maintaining viable populations of all native species, as well as eradication of troublesome exotic species (Noss, 1990). Bonar *et al.* (1990) attempted the former by choosing some 30 species across a wide range of habitat requirements, determining minimum viable populations for each, and then performing habitat supply analyses to determine if forest management could maintain sufficient habitat to meet the needs of the minimum viable populations.

What about within-stand diversity?
Diversity within forest stands is widely recognized to be an important determinant of overall biodiversity (e.g. Hunter, 1990). Some key variables include overstorey composition, vertical and horizontal structure of the stand, understorey composition, and the quantity and quality of dying, dead, and down trees.

As part of the among-stand approach, within-stand diversity can be brought in as an additional layer of detail in defining the forest stand inventory. For example, the Ontario Forest Resource Inventory includes an estimate of the species composition of the overstorey, in 10% classes. This information might be used to refine stand typing for among-stand diversity analysis, going beyond the inventory's 'working group' designation for each stand (based on predominant overstorey species). As another example, once stands are classified using information in the standard forest inventory, subclassifications based on amounts of snags or coarse woody debris might be created, and a diversity of these conditions, consistent with the natural range of variation, pursued among stands of the same type.

Benchmarks

For some facets of forest biodiversity, the benchmarks to manage toward seem clear. For example, species extinctions, or even extirpations, are to be avoided, even at high costs. Another example is no net loss of rare forest types, such as the Carolinian Deciduous Forest of southern Ontario, the area of which has been reduced to a very small fraction of its extent before European settlement. Most people today would agree that there is much too little Carolinian forest in southern Ontario for a balanced environment.

For other facets of forest biodiversity, the benchmarks are less clear, or subject to much debate. For Canada's boreal forests, should the pre-European-settlement state of forest biodiversity be the target? If so, many questions are raised. Can the presettlement biodiversity conditions be unambiguously defined (Perhaps this is a key role for large forest reserves or parks.) If they can, might we have confidence that survival of the species we have come to expect to inhabit forests will thus be guaranteed? What influences did Aboriginal peoples have on forest biodiversity? How broadly did key forest biodiversity variables fluctuate over time (say, centuries) under purely natural influences such as fires, insects, diseases, and windthrow?

Other benchmarks are possible. Notice that the strategic objective for forest biodiversity in Ontario's policy framework for sustainable forests (see above) calls for current forest biodiversity to be maintained. Perhaps this is a reasonable interim objective while people figure out just what types and amounts of biodiversity are desirable and attainable. Given that forest biodiversity in some areas may change markedly in response just to the forces of nature, forest managers may be faced with a situation where they are called upon to ensure that biodiversity remains between some widely set goalposts, and they may pursue a range of other objectives and values within that frame of reference.

Pressures on Forest Biodiversity

Humans put a number of pressures on forests which may alter biodiversity in undesirable ways. The main pressures on Canada's publicly owned forests, past and future, include: (i) timber management; (ii) harvest of forest fauna; (iii) air pollution; and (iv) climate change. Conversion of land cover from forest into other ecosystems was certainly a strong pressure on forest biodiversity in the past. It is still a pressure on privately owned forests in urbanizing areas.

Timber management

An estimated 29%, or about 119 million ha, of Canada's forest land is managed for timber production (Canadian Forest Service, 1994). Some of these forests have been subject to industrial timber harvest for a couple of centuries (e.g. the forests of Canada's eastern provinces), but others have been commercially harvested only in the last few decades.

Timber management can have profound effects upon forest biodiversity in the following ways.

1. Roads: previously unroaded territories across much of Canada must be roaded to access timber. In addition to turning forest ecosystems into road ecosystems, roads provide for human traffic and frequently other activities (e.g. settlement, industrial development, hunters, trappers) which may potentially be hostile to the conservation of forest biodiversity. Poorly built and maintained roads often erode into streams and adversely affect aquatic life.

Given the high cost of roads, improving standards for environmentally sound road building, the public appeal for more roadless areas in forests (e.g. Koven and Martel, 1994), and the already expansive road network throughout much of the commercial forest land in Canada, I believe that the effects of roads on forest biodiversity will be much smaller in the future than they were in the past. However, given the paucity of roadless areas in many forest regions of Canada, new roads into them may have proportionally more impact on overall forest biodiversity than an equivalent amount of roads when the regions were first accessed.

2. Timber harvest: all forms and styles of timber harvest have effects on forest biodiversity, some subtle and some dramatic. Clearcutting receives the most attention (e.g. Kimmins, 1992b; Duinker, 1995), particularly because it removes the entire tree overstorey in one operation, it is widely used in Canada, and it has often occurred over large contiguous areas. In a very short time, clearcutting a forest stand resets the successional clock from 'mature' forest to regenerating forest. Many forest types experience similar resetting through fire, insect infestations, and blowdowns, but other types experience much more localized disturbances.

Clearcutting is not the only harvest method that may have strong effects on forest biodiversity. Throughout Canada's development of European settlement, forests have been highgraded repeatedly. This means that, in response to market conditions and what the forests could offer by way of quality raw materials, cutters moved through the forests taking only the best logs. Such indiscriminate harvesting has not only changed the species composition of many of Canada's forests, but it has also changed the genetic stocks of the natural species.

Timber harvest practices in Canada have changed markedly in the past decades, and will continue to change dramatically as foresters discover new possibilities in response to public demands. All the changes point in one direction – trying to remove trees in a manner much closer to the way natural disturbances operate. Clearcut sizes, shapes and residuals (unharvested trees) are following much more natural patterns than in the past. Many of these changes in harvest practices are being made specifically for the purpose of conserving forest biodiversity.

3. Regeneration: while natural regeneration predominates in Canadian forest management, artificial regeneration is a vital component of the overall

picture. In recent decades, significant federal monies were made available to treat hundreds of thousands of hectares of provincial forest land that had not adequately regenerated naturally. Key ways in which artificial regeneration treatments can adversely affect forest biodiversity include: (i) conversion of stands from a relatively uncommon type to a common, industrially more-useful type; (ii) planting of populations of seedlings with narrow genetic amplitude; (iii) planting stands of a single species where mixedwoods formerly grew; and (iv) bringing on stands of commercial tree species more rapidly than might occur with natural regeneration following natural disturbance.

As with timber harvest practices, regeneration practices are tending towards gentler intrusions into forest ecosystems, and taking advantage of natural regeneration whenever possible. Foresters are much more inclined today than in the past to assist replacement of the kind of stand that was harvested.

4. Pesticides: many people have been concerned about the silvicultural use of chemical pesticides to combat insect infestations and weed competition. The key issue has been mortality of non-target organisms. Whether justified from a technical point of view or not, public sentiment in Canada has led to strong reductions in the application of chemical pesticides in forests. Some provinces have banned their use, favouring biological insecticides and mechanical weed control. The future of chemical pesticides in Canada is likely to see foresters using them only when all other means of controlling insects and weeds in crisis situations are ineffective or prohibitively expensive (e.g. Ontario Forest Industries Association, 1993).

5. Fire suppression: across the commercial forests of Canada, the general fire-suppression policy has been to suppress all fires as soon as possible following detection. Such a policy may protect commercial interests and human safety in the short term, but for many forests it means a gradual buildup of fuel loads which increases the intensity of fires when they do occur. In terms of areas burned, most of the fire in Canada's commercial forests is of lightning origin, and many forests are adapted to repeated fires every few to several decades.

Despite attempts to suppress all wildfires, they remain a significant agent of change in Canada's forests. More and more people are beginning to recognize the important role of fire in Canada's forests, and the influence that fire suppression has had on forest biodiversity. There are calls for both a stronger emulation of fire in timber harvest and regeneration practices where fire will be suppressed, and a balanced fire management programme, particularly in parks, where some fires are allowed to take their own course (Addison, 1994), provided they do not jeopardize forest values outside the parks.

I have reflected on these agents of changing forest biodiversity as if they were all undesirable. In fact, this same slate of tools – timber harvest, regeneration, pesticides, fire – can play a vital part in improving forest biodiversity. The conservation of biodiversity – either restoring it or maintaining it – is not necessarily best accomplished by walking away from ecosystems once they have been used or abused. Forest managers in Canada will increasingly be called upon to use their knowledge and management tools to shape forest biodiversity in desirable directions.

Harvest of forest fauna

Hunting of forest game animals, trapping of furbearers, and fishing in lakes and streams surrounded by forest are all important activities across Canada (Filion et al., 1988). Hunted and trapped species can have substantial influences on forest biodiversity (Naiman et al., 1988; Johnston et al., 1993), so harvest activities that strongly reduce the faunal species populations can alter overall forest biodiversity.

In Canada's public forests, forest managers are rarely in control of the harvest of forest fauna. Usually, such control is vested in fish and wildlife resource managers. Two moves are at present underway in Canada: (i) stronger controls on the rates of harvest of many forest fauna, so that populations remain at higher levels; and (ii) stronger coordination between habitat management and wildlife population management.

Air pollution

Some parts of Canada's forest lands, particularly in eastern Canada, are subject to regional depositions of airborne pollutants (e.g. sulphur dioxide) that some experts believe to be stressful to forests. Although there is uncertainty over the role of air pollution in contributing to the forest declines of eastern North America in the 1980s (Freedman, 1989), few would deny that regional air pollution may be stressing forests to some degree, and that near-ground point sources of air pollution – including oxides of sulphur and nitrogen, metals, organic compounds, ozone, and other – can give rise to local forest declines (Duinker, 1990a; Nilsson, 1991; Nilsson et al., 1992). Given the increased control of air pollution now practised in North America, local forest declines due to pollution are likely a thing of the past. However, regional air pollution may be having subtle and poorly understood effects on forest biodiversity (Freedman, 1989).

Climate change

If the global climate changes in ways that many experts predict, during the next century Canada may experience dramatic warming, especially in winter, and in many forests a general drying of the soil. Forest ecosystems will be in a state of confusion as climate resources shift at rates that plants, animals and soils cannot keep up with (Duinker, 1990b). Biotic communities and soils will have to respond to climates that have rapidly become much different from the climates they had evolved with over centuries. Given the potential for increased forest fires and insect/disease infestations, more difficult regeneration, and general losses of tree vigour, some ecologists believe that Canadian forests will experience a general state of decline until the vegetation catches up with the new climate (e.g. Solomon and West, 1985). Although there is still much uncertainty about the effects of climate change on forest biodiversity in North America, initial explorations of the topic (e.g. Peters and Lovejoy, 1992) suggest the potential for large undesirable effects. It is far from clear whether forest managers' efforts can have significant influence in mitigating the effects of climate change on forest biodiversity.

Summary

My assessment of the relative threats to forest biodiversity in Canada leads me to conclude that, during this century, timber-management practices and harvest of forest fauna have been the most influential factors degrading forest biodiversity in Canada's publicly owned forests. I believe climate change to be the most serious potential future threat to conservation of forest biodiversity, and careful timber-management practices to be the main means of maintaining or restoring forest biodiversity.

Research and Development Needs

I believe that the forest community has access to sufficient knowledge about forest biodiversity conservation to begin implementing it today in all public forests in Canada. However, many knowledge gaps need to be addressed, and technology needs to be developed. I identify below some areas where research and development are especially needed.

Public attitudes

We know that the Canadian public is demanding the conservation of forest biodiversity on public lands, but we do not have clear ideas on what the public means by this. Studies are needed to determine more precisely just how the public perceives forest biodiversity, what levels of forest biodiversity are desired, and what tradeoffs in terms of providing economic opportunities from the forests the public might be willing to make to obtain the desired levels of biodiversity conservation. Work is also needed to determine how to reconcile the public's appetite for biodiversity conservation with its equally strong desire to curb foresters' use of some biodiversity-conserving tools (e.g. timber harvest techniques, controlled burning).

Indicators

It is imperative in adaptive management of forest biodiversity to have unambiguous measures indicating its quantities at specific times and places (Ontario Forest Policy Panel, 1993). Much work in Canada is now being devoted to the general topic of indicators of sustainable forest development (e.g. Forestry

Canada, 1992; 1993; Anon., 1993; Canadian Forest Service, 1994; Plinte et al., 1995). Regarding forest biodiversity, there has been work both in Canada (e.g. Duinker, 1993) and abroad (e.g. Reid et al., 1993). A key initiative which is presently developing indicators for a range of sustainable forest development criteria, including biodiversity, is the project led by the Canadian Forest Service called the 'Canadian Criteria and Indicators Initiative for Sustainable Forest Management'. Its results will require operational testing at forest, regional, provincial and national levels.

Inventories

Once specific forest biodiversity indicators are chosen, forest managers will need to determine the state of their forests against the chosen indicators. Many of the indicators will have never before been inventoried. Some may be interpretable from standard forest resource inventories, but others will not. New classification systems for forest biodiversity will be needed, and inventories taken according to the new systems.

Effects of human activities

Forest managers need to know how the tools available for forest management might affect forest biodiversity, not only at the site level in the short term but more importantly across large forest landscapes over the long term. They also need better understandings of how stressors such as air pollution and climate change may affect forest biodiversity, and how their tools might be used to mitigate any such undesirable effects. Although some of this knowledge exists at present, work is urgently needed to synthesize existing information into models to help managers make credible dynamic forecasts of potential effects.

Decision support for adaptive forest ecosystem management

Forest managers are being called upon to manage forests for an ever-expanding array of values and benefits, one of which is biodiversity conservation. It is not possible for managers to determine efficient and effective ways to arrange human activities on broad forest landscapes for so many values and benefits without computational assistance. Decision-making now requires long-term, spatially explicit forecasts for the delivery of all quantifiable forest values (e.g. timber, recreation, biodiversity). Biodiversity presents a formidable planning chore to forest managers because there are so many ways of quantifying its elements. Much work will need to be done to assist managers in illuminating for them the myriad possible futures for forest biodiversity on their respective territories.

Natural disturbances

A key obstacle facing analysts in making dependable forecasts of forest biodiversity under a range of management strategies relates to broadscale disturbance. Events such as fire, windthrow, insect/disease infestations, and global atmospheric change could all have strong effects on forest biodiversity in reality, yet these forces are very difficult to account for properly in forest inventory simulations. The research and development community is urged to develop appropriate ways to build these factors into biodiversity forecasting models.

Conclusions

Some may liken biodiversity to love or money – the more, the better. However, it is perhaps more like colour – its appropriateness depend on many elements of the specific context. Enlightened forest managers and stakeholders are unlikely to want to maximize biodiversity, but rather to get it at approximately right. Getting biodiversity right in most cases will mean trying to conserve it at 'natural' levels, but this will be difficult to define precisely. Thus, the zone of 'rightness' is likely to be broad for Canada's public forests. Perhaps it will at first be necessary to define clearly what are undesirable states of biodiversity, i.e. what is to be avoided. As scientists, managers and the public begin to understand more and more about forest biodiversity, they can become increasingly better at defining just what constitutes the right kinds and amounts of biodiversity for specific forests.

Among-stand diversity and provision of habitats for special species are two promising management approaches for the conservation of forest biodiversity. Using these approaches, the guidance emerging in the literature, and the concepts of hierarchy of diversities (Perera, 1992) and diversity of diversities (Hunter, 1990), forest managers and stakeholders in Canada can aggressively pursue sustainable forest development through biodiversity conservation.

Acknowledgements

I benefited greatly from collaborations related to this chapter with R. Plinte, and from comments on an earlier draft from R. Bonar, D. Daust, G. Eason, D. Neave and S. Nilsson.

References

Addison, B. (1994) Born to burn. *Seasons* 34(2), 20–25.
Anon. (1992) *Sustainable Forests: A Canadian Commitment.* National Forestry Strategy, Canadian Council of Forest Ministers, Hull, PQ, 51 pp.
Anon. (1993) *Indicators of Sustainable Development Workshop: Draft Proceedings.* Model Forest Program, Canadian Forest Service, Hull, Quebec, 149 pp.
Baskerville, G.L. (1985) Adaptive management: wood availability and habitat availability. *Forestry Chronicle* 61, 171–175.

Bonar, R., Quinlan, R., Sikora, T., Walker, D. and Beck, J. (1990) *Integrated Management of Timber and Wildlife Resources on the Weldwood Hinton Forest Management Agreement Area*. Weldwood of Canada Limited, Hinton Division, Hinton, Alberta, 44 pp. and appendices.

Booth, D.L., Boulter, D.W.K., Neave, D.J., Rotherham, A.A. and Welsh, D.A. (1993) Natural forest landscape management: a strategy for Canada. *Forestry Chronicle* 69, 141-145.

Bunnell, F.L. and Kremsater, L.L. (1990) Sustaining wildlife in managed forests. *Northwest Environmental Journal* 6, 243-269.

Canadian Forest Service (1994) *The State of Canada's Forests 1993*. Fourth Report to Parliament. Minister of Supply and Services Canada, Ottawa, Ontario, 112 pp.

Duckworth, G. and Fleming, M. (1993) *Interim Strategy for Biodiversity Considerations in Timber Management Planning*. Strategic Planning Statement No. 1, Site Region Planning Group, Northeast Region, Ontario Ministry of Natural Resources, Timmins, Ontario, 8 pp.

Duinker, P.N. (1986) A systematic approach for forecasting in environmental impact assessment: a deer-habitat case study. Unpublished PhD thesis, Faculty of Forestry, University of New Brunswick, Fredericton, New Brunswick.

Duinker, P.N. (1987) Forecasting environmental impacts: better quantitative and wrong than qualitative and untestable! In: Sadler, B. (ed.) *Audit and Evaluation in Environmental Assessment and Management: Canadian and International Experience*. Environment Canada, Ottawa, and Banff Centre School of Management, Banff, Alberta, pp. 399-407.

Duinker, P.N. (1989a) Wildlife in forest management: from constraint to objective. In: Symposium Proceedings, *Forest Investment: A Critical Look*. Great Lakes Forestry Centre, Forestry Canada, Sault Ste Marie, Ontario, pp. 133-144.

Duinker, P.N. (1989b) Ecological effects monitoring in environmental impact assessment: what can it accomplish? *Environmental Management* 13, 797-805.

Duinker, P.N. (1990a) Biota: forest decline. In: Solomon, A.M. and Kauppi, L. (eds.) *Toward Ecological Sustainability in Europe: Climate, Water Resources, Soils and Biota*. Research Report RR-90-6, International Institute for Applied Systems Analysis, Laxenburg, Austria, pp. 115-154.

Duinker, P.N. (1990b) Climate change and forest management, policy and land use. *Land Use Policy* 7, 124-137.

Duinker, P.N. (1993) Indicators and goals for biodiversity in Canada's Model Forests. In: *Draft Proceedings, Indicators of Sustainable Development Workshop*. Model Forest Program, Canadian Forest Service, Hull, Quebec, pp. 51-61.

Duinker, P.N. (1995) Clearcuts. In: Paehlke, R. (ed.) *The Encyclopedia of Conservation and Environmentalism*. Garland, New York.

Duinker, P.N. and Baskerville, G.L. (1986) A systematic approach to forecasting in environmental impact assessment. *Journal of Environmental Management* 23, 271-290.

Duinker, P.N., Higgelke, P. and Koppikar, S. (1991) GIS-based habitat supply modelling in Northwestern Ontario: moose and marten. In: Proceedings, *GIS'91: Applications in a Changing World*. Forestry Canada, Victoria, BC, pp. 271-275.

Duinker, P.N., Higgelke, P. and Bookey, N. (1993) Future habitat for moose on the Aulneau Peninsula, Northwestern Ontario. In: Proceedings, *GIS'93: Eyes on the Future*. Forestry Canada, Vancouver, BC, pp. 551-556.

Federal-Provincial-Territorial Biodiversity Working Group (1994) *Draft Canadian Biodiversity Strategy*. Biodiversity Convention Office, Hull, Quebec, 69 pp.

Filion, F.L., Parker, S. and DuWors, E. (1988) *The Importance of Wildlife to Canadians: Demand for Wildlife to 2001*. Canadian Wildlife Service, Ottawa, Ontario, 29 pp.

Forestry Canada (1992) *The State of Canada's Forests 1991*. Second Report to Parliament. Minister of Supply and Services Canada, Ottawa, Ontario, 85 pp.

Forestry Canada (1993) *The State of Canada's Forests 1992*. Third Report to Parliament. Minister of Supply and Services Canada, Ottawa, Ontario, 112 pp.

Freedman, B. (1989) *Environmental Ecology: The Impacts of Pollution and Other Stresses on Ecosystem Structure and Function*. Academic Press, San Diego, California, 424 pp.

Greig, L., Duinker, P.N., Wedeles, C.H.R. and Higgelke, P. (1991) *Habitat Supply Analysis and Modelling: State of the Art and Feasibility of Implementation in Ontario*. Report prepared for Wildlife Branch, Ontario Ministry of Natural Resources. ESSA Ltd, Richmond Hill, Ontario, 81 pp.

Holling, C.S. (ed.) (1978) *Adaptive Environmental Assessment and Management*, no. 3, Wiley International Series on Applied Systems Analysis. John Wiley, Chichester, 377 pp.

Hosie, R.C. (1969) *Native Trees of Canada*. Canadian Forestry Service, Ottawa. 380 pp.

Hunter, M.L. Jr (1990) *Wildlife, Forests and Forestry: Principles of Managing Forests for Biological Diversity*. Prentice Hall, Englewood Cliffs, New Jersey, 370 pp.

Johnston, C.A., Pastor, J. and Naiman, R.J. (1993) Effects of beaver and moose on boreal forest landscapes. In: Haines-Young, R., Green, D.R. and Cousins, S.J. (eds) *Landscape Ecology and Geographic Information Systems*. Taylor and Francis, London, pp. 237-254.

Kimmins, J.P. (1992a) Biodiversity: An environmental imperative. In: *Forestry on the Hill*, Special Issue no. 3. Canadian Forestry Association, Ottawa, Canada, pp. 3-12.

Kimmins, J.P. (1992b) *Balancing Act: Environmental Issues in Forestry*. UBC Press, Vancouver, British Columbia, 244 pp.

Koven, A. and Martel, E. (1994) *Reasons for Decision and Decision: Class Environmental Assessment by the Ministry of Natural Resources for Timber Management on Crown Lands in Ontario*. Environmental Assessment Board, Toronto, Ontario, 561 pp.

Lee, K.N. (1993) *Compass and Gyroscone: Integrating Science and Politics for the Environment*. Island Press, Washington, DC. 243 pp.

McGarigal, K. and Marks, B. (1994) *Fragstats: Spatial Pattern Analysis Program for Quantifying Landscape Structure*. Forest Science Department, Oregon State University, Corvallis, Oregon, 67 pp.

Miramichi Pulp and Paper Inc. (1992) Report on the task force on biodiversity. In: *Forestry on the Hill, Special Issue no. 3*. Canadian Forestry Association, Ottawa, Canada, pp. 23-26.

Moore, T.G.E. and Lockwood, C.G. (1990) *The HSG Wood Supply Model: Description and User's Manual*. Information Report PI-X-98. Petawawa National Forestry Institute, Chalk River, Ontario, 31 pp.

Naiman, R.J., Johnston, C.A. and Kelley, J.C. (1988) Alteration of North American streams by beaver. *BioScience* 38, 751-762.

Nilsson, S. (ed.) (1991) *European Forest Decline: The Effects of Air Pollutants and Suggested Remedial Policies*. International Institute for Applied Systems Analysis, Laxenburg, Austria, 228 pp.

Nilsson, S., Sallnäs, O. and Duinker, P.N. (1992) *Future Forest Resources of Western and Eastern Europe*. Parthenon Publishing Group, Carnforth, UK 496 pp.

Noss, R.F. (1990) Indicators for monitoring biodiversity: a hierarchial approach. *Conservation Biology* 4, 355–363.

Ontario Forest Industries Association (1993) *Ontario Forest Industries Association Guiding Principles and Code of Forest Practices*. Ontario Forest Industries Association, Toronto, Ontario, 12 pp.

Ontario Forest Policy Panel (1993) *Diversity: Forests, People, Communities – A Comprehensive Forest Policy Framework for Ontario*. Queen's Printer for Ontario, Toronto, 147 pp.

Ontario Ministry of Natural Resources (1988) *Timber Management Guidelines for the Provision of Moose Habitat*. Wildlife Branch, Ontario Ministry of Natural Resources, Toronto, Ontario, 33 pp.

Ontario Ministry of Natural Resources (1994) *Policy Framework for Sustainable Forests*. OMNR, Toronto, Ontario.

Perera, A.H. (1992) The enigma of biodiversity. In: *Forestry on the Hill, Special Issue no. 3*. Canadian Forestry Association, Ottawa, Canada, pp. 73–76.

Peters, R.L. and Lovejoy, T.E. (eds) (1992) *Global Warming and Biological Diversity*. Yale University Press, New Haven, Connecticut, 386 pp.

Plinte, R.M., Duinker, P.N. and Bookey, N.A. (1995) *Measuring Up: Indicators of Sustainability for Ontario's Boreal Forests*. Report to the Canadian Forest Service, Sault Ste Marie, Ontario.

Probst, J.R. and Crow, T.R. (1991) Integrating biological diversity and resource management. An essential approach to productive, sustainable ecosystems. *Journal of Forestry* 89(2), 12–17.

Reid, W.V., McNeely, J.A., Tunstall, D.B., Bryant, D.A. and Winogard, M. (1993) *Biodiversity Indicators for Policy-makers*. World Resources Institute, Washington, DC. 42

Salwasser, H., Thomas, J.W. and Samson, F. (1986) Applying the diversity concept to national forest management. In: Cooley, J.L. and Cooley, J.H. (eds) *Natural Diversity in Forest Ecosystems*, Proceedings of a Workshop. Institute of Ecology, University of Georgia, Athens, Georgia, pp. 59–69.

Society of American Foresters Task Force (1991) *Task Force Report on Biological Diversity in Forest Ecosystems*. Report SAF 91-03. Society of American Foresters, Bethesda, Maryland, 52 pp.

Solomon, A.M. and West, D.C. (1985) Potential responses of forests to CO_2-induced climate change. In: White, M.R. (ed.) *Characterization of Information Requirements for Studies of CO_2 Effects: Water Resources, Agriculture, Fisheries, Forests and Human Health*. DOE/ER-0236, Carbon Dioxide Research Division, United States Department of Energy, Washington, DC, pp. 145–169.

Thompson, S. and Webb, A. (eds) (1994) *Forest Round Table on Sustainable Development: Final Report*. National Round Table on the Environment and the Economy, Ottawa, Ontario, 44 pp.

US Office of Technology Assessment (1987) *Technologies to Maintain Biological Diversity*. OTA-F-330. US Government Printing Office, Washington, DC.

Walters, C.J. (1986) *Adaptive Management of Renewable Resources*. MacMillan Publishing Company, New York, 374 pp.

Welsh, D.A. (1992) Biodiversity and timber production are compatible. In: *Forestry on the Hill, Special Issue no. 3*. Canadian Forestry Association, Ottawa, Canada, pp. 28–29.

Some Canadian Approaches to Partnership in Agricultural Biodiversity

B. Fraleigh
Biodiversity Convention Office, Environment Canada, 351 St. Joseph Blvd., Hull, Quebec, Canada, K1A OH3

Introduction

Although this chapter is not exclusively focused upon activities within Canada, it seeks to provide a description of some approaches and perspectives taken on sectoral, national and international levels relating to issues in crop plant genetic resources. They are linked by one common theme: the search for partnership.

Biodiversity in Agriculture: Problems and Opportunities

Analysis of the relationship between agriculture and biodiversity reveals two major themes: conservation of biological resources important to the agriculture sector itself, and the impact of agricultural practices on biological diversity beyond that sector.

Conservation of agricultural biodiversity

Agricultural products are biological in nature, so the sector itself is ultimately dependent on biological diversity. Genetic resources are used to breed new crop varieties, to improve races of domesticated animals and to obtain adapted microorganisms. The use of genetic resources broadens the genetic base in agriculture, thereby enhancing farmers' competitiveness in domestic and international markets. The loss of genetic diversity limits the supply and range of genetic material available for future crop and animal improvement and industrial application.

Biological resources important to agriculture which are not preserved *de facto* in pastures, farmers' fields or protected areas, must be deliberately maintained as part of a conservation programme. In Canada, the conservation of biological resources important to agriculture will continue to be characterized by *ex situ* rather than *in situ* conservation methods, and will emphasize within-species genetic diversity as compared to between-species and ecosystem diversity.

Agrologists also recognize the need to identify and preserve certain wild species *in situ* such as beneficial insects and microorganisms, and wild relatives of cultivated plants (see especially Hoyt, 1988). Since agricultural authorities in Canada rarely have the mandate or the means to implement *in situ* conservation measures, this is an important opportunity for partnership with environmental conservationists and non-governmental organizations.

One example of partnership in Canada in the domain of plant genetic resources is a Memorandum of Understanding signed, in October 1991, between the federal government's Department of Agriculture, represented by Plant Gene Resources of Canada (PGRC), and a non-governmental organization, the Heritage Seed Program (HSP). Under the terms of this agreement, PGRC and HSP decided to establish conservation plots for the preservation of the genetic diversity of crop plants. HSP growers have already rejuvenated dozens of seed samples preserved in the national crop genebank during the three growing seasons over which the agreement has been in effect. The HSP growers have the opportunity to enjoy a greater selection of crop diversity, because by taking advantage of the genebank's storage capacities, they are not obliged to regrow the same variety so often. The HSP growers have unrestricted access to samples of genetic resources for non-commercial purposes. Choice of the varieties to grow is guided by the need to rejuvenate accessions which have a small amount of seed left or a low percentage of viability. Similar agreements are being considered for the conservation of clonal crops such as fruit trees, berries and potatoes.

The impact of agriculture on biodiversity

Agriculture's impact on biodiversity extends well beyond the farm gate. A Federal–Provincial Agriculture Committee on Environmental Sustainability (1990) reported that, with some exceptions, the availability and quality of wildlife habitat was seriously affected by settlement and agricultural development in Canada. While farmland and rangelands do provide habitat for certain species of plants, animals and microorganisms, populations of many others have declined as a direct result of agricultural expansion and agricultural production practices. Pollution and degradation of soil, water and air resources pose major challenges to maintaining biological diversity. An assessment by Environment Canada (1994) identified impacts of specific agricultural practices, including tillage, fertilizers, pesticides, low-input farming, drainage, and grazing.

The expansion of agricultural areas reflects society's need to produce food. However, agricultural expansion has resulted in a growing concern for the continued viability of wildlife species. The loss of native habitat converted into agricultural land has been significant; Canada has less than 13% of original short grass prairie, 19% of mixed grass prairie, 16% of aspen parkland and almost none of the tall grass prairie remaining in their native state (Millar, 1986). In the Prairie provinces, agricultural drainage has eliminated 1.2 million hectares, or 40%, of wetland habitat, resulting in significant declines in waterfowl populations. Some 35 species of birds, fish, mammals and reptiles are threatened as a consequence of Prairie habitat loss due to agriculture (Federal-Provincial Agriculture Committee on Environmental Sustainability, 1990).

Discussion continues regarding the benefits of reintroducing heterogeneity into existing cropland through measures such as rotation, crop diversification, intercropping and reductions in field size. The Federal-Provincial Agriculture Committee on Environmental Sustainability (1990) reported that there is now a general consensus that wildlife habitat within agricultural areas in Canada should be conserved and enhanced to accommodate a wide diversity of plants, animals and microorganisms. There is growing recognition within the farm sector that agriculture can not only coexist with wildlife, but also benefit from maintaining and enhancing wildlife habitat.

Paradoxically, excellent agricultural land is being sacrificed to the social pressures of increased urbanization. The tendency to push agriculture onto ever-more marginal lands is a danger that both agriculturalists and environmentalists want to resist.

National Biodiversity Planning and Capacity Building

In November 1992, Canadian federal, provincial and territorial ministers of environment, parks, wildlife and forestry departments met. They unanimously supported ratification of the Convention on Biological Diversity and adopted a process for the development of a National Biodiversity Strategy to ensure that Canada's obligations are implemented at local, provincial, national and international levels.

The consultation process leading to the national strategy has been deliberately open and inclusive. Three advisory groups played important roles in the development of the Canadian Biodiversity Strategy: the Biodiversity Working Group, the Biodiversity Convention Advisory Group and a federal government interdepartmental committee. Environment Canada's Biodiversity Convention Office was named to coordinate the overall development of the Strategy.

The Biodiversity Working Group is a federal-provincial-territorial intergovernmental body. In Canada, provinces have important constitutional responsibilities in the management of natural resources. Most provincial governments formed provincial interdepartmental biodiversity working groups and/or named a biodiversity contact. Although the provincial groups are mostly led by an environment or natural resources department, the provincial agriculture department is usually encouraged to participate. The Biodiversity Working Group was given the task of producing the Canadian Biodiversity Strategy for ministerial approval (including by sectoral ministers), to formulate a public awareness programme for biodiversity conservation, and to advise on positions relating to the Convention on the international level.

Membership of the Biodiversity Convention Advisory Group (BCAG) was drawn from business, non-governmental organizations and scientific societies. Several agricultural organizations were represented: the Canadian Federation of Agriculture, the Agricultural Institute of Canada, the seed sector, and the Rare Breeds Conservancy. In addition, the Canadian Federation for Biological Sciences was represented by an agriculturalist. Other organizations participating in the Advisory Group included the Canadian Pulp and Paper Association, the Mining Association of Canada, the Canadian Institute of Biotechnology, the Canadian Wildlife Federation, the Canadian Nature Federation, Cultural Survival (Canada), the National Round Table on the Environment and the Economy ... this is a group with wide-ranging expertise and interests!

Agriculture Canada, the federal department, participates in the federal interdepartmental committee on biodiversity. In addition, the author was loaned by Agriculture Canada to Environment Canada's Biodiversity Convention Office.

A Draft Canadian Biodiversity Strategy was distributed for wider discussion in June (Federal–Provincial–Territorial Biodiversity Working Group, 1994). Once the views of stakeholders are compiled and considered, the Canadian Biodiversity Strategy will be drafted for the approval of all governments. Each jurisdiction will determine its own priorities for action.

An important result of the consultative process was an emphasis on integrated, participatory decision-making. The Draft Strategy stated that 'successful implementation of Canada's biodiversity strategy will require a coordinated approach based on cross-sectoral cooperation, and partnerships among governments, non-government organizations, private sector interests and individuals. The capacity to determine how Canada's biodiversity is managed is not limited to governments. Indigenous communities, business and industries, local communities and individuals must be involved with the implementation of the Canadian Biodiversity Strategy'.

It is worth noting that the Global Environment Facility offers financial support to developing countries for the development of their national biodiversity strategies and action plans. In Canada, we have found that the development of national strategies and action plans is in itself an excellent tool to build our capacities for strategic partnerships.

Complementarity of International Agencies and Forums

Countries depend on each others' biodiversity, and threats to biodiversity are the common concern of all. Countries need to pool resources and organize their activities in order to ensure that agricultural biodiversity is protected and used sustainably.

In recent years, the Convention on Biological Diversity and the Food and Agriculture Organization have been active fora for international discussion concerning agricultural biodiversity. These intergovernmental forums have had to grapple with difficult and complicated issues of national priorities and sovereignty.

The Convention on Biological Diversity provides a biological foundation for sustainable development. Its objectives, stated in the first Article, are 'the conservation of biological diversity, the sustainable use of its components and the fair and equitable sharing of the benefits arising out of the utilization of genetic resources, including by appropriate access to genetic resources' (Convention on Biological Diversity, 1992). Agricultural genetic resources are clearly within its terms of reference. Numerous references are made to genetic resources, which are defined as 'genetic material of actual or potential value'. Article 9 treats *ex situ* conservation.

Article 15 of the Convention addresses the controversial issue of access to genetic resources. Essentially, this Article provides a set of principles. It does not spell out how these principles are to be applied. In this important respect, the Convention is a framework, requiring completion through supplementary agreements.

Throughout the negotiation of the Convention on Biological Diversity, Canada took an approach which respected the specificity of the economic sectors which use biological resources. Different sectors have developed different practices, and therefore require different rules for access.

The Intergovernmental Committee for the Convention on Biological Diversity, which had the responsibility of preparing the work of the first Conference of Parties to the Convention, addressed the issue of ownership of and access to *ex situ* genetic resources, at its second session held in June 1994. It reported (UNEP, 1994, paragraph 233) that:

> ... any multilateral agreement on access to genetic resources should take into account *ex situ* collections which existed before the Convention entered into force [*Comment: this presumably does not exclude taking into account collections existing after entry into force*]. Such an agreement should be in accordance with the objectives of the Convention on Biological Diversity ... and should be properly examined by Governments ... Many representatives supported the work of the Commission on Plant Genetic Resources, which is an intergovernmental forum established within the FAO, in addressing the issue of plant genetic resources for food and agriculture. Reference was made to the negotiations among Governments for the adaptation of the International Undertaking on Plant Genetic Resources in harmony with the Convention on Biological Diversity, for consideration of the issue of access on mutually agreed terms to plant genetic resources ... the Conference of the Parties should provide guidance to the interpretation and further development of these issues. There was strong general support for the renegotiation process of the International Undertaking on Plant Genetic Resources, bringing it within the framework of the Convention on Biological Diversity, possibly in the form of a protocol developed on a step by step basis.

These statements reflect the desire of the intergovernmental biodiversity community to engage the appropriate forum in order to come to an agreement on an issue which is of primary interest to the agricultural sector. It is a measure of the maturity of participating governments that they were able to agree that more specialized expertise can be found in a different forum to resolve a particular biodiversity issue. One can hope that this attitude sets a precedent!

References

Convention on Biological Diversity (1992) United Nations Environment Program, Nairobi, Kenya.

Environment Canada (1994) *Biodiversity Science Assessment* Final Draft, Environment Canada, Canada, February 1994.

Hoyt, E. (1988) *Conserving the Wild Relatives of Crops*. IBPGR, Rome, IUCN and WWF Gland.

Federal-Provincial Agriculture Committee on Environmental Sustainability (1990) *Growing Together*. Report to Federal and Provincial Ministers of Agriculture, Canada.

Federal-Provincial-Territorial Biodiversity Working Group (1994) *Draft Canadian Biodiversity Strategy For Discussion*. June 1994. Biodiversity Convention Office, Enviroment Canada, Canada.

Millar, J.B. (1986) *Estimates of Habitat Distribution in the Settled Portions of the Prairie Provinces in 1982*. Canadian Wildlife Service, Saskatoon, Saskatchewan, Canada.

UNEP (United Nations Environment Programme) (1994) *Report of the Intergovernmental Committee on the Convention on Biological Diversity on the Work of its Second Session*. UNEP/CBD/COP/1/4, September 1994.

Three Levels of Conservation by Local People   30

A. Gómez-Pompa

Department of Botany and Plant Sciences, University of California at Riverside, Riverside, CA 92521, USA

Introduction

There is no longer any doubt in our minds on the importance of tropical forests and about the need to understand them better. Hardly anyone objects to the overwhelming need to know and protect as much of the planet's biodiversity as possible, especially in the tropics. One question often raised is on the past and present role of humans on the management and conservation of biodiversity.

Management usually involves the sense of 'improvement' from a previous stage and conservation always involves maintenance of what exists. However, the meaning of management may be very different to different people, and in fact, may include conservation and even preservation (Gómez-Pompa and Burley, 1991). Our modern society has mainly been concerned with the issue of the preservation side of conservation and has paid little attention to the management side of conservation.

The biodiversity we have today is in great part the product of the actions of thousands of generations of humans on earth. They have tried all kinds of management approaches to 'improve' the production of the natural ecosystems of its basic needs for food, fibre, and housing materials.

History is full of examples of the successes and failures of past civilizations. Their attempts to modify their natural habitat to produce more and better goods drastically modified the ecological landscape of where they lived. Deforestation was a common trademark of most civilizations, including ours. The search for new fields to plough or wood to harvest has been the common denominator of humans. In doing so, the plants and animals of the 'natural' environment were pushed to other areas and most of their populations were drastically reduced, except for those species that found new niches in the human-modified environments. The original ecosystems were replaced in great quantities by human-made agroecosystems and large areas of disturbed ecosystems.

Humans in their search for better food production systems were able to find and select a certain number of species that were nutritious, taste better, were easy to harvest and most important, were prone to cultivation. These species in former times came mainly from the wild. The wild lands were the providers of food, medicine, and fibres. Humans since ancient times have been acknowledging the sources of their subsistence. Forests and other wild lands were frequently revered, respected and protected. Sacred forest groves are known from all over the world. Deities based on trees and wild lands are common in all cultures.

Biological conservation was an important component in the management of landscape in many cultures. Alcorn (1991) refers to this type of conservation as archaic conservation: 'Archaic conservation operates in subsistence economies by placing limits on the extraction of natural resources; local rules therefore tend to prevent destruction of nature because raw materials inputs derived from nature comprised the pre-eminent resource base'. This view contrasts with the prevailing view in some quarters of native people as destroyers of nature.

Our only option for the future is to find the precise role the human species plays and has played in the conservation, depletion, and enrichment of biodiversity. Humans should not be seen only as the destroyers of biodiversity but also as the managers of it (Gómez-Pompa and Kaus, 1992).

In order to properly manage anything we must know it well. Yet at the present time, we do not know enough about the processes that affect, reduce, increase, and maintain different kinds and levels of biodiversity in the world.

Unfortunately the issue of our ignorance is frequently misunderstood and misused. Questions raised concerning the estimates of alarming extinction rates of taxa are valid ones, and need to be addressed seriously; however, if we lose one species per year, or one species per day or one per minute is not the issue. The issue is that with the little information available we do know that we are losing them, and that the tropics in particular are especially vulnerable because of the high biological richness, the fast conversion rates of its ecosystems and the economic priorities.

We also know that major causes of environmental deterioration are human actions that can be changed. But also we know that there are human actions that not only prevent biological losses but in fact may be responsible for its enrichment. It seems that the study of human activities that conserve and may even enrich biological diversity is an utmost priority.

In this chapter I will discuss three levels of the role of humans in managing and conserving biodiversity in the tropics of Mesoamerica: (i) the management for conservation of large areas; (ii) the creation of new agroecosystems; and (iii) their role in the evolution and domestication of trees.

The three examples are from the Maya area of Mesoamerica but similar examples can be mentioned from many other areas of the world.

Level 1: The Conservation and Management of Large Areas: The Maya Forests

Much of our research is based on the puzzlement over how the Maya culture developed and flourished in the tropical lowlands of Mexico and Central America. The ancient Maya reached population densities estimated to be much higher than those which the same areas support today. Though there is some disagreement among archaeologists, an estimate of 200–500 people per km^2 in the rural areas for several centuries is widely accepted. Much higher numbers have been estimated for some specific sites (Turner, 1976).

What is intriguing is that this population density was apparently reached, without diminishing the biological diversity of the region. The Maya region is considered to be one of the most important centres of biodiversity in our continent. A new family of flowering plants, the Lacandoniaceae was discovered a few miles from the known archaeological sites of Yaxchilan and Bonampak, that were heavily populated for several centuries (Martínez and Ramos, 1989).

The vegetation and flora of the region have given us an understanding of the diversity that these areas once had and a rough estimate of the amount and rate of change. The study areas were also the sites of long-past human habitation.

Based on these studies the conclusion was reached that most of the mature forests of Mesoamerica were areas inhabited by humans; evidence can be found almost anywhere one looks.

There is very little doubt that most of the present day mature forests of the Maya area are the result of past management. The abundance and unexplainable distribution of useful trees, many of them with edible fruits, can only be explained by human intervention (Barrera-Marín et al., 1977).

What kind of intervention?

In the process of studying these sites I also became aware of the extraordinary amount of traditional knowledge the farmers have on the plants, the vegetation, crops, and the diversity of their land use practices. We began noticing the abundance and dominance of useful trees in what were considered in our surveys as 'primary' vegetation in 'primeval' areas of the humid tropics of Mesoamerica. In addition, we found that the many species that were dominant in the forests were also abundant in the home gardens.

Several authors have suggested the influence of the old Maya in the present day vegetation in the archaeological sites. Puleston (1968) even suggested that a tree, *Brosimum alicastrum*, the ramon tree was in fact an important alternative staple food of the Maya. New evidences from present day Maya support that hypothesis (Atran, 1993).

Others have proposed that the ancient Maya had complex forest gardens similar to the present day home gardens where many species were encouraged, attracted, relocated, selected, protected, transplanted, domesticated, semidomesticated, eliminated, or introduced. In these forests wildlife was also managed, by favouring certain game-food plants (Wiseman, 1978). There are early descriptions of very sophisticated breeding methods for rare birds (Hamblin, 1984), including the famous Quetzal in the Maya forests (M. Aliphat, personal communication, Mexico City).

I have supported and contributed to this intriguing idea of the existence of extensive ancient forest gardens by proposing a silvicultural system of the ancient Maya based on the food production systems available and the resource management methods practised by Maya farmers today (Gómez-Pompa, 1987).

This mega-management of biodiversity by the ancient Maya was not the result of a centralized mandate. The explanation lies in the presence of a series of techniques and methods for their agriculture, silviculture, and wildlife management. These techniques were selected over time and practised by all. Something similar today is the worldwide acceptance of the modern techniques and methods for agriculture (green revolution), resource management, and conservation.

The large scale management of resources has been reported from some present day indigenous groups. One extraordinary example has been described from the Kayapo Indians of Brazil who have managed extensive areas for centuries and have influenced the composition of forests and in fact created what have been called 'anthropogenic forests' (Posey, 1985).

As in the case of the Maya, the findings for the Kayapó are not unique. According to Baleé (1989) 'at least 11.8% of terra firma forest in Amazonian Brazil is anthropogenic'.

Level 2: The Creation of New Agroecosystems: Home Gardens

This section briefly discusses this remarkable anthropogenic forest: the home gardens. These are a key component of the biodiversity conservation strategy by local people. They are known to be very efficient agroforestry systems very widely used by many traditional cultures of the world.

The home garden has been considered as the site of experimentation on plant introduction, plant cultivation, and animal breeding. Probably it was the site where the domestication of many plants and animals occurred and is occurring now. It is also the site par excellence of crop diversity conservation.

Each garden is an experiment in the design of a multispecies ecosystem in space and time. The kinds, number and individuals of each species, as well as the landscaping is done by each gardener. It is based on his/her tastes, needs, knowledge and curiosity, and previous experiences. The gene pool available comes from the local biota, other home gardens, or from local markets. The home gardens are composed by a mixture of the selected species with the local flora that is allowed to coexist in the garden.

Another important factor is the zonation found in their gardens. These vary from intensively managed areas to low management areas. In the latter 'wilder' areas many native species are spared and are able survive in the gardens.

Home gardens may comprise more than 10% of the total forested area of the State of Yucatan in Mexico. In a study of the home gardens at Xuilub 339 species of flowering plants were found living in 52 gardens. These gardens have an average size of 3800 m^2. This means that in 19.76 ha of home gardens comprise 30% of the flora of Yucatan (Herrera et al., 1993).

Level 3: The Domestication or Semidomestication of Trees

In the studies on early domestication there are few examples on how some trees were domesticated. We know that many early pre-agriculture human groups lived in tropical forest environments (Hladik et al., 1993). They were hunter–gatherers, whose survival depended on their knowledge of where to find abundant game and also areas where edible plants were abundant. The reliability of food sources was of utmost importance.

Fruit trees were without doubt a staple food source. They were the reliable sources of food. The knowledge of their phenology became important to ensure food through the year.

The gathering of fruits and seeds of trees became a routine activity for centuries and maybe millennia. The knowledge of germination of seeds was a logical event that may have occurred in their temporary home sites or even in the forest. Seeds of many trees from the rain forest have no dormancy, they germinate very soon after they fall.

It is not difficult to believe that an intelligent hunter–gatherer might have learned the advantage of planting seeds of edible tree species or even selecting seeds from individual trees or populations of a favourite species. Nor is it difficult to visualize the initial management of forests by the keen observers of the forests. It does not take much to realize that if some inedible trees are eliminated, the edible ones will have an advantage and in the near future more food could be gathered. If the useful edible trees were not there, they could be planted. This is the most logical scenario as to how the 'anthropogenic forests' were born. This was in fact the beginning of a domestication process by hunters and gatherers. Even though hunter–gatherers may not have domesticated plants or animals, they domesticated the environment in which these species grow (Yen, 1989).

The result of all these activities created the original forest home gardens and also the initiation of what I called the mega-management of forests.

In this scenario, the trees introduced to home gardens were not very different from their wild relatives. However, over time, certain genotypes were selected by different human groups, based on taste, cooking preferences and ecological advantages (drought and pest resistance, etc.). These genotypes have been inherited, maintained, and developed further by the many ethnic groups around the world that have conserved the cultural traditions and the genetic and biotic patrimony from their ancestors.

The cacao story

The cacao story is an example of a sophisticated domestication of a tropical rain forest tree (Gómez-Pompa et al., 1990).

Landa in 1566 wrote about the Maya in Yucatán in his *Relaciones de Valladolid* the following: 'They have sacred groves where they cultivate certain trees, like cacao'. Gaspar Antonio Chi (1582) wrote: 'the lands were in common ... except between one province and another because of wars, and in the case of certain hollows and caves, (plantations of fruit trees and) cacao trees'. Several other early chroniclers mentioned the presence of cacao trees in northern Yucatan.

These reports of cacao from the state of Yucatan, Mexico, have intrigued scientists for a long time, since the regional soil and climate (1300 mm of rainfall per year) are not appropriate for the cultivation of such a species so well-known for its high demands for humidity. Obviously the only way cacao could be grown in this area would be in areas where the soil humidity could be kept constantly high by natural conditions or by irrigation.

This contradictory information made us start an intensive search for microhabitats where cacao could be grown in this northern region of the Yucatan Peninsula. Our search was based on the belief that cacao or a cacao-related species was grown there and remnant trees would still be found in special microenvironments.

Two possible sites were chosen. One was the home gardens of the present day Maya, these sites are very rich biologically and irrigation is a common practice. Unfortunately, no living specimens have yet been found. This environment is still a possible site, since soils and humidity are controlled.

Another place we looked for cacao was in the sinkholes and large cenotes with deep soils. We suspected that these sites could have been the place for ancient cacao cultivation in northern Yucatan. These microhabitats have high humidity and rich soils available and frequently have water in the bottom. Those with water are the famous 'cenotes' of the Maya area, known in Maya as 'co'op'.

Our hunch and subsequent efforts were rewarded and in the town of Yaxcabá, an old informant told us that in a sinkhole named 'kuyul' in his community there were some trees of cacao. We visited the place and we found cacao trees, one of which was in flower. After that discovery we have been able to find more sinkholes in the same region with cacao trees.

Our informants told us that those trees have been there ever since they can remember, and they protect them as they always do for useful trees found in the forests of the cenotes.

The cacao collections were identified initially as *Theobroma cacao* subsp. *cacao* forma *lacandonense*. Is this a wild cacao? Wild cacao is difficult to distinguish from the cultivated ones. In fact, most of the collections of wild cacao we have been studying are called wild only because they were found in a forest and not in a plantation. But most of the wild cacaos are very similar to the cultivated ones.

The cacao populations of Yucatan are the relics of ancient forest gardens in sinkholes mentioned by the chroniclers. They have been reproducing there through the centuries, providing the genetic continuity through time. These trees have been managed and introduced into human-made environments since remote times. Recent studies (de la Cruz et al., 1995) using random amplified polymorphic DNA (RAPD) genetic markers were used to determine the genetic diversity and relationships between and within wild and cultivated cacao trees. The results of this study indicate that wild accessions are genetically different from the cultivated varieties. The cacao from Yucatan have diverged genetically from domesticated cacaos, and from wild cacaos from Mesoamerica and from the Amazon region. This is a remarkable example of an ancient conservation practice of great potential economic value.

What is the relation of the cacao to the forest management of the Maya? It is one additional forest system managed by the Maya in the past and kept by the present-day Maya. The species in these groves are a mixture of three groups of species: wild species (for example, species of *Ficus, Melicoccus, Bursera*), native useful species (species of *Achras, Pouteria, Annona, Brosimum, Castilla, Quararibea*), and useful exotic species (coconut, *Citrus*).

Three Levels of Conservation

There is a remarkable similarity between the species composition of the cacao groves to the home gardens of the Maya and to the anthropogenic forest gardens of the Maya area.

We are uncovering a continuum that goes from the protected species, to the small home garden to the large home garden and from the forest gardens away from home to the 'natural' forests. The dominant tree flora is very similar in all cases.

These findings help to explain the abundance of useful species in mature forests all over the tropical lowlands of Mesoamerica. If a human occupied area is abandoned, as we know happened several times in the past in that region, the species with the most ecological advantage are those that are already there (Gómez-Pompa and Burley, 1991). Native species will have an additional advantage, and abundant species would have even more of an advantage. The forest dominants will be the tallest and with the longest life span.

In the recovery process that occurs over the centuries, the same trees may not remain, but the populations that will replace them will be direct descendants of the original trees (as the case of the cacao from Yucatan). Many other trees may come from distant places by long-distance dispersal, but the most abundant propagules will come from managed trees from seminatural forests nearby.

The regeneration of the abandoned area will be strongly dominated by the available tree flora which today, as in the past, comes from forest fallows, forest gardens, and other managed or semimanaged areas. If the tree flora is diversity-poor, a poor forest may regenerate. But if the number of tree species is large the forest may be richer.

This is the type of forest Landa (1566) saw and wrote about so eloquently when he said that the Maya had so many kinds of trees that they use and protect that it is frightening. Not so surprisingly, this human-influenced biological richness in mature forests is widespread throughout the tropics of the world.

Final Comments

The three examples from the Maya region are only the 'tip of the iceberg' on the role of traditional people in biological and genetic conservation. The traditional systems of agriculture, and pastoral systems have been based on the diversity of options available and have produced an impressive number of traditional cultivars and land races that are part of a strategic patrimony of humanity as a whole: the thousands of land races of rice in the paddies of Asia, the races of corn and their relatives in the Americas, are examples of this biological wealth. The coexistence of traditional cultivars with their wild relatives is a well-known fact; the gene flow in both directions is an ongoing process that needs to be protected.

However, the conservation of this biodiversity has been left to the local people, without a major effort to support and reward their activities. Local people are indeed the custodians of the most important gene pool for the future of humankind, their archaic methods of conservation have proven successful.

Unfortunately the future does not look very promising. The promises of a better life through intensification of production via green-revolution methodologies are producing a dangerous change in many areas of the world. The change from traditional to modern has had its toll by the loss of many traditional cultivars and land races. 'The principal cause of genetic erosion has been the widespread adoption of modern crop cultivars within areas of ancient agriculture' (Oldfield and Alcorn, 1987).

It is obvious that we urgently need a strategy for the conservation of the genetic pool of plants and animals used by local people all over the world. We have to preserve also the systems of resource management that have provided the environment for the production and conservation of the human-made biodiversity. Unfortunately we lack proven approaches to do this.

Oldfield and Alcorn (1987) described a few approaches that have been proposed that include a system of land village custodians, freezing the genetic landscape, the setting up of a world network of large strips of land under traditional agriculture, subsidies to traditional farmers, scientifically controlled *in situ* reservoirs of crop populations in experimental stations located in the original environment, *in situ* conservation and research stations independent of private farmers and *ex situ* reserves.

In Mexico we are experimenting with a new strategy called the 'Tripartite Alliances for Conservation and Development' that consist of a network of traditional conservation projects of farmers funded through the Mexican

NGO (PROAFT A.C.) which combines several suggested approaches (del Amo et al., 1993). The farmers are encouraged to continue their old conservation methods and PROAFT provides resources to increase the quality of life of the community (land use planning, local reserves, marketing advice, development of local products, education, health services, etc.).

However, this can not be accomplished if local knowledge is lost. This knowledge represents a bank of potential alternatives to be used in combination with present day environmental, economic, and social conditions. Any effort to conserve the genetic resources of traditional systems has to be complemented with research that documents this knowledge.

References

Alcorn, J.B. (1991) Ethics, economics and conservation. In: Oldfield, M.L. and Alcorn, J.B. (eds) *Biodiversity. Culture, Conservation and Ecodevelopment*. Westview Press, Boulder, Colorado.

Atran, S. (1993) Itza Maya tropical agro-forestry. *Current Anthropology* 34(5), 633–700.

Baleé, W. (1989) The culture of Amazonian forests. *Advances in Economic Botany* 95, 1–21.

Barrera-Marín, A., Gómez-Pompa, A. and Vázquez-Yanes, C. (1977) El manejo de las selvas por los mayas: sus implicaciones silvícolas y agrícolas. *Biotica* 2(2), 47–60.

Chi, Gaspar Antonio (1582) Relación. In: Tozzer, A.M. (ed. and translator), 1941. Appendix in the translation of Landa's *Relaciones de las Cosas de Yucatan*. Harvard University Peabody Museum of American Archaeology and Ethnology. Paper 18. Cambridge, USA, pp. 230–232.

de la Cruz, M., Whitkus, R., Gómez-Pompa, A. and Mota-Bravo, L. (1995) Origins of cacao cultivation. *Nature* 375, 542–543.

del Amo, R., S., Gómez-Pompa, A., Roldán, A. and Kaus, A. (1993) Tripartite alliances: lessons for conservation and sustainable development. In: Ferrera Cerrato, R. and Quintero Lizaola, R. (eds) *Agroecologia, Sostenibilidad y educación*. Centro de Edafología, Colegio de Poostgraduados, Montecillo, México, pp. 8–18.

Gómez-Pompa, A. (1987) On Maya silviculture. *Mexican Studies* 3(1), 1–17.

Gómez-Pompa, A. and Burley, F.W. (1991) The management of natural tropical forests. In: Gómez-Pompa, A., Whitmore, T.C. and Hadley, M. (eds) *Rain Forest Regeneration and Management*. MAB Series. Vol. 6. Parthenon Publishing Group, Carnforth, UK, pp. 3–18.

Gómez-Pompa, A. and Kaus, A. (1992) Taming the wilderness myth. *Bioscience* 42(4), 271–279.

Gómez-Pompa, A., Flores-Guido, J.S. and Aliphat, M. (1990) The sacred cacao groves of the Maya. *Latin American Antiquity* 1, 247–257.

Hamblin, N.L. (1984) *Animal Use by the Cozumel Maya*. University of Arizona Press, Tucson.

Herrera, N., Gómez-Pompa, A., Cruz Kuri, L. and Flores, J.S. (1993) Los huertos familiares mayas de X'uilub, Yucatán, México. Aspectos generales y estudio comparativo entre la flora de los huertos familiares y la selva. *Biotica. Nueva Epoca* 1, 19–36.

Hladik, C.M., Linares, O.F., Hladik, A., Pagezy, H. and Semple, A. (1993) *Tropical Forests, People and Food: an Overview*. MAB Series 13. Parthenon Publishing Group, Cornforth, UK, pp. 3–14.

Landa, Diego de (1566) *Relación de las Cosas de Yucatán*. Ms. en la Real Academia de la Historia, Madrid.

Martinez, E. and Ramos, C.H. (1989) Lacandoniaceae (Triuridales). Una nueva familia de México. *Annals Missouri Botanical Garden* 76, 128–135.

Oldfield, M.L. and Alcorn, J.B. (1987) Conservation of traditional agroecosystems. *Bioscience* 37(3), 199–208.

Posey, D.A. (1985) Indigenous management of tropical forest ecosystems: the case of the Kayapó indians of the Brazilian Amazon. *Agroforestry Systems* 3, 139–158.

Puleston, D.E. (1968) *Brosimum alicastrum* as a subsistence alternative for the classic Maya of central southern lowlands. MA Thesis, Ann Arbor: University Microfilms International.

Turner II, B.L. (1976) Prehistoric population density in the Maya lowlands: new evidence from old approaches. *Geographic Review* 66, 73–82.

Wiseman, F.M. (1978) Agricultural and historical ecology of the Maya lowlands. In: Harrison, P.D. and Turner II, B.L. (eds) *Pre-Hispanic Maya Agriculture*. University of New Mexico Press, Albuquerque, pp. 63–115.

Yen, D.E. (1989) The domestication of the environment. In: Harris, D.R. and Hillman, G.C. (eds) *Foraging and Farming: the Evolution of Plant Exploitation*. Unwin Hyman, London, pp. 55–75.

Some Current Issues in Conserving the Biodiversity of Agriculturally Important Species

T. Hodgkin
International Plant Genetic Resources Institute, Via delle Sette Chiese 142, 00145 Rome, Italy

> In the hinterlands of Asia there were probably barley fields when man was young. The progenies of these fields with all their surviving variations constitute the world's priceless reservoir of germplasm.
>
> H.V. Harlan and M.L. Martini (1936)

Summary

The dangers of depending on uniform genetic material in crop production have been amply demonstrated and the need to maintain the genetic diversity of crop and pasture species, and their wild relatives, firmly established. The last 30 years have seen a steady increase in national and international activities to ensure that crop diversity is conserved for the benefit of future generations. Over 100 countries now have some type of plant genetic resources programme and it is estimated that over 4 million crop samples or accessions are maintained in gene banks throughout the world. Most recently the importance of conserving plant genetic resources of crop and pastures species was confirmed by the development of the programme outlined in Agenda 21 and the entry into force of the Convention on Biological Diversity.

The effective conservation of plant genetic resources depends on continuing research to improve our ability to determine what material should be conserved and how it should be maintained. Areas of particular interest include: investigating the relationships between crop species and their wild relatives; determining the extent and distribution of genetic diversity within crop genepools; and exploring how *ex situ* and *in situ* can complement each other to provide effective coverage of the whole genepool. The knowledge gained from these studies will not only improve the effectiveness of conservation work, it will also enable crop breeders to identify the variation they require for crop improvement more easily.

© 1996 CAB INTERNATIONAL. *Biodiversity, Science and Development: Towards a New Partnership* (eds F. di Castri and T. Younès)

Introduction

The need to conserve the genetic diversity of crop and pasture species is widely recognized. Plant breeders make extensive use of the characters found in traditional crop varieties and in their wild relatives. A notable example is the gene for resistance to barley yellow dwarf virus, found in Ethiopian barley landraces, which has been estimated to be worth several million dollars to the Californian barley industry. Similarly, the wild relative of rice, *Oryza nivara*, has provided the only known source of resistance to grassy stunt virus in the crop. This resistance has so far been found in only one accession of *O. nivara* and now forms an essential character in newly released rice cultivars. Many other examples from all our major crops could be given (e.g. see Holden et al., 1993).

The dangers of depending on uniform genetic material in agricultural production have also been widely appreciated. Many examples can be cited where the genetic uniformity of the crop grown has been deleterious to production. Certainly, varietal uniformity was a significant factor in the Irish famines of the 19th century which resulted from the devastating attacks of blight (*Phytophthora infestans*) on the one variety of potato that was grown (Oldfield, 1989). More recently, an epidemic of southern corn leaf blight, caused by the pathogen *Helminthosporium maydis*, resulted in a reduction in production of approximately 710 million bushels of maize in the United States in 1970. Again, many further examples could be cited where the genetic uniformity of the crop grown has resulted in the development of a disease epidemic or other similar event and led to a dramatic loss of production with resultant human suffering and dislocation.

The role that diversity plays in traditional crop production systems should also be recognized. Such diversity continues to be an essential component of production for most of the world's farmers. Approximately 60% of global agriculture still involves subsistence farming using traditional methods. Wood and Lenné (1993) suggest that this provides some 15–20% of the world's food, a figure that may well underestimate the nutritional contribution made by such farming, given the importance of traditional methods in vegetable and fruit production. Traditional production depends on diversity with respect to the number of crops and species grown, which is high, the way in which they are grown, using methods such as multicropping and intercropping, and the variation maintained within the individual crops. The importance to farmers of maintaining high levels of intracrop genetic variation is becoming increasingly well documented by research workers. It is now clear that, even when farmers make considerable use of newly available varieties, they will frequently maintain traditional varieties with high levels of diversity (Brush, 1995).

The conservation of diversity is therefore important, both for the contribution that the range of genetic variation found within crops can make to development of improved cultivars, and to ensure that there is a sufficient variation in current agricultural production systems to prevent disease epidemics and other disasters. In addition, diversity in cropping systems is inherently important as a component of sustainable cropping systems both now and in the future. In response to this need national and international plant genetic resources conservation programmes have been developed. The development of these programmes has involved continuing biological studies on the nature, extent, and distribution of genetic diversity in crop species and their wild relatives and on the ways in which it can best be conserved. In this chapter the growth of national and international conservation work is briefly described and some of the areas identified in which further research will be important to the further growth of these programmes.

The Development of Genetic Resources Conservation Programmes

The value and worldwide significance of the genetic variation present in traditional varieties of crops and their wild relatives was clearly established by Vavilov and his collaborators as a result of their extensive studies and collecting programmes in the 1920s and 1930s. In detailed studies (e.g. Vavilov, 1926, 1951) on a wide variety of major crop plant species, Vavilov and his collaborators described the variation present in different crops and in different areas of the world and established that there were certain areas of the world which contained remarkably large amounts of genetic diversity of particular crops and species. These centres of diversity (of which Vavilov recognized eight major ones and three subcentres) were also considered by Vavilov to be the probable centres of origin of our crop plants. The significance of the centres of diversity described by Vavilov has been the subject of considerable discussion. However, the fact that there are specific areas of the world in which very large amounts of crop genetic diversity may be found is of enormous significance for conservation work (Table 31.1). It is also clear that, in a number of these areas, crop domestication occurred and that the juxtaposition of crops and their wild progenitors is important in the study of crop evolution and the detailed planning of conservation work.

The work by Vavilov and his colleagues was paralleled by collecting and research by a number of other important early workers such as Harlan in the United States (Harlan, 1951). The approach was highly practical in that the primary objective was to obtain material for the production of new cultivars (in the case of the USSR for the widely differing ecological conditions in that vast new country). This practical objective was combined with research of fundamental significance, aimed at providing a more general understanding of the nature and distribution of variation. A further consequence of the work was to establish the importance of adopting an agroecological approach in describing and understanding the observed distribution of genetic variation within a crop.

The international distribution of our major crop species (often far outside their centre of origin or centre of diversity) has been a feature of profound significance in the development of conservation work. Thus, wheat, barley,

Table 31.1. Examples of the crops found in two world centres of diversity of cultivated plants (adapted from Hawkes, 1983).

China	Ethiopia
Avena nuda, naked oat (secondary centre of origin)	*Triticum durum*, durum wheat
	Triticum turgidum, Poulard wheat
Glycine hispida, soybean	*Triticum dicoccum*, emmer
Phaseolus angularis, adzuki bean	*Hordeum vulgare*, barley
Phaseolus vulgaris, bean (recessive form - secondary centre)	*Cicer arietinum*, chickpea
	Lens esculentum, lentil
Phyllostachys spp., small bamboos	*Eragrostis abyssinica*, teff
Brassica juncea, leaf mustards	*Eleusine coracana*, African millet
Prunus armeniaca, apricot	*Pisum sativum*
Prunus persica, peach	*Linum usitatissimum*, flax
Citrus sinensis, orange	*Sesamum indicum*
Sesamum indicum, sesame, (endemic group of dwarf varieties - secondary centre)	
Camellia sinensis, China tea	

maize, and rice are important crops throughout the world despite their origins in the Middle East and in Asia. Similarly, soybean, originally from Northern China, is now a major export of the USA. In some cases, crops may develop important new centres of diversity in areas to which they have been introduced and in which they are successful. Examples of this would include *Phaseolus* beans and cassava from S. America which now have secondary centres of diversity in Central Africa and West Africa, respectively. The importance of international collaboration was therefore recognized by plant genetic resources workers in a way that was often quite different from workers in other areas of science or agriculture. It was not considered just as a valuable additional dimension to the work but as an essential component of it.

During the 1950s and 1960s crop geneticists and plant breeders became increasingly concerned at the loss of traditional crop varieties (landraces) and the genetic diversity they contained. FAO had initiated work on genetic resources very early in its life but, at first, this work was largely concerned with germplasm exchange. The emphasis on exploration, collection and conservation came later. During the 1960s FAO sponsored a number of meetings and other initiatives on plant genetic resources which resulted in key publications on the subject (e.g. Frankel and Bennett, 1970), the development of agreed international priorities and a number of important collecting and conservation initiatives.

In 1972, the United Nations Environment Conference emphasized the importance of conserving crop genetic resources and, at the about the same time, the Technical Advisory Committee of the Consultative Group on International Agricultural Research (CGIAR) developed a blueprint for world plant genetic resources work. Building on this initiative, the CGIAR agreed with

FAO to establish the International Board for Plant Genetic Resources (IBPGR) which was founded in 1974 to stimulate and coordinate plant genetic resources conservation and use throughout the world. An important aspect of its work was considered to be the development of a global network of conservation activities.

The period since the formation of IBPGR has seen a dramatic increase in plant genetic resources programmes throughout the world. In 1974 there were fewer than 10 recognizable national plant genetic resources programmes, whereas today there are over 100. Similarly, the number of storage facilities able to maintain seed safely for long periods at temperatures below $-20°C$ has grown to over 100 and there has been a similar increase in the number of collections and in the numbers of short-term storage facilities and field genebanks. Today, it is estimated that over 4 million plant accessions are maintained in some facility somewhere in the world.

Much of IBPGR's early concerns were with collecting and *ex situ* maintenance of traditional farmers' varieties or landraces that were disappearing with great speed. It was also concerned with the technologies of conservation, stimulating, with FAO, work on seed storage procedures (Roberts, 1975) and on collecting procedures (Marshall and Brown, 1975). The need was to identify material most at risk, to collect it and then to maintain it in whatever way was possible.

Another function of IBPGR was the promotion of a worldwide network of genetic resources centres (IBPGR, 1979). As suitable seed storage facilities became available and germplasm was collected, IBPGR began to organize a register of base collections for a given crop, maintained under specified conditions by an Institution which had made certain specific commitments such as the unrestricted availability to *bona fide* users of the germplasm conserved.

Although IBPGR was largely concerned with the scientific and technical aspects of conservation, FAO, as a UN Agency, was able to address some of the important political dimensions of plant genetic resources conservation. In 1983 the FAO Commission on Plant Genetic Resources was established and, as a result of its work, the International Undertaking on Plant Genetic Resources became the first comprehensive international agreement on genetic resources. The Undertaking was born out of the demands of developing countries for international regulation of the exchange of genetic resources and sought to put all plant genetic resources on an equal footing.

Major elements of the Undertaking which have had a significant impact on international aspects of plant genetic resources conservation include the belief that plant genetic resources are a heritage of mankind and the concept of farmers' rights. The common heritage concept has always been open to rather variable interpretation and the Convention on Biological Diversity which came into force at the end of 1993 lays much greater emphasis on the recognition of national sovereignty in respect of such resources. Of course, a recognition of national sovereignty clearly implies a recognition of both the rights of a country over resources found within its borders and a responsibility towards their conservation.

The Undertaking as originally drafted was amended by a number of specific resolutions which dealt with key issues. Of these, the first recognized that plant breeders' rights were compatible with the Undertaking and simultaneously recognized farmers' rights. These were defined in a second resolution as 'rights arising from the past, present and future contributions of farmers in conserving, improving and making available plant genetic resources . . .'. This concept is of key importance in providing a framework to recognize the importance for conservation of the countries and communities which are in the major centres of crop diversity.

The work involved in the preparations for the UN Conference on the Environment and Development and in drafting Agenda 21 gave work on the conservation of crop plants and their wild relatives a new impetus. With the entry into force of the Convention on Biological Diversity a new political framework is created which provides much clearer commitments by individual countries to conservation work. This framework is likely to create conditions in which international collaboration on conservation can be further strengthened.

The increasingly developed political framework for the conservation of plant genetic resources makes new demands on the technical capabilities of those involved in carrying out conservation work. As noted above, this has been reflected in the growth of conservation facilities and in the numbers of accessions conserved by national programmes and by international institutions. IBPGR, with its international mandate to further conservation of plant genetic resources, has also grown and changed. In 1994 it became the International Plant Genetic Resources Institute (IPGRI), the CGIAR Centre responsible for plant genetic resources work with concern also for the conservation of forestry genetic resources and an increasing interest in *in situ* conservation and the socioeconomic and cultural aspects of conservation.

Issues in Conserving Plant Genetic Resources of Crop and Pasture Species

Ever since Vavilov's work, research has been an essential component of plant genetic resources conservation. A considerable research effort will continue to be needed to translate into reality the commitments in the Undertaking on Plant Genetic Resources or the Convention on Biological Diversity. This is recognized in Agenda 21 and in the current efforts to summarize the status of biodiversity knowledge and conservation. It is also reflected in the work being undertaken for the International Conference and Programme for Plant Genetic Resources. The rest of this section surveys three areas where further research will be particularly important and where there are opportunities for exciting new discoveries. The areas discussed are not the only ones of importance and other authors might well choose different topics. However, research in these areas will be an essential component of the work needed to answer the primary conservation questions: what should we conserve? and how should we conserve it?

As part of the response to the Convention on Biological Diversity, countries need to develop national conservation plans, to identify the resources present in a county and decide which should have priority in conservation work. The issue of how to select what should be conserved certainly has a very high priority at present, as does the way in which different conservation methods may best be used for that which is selected. Three different areas may be highlighted.

Describing the genepool

In conserving biodiversity of importance for agriculture, it is necessary to conserve not only the variation found in the crop species itself, but also that found in closely related species – its wild relatives. A practical and useful way of describing the relationship between a crop species and its close relatives was developed by Harlan and de Wet (1971). They distinguished three kinds of relationship between a crop species and its wild relatives. The primary genepool includes the crop species itself and all those wild taxa with which it is fully interfertile – the biological species – whereas the secondary and tertiary genepools include species of increasing incompatibility with respect to seed production from interspecific crosses. Thus, the secondary genepool includes species which, when crossed with the crop species, produce few seeds, and the tertiary genepool includes species with which the crop species is normally infertile and with which progeny can only be produced if artificial methods such as embryo culture are used.

The study of species relationships and the investigation of species evolution has been of great practical significance for both conservation and use of plant genetic resources. Studies of crossing relationships together with investigation of similarities in seed protein composition or in isozymes have enabled workers to identify unique variation in taxa that can be easily crossed with crop species in crops as varied as lentils, watermelon, peanut, banana, foxtail millet and tomato (Doebley, 1989; Gepts, 1990).

Molecular genetic techniques have greatly increased the power of these investigations and permitted them to be more effectively extended to include species of the secondary and tertiary genepools, particularly through examining variation in plastid DNA. In the Andean potato, *Solanum tuberosum* subsp. *andigena*, similarities in cpDNA sequences have confirmed the earlier assumption that forms of *S. stenotonum* constitute one of the ancestors of the cultivated species (Hosaka and Hanneman, 1988). RFLP analysis of nuclear DNA in *Lycopersicon* species has confirmed the close relationship between *L. esculentum*, *L. pimpinellifolium* (both very closely related probably due to introgression where they overlap) and *L. cheesmanii* and apart from the other group of green fruited species (Miller and Tanksley, 1990).

RFLP analysis has also revealed striking similarities in genome organization between *L. esculentum* and *S. tuberosum*, demonstrating that for 9 of the 12 chromosomes in the two species the order of identified loci was identical (Bonierbale *et al.*, 1988). Studies of genome organization in the *Gramineae* which include the more closely related species such as wheat, barley, rye as well as the distantly related species such as rice and sugar cane are beginning to give a fascinating picture of the way in which genome evolution has occurred.

These studies are important because they enable decisions on conservation and use to be more effectively targeted to taxa, species, and genera which contain unique variation and to develop utilization strategies based on a knowledge of the way in which different genomes are organized and where important characters may be found on particular chromosomes.

The distribution of diversity

Variation does not occur at random in nature. There are great differences in the amounts of genetic variation present in different species and in the way in which the variation present is distributed. Some of the general factors that affect the distribution of variation are well known and others are becoming clearer as we extend our studies of the distribution of genetic diversity for different character sets. But, in many cases we remain rather unclear about the amount of genetic variation that we can expect to find in a particular taxon, the way it is distributed in the taxon as a whole or individual populations, or the factors that will most affect its distribution.

Studies on the distribution of isozyme variation allow us to make some useful generalizations about the relationship of such variation to plant life cycle and reproductive behaviour (Hamrick and Godt, 1990). Schoen and Brown (1991) have shown that there is much more variation in the distribution of genetic diversity in self-pollinated species than there is in cross-pollinated species. These kinds of investigations are of great significance in designing conservation strategies, as are results such as those of Olivieri (1993) which provide an insight into the distribution of resistance genes in plants and virulence genes in pathogens and thus provide a basis for a much more targeted selection of populations for conservation.

As with research on species relationships, molecular genetic techniques are capable of providing important new information in this area. In *Lycopersicon*, molecular genetic studies of single copy nuclear RFLPs have shown that both within accession and between accession variation is greatest in the wild self-incompatible species *L. hirsutum*, *L. penellii* and *L. peruvianum* (Miller and Tanksley, 1990). These kinds of studies are clearly important when selecting what to conserve and need to be extended to much wider ranges of material. For crop species, a particular need is to increase the range of species subject to studies of the distribution of genetic variation. Although we begin to have a clear picture of the distribution of variation in some major crop species such as wheat, barley, rice, maize, *Phaseolus* beans, tomato, etc. we remain largely ignorant of the distribution of variation in many other crops including such crops as sesame, safflower, taro, coconut and many others.

It is important that such studies include a range of characters and are not only restricted to particular classes of characters such as isozymes or specific types of molecular markers. Partly, this reflects our continuing interest in the use of plant genetic resources and our need to know about characters of direct interest to plant breeders such as yield-related characters or disease resistance. Partly, it reflects the need to determine whether different types of character are distributed in different ways. For example, Brown and Schoen (1992) have suggested that some molecular markers show much more between-population diversity that do many isozymes. Conservationists who make decisions on whether to concentrate on conserving a few seeds of many accessions or many seeds of a few accessions would clearly welcome further information on this point.

A major feature of much recent work on the distribution of genetic diversity within crop species has been the emphasis on an ecogeographic approach. Detailed studies of the distribution of variation have tended to confirm Vavilov's approach and the importance of an agroecological approach is firmly established. Much interest therefore focuses on the identification of evidence for the existence of adaptive gene complexes and the ways in which they have developed and might be maintained.

In situ *and* ex situ *conservation*

Plant genetic resources conservation work has tended to place greatest emphasis on the development of satisfactory *ex situ* conservation facilities. It was considered vital that national programmes develop satisfactory cold stores for long-term seed storage and that adequate field genebank facilities became available for crops whose seed could not be stored (species which are clonally propagated or which possess recalcitrant seed). For crop species there are strong arguments for ensuring that good *ex situ* storage facilities exist within a country. They provide an economic way of maintaining large numbers of seed samples and ensure that material that might be required by plant breeders is reasonably accessible. They also provide a ready supply of material for research on a wide range of conservation and use questions.

Crop wild relatives (and most forestry species) are probably better conserved *in situ* where this is possible. *In situ* conservation allows evolution to continue and the species to adapt to changing environmental conditions. Many wild species are particularly difficult to conserve *ex situ* because it is difficult to obtain sufficient seed for *ex situ* storage and to provide appropriate conditions for plant growth and reproduction. However, *in situ* conservation of crop relatives raises specific questions which require investigation. Appropriate methods are needed to locate populations of the taxa that should be protected; the nature of the protection required needs to be determined and ways of accessing the diversity conserved for plant breeding programmes need to be developed.

Just as *in situ* conservation of crop relatives has been attracting more interest, there has been increasing interest in *in situ* conservation of the crop species themselves. The desirability of *in situ* conservation of traditional varieties has always been recognized (Frankel and Soulé, 1981). However, until recently, many involved in conserving such material doubted its feasibility. It is now clear that many traditional landraces continue to be maintained by farmers even when new cultivars have been introduced (Brush, 1995) and that communities may wish to ensure that these traditional varieties are conserved for the future.

The operation of *in situ* conservation strategies (both for wild and cultivated materials) raises many important new questions for workers in this area. How many populations? what size? what distribution? how much human interference? of what type? In practice, the approach should clearly be to integrate *ex situ* and *in situ* methods together and develop a complementary approach whereby the different populations, taxa and species within a genepool are each maintained in the most appropriate way. This will differ depending on the genetic, biological, ecological, and other characteristics of a genepool and quite different solutions will be needed for example for the *Brassica* genepool than for the *Musa* one.

Concluding Comments

The issues identified above concern the related questions of how we identify that variation which we should be most concerned to conserve and what methods will be most effective in ensuring its secure conservation. They are by no means the only questions that could be raised and each conservationist will probably have their own specific set. However, all such questions share certain features. They concern the development of an improved appreciation and understanding of the wealth of variation which we observe in useful plant species and they reflect the need to carry out research (often at a fairly fundamental level) in order to provide practical answers to those who make decisions on what to conserve and how to do it.

Although conservation may require (and benefit from) biological science of a high order, such issues are not the only ones that should determine conservation decisions. Indeed they may not even be the most important ones. Increasingly it is being realized that much more emphasis should be given to the needs and interests of communities in developing our conservation programmes. This has significant effects for example with respect to our approach to *in situ* conservation of crop plants and the way in which *in situ* conservation areas are managed (by whom and for whom). It is likely that these developments will increase support for work on conservation and increase demand for knowledge of what to conserve, how to conserve it and how it may be best used. Our ability to respond to this kind of demand will depend on the biological research we have carried out on relevant issues.

References

Bonierbale, M.W., Plaisted, R.L. and Tanksley, S.D. (1988) RFLP maps based on a common set of clones reveal modes of chromosomal evolution in potato and tomato. *Genetics* 120, 1095–1103.

Brown, A.H.D. and Schoen, D.J. (1992) Plant population genetic structure and biological conservation. In: Sandlund, O.T., Winder, K. and Brown, A.H.D. (eds) *Conservation of Biodiversity for Sustainable Development*. Scandinavian University Press, Oslo, Norway, pp. 88–104.

Brush, S.B. (1995) *In situ* conservation of landraces in centers of crop diversity. *Crop Science* 35, 346–354.

Doebley, J. (1989) Isozymic evidence and the evolution of crop plants. In: Soltis, D.E. and Soltis, P.S. (eds) *Isozymes in Plant Biology*. Dioscorides Press, Portland, Oregon, pp. 165–191.

Frankel, O.H. and Bennett, E. (eds) (1970) *Genetic Resources in Plants – Their Exploration and Conservation*. Blackwell, Oxford.

Franker, O.H. and Soulé, M.E. (1981) *Conservation and Evolution*. Cambridge University Press, Cambridge.

Gepts, P. (1990) Genetic diversity of seed storage proteins in plants. In: Brown, A.H.D., Clegg, M.T., Kahler, A.L. and Weir, B.S. (eds) *Plant Population Genetics, Breeding and Genetic Resources*. Sinauer Associates, Sunderland, Massachusetts, pp. 64–82.

Hamrick, J.L. and Godt, M.J.W. (1990) Allozyme diversity in plant species. In: Brown, A.H.D., Clegg, M.T., Kahler, A.L. and Weir, B.S. (eds) *Plant Population Genetics, Breeding and Genetic Resources*. Sinauer Associates, Sunderland, Massachusetts, pp. 43–63.

Harlan, H.V. and Martini, M.L. (1936) *USDA Yearbook of Agriculture*. USDA, Washington, DC, pp. 303–346.

Harlan, J.R. (1951) Anatomy of gene centres. *American Naturalist* 83, 97–103.

Harlan, J.R. and de Wet, J.M.J. (1971) Toward a rational classification of cultivated plants. *Taxon* 20, 509–517.

Hawkes, J.G. (1983) *The Diversity of Crop Plants*. Harvard University Press, Cambridge, Massachusetts.

Holden, J., Peacock, J. and Williams, T. (1993) *Genes, Crops and the Environment*. Cambridge University Press, Cambridge.

Hosaka, K. and Hanneman, R.E. (1988) Origin of chloroplast DNA diversity in Andean potatoes. *Theoretical and Applied Genetics* 76, 333–340.

IBPGR (1979) *A Review of Policies and Activities 1974–78 and the Prospects for the Future*. International Board for Plant Genetic Resources, Rome.

Miller, J.C. and Tanksley, S.D. (1990) RFLP analysis of phylogenetic relationships and genetic variation in the genus *Lycopersicon*. *Theoretical and Applied Genetics* 80, 437–448.

Marshall, D.R. and Brown, A.H.D. (1975) Optimum sampling strategies in genetic conservation. In: Frankel, O.H. and Hawkes, J.G. (eds) *Crop Genetic Resources for Today and Tomorrow*, IBP synthesis, vol. 2. Cambridge University Press, Cambridge.

Oldfield, M.L. (1989) *The Value of Conserving Genetic Resources*. Sinauer, Associates, Sunderland, Massachusetts.

Olivieri, M. (1993) Spatial, genetic and demographic structure of co-evolving organisms. *Proceedings of Council of Europe Workshop on Conservation of the Wild Relatives of European Cultivated Plants*, Neuchatel, October, 1993 (in press).

Roberts, E.H. (1975) Problems of long-term storage of seed and pollen for genetic resources conservation. In: Frankel, O.H. and Hawkes, J.G. (eds) *Crop Genetic Resources for Today and Tomorrow*, IBP synthesis, Vol. 2. Cambridge University Press, Cambridge, pp. 269–295.

Schoen, D.J. and Brown, A.H.D. (1991) Intraspecific variation in population gene diversity and effective population size correlates with making systems in plants. *Proceedings of the National Academy of Sciences* **88**, 4494–4497.

Vavilov, N. (1926) Studies on the origin of cultivated plants. *Bulletin of Applied Botany, Genetics and Plant Breeding* **16**, 1–248 (in Russian).

Vavilov, N. (1951) The origin, variation, immunity and breeding of cultivated plants. *Chronica Botanica* **13**, 1–366. (Translated from Russian by K. Starr Chester.)

Wood, D. and Lenné, J. (1993) Dynamic management of domesticated biodiversity by farming communities. In: *Proceedings of the UNFP/Norway Expert Conference on Biodiversity*, Trondheim, Norway, May 1993.

Ecosystem Management: An Approach for Conserving Biodiversity

R.C. Szaro, G.D. Lessard and W.T. Sexton
USDA Forest Service, Ecosystem Management, PO Box 96090, Washington, DC 20090-6090, USA

Introduction

One of today's most pressing environmental issues is the conservation of biodiversity (Wilson and Peter, 1988; Szaro and Johnston, 1995). The challenge is for nations, government agencies, organizations, and individuals to protect and enhance biodiversity while continuing to meet people's needs for natural resources. Ecosystem management is an emerging ecological philosophy and approach to this challenge that looks at ecosystems as functioning systems rather than through a single species or single function view. Rather than simply assuring we meet the needs of individual species, we now need to make decisions based on how the function of the system is affected, knowing that more species than we may even recognize depend upon it. An ecological approach is the means to ensure the sustainability of the environment while meeting human needs and interests. If the challenge is not met, future generations will live in a biologically impoverished world and perhaps one that is also less capable of producing desired resources.

Developing An Ecological Approach

Clearly every effort should be made to conserve biodiversity (Szaro and Shapiro, 1990; Szaro, 1994, 1995). The conservation of biodiversity encompasses genetic diversity of species populations, richness of species in biological communities, processes whereby species interact with one another and with physical attributes within ecological systems, and the abundance of species, communities, and ecosystems at large geographic scales (Harrington *et al.*, 1990; Salwasser, 1990). Current programmes on conservation focused on populations of particular species contribute to the welfare of selected components of biodiversity but can not possibly be expanded to deal effectively with the ever-expanding list of threatened and endangered species (Reid et al., 1992; Miller, 1995). Conserving biodiversity involves restoring, protecting, conserving, or enhancing the variety of life in an area so that the abundances and distributions of species and communities provide for continued existence and normal ecological functioning, including adaptation and extinction (Szaro, 1995). This does not mean that all things must occur in all areas, but that all things must be cared for at some appropriate geographic scale.

It is easy to understand why threatened and endangered species have received the focus of attention. Many are large, easily observable, and oftentimes aesthetically pleasing. This has resulted in most efforts at restoration and rehabilitation being directed towards endangered as well as harvested species (Bridgewater et al., 1995). The best way to minimize species loss is to maintain the integrity of ecosystem function. The important questions therefore concern the kinds of biodiversity that are significant to ecosystem functioning. To best focus our efforts we need to establish how much (or how little) redundancy there is in the biological composition of ecosystems. Functional groups with little or no redundancy warrant priority conservation effort (Walker, 1992). It is axiomatic that conservation of biodiversity cannot succeed through 'crisis management' of an ever-expanding number of endangered species. The best time to restore or sustain a species or ecosystem is when it is still common. And for certain species and biological communities, the pressing concern is perpetuation or enhancement of the genetic variation that provides for long-term productivity, resistance to stress, and adaptability to change. A biologically diverse forest holds a greater variety of potential resource options for a longer period of time than a less diverse forest. It is more likely to be able to respond to environmental stresses and adapt to a rapidly changing climate. And it may be far less costly in the long run to sustain a rich variety of species and biological communities operating under largely natural ecological processes than to resort to the heroic efforts now being employed to recover some endangered species. Resource managers know from experience that access to resources is greater when forests and rangelands are sufficiently healthy and diverse. Management strategies for biodiversity must be planned for and applied at several scales.

For example, a strategy to maximize species diversity at the local level does not necessarily add to regional diversity. In fact, oftentimes in our haste to 'enhance' habitats for wildlife we have emphasized 'edge' preferring species at the expense of 'area' sensitive ones and consequently may have even decreased regional diversity. It is important to realize that principles that apply at smaller scales of time and space do not necessarily apply to longer time periods and larger spatial scales (Crow, 1989). Long-term maintenance of species and their genetic variation, will require cooperative efforts across entire landscapes (Miller, 1995). This is consistent with the growing scientific sentiment that biodiversity should be dealt with at the scale of habitats or ecosystems rather than species (Hunter et al., 1988). If context is ignored in conservation decisions and the surrounding landscape changes radically in pattern and structure, patch content too will be altered by edge effects and other external influences (Noss, 1995). For example, landscape connectivity is a direct consequence of the abundance of suitable habitat, its spatial patterning in the landscape, and the organism's scale of resource utilization (Pearson et al., 1995). Moreover, the scale and scope of conservation has been too restricted and steps must be taken to incorporate the benefits of biodiversity and the use of biological resources into local, regional, national and international economies (WRI/IUCN/UNEP, 1992; Miller, 1995). The maintenance of biodiversity requires attention to a wider array of components in determining management options as well as the management of larger landscape units.

Significant biological responses and cumulative management effects develop at the landscape level. Planners and managers are increasingly aware that adequate decisions can not be made solely at the site level. The opportunities associated with a particular management unit are determined not only by the content of that unit, but also to a great extent by the landscape context in which it exists. Regardless of whether the primary management objective is producing food or fibre, creating suitable habitats for animals and plants, protecting watersheds, or providing wilderness experiences, assessing these opportunities requires consideration beyond the boundaries of a particular planning or management unit. Landscape ecology is an emerging field, but it lacks a strong conceptual framework and established research methodology. Research is needed to better define how natural resource management can be enhanced through a landscape ecological approach.

Endangered species are fundamental indicators of environmental disturbance. Since extinction is a process, not a simple event, the recognition that a species is endangered is little more than a snapshot of a moving vehicle. Attempts at therapy most often address symptoms rather than causes. We have failed to communicate successfully why rehabilitation and restoration beyond the narrow focus of the endangered and harvested, are essential. The environmental variables which affect the health and welfare of all the flora and fauna also affect people through water and air quality, recycling of organic and inorganic substances, microclimate, etc. Loss of biodiversity means loss of ecological services and options for the future. The cost of replacing ecological services, already great, will increase to staggering proportions. The real and potential wealth represented by conserved biodiversity cannot be replaced (Bridgewater et al., 1995).

The Human Dimension

The reason a multi-faceted or ecosystem perspective is needed for conserving biological diversity through multiple-use management is simple. Continued growth in human populations and increases in their production, use and disposal of resources are not matched by corresponding growth in the land base available to meet those demands under traditional resource management approaches while sustaining desired levels of environmental quality. The

USDA Forest Service translated these issues into four reasons for exploring ecosystem management (Overbay, 1992).

1. People need and want a wider array of uses, values, products, and services from public lands than in the past, especially, but not limited to, the amenity values and environmental services of healthy, diverse lands, and water.

2. New information and a better understanding of ecological processes highlight the role of biodiversity as a factor in sustaining the health and productivity of ecosystems and the need for integrated ecological information at various spatial and temporal scales to improve management.

3. People want more direct involvement in the process of making decisions about public resources.

4. The complexity and uncertainty of natural resources management call for stronger teamwork between scientists and resource managers than has heretofore been practised.

Research Needs

Research on ecosystem management should focus on providing the technological advances and new scientific information essential to meeting current and future resource needs. Research will (i) collect, assess and analyse the information needed to advance the social goals of protecting, restoring, and managing the environment; (ii) improve our knowledge of the fundamental processes that shape the natural world and the human behaviour that affects that world; and (iii) apply the knowledge to solving environmental problems with a comprehensive management strategy in the context of economic and social goals.

Recommended major research programme components include the following.

1. *Understanding ecosystems* Understanding the structure and function of forest and range ecosystems includes study of the mechanisms that control ecosystem processes and the interdependency of ecosystem components at multiple scales. Ecosystem research provides the scientific basis for addressing resource health and productivity issues.

2. *Understanding people and natural resource relationships* Understanding how people perceive and value the protection, management, and use of natural resources, including agriculture, is critical to managing our natural resources today. Increased diversification of resource uses and differing perceptions and values among user groups make finding a consensus on resource protection, management and use a challenge.

3. *Understanding and expanding resource options* This component focuses on determining which protection and management practices and utilization systems are most suitable for the sustainable production and use of natural resources. Special attention is on management options for water, fish, wildlife, recreation use, and recycling wood fibre for new products.

4. *Understanding the impacts of land use* Nearly all humans depend on agriculture for their food. Misuse of land for agriculture, even more than for forest and range, leads to irreversible degradation of the resource on which civilization depends. We need to understand how various options of crop and soil management affect sustainability as well as the short-term yield and economic return.

Inventory, Monitoring and Evaluation

Implementing sustainable management goals and assessing the impacts of management activities will require resources and knowledge. Surveys and inventories are critical in helping determine which areas are in greatest need of preservation or management, and in identifying threatened species and habitats. More research to improve methodologies, distributional and status information, and strategies based on sound information will ultimately provide the basis for all sound policy and management decisions. Current scientific understanding of ecological processes is far from perfect. Better inventories and assessments are needed of current conditions, abundances, distributions, and management direction for genetic resources, species populations, biological communities, and ecological systems.

Monitoring should provide sufficient information about the abundance of animals or plants targeted for monitoring to ensure that current management practices are not threatening the long-term viability of their populations. Effort should not be wasted on monitoring systems that fail to give the level of confidence needed to deduce the most likely effects of management activities on natural resources. In an era when mankind's activities are the dominant force influencing biological communities, proper management requires understanding of pattern and process in biological systems and the development of assessment and evaluation procedures that assure protection of biological resources. It is essential that appraisals of these resources give us the ability to forecast the consequences of human-induced environmental changes accurately. Ideally, the results from monitoring should feedback into the system to correct or fine-tune management activities. But we have a long way to go in this process.

It is a challenging task to formulate an integrated, concise and relevant approach to inventory, monitoring and evaluation. The need for more specific data and more efficient ways for collecting and managing data will lead to significant changes in inventory processes. These include the development of methods and technology that will: (i) provide resource estimates for specific geographic units and evaluate the reliability of such estimates; (ii) display estimates and units spatially; (iii) make maximum use of existing information and new technology, such as remote sensing and geographic information systems; (iv) provide a baseline for monitoring changes in the extent and condition of the resource; (v) eliminate redundant data collection, develop common terminology, and promote data sharing through corporate databases; (vi) utilize information management systems to provide maximum flexibility for data integration, manipulation, sharing, and responding to routine and special requests; and (vii) provide up-to-date databases using modelling techniques, accounting procedures, and re-inventories.

Sustaining the Environment – into the Next Century

The demands and expectations placed on biological resources are high and widely varied, calling for new approaches that go beyond merely reacting to resource crises and concerns (Szaro and Salwasser, 1991; Szaro, 1993). New approaches must incorporate fundamental shifts in the scale and scope of conservation practice (Miller, 1995). These include the shift of focus from the more traditional single-species and stand level management approach to management of communities and ecosystems (Reynolds et al., 1992).

Public concern about the environment, together with new thinking by scientists and resource managers, has led to a new philosophy about how to manage land and resources. This philosophy says that we can manage forest and grasslands to sustain their full array of values and uses for current and future generations. But this calls for changes in the traditional ways of managing resources. It is an attitude, an approach, and a philosophy about how we carry out multiple-use, sustained-yield management. In short, it is enviromentally responsible resource management.

In order to respond to this challenge in the United States, we looked at the dominant social trends affecting future protection, management, and use of natural resources, and we evaluated how the trends might affect the content and conduct of our natural resource and research programmes. Although numerous significant trends were evaluated, three emerged as being the most important in determining future direction. They are not surprising. First is the expanding world population and associated demographic changes. Second is the increasing competition for the many uses of natural resources – more people are demanding ever-increasing amounts of products and services from a finite resource base. And, third is the increasing public awareness and concern for natural resources and for national and global environmental issues. In addition to the traditional forest products – timber, big game animals, fish and water – society is now interested in how forests and climate affect each other; loss of biological diversity especially reflected in threatened and endangered species; growing demand for wood production integrated with the protection of other resources values such as clean water; and maintaining of forest health.

The United States is moving forward with an ecosystem management approach, one that is scientifically sound, ecologically based and totally integrated. Commonsense dictates that this approach, one that considers the sum of the parts rather than each resource in isolation, is the proper and practical way to head. It uses as its foundation principles derived from conservation biology theory for conserving biodiversity and maintaining ecological systems (Soulé and Wilcox, 1980; Soulé, 1986, 1987; Salwasser et al., 1995). These principles include:

- Recover and conserve formally listed threatened or endangered species;
- Provide for viable populations of native plant and animals species;
- Maintain a viable network of native biological communities and ecosystems;
- Maintain structural diversity;
- Sustain genetic diversity;
- Produce and conserve resources needed by people;
- Protect ecosystem integrity – soils, waters, biota and ecological processes;
- Restore and renew degraded ecosystems.

Ecosystem management responds to a significant shift in social values, scientific understanding and land management interests from that of the past. Ecosystem management is an identifying name tag for a new and evolving approach to land management. For practical purposes it is generally synonymous with sustainable development, sustainable management, sustainable forestry and a number of other terms being used to identify an ecological approach to land and resource management. Ecosystem management is a goal-driven approach to restoring and sustaining healthy ecosystems and their functions and values. It is based on a collaboratively developed vision of desired future ecosystem conditions that integrates ecological, economic, and social factors affecting a management unit defined by ecological, not political boundaries. Its goal is to restore and maintain the health, sustainability, and biodiversity of ecosystems while supporting communities and their economic base.

There are four basic operating tenets that provide an 'umbrella' for an ecosystem management approach. Under this umbrella are a number of components all driven or related in some degree to participation, collaboration, using the best science and following an ecological approach. These tenets are as follows.

1. *Partnerships* Sharing responsibility for land management is fundamental. Ecosystems cross boundaries, making the need for cooperation, coordination and partnerships a must for managing the entire ecosystem.

2. *Participation* Get people involved in all aspects of public resource decision-making so that managers will know their needs and views. It is essential to use a highly participatory process, from beginning to end, before deciding on a course of action by involving all those interested in formulating alternatives, evaluating those alternatives and describing the process used to select one. The focus should be on end results – desired future ecological and social conditions and the landuse classes and management actions that will best attain them.

3. *Scientific knowledge* Use the best scientific information and the most appropriate technologies available to understand the range of choices of actions and the consequences of each. Integrate information and technology, such as ecological classifications, inventories, data management systems, and predictive models, and use them routinely in landscape-scale analyses and conservation strategies. This includes strengthening teamwork between researchers and resource managers to improve the scientific basis of ecosystem management (Soulé and Kohm, 1989; Solbrig, 1991; Szaro, 1994).

4. *Ecological approach* This means in the simplest terms looking at many factors, across a broad landscape, using several scales, addressing linkages between landscape elements and ecological processes, and a number of other activities. The science of ecology will be applied to multiple-use management recognizing that people are part of the ecosystems we manage. Landscapes should be used as the basic unit for planning and managing ecosystems to meet specific objectives – both desired future ecological conditions and desired economic and social goals while reconciling conflicts between competing uses and values.

Evolving from these four principles are a set of methods and tools that compose the basic elements of any ecosystem management approach. The following represent key elements of such an approach.

1. Address activities and information across several geographic scales. For aquatic information use a range of nested watersheds, for terrestrial information use the levels described in Ecoregions of the Country.

2. Use landscapes as a basic unit for planning and managing ecosystems to meet specific objectives – both desired future ecological conditions and desired economic and social goals while reconciling conflicts between competing uses and values.

3. Select scales/boundaries appropriate for highly mobile species.

4. Adopt means to deal with the complexity that comes with using multiple scales and multiple boundaries across scales for organizing and using information necessary for sound analyses.

5. Conduct information collection, analyses and planning across administrative and jurisdiction borders to coincide with useful ecological boundaries.

6. Address biotic information across levels of biological organization (cell, organism, population, community, ecosystem, landscape, biome, biosphere).

7. Develop and use methods to recognize and address patterns and change over time and space for key elements at multiple scales.

8. Define major disturbance factors and their range of historic variation.

9. Develop common approaches to ecological classification.

10. Develop, seek out, utilize and transfer the very best available scientific knowledge.

11. Conduct analyses over large geographic areas that encompass smaller project areas.

12. Cooperatively develop desired conditions.

13. Address effects at the project level and at least at one scale above and below.

14. Develop approaches to share information across many borders, including integrated resource inventories and information provided for national uses.

15. Develop decision support technologies and methods to support the complexities of ecosystem management. Build recognition of uncertainty into those processes including the fact that most questions will probably never be answered and major mistakes can take a long time to heal.

16. Integrate information and technology, such as ecological classifications, inventories, data management systems, and predictive models, and use them routinely in landscape-scale analyses and conservation strategies.

17. Develop information about a variety of species habitat needs.

18. Develop information about ecological processes; carbon cycle, nutrient cycle, hydrologic cycle, succession, biological diversity, population dynamics.

19. Develop knowledge of linkages within and between systems and processes.

20. Work within the scope of natural processes that shape landscape and ecosystem conditions.

21. Develop knowledge about the human dimensions of ecosystem management.

22. Use highly participatory processes, involve all the public that want to be from basic data collection through to monitoring.

23. Seek and form as many partnerships as possible in doing ecosystem management, with federal, state, local and other organizations.

24. Use an adaptive management process as an integral part of monitoring and evaluation.

25. Focus on end results – desired future ecological and social conditions and the landuse classes and management actions that will best attain them.

26. Develop, monitor, and evaluate vital signs of ecosystem health.

These are some of the key tools and methods that must be in place to support ecosystem management. There should be independent and unique decisions on individual projects and plans, but there should be a general approach towards an ecosystem management process. Many of the tools and methods noted require sharing and cooperation across administrative boundaries. Much of the information needed at each unit to conduct ecosystem management, especially information at the higher geographic scales, is useful to many units, and many other organizations also interested in ecosystem management.

Adaptive Management

A formal process of adaptive management will be required to maximize the benefits of any option for land and natural resource management and to

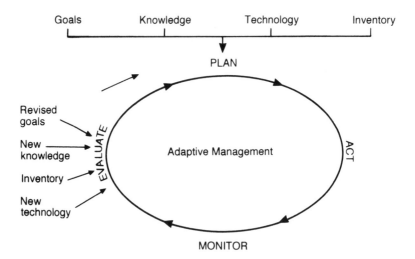

Fig. 32.1. Basic adaptive management model (from ROD, 1994).

achieve the long-term objective of ecosystem management. The process itself is straightforward and simple: new information is identified, evaluated, and a determination is made whether to adjust strategy or goals. Adaptive management is a continuing process of action-based planning, monitoring, researching and adjusting with the objective of improving the implementation and achieving the desired goals and outcomes (ROD, 1994). The general planning model illustrated in Fig. 32.1 illustrates the adaptive approach to management (FEMAT, 1993). This approach (Walters and Holling, 1990) provides a basis for immediate implementation of ecosystem management. In this process goals and objectives are clearly stated, an initial hypothesis of ecosystem behaviour is described, and monitoring is conducted to provide rapid feedback for redirection of management experiments.

In an adaptive management approach, Watters and Holling (1990) offer challenges for justifying and designing experimental management programmes. These are:

1. To demonstrate that a substantial, deliberate change in policy should even be considered, given the alternative of pretending certainty and waiting for nature to expose any gaps in understanding.

2. To expose uncertainties (in the form of alternative working hypotheses) and management decision choices in a format that will promote both intelligent choice and a search for imaginative and safe experimental options, by using tools of statistical decision analysis.

3. To identify experimental designs that distinguish clearly between localized and large-scale effects, and hence, make the best possible use of opportunities for replication and comparison.

4. To develop designs that will permit unambiguous assessment of transient

responses to policy changes, in the face of uncontrolled environmental factors that may affect treated and reference experimental units differently.

5. To develop imaginative ways to set priorities for investments in research, management, and monitoring, and for design of institutional arrangements that will be in place for long enough to measure large-scale responses that may take several decades to unfold.

Although the concept of adaptive management is relatively straightforward, applying it to complex management strategies requires answers to several critical questions (ROD, 1994). What new information should compel an adjustment to the management strategy? What threshold should trigger this adjustment? Who decides when and how to make adjustments? What are the definitions and thresholds of acceptable results? Adaptive ecosystem management depends on a continually evolving understanding of cause-and-effect relationships in both biological and social systems (Everett et al., 1994). The key features in an adaptive approach are (McConnaha, Alaska, 1994, personal communication) as described in the following paragraphs.

First, there should be an *experimental design* for implementation. Adaptive management demands taking a rigorous scientific approach to management. Holling (1978) noted that the 'heart of adaptive environmental management' is 'an interactive process using techniques that not only reduce uncertainty but also benefit from it. The goal is to develop more resilient policies.' Holling (1978) recommended that in an adaptive management context 'environmental dimensions should be introduced at the very beginning of the development, or policy design process, and should be integrated as equal partners with economic and social considerations, so that the design can benefit from, and even enhance, natural forces.' Experiments designed to produce information should be an integral part of the actual management activities. Managers and scientists can learn from changes over time. This can only be effective if monitoring and evaluation procedures are an integral part of the design that should be incorporated at the beginning of the process and not simply added post hoc after implementation (Holling, 1978).

Secondly, there should be an explicit *description* of the system. In order to answer critical questions related to resource management, there is a need to understand the current ecological conditions of ecosystems of interest, changes in the ecosystem components over time, and the likely ecological trends within the ecosystems. That is, we need to know where we are today, how we got here, and what the future scenarios are likely to be under different management regimes. Therefore, it is necessary to develop a scale-relevant assessment to put the social, biological, chemical, and physical components of the management area for which we have stewardship into the larger ecological context.

Thirdly, well-defined *goals and objectives* are necessary. The planning process, a collaborative and cooperative approach, uses the assessment to assign values to the current condition and describes the 'desired future condition' of the resources. An inappropriate decision is likely to occur without widespread understanding, acceptance, and support for the desired future condition.

Goals and objectives provide the guidance for managing towards the agreed upon desired future condition. An overarching goal explicit in taking an ecological approach to management is that of sustaining ecosystems. Two working hypotheses formulated by Everett *et al.* (1994) are: (i) human values and expectations can be incorporated into ecosystem management by identifying landscape patterns that are representative of these values, and (ii) sustainable ecosystems can be achieved by integrating people's expectations with the ecological capacities of ecosystems.

Fourthly *critical uncertainties* should be identified. Because there is no such thing as perfect information, there is always a level of uncertainty in any decision. The concept of adaptive management acknowledges the need to manage resources under circumstances that contain varying degrees of uncertainty, and the need to adjust to new information (ROD, 1994). Even the best decisions may have unacceptable outcomes. To reduce the level and impact of unacceptable outcomes the decision-maker needs to focus on critical uncertainties. Obtaining the best information on these uncertainties and designing a monitoring and evaluation process to track decisions is therefore an essential step in the process. Holling (1978) notes that 'some systems are inherently more capable than others of absorbing insults and changes without losing their integrity.' He describes an 'axiom that underlies any design for uncertainty ... There exists a serious trade-off between designs aimed at preventing failure and designs that respond and survive when that failure does occur.' It is the latter that needs to be incorporated into an adaptive approach.

Fifthly, there needs to be a *monitoring and evaluation* programme. What new information should trigger an adjustment of strategy or direction? New information can come from monitoring, statutory or regulatory changes, organizational or process assessments, or a variety of other sources (ROD, 1994). There are two distinct phases in a monitoring and evaluation programme. The first phase, traditionally, monitors and evaluates the consequences of our management actions. It constantly measures our progress towards our stated desired future ecological condition and provides for mitigation when our actions steer off course. During the evaluation process, information is analysed to determine the nature, scope, and importance of that information (ROD, 1994). Adaptive management depends on both negative and positive feedback in the reiterative evaluation of both the continued desirability of previously selected management goals and progress toward their achievement (Everett *et al.*, 1994). If the impacts detected during monitoring and evaluation are of little significance, then the management regime can be adjusted and refined without adjusting the long-term plan. If, however, the impacts are significant the plan will need to be revised. This can be done directly or, if needed, by modifying the assessment to bring it into line with new knowledge, changed social values, impacted environments, etc.

The second, and less traditional, phase monitors and evaluates social needs in relation to changing societal values and settings. Here social values and needs are constantly assessed in relation to the validity of the desired future ecological condition. Societal values often change more rapidly than ecological conditions. Also, in this second phase, unexpected disturbance events may significantly deflect our trajectory towards the desired future ecological condition. In this case the supply of goods and services may be disrupted and ecological restoration may become an immediate need. Lastly, in phase two, technology transfer of new knowledge is formally incorporated into the adaptive management system. This is where research and management join to ensure that best science is incorporated into the adaptive management process.

Sixthly, there should be an aggressive approach to *learning*. As noted above the less traditional approach to monitoring and evaluation requires active learning in all aspects of ecosystem management. Resource managers and scientists must learn more about the social, biological, and physical attributes of ecosystems and adapt more quickly to new knowledge. Technology transfer must be an inherent and valued part of ecosystem management.

Lastly the structure must be *adaptable*. The adaptive management structure itself must be inherently adaptive. This includes the organization responsible for implementing the process. The key element in this flexibility is the development of a management system that operates strategically and not functionally. This system emphasizes functional skills and eliminates functional organizational barriers.

Conclusions

The global focus on issues related to the conservation of biodiversity will continue to increase and with it, highlight serious and complex problems not likely to be easily resolved. A broad understanding of the significance of managing biological resources currently exists in the United States across the social and political spectrum. In fact, most areas of the United States, and levels of government, have experienced first hand the difficulty of understanding and managing for the conservation of biodiversity once species or ecosystems have been put in jeopardy.

The national paradigm of acceptable land management in the United States and provisions for associated values have changed dramatically over the last 200 years. Social, cultural, economic, and environmental views and values have continued to adjust based on perceptions of scarcity, national security and development, scientific understanding and a desired condition for the country's health and well-being. This has required dramatically different responses from all levels of government, economic sectors, educational systems, and non-governmental organizations.

The current framework for conservation of biodiversity has evolved as a mix of related individual laws and regulations over the last 125 years. The majority of these were put in place within the last 30 years, with a variety of relationships to federal, state or private lands. Specific direction for conservation of biodiversity resources in the United States remains primarily aimed at federal lands and agencies. Designation of a particular species as Threatened or Endangered creates responsibilities and constraints for all ownerships, public, and private. Improving scientific awareness and shifting societal values and priorities have resulted in a new approach to managing lands and resources. This approach is focused on looking at large systems and landscapes, as opposed to the individual component parts. The term used to describe this philosophy and approach on public lands in the United States is ecosystem management. The fundamental focus of ecosystem management is on the maintenance of biodiversity.

Public lands and resources will continue to be a focal point for divisive opinions, interests and values. Ecosystem management will not remove controversy. It is an approach that is based on using the very best information, in a very professional manner to determine the 'sustainable' decision space. The selection of alternatives will continue to be a mix of resource, social, cultural and political interests. The key is to apply ecosystem management in a manner that provides the very best information on which to examine sustainable options and make decisions.

References

Bridgewater, P., Walton, D.W. and Busby, J.R. (1995) Creating policy on landscape diversity. In: Szaro, R.C. and Johnston, D.W. (eds) *Biodiversity in Managed Landscapes*. Oxford University Press, Oxford.

Crow, T.R. (1989) Biological diversity and silvicultural systems. In: *Proceeding of the National Silvicultural Workshop: Silvicultural Challenges and Opportunities in the 1990s*, Petersburg, Alaska, July 10–13, 1989. USDA Forest Service, Timber Management. Washington, DC, pp. 180–185.

Everett, R., Oliver, C., Saveland, J., Hessburg, P., Diaz, N. and Irwin, L. (1994) Adaptive ecosystem management. In: Jersen, M.E. and Bourgeron, P.S. (Tech. eds) *Eastside Forest Health Assessment*, Volume II: Ecosystem Management: Principles and Applications. Gen. Tech. Rep. PNW-GTR-318. USDA Forest Service, Pacific Northwest Research Station, Portland, Oregon, pp. 361–376.

Forest Ecosystem Management Assessment Team (FEMAT) (1993) *Forest Ecosystem Management: An Ecological, Economic, and Social Assessment*. US Government, Portland, Oregon.

Harrington, C., Debell, D., Raphael, M., Aubry, K., Carey, A., Curtis, R., Lehmkuhl, J. and Miller, R. (1990) *Stand-level information Needs Related to New Perspectives*. Pacific Forest and Range Experiment Station, Olympia Forestry Sciences Laboratory, Olympia, Washington.

Holling, C.S. (ed.) (1978) *Adaptive Environmental Assessment and Management*. John Wiley, New York.

Hunter, M.L., Jr, Jacobson, G.L., Jr and Webb, T. (1988) Paleoecology and the coarse-filter approach to maintaining biological diversity. *Conservation Biology* 2, 375–385.

Miller, K.R. (1995) Conserving biodiversity in managed landscapes. In: Szaro, R.C.

and Johnston, D.W. (eds) *Biodiversity in Managed Landscapes*. Oxford University Press, Oxford.

Noss, R.F. (1995) Conservation of biodiversity at the landscape scale. In: Szaro, R.C. and Johnston, D.W. (eds) *Biodiversity in Managed Landscapes*. Oxford University Press, Oxford.

Overbay, J.C. (1992) Ecosystem management – keynote address. In: *Proceedings National Workshop: Taking an Ecological Approach to Management* 27–30 April 1992, Salt Lake City, Utah. WO-WSA-3. USDA Forest Service, Washington, DC, pp. 3–15.

Pearson, S.M., Turner, M.G., Gardner, R.H. and O'Neill, R.V. (1995) Creating policy on landscape diversity. In: Szaro, R.C. and Johnston, D.W. (eds) *Biodiversity in Managed Landscapes*. Oxford University Press, Oxford.

Reid, W., Barber, C. and Miller, K. (eds) (1992) *Global Biodiversity Strategy: Guidelines for Action to Save, Study, and Use Earth's Biotic Wealth Sustainably and Equitably*. World Resources Institute, Washington, DC.

Record of Decision (ROD) (1994) *Record of Decision for Amendments to the Forest Service and Bureau of Land Management Planning Documents within the Range of the Northern Spotted Owl. Standards and Guidelines for Management of Habitat for Late-successional and Old-growth Forest Related Species Within the Range of the Northern Spotted Owl*. US Government, Portland, Oregon.

Reynolds, R.T., Graham, R.T., Reiser, M.H., Bassett, R.L., Kennedy, P.L., Boyce, D.A., Jr, Goodwin, G., Smith, R. and Fisher, E.L. (1992) *Management Recommendations for the Northern Goshawk in the Southwestern United States*. Gen. Tech. Rep. RM-217. USDA Forest Service, Rocky Mountain Forest and Range Experiment Station. Fort Collins, Colorado.

Salwasser, H. (1990) Conserving biological diversity: a perspective on scope and approaches. *Forest Ecology and Management* 35, 79–90.

Salwasser, H., Caplan, J.A., Cartwright, C.W., Doyle, A.T., Kessler, W.B., Marcot, B.G. and Stretch, L. (1995) Conserving biological diversity through ecosystem management. In: Szaro, R.C. and Johnston, D.W. (eds) *Biodiversity in Managed Landscapes*. Oxford University Press, Oxford.

Solbrig, O.T. (ed.) (1991) *From Genes to Ecosystems: A Research Agenda for Biodiversity*. International Union of Biological Sciences, Cambridge, Massachusetts.

Soulé, M.E. (1986) Conservation biology and the real world. In: Soulé, M.E. (ed.) *Conservation Biology: the Science of Scarcity and Diversity*. Sinauer Associates, Sunderland, Massachusetts, pp. 1–12.

Soulé, M.E. (ed.) (1987) *Viable Populations for Conservation*. Cambridge University Press, New York.

Soulé, M.E. and Kohm, K.A. (eds) (1989) *Research Priorities for Conservation Biology*. Island Press, Washington, DC.

Soulé, M.E. and Wilcox, B.A. (eds) (1980) *Conservation Biology: an Evolutional-Ecological Perspective*. Sinauer Associates, Sunderland, Massachusetts.

Szaro, R.C. (1993) The status of forest biodiversity in North America. *Journal of Tropical Forest Science* 5, 173–200.

Szaro, R.C. (1994) Research needs and opportunities: the response of forest biodiversity to global change. In: Boyle, T.J.B. and Boyle, C.E.B. (eds) *Biodiversity, Temperate Ecosystems and Global Change*, NATO Advanced Science Workshop, 15–19 August 1993, Montebello, Canada, Springer-Verlag, Berlin, pp. 399–416.

Szaro, R.C. (1995) Biodiversity maintenance. In: Bisio, A. and Boots, S.G. (eds)

Encyclopedia of Energy Technology and the Environment, Wiley-Interscience, John Wiley, New York.

Szaro, R.C. and Johnston, D.W. (eds) (1995) *Biodiversity in Managed Landscapes*. Oxford University Press, Oxford.

Szaro, R.C. and Salwasser, H. (1991) The Management Context for Conserving Biological Diversity. 10th World Forestry Congress, Paris, France, September 1991. *Revue Forestiere Française, Actes Proceedings* 2, 530–535.

Szaro, R. and Shapiro, B. (1990) *Conserving Our Heritage: America's Biodiversity*. The Nature Conservancy, Arlington, Virginia, 16 pp.

Walker, B.H. (1992) Biodiversity and ecological redundancy. *Conservation Biology* 6, 18–23.

Walters, C.J. and Holling, C.S. (1990) Large-scale management experiments and learning by doing. *Ecology* 71, 2060–2068.

Wilson, E.O. and Peter, E.M. (eds) (1988) *Biodiversity*. National Academy Press, Washington, DC.

WRI/IUCN/UNEP (1992) *Global Biodiversity Strategy: Guidelines for Action to Save, Study and Use of Earth's Biotic Wealth Sustainably and Equitably*. World Resources Institute, Washington, DC.

Biological Diversity and Agrarian Systems

B. Vissac

Département de Recherches sur les Systèmes Agraires et le Développement, INRA, 147, rue de l'Université, 75338 Paris Cedex 07, France

Introduction

The conflicts underlying the implementation of the 'Convention on Biological Diversity' highlight the essential links that exist between this Convention's objectives and the development models intrinsic to government policies and the rules of international trade. These links find expression at widely differing spatial and social levels, such as regions of the world, communities of States, provinces and localities.

It is appropriate to compare the developed nations, which have based their development on a limited selection of species and clones in simplified, highly controlled systems, with developing countries. The developed world has often set itself as an example for developing countries, while seeking to benefit at low cost from the biological reserves that have been protected by the lag in development of the poorer nations. However, we might also take a look on our own doorstep: behind a renewed interest in local products and models, there often appears to be a misuse of the image of rare genetic material (local breeds shown on television, for instance) to promote products by usurping names drawn from places of origin, and fabricated, here or elsewhere, by widespread but ordinary breeds (Vissac, 1995).

One cannot therefore discuss biodiversity of domestic livestock without referring: (i) to its relations with ancient human cultures within agrarian systems (the set of items and relationships both induced by the environment and developed by a society in order to put the resources of a territory to use through agriculture); and (ii) to the relations maintained between existing vestiges of those cultures via biodiversity.

Biodiversity and Human Evolution: the Confusion Resulting from Overlapping Forms of Livestock Farming

If we consider the example of livestock farming, the situation at a given moment in time corresponds to a stratification of the development models and types of biological material elaborated during the course of history. Some situations have enabled social groups to profoundly transform their husbandry practices and animal material, whereas other highly restrictive situations are the reflection of deeply ingrained practices in relation to biodiversity (pastoral herding, for example).

Nowadays, when discussing biodiversity it may seem inappropriate to begin by referring to issues of domestication. However, in animal husbandry, situations exist where exchange takes place between wild and domesticated species. In Corsica, for example, increasingly extensive forms of local pig farming facilitate reproduction with the Corsican boar, whose chromosome number is equivalent to that of the Corsican domestic pig (Franceschi, 1980). Making a distinction between genetic materials is difficult in such a context, as can be illustrated by the hunting tallies (Franceschi, 1984).

Until the end of the eighteenth century, cattle farming in Europe corresponded to local agrarian forms adapted to a large range of uses through controlled selection while at the same time subject to the forces of natural selection often linked to catastrophic events (epidemics, wars, famines, etc.). The relatively general biological opposition between these two selective forces led to fluctuations around states of equilibrium which characterize local populations. Contrary to crop-growing, for which ancient agronomic rules had been laid down, it would seem that the practice of livestock rearing in France was linked to a certain way of life and was, in variable forms, a vital contribution (as a source of power for work and a source of organic fertilization) to the growing of cereal crops (Bloch, 1976). Whatever the case, the confusion that existed in the ideas of 'animal generation' before the nineteenth century could not have made the work of breeders any easier. According to Russell (1986) this term included simultaneously reproduction, heredity (the resemblance between relatives), and the environment (confusing the effects of genes and of the environment on the phenotypes observed).

'Modern' forms of cattle livestock farming appeared in England where climatic conditions favoured the development of grasslands. The concurrence of the first technical advances with the onset of a market economy led to the creation of enclosures on large estates of 'wool-bearing animals', the control of which had been seized by an aristocratic class under the influence of a powerful monarchy, the Tudors (Crotty, 1980). This occurred at the same time as the outbreak of famines and epidemics, which not only limited the human population but also weakened any form of social resistance. It was against this background, just when the influence of global trade and industry (1700) was beginning to be felt, that techniques were developed with cattle in Lancashire for the intensive rearing of bull calves (suckled by several cows in succession) capable of mating the cows of the herd in order to obtain grouped calving when put out to graze. A correspondence was thus established between the seasonal dynamics of animal needs and those of grazing resources. The economic value of these bulls, whose virility was presumed to be transmissible, went far beyond that of the value of their meat production (Russell, 1986). This is the first mention of capital gains in cattle, which was presumed to be due to selection!

It was based on these breeding practices and somewhat extravagant suppositions that around 1780, Bakewell started working with sheep in the Midlands, as did Collins with cattle in the Tees valley. Both corresponded to the general trend close to Darwinian thought, and Bakewell, in particular, using an animal species (sheep) whose biological characteristics were more favourable, was for many years considered as the father of the pre-modern selection of livestock. These pioneers proceeded by morphological observation of the breeder bulls, their ancestry and descendants, in the herds of customers to whom they leased the animals, in order to keep the best for later use in their own herd. Practices such as these, associated with the type of animal husbandry then in use, favoured the promotion of fat animals that matured early, and whose tallow was used by industry. These practices were extended towards the end of the nineteenth century by the setting up of breeder societies which overemphasized the ancestry of the animals for the purposes of selection, an attitude stemming from their respect for the privileges granted to the nobility, which helped in maintaining the power of the aristocracy. The breeders had the enormous advantage of combining the functions of both judge and merchant. Until the advent of neo-Darwinism and genotypic selection (Lush, 1935), these were the bases for livestock selection and breeding. This was true of France, in regions where cattle used for draught purposes had remained 'muscular', and also of the United States, where, lacking land to enclose in Europe, the English emigrants spread out over a limitless land, taking their breeds and the ideas of Robert Bakewell with them as they went.

After World War II, in line with neo-Darwinian thought (combining population genetics and mathematical statistics), the discovery of artificial insemination of cattle and the advent of powerful, user-friendly computing systems acted jointly with the spread of specialized, intensive production systems favoured by the agricultural modernization policy. The conditions for controlling the various aspects of animal generation (the container, the content and the context) were now established:

1. Artificial insemination facilitated the application of the 'male approach' for controlling large-scale genetic progress;

2. It permitted statistical estimation of the genotypes that had hitherto been confused with phenotypes and the effects of the environment;

3. The policy of favouring specialized intensive breeding, in which the profitability of the animals closely followed the income of the breeders, facilitated the adoption of consistent objectives and the comparison of animals.

The generation of animal populations thus became a matter for collective management using analytical factors combined on the basis of pre-established methods. Developed countries have applied these procedures in a liberal or cooperative fashion with a few large-scale breeds suited to these productive principles. The latter approach has been particularly effective in France where a relatively unproductive livestock has become the subject of selection methods which are among the most efficient in the world. The reputation of French beef cattle has superseded that of the British breeds regarding the search for productivity and lean meat. The quantitative results have exceeded all expectations. However, the following points should be noted.

1. Potential outlets have been limited by the insolvency of the very poor populations in the developing world for whom the breeding methods were supposed to set an example.

2. The energy and protein crises (e.g. the soya slump) have drawn our attention to the risks of dependence on intensive forms of production and on the strategies resulting from them, hence the term 'food weapon' in use during the 1970s.

3. The damage caused to the environment by lack of, or excessive (pollution) farming activity increased during the 1980s.

At the same time, experiments have highlighted the multiplicity of interactions and oppositions between selected traits, between direct and maternal effects linked to genes and to the environment. In mammal species in particular, these interactions play a buffer role, and act as a counter force to the effects of directional selection. These observations have led to the separate selection of specialized male and female lines of livestock, either for the production of female breeders, or for the production of meat. The lines are also systematically used in cross-breeding in order to exploit heterosis effects. Developments of this type, which are dominant in the case of the most prolific animal species (poultry and pigs), have resulted in the development of animal production methods whereby the various stages of the process have been integrated within a single industrial firm (reproduction, selection, feeding and marketing). For the farmers, this results in a high degree of dependency and restrictions, and in major environmental effects.

In parallel, and following on the progress made in molecular biology, there has been a segmentation of disciplines such as genetics, and a more critical analysis of the neo-Darwinian approach which provided the basis for quantitative selection (Kimura, 1990). As a result, researchers in applied genetics are aiming at control of the genome, by limiting their work to a few QTL (quantitative trait loci), given their known effects on performance. In order to achieve this, they have established an order to validate relationships between the fractionated genome and the productive genotype. The speed with which the analysis of the genome has been implemented contrasts, however, with the delays in its evaluation on the basis of performance. The databases available for that evaluation concern controlled animal production conditions and juvenile performances that are given preference in order to accelerate

Table 33.1. The stages in livestock production as managed through human history.

Types of livestock use and farming	Degree of mobility	Reproduction (the 'container')	Selection (the 'content')	Management of forage resources (the 'context')	Management of sanitary conditions (the 'context')
Hunting	+++	Natural	Natural	None	None
Domestic	++	Natural (collective)	Artificial (phenotypes)	Extensive	Protection (vaccination)
Elitist (breeders' societies)	+	Natural + Artificial insemination	Artificial (phenotypes)	Semi-intensive (grazing)	Protection and culling
State- or company-managed	+	Artificial insemination	Artificial (genotypes)	Intensive feeding	Culling
Industrially integrated	0	Artificial	Artificial (genotype + gene)	Confined rearing	Germless

genetic progress (i.e. the first production cycle). Yet the growing importance of the various forms of extensive types of animal production are drawing increasing attention to performance profiles throughout the career of the animals. One final point is that the emergence of a slump in livestock farming and the resulting uncertainty present a number of disadvantages, also of concern to the geneticist. They notably limit the definition of precise objectives which are necessary for the collective selection of genotypes and for the validation of relations between genotypes and molecular variants, contrary to the situation in medical genetics.

Biodiversity, in relation to agricultural development, ultimately implies referring to successive forms of animal production during the course of evolution of humans. During this evolution, each new phase seemed to necessarily replace the previous ones, rejected as archaic. We know this can come about only if made possible by the natural and socioeconomic context. The new techniques either partially transformed the older techniques or replaced them totally to the point where the old breeds disappeared. Linked with the development of trade, this can result in the coexistence on neighbouring and even contiguous territories, of forms of animal production which correspond to the whole range of previous techniques (Table 33.1).

It thus becomes clear that when speaking of both diversity and agricultural development in relation to livestock farming, one is obliged to refer simultaneously to the three types of factors in animal generation, commonly designated as the 'container' (sperm and embryos), the 'content' (genetic information), and the 'context' (nutritional and pathological effects and their determinant socioeconomic and cultural factors). Extending beyond the framework of the domesticated animal section of the biosphere, we must therefore include factors concerning other components of the biosphere, chiefly those linked to forage resources and to pathogens affecting all of these elements.

It therefore seems absolutely essential to link forms of livestock farming and factors of generation. The items targeted for use in biodiversity (sperm, embryos, living animals) and the way in which they are combined or controlled in fact vary considerably from one form of livestock farming to another. Given their possible territorial contiguity and the significant pre-eminence of the more recent forms of animal production, the wide array of regulatory and economic incentives create friction and even conflicts between the different forms of livestock farming with regard to the management of biodiversity. Table 33.1 highlights these underlying conflicts.

The Effects of Reductionism on Biodiversity Reasoning

Research on biodiversity in livestock farming ultimately comes up against major limits:

1. biodiversity in general refers to domesticated livestock populations and not to the habitat in which they have developed;
2. the subject is treated above all in terms of genetics (populations, breeds, strains);
3. from this restrictive stand-point, biodiversity refers to a range of entities (animals, sperm, embryos) which do not have the same meaning depending on which form of livestock farming is under consideration, and whose qualification as far as improvements are concerned encounters difficulties due to the uncertainty of farming aims and environments.

It is clear that this development, which resulted in the application of Taylorism in productive activities, is largely concomitant with the analytical and Cartesian mode of thought from which it stems. From this stand-point, research discoveries have been applied separately to individual parts in the long productive process on which livestock farming is based, such as feed, animals, and processed products. As a result, various disciplines, fields or disciplinary subfields have gradually emerged, dealing with each item, and with each component of the generation it concerns.

These disciplines have accumulated their contributions in various combinations partially linked to the thought processes and strategies of the complex under construction (i.e. Research, Training and Development), motivated by socioeconomic and cultural change. The development of these contributions is also contingent on the past, to an often excessive degree. The emergence of a new scientific field takes place within an existing generative framework (container, content and context) that is to say, it is primarily based on the most advanced types of genetic material and forms of animal production in the previous stage of human evolution. The socioeconomic pressure of the time, and the fact that in order to progress and have that progress evaluated, research needs the most up-to-date information, are factors used as reasons to explain and even justify this evolution.

These limits are inherent to the way in which knowledge about domesticated livestock has progressed since the beginning of the century. In this respect, the United States is an interesting example, having inspired a development which the rest of the world and Western Europe in particular experienced in an accelerated fashion. It was necessary to fill the 'technology gap' just after World War II, and this caused the acceleration to be even faster as there was more ground to be made up.

At the beginning of the century, biodiversity was evaluated in terms of the taxonomy of species and varieties, according to 'invariable' morphological and aesthetic criteria, such as the shape of the skullbone, the colour of the coat, of the mucous membrane and of the hooves in cattle. The animal was evaluated in its regional living area, and limits for its habitat were established.

The first research carried out on animal performance in the United States, a country which was a pioneer in the field, dates from the beginning of the present century (Willham, 1986). In addition to pathology, research focused on the nutrition of dairy cows and poultry, pioneer species in terms of production specialization (1908), at a time when a specialized organization, the 'American Feed Manufacturers', was being set up. Nutritional research formed the crucible for animal production research and was therefore much earlier than the work on product processing and on selection, whose direction it determined. Research on produce processing developed after World War I, notably due to pressures and needs related to the organization of distribution circuits for produce (the meat circuit in particular).

Animal breeding research only took off after 1935, with the publication of the book by Lush, which applied quantitative genetics to the breeding of domestic livestock. The emergence of scientific disciplines in animal production and their integration into agronomic research often occurred after this period; indeed, the term 'animal husbandry' disappears from the literature on applied research, to be replaced by that of 'animal production'. This integration took place in experimental research stations, where biodiversity was characterized by experimental arrangements for comparing animal samples that deviated little from the standards of maximum profitability per head, which in Europe continued to be linked to farmer income, in particular during the 1960s. Systems for recording animal performance were developed for a fraction of the livestock, which in France were entitled 'selection basis', and were supposed to meet the aims of modernization. They focused on economically interesting traits, with a basic emphasis being given to the juvenile phase of the animals. The livestock were generally reared under controlled conditions. Testing stations where the animals were kept under standard rearing and nutritional conditions completed the system. In short, what they were trying to improve was known (the future was certain), the various component factors of generation were analysed and identified (reproduction, heredity, and environment) and were then recombined to obtain the technological optimum, but the genetic factors affecting performance remained unknown. Research work was based on their probable effects.

Research on blood groups dates from the beginning of the century, but

Table 33.2. The main trends in livestock farming and production research (Willham, USA).

Period	Livestock farming ⟶ Production	Research focus
1910	Specialization in animal production	Animal nutrition (maize, etc.)
1925	Organization of production-processing systems (qualification)	Product technology
1935	Organization of animal selection based on productive traits	Quantitative genetics
1950	Artificial insemination	Animal physiology
1980-90	Selection based on genetic variants	Molecular genetics
	Embryo transfer, sexing, cloning	Embryo physiology
2000?	Transgenosis	

only really got under way after the war, in order to lend a moral tone to the market for high value breeders in which the outlay for individual (horses) or collective (cattle) selection was concentrated. The latter area of selection also stimulated interest in genes with visible effects, such as hereditary anomalies affecting the viability of the animals, and above all muscular hypertrophy of genetic origin, the value of which was heightened by the need for lean meat. Towards 1960 this focus led to research of the same type as that undertaken at the turn of the century, particularly in the United States, on the hereditary dwarfism of cattle. Despite the multiple problems in female reproduction associated with this factor, muscular hypertrophy of genetic origin turned out to be particularly useful for producing specialized sire lines in cattle and pigs. Towards 1970, work was started on cytogenetics applied to animals. Their development owes a great deal to the discovery of a translocation which affects reproduction in various breeds of cattle in which the risks of extension seem to be linked to the small size of the male breeder population used in artificial insemination. It was in the field of milk protein polymorphism that genetic research developed the most, however. This development in research and its applications were, as we saw above, accompanied by complementary progress in work on animal physiology, leading ultimately to proficiency in the use of embryos (transfer, sexing, cloning) in a situation of increasing integration of research work in genetics and in physiology. The possibilities for genetic manipulation are now in hand and announced for the end of the century (Table 33.2).

Research on forage plant resources that accompanies livestock farming first of all concerned itself with the selection of cultivated varieties. It is true that ecologists were at work beforehand but their research was limited to the description of the large-scale man-induced developments of plant cover (Kunholz-Lordat in the Mediterranean region). It was in the early 1960s that the concept of ecosystem was defined, and towards 1970, under the auspices of UNESCO's MAB programme, that their evolution and the role of interfaces between ecosystems were first tackled. In France, the ecological movement stimulated debate about the relations between the different parts of the domesticated biosphere and about the management of the interfaces between areas of different but complementary potential by the use of adapted forms of livestock farming (the soya slump at that time leading in France to the expression 'prospecting for green fuel'). Various committees of the General Delegation for Research in Science and Technology (DGRST) encouraged French researchers to adopt a multidisciplinary approach to agricultural and rural planning and development, in situations where the approach by separate disciplines, used in agricultural research, seemed either insufficient or unsuitable. Researchers engaged in this localized multidisciplinary work then attempted to make a comparative analysis of their approach to biodiversity, depending on the subject of their study and specialized field, in a context of complexity.

Research on biodiversity is thus based on the development of science in relation to livestock farming. Since the turn of the century, animal husbandry has become a means of production. Knowledge generated by research has been applied to livestock farming against a background of segmented and sub-segmented disciplines. This persists despite recent efforts to establish links between the different parts of the biosphere concerned with the management of the productive processes (production flows) and the areas in which they occur.

Despite these efforts, this form of segmented research has diminished the overall knowledge of researchers and is distancing them from that or the practitioners. As a result, it is not surprising that the reactions of researchers who are aware of the issues surrounding the preservation of biodiversity are expressed in various ways, from those who analyse increasingly elementary items while trying, with growing difficulty, to link them to a humanized whole which itself is increasingly complex, to those who attempt to embrace the whole without being able to accept fully the changes taking place at high speed in our knowledge of the biologically infinitely small.

When the report which led to the creation of the French Bureau of Genetic Resources, was drafted in 1980 we recalled the importance of not breaking the system of relations existing between the component parts of domesticated livestock generation (Cassini and Vissac, 1980). Those parts are managed in terms of:

1. visible genetic structures, which are therefore usable by practitioners, or invisible ones (such as laboratory measurements by sequencing techniques);
2. genotypic entities selected in schemes according to indices of economic performance;
3. living animals or phenotypes kept under conditions outside the agricultural context;
4. herds of animals used by livestock farmers, the exploitation and component parts of which are interlinked;
5. continuous territories used for production purposes in which herds are raised with specifically defined objectives associating natural and cultural conservation concerns.

At each of these levels, the items concerned by the management methods, namely sperm, embryos and seeds, and living beings, reflect the different relations that exist between biological material and its environment, and have different meanings in relation to their use in livestock production. Biological organisms, genetic data, the pathological and nutritional context – in short, the constituent parts of population generation – are associated in different ways and jointly controlled in forms that differ according to the level considered. In one area, natural reproduction may be linked to a genetic adaptation to diseases and to a pastureland context. In another, on land which is often nearby, an artificially controlled method of reproduction, of agronomic management of food resources, and of fodder and animal health management, may correspond to specific genetic lines which are clearly identified. Coexistent systems of this type can create problems, however, in relation to cultural combinations which are not only different but antagonistic in many ways (Table 33.1).

Ultimately, although we are making progress in the management of livestock farming on the one hand, and in the levels of analysis and the accuracy of our knowledge about biodiversity on the other, the aims and circumstances of livestock farming are becoming increasingly varied and uncertain. Where biodiversity is concerned, we are therefore justified in developing an approach that starts with livestock farming as a human activity and goes on to deal with management of its biodiversity.

Some Concepts for Systems Modelling in Livestock Farming

The following section of this chapter starts from the complex and from the actors involved in the practice of exploiting domesticated biodiversity, taking as a basis the recent progress in the sciences of complexity in their relations with action. The questions raised by biodiversity – as far as breeding units, such as herds, are concerned – involve not only production subsectors such as dairying, meat, etc., which research agronomists are understanding increasingly well, but also land, which has social implications frequently overlooked by agronomic and zootechnical research (Fig. 33.1). The notion of quality, which involves both of the above-mentioned levels, may be appreciated in different, even diametrically opposed ways which have become the subject of a general debate that goes beyond the agricultural world, but a debate which research should help clarify. What is a good food product, what is a beautiful landscape, and how compatible are they? In fact, the answers to the excesses or to the shortcomings of production subsectors may perhaps be found in appropriate forms of land use, and vice versa. In one region, a poor society will seek to obtain a hypercalorific source of food in order to survive or to make better use of its physical strength. In an inactive sedentary society, threatened by a surplus of cholesterol, its members will seek out hypocalorific products. In this respect, Willham (1986), notes that cultural evolution is much faster than biological evolution.

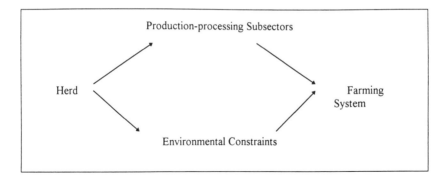

Fig. 33.1. Problem complexity and structure.

Table 33.3. Modes of reasoning.

Well-structured	Problem	Ill-structured
Substantive (positive)	Rationality	Procedural
Defined	Objective and conditions	non defined
Optimum (final)	Solutions	Satisfactory (partial)
Model of object	Representation	Modelling of project

This dual complexity cannot be tackled jointly without basing one's approach on the notion of an agrarian system. This makes it impossible to use approaches based on substantive or positive rationality (Table 33.3), which would enable, on the basis of a model, an optimum solution to be found for well-structured problems. In such a situation, which has become commonplace, with ill-structured problems, a procedural form of rationality is used. No objective is indicated, nor are there any paths laid down in order to attain it; the aim is gradually constructed in parallel with the development of negotiated management procedures between social groups leading at each stage to satisfactory solutions, given the perception that those involved have of their situation and of their aims at that moment in time.

The first French research work on a systems representation of livestock farming (Fig. 33.2) was based on the functioning of a farm, taken as the basic agricultural unit in Europe (Osty, 1978). A farm is considered to be a steered system. The concept of practice (Teissier, 1978; Landais *et al.*, 1989) enables farming operations to be defined in terms of the use of the resources of land, finance, production, skills, and aims of the farmer involved. The practices, which organize themselves into a system, reveal the importance of the herd entity (which is not just a group of animals as is usually considered the case in most research), and of the organization of farm land into functional land units whose resources are used by the animals (Hubert, 1994). The practices can be linked to the build-up of animal and plant cover performances. They

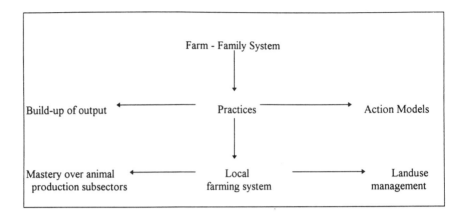

Fig. 33.2. Livestock farming viewed as a complex managed entity.

thus meet up with and put the finishing touches to the work of agronomists and ecologists: the concept of disturbance adopted by the latter thus takes on an operational meaning. The practices can also be linked to decisions through 'action models' (Sebillotte and Soler, 1990), the use of which by the livestock farmer was notably illustrated by Hubert (1994). These concepts can be used to describe the variety of farm types, and provide a way to conceptualize the characteristics and methods of management used for the plant and animal material they comprise (Capillon, 1985; Perrot, 1991).

These concepts can also be applied at the level of continuous land, by integrating them into a local agrarian system (Deffontaines and Osty, 1977), in relation to frequently controversial questions about land planning and (or) development which rural communities have to face:

1. the management of mineral or drinking water resources, given the danger of pollution by nitrates from agriculture (research in Vittel and at other sites in Lorraine) (Deffontaines et al., 1994);
2. protecting the land against large-scale fire through the organization of forms of pastoral farming to create and maintain a heterogeneous plant cover (Provence).

In these cases, questions related to biodiversity fit into a framework which makes them easy to understand and to be seen in their social context concerning land management and production. Tell-tale signs such as the landscape, which can be interpreted both from an ecological and from an agronomic point of view, can be used to help those involved to discuss and negotiate their practices and relations when there is disagreement about their use. In this respect, 'contractual economics' sets down some general guidelines, thereby facilitating and organizing the corresponding relationships.

Management of Biodiversity Through Animal Population

This type of applied research generally lacks an integrating concept that can be recognized by human communities and which has a territorial base sufficiently widespread and temporal foundations sufficiently deep to serve as a collective pilot light or guideline. The animal population, the breed in particular, can fulfil this type of requirement, with the dual definition it has been given (Vissac, 1993), i.e. as a biological entity defined by its genetic informational content, and as an anthropological entity corresponding to the individual and collective practices of its management, including selection (Fig. 33.3).

The developments which have appeared since the multidisciplinary work in Aubrac (1964) involving agronomists, animal scientists and ethnologists, are a perfect illustration of how the modernization of forms of livestock farming, which was in danger of leading to the disappearance of the Aubrac breed, has combined tradition, in the form of its genetic potential and image, with modernity brought in from the outside (Charolais crossing; Rouquette, 1994).

Similarly, in the Franche-Comté region, the struggle by the upholders of the pure Montbèliard breed, linked with the defence of Comté cheese, and the individual and collective production and processing practices associated with them, has done a great deal to prevent the adulteration of the Montbéliard breed by the American Holstein. The different ways by which social groups integrated modernity and tradition resulted in this instance in violent disputes about the application of the French law on animal breeding (Jacques, 1985).

In the Northern Alps, the Tarine breed, protected by the 'Appellation d'Origine Contrôlée' granted to the Beaufort cheese made from their milk, is also vital for maintaining, by grazing, the steeply sloping land used for internationally renowned summer tourism and winter skiing. The argument in this case is between those in favour of improving yield and a form of livestock farming which uses imported hay (Crau hay), and those who defend the cheese characteristics of the milk produced from the native alpine grasslands (Vissac, 1994). Different ways of integrating endogenous and exogenous cultures also concern the breed (as a genetic entity), and the ways in which it is used to up-grade land and its produce (anthropological entities), based on a conceptual framework that changes with developments in the sociocultural and economic environment (for example, the emergence of 'urbanization' in industrial societies).

These facts tend to broaden farmers' individual practices and give them a collective sense (collective practices). Breeds, like the landscape, in their hybrid genetic and anthropological form, become an indicator of the way in which an agrarian system or an association of agrarian systems function in seeking common qualities for both the land and its produce. Animal biodiversity, seen in this light, becomes a concept that can organize development, whose conflicts it integrates over the long term and within a spatial and social framework suited to its management and conservation.

It was a group of research workers in Corte (Casabianca and Vallerand,

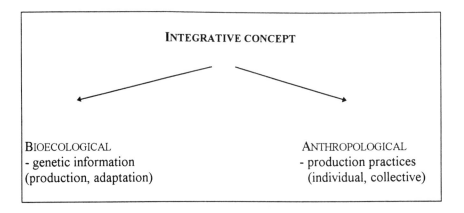

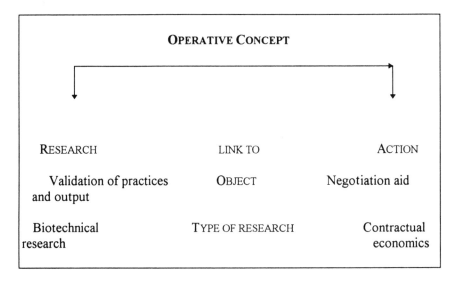

Fig. 33.3. A dual definition of the animal population.

1994) who first clearly formulated a project for the auto-eco-reorganization (in the sense given to the term by the French sociologist Edgar Morin) of an agrarian system based on farming of regional breeds of cattle. Breed management in Corsica is currently based on procedures of 'domestic co-ordination'. These procedures run contrary to those set out in the French law on animal breeding, which, for the breeders themselves, is a form of 'administrative regulation' which does not take into account the aims and limitations of the island's breeding system. The breed is considered by the research workers of Corte as inalienable and jointly owned common property, an emblem recognized and defended by society, which integrates the relational practices and networks implemented at the level of production, processing and the effective use of the cultural benefits generated by the produce. The formulation has something of a global concept for regional development about it. Far from appearing marginal, this example may provide a principle for dealing with the animal populations of many other countries. It may also, however, provide a safety net for the management of widespread breeds affected by the restrictions resulting from the uncertainty of the aims of livestock farming which, given the variation of attitudes among breeders, weigh on the establishment and implementation of collective breeding schemes.

Animal breeds can therefore be used as an indicator linked to forms of livestock farming, which can be interpreted in terms of an agrarian system and of its component parts (the production subsectors and the landscape structures linked to their use). Although this observation does not apply generally, it can be more easily related to society and collectively defended, at the level of the domesticated biosphere, than a microorganism or a forage grass. It still has to be linked, however, to the other aspects of biodiversity which occur in livestock farming, such as the plant cover, and vegetable and animal parasites.

In this respect, the work carried out on landscape ecology converges with this type of activity-based thinking that leads on to the study of biodiversity in the broad sense of the term. In this way, research based on fractal geometry (Baudry, 1992) reveals the way in which the land is divided and how those divisions develop, and links the form of plant cover observed with their function and the ways in which they are being managed (types of husbandry and practices, breed). The use of botanical or zoological indicators (Balent and Courtiade, 1992), and the establishing of their dynamic relations with anthropic phenomena as described by practices, can give meaning to the study of biodiversity in relation to the scales and levels of organization thus described, where it appears jointly with the phenomena in question, i.e. plant cover serving as a food source for ruminants, managed agricultural landscapes, etc. (Balent and Courtiade, 1992).

On a more general level, it has been suggested (Vissac, 1993) that the relations between breeds and livestock farming practices that are associated with them to a greater or lesser extent should be more deeply examined, so as to better understand their productive efficiency and landscaping effects. A breed such as the Normandy cow, for instance, seems more adapted than the Holstein to quality production making optimum use of grazed pastures, and in the same way, the Holstein is more suited to production based on specialization and forage intensification on arable land, wherever that is still possible.

Conclusion

Ultimately, starting with the present situation of crisis in animal husbandry in order to describe it in all its complexity and cultural diversity does not run counter to research on particular features of biodiversity which develops spontaneously as the fields of disciplinary knowledge and forms of science multiply. In fact, this is far from being the case. In this context, this approach gives meaning to the work which, in a crisis situation, needs to be based on the attitudes of the people and communities involved in exploiting animals in order to link up with the aims of production (product quality) and the maintenance of the land allocated to livestock farming.

Some will object that this type of thinking runs counter to the inability of the human communities concerned by livestock farming to face the issue of long-term conservation based on the realities of action. They prefer to turn towards central government, whose incentive role could be turned to advantage, as was the case with the Animal Breeding Act in France, provided that its application does not just follow in the wake of the dominant models of the semen suppliers. The question of the respective role of individuals and communities is wide open; the answer seems to depend on the degree of certainty of the aims, central government being frequently poorly placed to deal with uncertainty related to land use. Perhaps we also no longer have the time to discuss the issue, given the urgency of the questions being raised.

This chapter has concerned itself solely with livestock farming, as its collective structures for selection are well suited to taking account of the collective cultures of those involved. The research done by a number of agronomists such as Haugerud and Collinson (1990) in Kenya expresses, in differing degrees, the same line of thought. Haugerud and Collinson note, for example, the unsuitability of maize hybrids to the practices of crop associations, to the increase in the number of harvests per year and to climatic uncertainty. They argue for a renewal of the selection of local ecotypes; apart from this remark, however, for the moment we shall not extend our comments to agrarian systems based on cultivated crops.

On a more general level, considerations such as these raise questions about the ethics of research and the need to diversify approaches to the management of biodiversity. They query the postulates and operating rules of what Gruson (1987) calls the 'heavy structures', mostly resulting from western technology, that often restrict human freedom and place agrarian societies in situations of dependence which are often irreversible.

To conclude, as far as livestock farming is concerned and, more generally, the role it plays in human cultures, we can consider what Haudricourt (1962) said on 'the relations between the domestication of animals, the growing of crops, and the ways in which we treat others'. Haudricourt is concerned with the evolution of the Eurasian continent, where many of the roots of animal domestication are to be found. He contrasts Asia with its monsoons to Europe and mountainous regions. In the former case, the two basic climatic features for agricultural production, heat and humidity, are in seasonal harmony. The cultivation of annual plants dominates, and oxen, reared individually, are an aid to crop-growing. In the second case, the seasonal contrasts between heat and humidity (it is cold when water is available, and droughts dominate in summer), combined with those imposed by the relief, force farmers to keep animals in herds, to adopt forms of pastoral farming and winter feed storage strategies in order to be secure against uncertainty. The same contrast can be found in the Bible where the 'wicked farmer' is contrasted with Abel, the 'good shepherd'. Haudricourt (1962) wonders whether this contrast in the attitudes of people towards animals may not have led to differences between the great civilizations in the ways in which people conduct themselves and their affairs. One can at least ask the question with Haudricourt and show greater circumspection and respect for the actors involved in the ways in which we treat domesticated biodiversity in research.

References

Balent, G. and Courtiade, B. (1992) Modeling bird communities/landscape patterns relationships in a rural area of South-Western France. *Landscape Ecology* 3, 195–211.

Baudry, J. (1992) Approche spatiale des phénomènes écologiques: détection des effets d'échelle. In: Auger, P., Baudry, J. and Fournier, F. (eds) *Hiérarchies et Échelles en Écologie*. Publ. Min. Environnement/SCOPE, pp. 157–171.

Bloch, M. (1976) *Les Caractères Originaux de l'Histoire Rurale Française*. Librairie Armand Colin, Paris.

Capillon, A. (1985) Connaître la diversité des exploitations: un préalable à la recherche des différentes techniques régionales. *Agriscope* 6, 31–40.

Casabianca, F. and Vallerand, F. (1994) Gérer les races locales: une dialectique entre ressources génétiques et développement régional. *Génétics Selection, Evolution* 26, supp. 1, 343s–357s.

Cassini, R. and Vissac, B. (1980) *Conservation des Ressources Génétiques*. Report presented to the Minister of Agriculture, Doc. Roneo, 25 pp.

Crotty, R. (1980) *Cattle, Economics and Development*. CAB, The Gresham Press, Surrey.

Deffontaines, J.-P. and Osty, P.L.O. (1977) Des systèmes de production aux systèmes agraires. *L'Espace Géographique* 3, 195–199.

Deffontaines, J.P., Brossier, J., Benoît, M., Chia, E., Gras, F. and Roux, M. (1994) Agricultural practices and water quality. A research-development approach. In: Brossier J., Bonneval, L. de, and Landais, E. (eds) *Systems Studies in Agriculture and Rural Development*. INRA Editions, Versailles. Coll. Science Update. pp. 31–61.

Franceschi, P. (1980) Essai de caractérisation génétique du porc corse. Aspects cytogénétiques et polymorphismes biochimiques. Unpublished PhD Thesis, Université de Paris IV.

Franceschi, P. (1984) Quelques caractéristiques de la population de sangliers de Corse et analyses de ses échanges avec le porc domestique. *Bulletin Mensuel Office National de la Chasse* 85, 25–35.

Gruson, C. (1987) In *Vers une Éthique politique*. Editions de la Maison des Sciences de l'Homme, Paris.

Haudricourt, A.G. (1962) Domestication des animaux, culture des plantes et traitement d'autrui. *L'Homme*, T. I, 1.

Haugerud, A. and Collinson, M. (1990) Plants, genes and people: improving the relevance of plant breeding in Africa. *Experimental Agriculture* 26, 341–362.

Hubert, B. (1994) Pastoralisme et territoire. Modelisation des pratiques d'utilisation. *Agricultures* 3(1), 9–22.

Jacques, D. (1985) La défense de la race Montbéliarde: une action collective pour la défense d'une identité. Unpublished PhD Thesis, Université de Paris X Nanterre.

Kimura, M. (1990) *Théorie Neutraliste de l'Évolution*. Nouvelle bibliothèque scientifique, Flammarion, Paris.

Landais, E., Deffontaines, J.-P., and Benoît, M. (1989) Les pratiques des agriculteurs. Points de vue sur un nouveau courant de la recherche agronomique. *Etudes Rurales* 109, 125-128.

Lush, J.L. (1935) *Animal Breeding Plans*. Iowa State University, Ames, Iowa.

Osty, P.L.O. (1978) L'exploitation vue comme un système. Diffusion de l'innovation et contribution au développement. *Bulletin Technique d'Information* 326, 43-49.

Perrot, C. (1991) Un système d'information construit à dire d'experts pour le conseil technico-économique aux éleveurs. Unpublished PhD Thesis, Institut National Agronomique Paris-Grignon, Paris.

Rouquette, J.-L. (1994) L'Aubrac: un pays, des hommes, des produits, une race. Chronique du développement agricole et rural de l'Aubrac. *Etudes et Recherches sur les Systemes Agraires et le Développement* 28, 195-204.

Russell, N. (1986) *Like Engend'ring Like: Heredity and Animal Breeding in Early Modern England*. Cambridge University Press, Cambridge.

Sebillotte, M. and Soler, L.G. (1990) Les processus de décision des agriculteurs. In: Brossier, J., Vissac, B. and Le Moigne, J.-L. (eds) *Modélisation Systémique et Système Agraire*, INRA, Versailles.

Teissier, J.H. (1978) Relations entre techniques et pratiques. Document Institut National de Recherche et d'Application Pédagogiques (INRAP), Dijon, 38, 1-13.

Vissac, B. (1993) Société, race animale et territoire. Entre les théories et l'histoire: réflexions sur une crise. *Natures, Sciences, Sociétés* 1(4), 282-297.

Vissac, B. (1994) Race animale et qualité des systèmes agraires. La race indicateur de pilotage de la qualité. *Etudes et Recherches sur les Systèmes Agraires et le Développement* 28, 241-248.

Vissac, B. (1995) Les sélectionneurs et le masque de la biodiversité. *Natures, Sciences, Sociétés* (in press).

Willham, R.L. (1986) From husbandry to science: a highly significant facet of our livestock heritage. *Journal of Animal Science* 62, 1742-1758.

34. Biodiversity, Science and Development: Towards a New Partnership in Aquaculture and Fisheries

E.A. Huisman

Wageningen Institute of Animal Sciences (WIAS), Department of Fish Culture and Fisheries, Wageningen Agricultural University, PO Box 338, 6700 AH Wageningen, The Netherlands

Grandfather Great Spirit

Fill us with the light.
Give us the strength to understand,
and the eyes to see.
Teach us to walk the soft Earth
as relatives to all that live

A Sioux Prayer

This chapter will discuss issues of biodiversity in aquaculture and fisheries. It should be kept in mind that the leading theme of the Forum which this book is based on was to seek out ways to promote new partnerships between science and development with a focus on biodiversity. If evolution is development, and I think it is, then biodiversity is in itself the key to development, especially to sustainable development.

As an introduction I would like to make a few – certainly not comprehensive – remarks on:

1. Aquaculture and fisheries development;
2. Threats and opportunities to conserve/enhance biodiversity in relation to aquaculture and fisheries development; and
3. Needs versus demands for a new partnership between aquatic sciences and development.

Some Trends in Aquaculture and Fisheries Development

FAO's last estimate of total world catch and culture of fish and shellfish amounts to some 97 million tonnes in 1992, which is equal to or slightly less than the figures for the 5 preceding years (FAO, 1993). There is ample evidence that over the years yields of capture fisheries stabilized (*The Economist*, 1994), whereas world aquaculture production showed a continuous increase of some 9% per annum over the last decade (Huisman *et al.*, 1993).

Most of this fish produce is disposed for human consumption. Although human consumption patterns vary considerably from region to region, in developing countries the share of fish protein intake relative to the total animal protein intake is often considerably higher than in developed countries.

Fish can, therefore, be regarded as 'poor man's animal protein'. But in 30 years from now, it is anticipated that there will be 3 billion more mouths to feed, of which 90% are expected to live in developing countries (Pinstrup-Andersen, 1994). From this, the question arises (respectively the challenge poses itself) whether fish will stay within reach of the poor man. This is a twofold challenge, e.g. taking present consumption level as a reference there should be some 50% more fish by then, and, this fish should be available at an affordable prize in the less endowed regions of the world. And here one may see a tension between such future needs of 'more and cheap' and present reality. In capture fisheries the vicious circle of increased fishing pressure and economical and biological over-fishing seems to be maintained and may lead to what is known as 'the tragedy of the commons', whereas in aquaculture the monetary value of the produce still indicates rather some emphasis on the culture of relatively expensive species.

Threats and Opportunities to Conserve/Enhance Biodiversity in Relation to Aquaculture and Fisheries Development

The 1994 calendar of the International Plant Genetic Resources Institute (IPGRI), an institute of the Consultative Group on International Agricultural Research (CGIAR) states on it's January page: 'Biodiversity is the total variability within all living organisms and the ecological complexes they inhabit'. It has four distinct but interlinked components – ecosystem, population, species and genetic diversity.'

Therefore, in addressing the issue of biodiversity in relation to aquaculture and fisheries, we should realize that:

1. the total number of fish species is estimated to be some 28,000 (including subspecies and strains the number becomes even more staggering); and
2. fish species inhabit, in complex interrelationships with their biotic and abiotic environment, an enormous variety of aquatic habitats (deep sea and mountain water bodies; (ant)arctic, moderate and tropical climatological ones; fresh, brackish and marine habitats, etc., and temporarily even terrestrial ones (walking catfish, climbing perch)).

So, total aquatic biodiversity on the genetic, species and ecosystem level is tremendously large, and for the greater part not (well)known (Greenwood, 1992), which in itself of course severely complicates determination of the extent of changes in aquatic biodiversity and their consequences.

With regard to aquaculture, it is my conviction that aquaculture sciences can contribute enormously to the conservation and enhancement of biodiversity, for instance vis à vis *ex situ* conservation, especially in the disciplinary fields of reproductive physiology and genetics, and also – more recently emerging – in the field of environmental stress research.

However, aquaculture development can also pose threats to biodiversity, as for instance:

1. aquaculture encroaching into the mangrove ecosystem (Bailey, 1988; Hamilton *et al.*, 1989); and
2. intensive aquaculture practices through their intensive use of environmentally produced goods (such as raw production materials as fish meal) and services (such as fish farm effluents being 'discharged' to the self-purification capacity of surface waters).

Folke and Kautsky (1992), advocating integrated aquaculture using principles of ecological engineering, state that aquaculture development 'is not only limited by what is happening in the market . . ., but also by an increasing demand for environmentally produced goods and services sustained by intricate ecological connections, which are more easily disrupted as the scale of aquaculture grows relatively to its supporting ecosystem'. From this, it also follows that decreased biodiversity, when leading to malfunctioning of the ecosystem will affect aquaculture directly or indirectly.

Another aspect I want to discuss briefly, is the possible negative effect of farm-produced fish stocked into open waters on the populations of their wild conspecifics (and others) through eventual genetic erosion and changes in composition of fish assemblages (Ryman and Laikre, 1991; Hilborn and Winton, 1993).

The first remark I want to make about fisheries is, that, although over the past some 30 years the rate of extinction of fish is estimated at one per year (which is similar for birds), there is no proof that fishing itself has led to extinction (Beverton, 1992). However, there is no doubt that present fishing technology and pressure have a pronounced effect on the structure of fish populations (collapse/recovery of stocks), thereby affecting aquatic biodiversity at the system level over time.

Other anthropogenic influences, which pose threats to biodiversity and ecosystem performance, thereby influencing the fishery, are related to different forms of water pollution, either acting directly on the fish (Alabaster and Lloyd, 1982) or indirectly by changing the habitat, as for instance changes of the aquatic vegetation (de Nie, 1987). Also water flow and water level control measures – which mostly occur in fresh water systems – can have a considerable impact on the availability and suitability of aquatic habitats (Alabaster, 1985). At this point, it must be emphasized that loss of habitat is regarded as the major cause for extinction.

Welcomme (1988) reviewed more than 1300 introductions, of which somewhat more than 300 were 'successful' ones. However, almost one third of these 'had sufficient impact on the local environment or fish stocks to cause serious concern'. It is clear that introductions, where successful, may affect biodiversity at all levels (cf. under aquaculture) either negatively or positively. It is promising that this topic attracts increasingly more attention (see for instance the contributions to the International Symposium on Biological Diversity in African Fresh and Brackish Water Fishes, Dakar, 15-20 November 1993), because introductions will continue to take place which urges the need to increase our knowledge base in this particular field.

Over the recent years global climatic change has become an issue of worldwide concern. This topic is discussed in later chapters, but it is worthwhile stressing that the effect of temperature on fish – as cold-blooded organisms – is profound and manyfold (on reproduction, growth, survival, susceptibility to diseases and toxicants, etc.). Therefore, climatic change may have a tremendous effect on especially the spatial dimension of biodiversity in aquatic ecosystems. Based on fish-specific physiological traits the impact is expected to be relatively more intense in the colder regions.

Apart from the above mentioned, and other, threats, we must realize and appreciate that fisheries science, and especially the underlying basic sciences such as autoecology and synecology, can contribute a wealth of knowledge about aquatic life which will prove instrumental, especially in *in situ* conservation of biodiversity, be it that there is a lot more to be known about the majority of the fish faunas.

Needs Versus Demands for a New Partnership

This Forum aims to demonstrate the important role of biodiversity for life as a whole, including all production sectors. However, when the question 'why do we need to conserve biodiversity?' is posed, the answer rather often has some degree of uncertainty: 'no prediction can be made of consequences of destroying so much of global life'; 'we may deprive ourselves and future generations of essential commodities (food, medicine, etc.)'. Moreover, the relation diversity–stability is to some extent argued on the basis of theoretical explorations. In this respect there are two recent publications that should be noted, namely Tilman and Downing (1994) and Naeem *et al.* (1994).

Tilman and Downing (1994) using long-term field studies on grass lands demonstrated 'that primary production in more diverse plant communities was more resistant to and recovered more fully from a major drought'. Naeem *et al.* (1994) published results from the ECOTRON at the NERC Centre for Population Biology (Silwood Park, Berkshire, UK). They used a series of balanced microcosms with decreasing numbers of plant and animal species representing the basic ecosystem functions of production, consumption and decomposition. In this elegant laboratory set-up they were able to experimentally provide evidence that 'reduced biodiversity may indeed alter the performance of ecosystems'.

I do hope that the aquatic ecosystem can also become an object of similar field and laboratory studies, to complement theoretical studies on the relation between biodiversity and ecosystem performance.

A partnership has its own costs, not only the combined costs originating from each individual partner's activities, but cooperation in partnership has its own transaction costs as well. However, the combined product can be of better quality and more fashioned to more commonly perceived needs.

It is my conviction that we will clearly demonstrate the tremendous needs for a strong positive partnership between science and development to combat the threats to and to foster opportunities for conservation and improving biodiversity.

Still, we live in an economy-based world, and there is a saying 'economy does not satisfy needs, it only can satisfy demands'. So, is there a demand? Here, I am inclined to say that fish and other aquatic life are in a somewhat disadvantageous position compared to, for instance, mammals and birds. One knows that fish exist, somewhere under the water surface, but one does not see them too often especially not too often suffering. And there is a danger that 'what exists but is not seen' is 'taken for granted' (in this case as an unlimited eternal renewable resource).

Indeed, the aquatic environment is an enormously rich resource offering an 'endless' – though for a greater part not known (yet) – array of goods and services to present and future humankind, but not limitless and not when it is 'taken for granted'. Making that message clear to the world audience can lead to a real actual demand for increasingly fostering a strong growing partnership between so much needed aquatic science and sustainable aquatic resources development of which biodiversity is a cornerstone.

In relating sustainability and biodiversity to each other, I may quote the often-cited phrase of the Brundtland report: 'sustainable development is a development that meets the need of the present generation without endangering the possibilities for future generations to also meet their needs'. I want to stress that this concept of sustainability also implies 'solidarity', on a global scale. Solidarity not only within the generation presently inhabiting this globe, but also solidarity between present and future generations. In this sense sustainability asks for 'equity'; not only 'equity' with respect to access to food and other commodities, but also 'equity' with respect to access to opportunities and possibilities.

If we are not succeeding in coming to grips with this 'solidarity and equity', we will continue to lose more biodiversity than strictly necessary.

References

Alabaster, J.S. (1985) *Habitat Modification and Freshwater Fisheries* (Proceedings of a Symposium of the European Inland Fisheries Advisory Commission). Butterworths, London.

Alabaster, J.S. and Lloyd, R. (1982) *Water Quality Criteria for Freshwater Fish.* Butterworths, London.

Bailey, C. (1988) The social consequences of tropical shrimp mariculture development. *Ocean and Shoreline Management* 11, 31–44.

Beverton, R.J.H. (1992) Fish resources: threats and protection. Proceedings of the 7th International Ichthyology Congress. *Netherlands Journal of Zoology* 42, 139–175.

de Nie, H. (1987) *The Decreace in Aquatic Vegetation in Europe and its Consequences for Fish Populations*. EIFAC/FAO Occasional Paper No. 19.

Economist, The (1994) The tragedy of the oceans. 19-23 March 1994, pp. 15, 16, 23, 24 and 28.

FAO (1993) *The State of Food and Agriculture 1993*. FAO Agriculture Series No. 26, Rome.

Folke, C. and Kautsky, N. (1992) Aquaculture with its environment: prospects for sustainability. *Ocean and Coastal Management* 17, 5-24.

Greenwood, P.H. (1992) Are the major fish faunas well known? Proceedings of the 7th International Ichtyology Congress. *Netherlands Journal of Zoology* 42, 131-138.

Hamilton, L.S., Dixon, J.A. and Miller, G.O. (1989) Mangrove forests: an under-evaluated resource of the land and of the sea. In: Mann Borgese, E., Ginsburg, N. and Morgan, J.R. (eds) *Ocean Yearbook 8*. University of Chicago Press, Chicago, pp. 254-288.

Hilborn, R. and Winton, J. (1993) Learning to enhance salmon production: lessons from the Salmon Enhancement Program. *Canadian Journal of Fisheries and Aquatic Sciences* 50, 2043-2056.

Huisman, E.A., Born, A.F. and Verdegem, M.C.J. (1993) Tropical aquaculture: its constraints, opportunities and development. In: Barnabé G. and Kestemont P. (eds) *Production, Environment and Quality*. Bordeaux aquaculture '92. European Aquaculture Society, Ghent, Belgium. Special Publication 18, pp. 385-406.

Naeem, S., Thompson, L.J., Lawler, S.P., Lawton, J.H. and Woodfin, M.R. (1994) Declining biodiversity can alter performance of ecosystems. *Nature* 368, 734-737.

Pinstrup-Andersen, P. (1994) *World Food Trends and Future Food Security*. Food policy report of International Food Policy Research Institute (IFPRI). Washington, DC.

Ryman, N. and Laikre, L. (1991) Effects of supportive breeding on the genetically effective population size. *Conservation Biology* 5, 325-329.

Tilman, D. and Downing, J.A. (1994) Biodiversity and stability in grasslands. *Nature* 367, 363-365.

Welcomme, R.L. (1988) *International Introductions of Inland Aquatic Species*. FAO Fisheries Technical Paper 294, Rome.

Biodiversity and Aquaculture 35

R.S.V. Pullin

International Center for Living Aquatic Resources Management (ICLARM), MC PO Box 2631, 0718 Makati, Metro Manila, Philippines

Aquaculture – Status and Definitions

Most commentaries on aquaculture refer to its ancient origins from China and Egypt and a variety of fish impoundment systems scattered across Africa, Asia, Europe, and the Pacific, and describe a wide range of operations (e.g. Bardach *et al.*, 1972). Global aquaculture production is currently growing at about 6–7% per year; faster than any other food production sector. However, aquaculture remains vastly underdeveloped and its development has been very uneven (Born *et al.*, 1994). Only about 13% of all aquatic produce comes from aquaculture, with Asia producing about 80% by weight and 75% by value.

Aquaculture is defined here as the farming of aquatic organisms for food and for other purposes, including hatchery and nursery operations for farms and for fisheries enhancement, and without any qualifications with respect to stock ownership. Defining aquaculture more narrowly, for purposes of separating the statistics for fish raised on farms from those harvested from capture fisheries, has been discussed by New and Crispoldi-Hotta (1992). The terms biodiversity (= biological diversity), domesticated, *ex situ* and *in situ* are defined here as in the Convention on Biological Diversity.

The main groups of aquatic organisms discussed here are those that are actually farmed or propagated: the marine macroalgae, molluscs, crustaceans, and finfish. Mass culture of microalgae and the farming of freshwater macrophytes, which include some of the rices, are not considered to be within the mainstream of aquaculture and require separate treatment. The food potential of aquatic macrophytes is, however, very great and remains largely unrecognized outside some Asian countries (Edwards, 1980).

In addition to the biodiversity of the farmed aquatic organisms, aquaculture depends heavily on a wide diversity of aquatic organisms for food and for maintenance of water quality: for example, the filter-feeding of bivalve molluscs and the grazing of finfish, such as some carps and tilapias, on microalgae and detrital aggregates in fertilized fishponds. This applies especially to aquaculture that is non-intensive in terms of externally derived nutrient and energy inputs, that uses net primary production effectively and therefore is usually more environmentally benign than intensive operations (Pullin, 1989). For example, microalgae, cyanobacteria, and other microbial populations in fertilized fishponds act as biological aerators and metabolite removers (especially for ammonia) in addition to their roles as foods for microphagous fish and in driving the foodweb (Moriarty and Pullin, 1985). Therefore, microbial and planktonic biodiversity is very important in many aquaculture operations.

Biodiversity in Aquaculture

Much of the world's aquatic flora and fauna have yet to be evaluated for aquaculture potential. Only a small minority of the aquatic species used by humans are presently farmed (Table 35.1), although accurate estimations of this are difficult because official statistics are sometimes aggregated into higher taxa or commodity groups. Moreover, the advent of new databases, that draw upon multiple sources, is revealing that more species are farmed than was previously recorded. For example, Pullin (1995) estimated from FAO and Taiwanese official statistics that 95% of farmed aquatic produce comprised the following numbers of species: seaweeds, 6; molluscs, 43; crustaceans, 27; and finfish, 105. The ICLARM-FAO database, Fishbase (Pauly and Froese, 1991; Froese and Pauly, 1994a), which draws upon FAO and other data sources, now has entries for significantly larger numbers of finfish species that are farmed for food and for various purposes. Fishbase, currently covers 12,000 (50%) of extant finfish species and concentrates on fish that are, in one way or another, important to humans.

In addition to the farmed aquatic organisms listed in official statistics, there are probably thousands more species that are used for human, animal and fish foods, for drugs and other chemicals, and for ecological services such as integrated pest management. Because such use is local and in small quantities, it goes largely unrecorded. For example, Trono and Ganzon-Fortes (1988) listed around ten species of macroalgae used in commercial quantities in the Philippines, of which three (*Caulerpa lentillifera*, *Eucheuma denticulatum* and *E. alvarezii*) are mainly farmed rather than gathered. Farming supplements gathering for another four species (mainly *Enteromorpha* spp.). These authors also listed over 200 spp. as being of economic importance but in limited use in the Philippines for food, medicine, etc.

Hence, official statistics may mask a wider use of aquatic biodiversity in aquaculture and it is probable that many more aquatic species could be farmed, on a larger scale, if proven technically feasible and economically viable. This would mean that aquaculture would use far more species than

Table 35.1. Use of finfish by humans: the totals for extant fishes are from Nelson (1994), the information on use is from the ICLARM-FAO FishBase.

	Species		Families		Freshwater[a] species	
	No.	%	No.	%	No.	%
In capture fisheries[b]	2575	10.5	243	50.4	339	3.4
In aquaculture[c]	178	0.7	50	10.4	91	0.9
As ornamental fish	1965	8.0	134	27.8	1228[d]	12.3
As sport fish	779	3.2	135	28.0	188	1.9
As bait[e]	134	0.5	41	8.5	20	0.2
Totals used by humans[f]	4638	18.8	295	61.2	1606	16.1
Totals for extant fish[f]	24,618		482		9966	

[a] Fish that do not enter brackish or marine waters.
[b] Industrial and artisanal fisheries, probably an underestimate.
[c] For food and for fisheries enhancement.
[d] Most (707) are bred in captivity.
[e] Probably an underestimate.
[f] The sums of the numbers and percentages differ from the totals because many species have more than one use. The totals count such species only once.

those exploited in agriculture: about 95% of livestock production derives from five species and for cultivated plants the figure is about 100 (Prescott-Allen and Prescott-Allen, 1990). Intraspecific biodiversity, well-studied in agriculture and documented as diverse breeds, strains, and cultivars, etc., has hardly been investigated at all in aquaculture, apart from that for common carp (*Cyprinus carpio*), some salmonids and catfish, and most recently tilapias (e.g. Eknath et al., 1993).

Domestication of most farmed aquatic organisms has not progressed far. Captive breeding of aquatic organisms, as for other organisms, involves genetic change, through natural selection to the farm environment in addition to any artificial selection or genetic management, and the natural selection pressures can be especially high on the early life history stages of very fecund aquatic organisms. However, the captive breeding and husbandry of most aquatic organisms have no histories comparable to those of crops and livestock.

There are a few exceptions; for example, the common carp has been bred in captivity for centuries and some salmonids for over a century. Apart from such exceptions, the breeding histories of most farmed aquatic organisms span only a matter of decades. Captive breeding of the Chinese and Indian major carps, species that now account for over half of the world's farmed freshwater finfish production, has been possible only since the 1960s, when induced spawning technology through injection of pituitary hormones became available. Previously, these huge aquaculture operations relied upon catching wild fry, as many milkfish (*Chanos chanos*) and mullet (Mugilidae) farmers still do.

Therefore, most farmed aquatic organisms are close to wildtypes and, with few exceptions, there is no comparable diversity of farmer-developed (domesticated) breeds in aquaculture, such as has been well-documented in agriculture. However, aquaculture shares with agriculture the needs to conserve not only the wild relatives of farmed species but also any breeds that are developed by farmers, and to define and protect farmers' rights. The importance of 'nurturing diversity' and of protecting farmers' rights as well as the intellectual property of corporations, have been thoroughly reviewed for plants (e.g. Crucible Group, 1994) but not yet for aquatic organisms.

There are now increasing efforts to breed and to improve the performance of farmed aquatic organisms (and hence farmers' profits) by applied genetics. A wide array of technologies for fish breeders, including gene transfer, is emerging before the classical conventional routes to genetic improvement (selective breeding and hybridization) have been applied to more than a few species.

Thorough documentation and evaluation of aquatic biodiversity have only just begun: for example, the ICLARM-FAO Fishbase project (Pauly and Froese, 1991; Agustin *et al.*, 1993; Froese and Pauly, 1994a); a proposal for a network approach to aquatic biodiversity databases (Froese and Pauly, 1994b); REEFBASE, a global database of coral reefs and their resources (ICLARM, 1994); and the initiation of scientifically based fish breeding programmes in developing countries (e.g. for tilapias (BFAR, 1993), based on the genetic resources collected, evaluated and selectively bred by the Genetic Improvement of Farmed Tilapias (GIFT) project (Eknath *et al.*, 1993), which followed the approach pioneered for Norwegian salmonids (Gjedrem, 1985)). A new International Network on Genetics in Aquaculture (INGA) (Seshu *et al.*, 1994) has to date seven Asian (Bangladesh, China, India, Indonesia, the Philippines, Thailand, and Vietnam) and four African members (Côte d'Ivoire, Egypt, Ghana, and Malawi).

Awareness of the importance of biodiversity for aquaculture is also growing rapidly. FAO and ICLARM have held recent consultations on the sustainable utilization and conservation of aquatic genetic resources (ICLARM, 1992; FAO, 1993) and there have been many recent reviews of biodiversity issues of relevance to aquaculture and fisheries (e.g., Ray, 1991; Smith *et al.*, 1991; Cataudella and Crosetti, 1993; Pullin, 1995).

Loss of Aquatic Biodiversity in Relation to Aquaculture

Aquatic biodiversity is being lost at an alarming rate. Moyle and Leidy (1992) suggest that at least 20% of freshwater fish species 'are already extinct or in serious decline' for reasons that can be summarized as environmental degradation and mismanagement of natural resources.

This loss will constrain efforts to evaluate aquatic organisms for their aquaculture potential and will limit the success of future breeding programmes. Just as plant breeders find it increasingly difficult to locate and to conserve for breeding purposes the remaining undisturbed populations of the wild relatives of crops, it is becoming difficult for would-be fish breeders to locate and to collect genetic material from undisturbed wild populations. A workshop to review the genetic resources of tilapia for aquaculture (Pullin, 1988), found alarmingly few African sites likely to have undisturbed populations of the major species of current importance in aquaculture. With the emergence of tilapia as a globally traded fish commodity (Pullin et al., 1994) and a consequent interest in farming more tilapia species and hybrids for aquaculture, threats to the remaining wild populations are serious. Some of these have a very limited natural distribution. For example, wild stocks of *Oreochromis urolepis hornorum* (the only tilapia subspecies that yields consistently 100% male progeny in certain hybrid crosses, without hormonal or genetic manipulation) occur only in the Wami river, Tanzania and possibly Zanzibar. Monosex male tilapia farming is important because males grow faster than females and breeding in production ponds is avoided.

Aquaculture can itself be a significant contributor to the loss of aquatic biodiversity. Like agriculture and forestry, and possibly even more so because of the connective nature of aquatic systems, it can have large impacts on adjacent habitats: gross environmental effects, through water abstraction, effluents, spreading diseases, and clearance or fragmentation of habitats (e.g. mangroves). Conversely, fishponds can be environmental assets that help the conservation of terrestrial and aquatic biodiversity by assisting soil conservation and water resources management (Pullin and Prein, 1995).

Fish frequently escape from aquaculture installations and releases of hatchery-reared juveniles are the basis of enhanced fisheries. When captive-bred fish mix with wild stocks and disperse through natural habitats, the possible environmental consequences can be categorized as follows:

1. Depletion or loss of wild fish stock, for example, by predation, competition for food or breeding sites, or the spread of parasites and pathogens;
2. Changes in natural aquatic habitats, for example, clearance of aquatic vegetation by herbivorous fish or increased turbidity by bottom-foraging fish;
3. Genetic change in wild fish stocks by their interbreeding with hatchery-reared fish of the same or similar species, thereby causing loss of natural biodiversity and sometimes establishing feral hybrids.

These and other environmental impacts of aquaculture have been reviewed by various authors in Pullin et al. (1993).

Risks of adverse environmental impact are generally higher with exotic species than with indigenous species. This has prompted much recent debate on the limited use of existing advisory codes of practice for introductions and transfers of aquatic organisms (Turner, 1988; Coates, 1992) and on the need for research and development organizations and professional fisheries societies to formulate and to implement policies that will minimize environmental risk, for example, ICLARM has prepared a position statement on the use of exotic species and genetically modified organisms (Pullin, 1994). It is probable that new legally binding instruments, within a general protocol towards improved biosafety, will be prepared under the Convention on Biological Diversity. At present, particularly in the developing countries where most aquaculture is practised, there is almost a *laissez-faire* situation with respect to introductions and stock transfers, especially by the private sector, and effective quarantine measures for aquatic organisms are generally lacking.

The genetic effects of escapees from aquaculture have been much debated, especially for salmonids (e.g. Gausen and Moen, 1991). Hindar *et al.* (1991) have provided a broad overview, though still largely dependent on salmonid examples, and show that genetic effects may progress to high levels of introgression and displacement. Several major groups of farmed finfish (e.g. cichlids, cyprinids and salmonids) contain species that hybridize readily and natural introgressive hybridization is widespread. Verspoor and Hammar (1991) cite 66 such cases.

Do such genetic effects, brought about entirely or accelerated by aquaculture development, constitute a serious loss of aquatic biodiversity? Most probably yes. When escapees or stocked fish not only hybridize freely but also produce fertile hybrids, the tendency is towards evolution of a single hybrid population (e.g. Bartley and Gall, 1991).

The legacies of escapes from aquaculture can sometimes be permanent and can affect farming operations as well as wild populations and habitats. For example, feral populations of *Oreochromis mossambicus*, formerly spread throughout the tropics and subtropics for aquaculture and now largely abandoned in favour of farming *O. niloticus*, have become established in open waters, have gained access to *O. niloticus* farms, and have hybridized with farmed stocks (Macaranas *et al.*, 1986). Another example is the golden snail (*Pomacea* sp.) which after introduction for farming now infests over 400,000 ha of ricefields in the Philippines (Acosta and Pullin, 1991). The eradication of such species would be difficult and costly, if not impossible. A decision was taken in 1991 to eradicate an exotic farmed fish (the common carp) from Malawi (Msiska and Costa-Pierce, 1993). The results are still awaited.

There have been fierce debates about the effects of hatchery programmes and of artificial propagation in general; for example, Pacific salmon (*Oncorhynchus* spp.). Hilborn (1992) argued that the hatchery programmes for these species have largely failed and advises against future attempts to add hatchery-reared fish to healthy wild stocks. There has been, however, a serious loss of genetic diversity in these salmonid stocks for many reasons (Nehlsen *et al.*, 1991); so what does 'healthy' really mean? compared to near-virgin wild stocks? (are there any?) or to the *status quo* for habitats, assuming better sustainable management in future? Martin *et al.* (1992) discussed this and emphasized the needs for the 'management and restoration of wild stocks ... within the productive potential of today's habitats, using existing genetic material and within the context of larger public resource management programs', and for public support of and belief in the scientists who devise such interventions. Stickney (1994), although emphasizing the genetic impacts that hatchery fish can have on wild populations, also pointed out that hatchery fish have sometimes been poorly bred to compete in the wild. Thus, there is a real dilemma here: breed to outcompete wild stocks and thus have a successful fishery, but with a narrow genetic base; or breed to let wild fish coexist or remain dominant?

It is therefore important that hatchery programmes, like all captive supportive breeding for enhancing wild populations, are undertaken in full awareness of their likely genetic effects. This has been rare for captive-bred aquatic organisms. Naevdal (1994) highlighted this and proposed genetic tagging and selective breeding for species used in sea ranching. Cowx (1994) discussed fish stocking strategies and their potential genetic, ecological and environmental impacts and recommended 'maintaining the genetic integrity of the indigenous stocks . . . stocking . . . fish derived from local populations, or failing that habitats which are environmentally similar, or fish that have not been held in captivity for more than one generation. Ryman and Laikre (1991) urged caution in captive breeding programmes to support wild populations because of the possibility of reducing the contributions from wild fish.

Similar caution is needed for all captive breeding in aquaculture (Tave, 1986; Padhi and Mandal, 1994). The most important provisions are to maintain adequate effective breeding numbers (N_e) to minimize inbreeding and to plan for long-term conservation of diversity. Eknath and Doyle (1990) found inbreeding between 2 and 17% in Indian carp hatcheries. Smitherman and Tave (1988) recommend an N_e of 390–500 per generation for reference populations of tilapias. Many hatcheries maintain too few broodstock and there is widespread ignorance of the dangers of inbreeding and genetic drift in aquaculture.

Genetically Engineered Aquatic Organisms

The Convention on Biological Diversity defines biotechnology as 'any technological application that uses biological systems, living organisms, or derivatives thereof, to make or modify products or processes for specific use'. All captive-bred organisms are somewhat 'genetically modified', but most authors restrict the use of this term to those organisms upon which actual gene manipulation has been performed – for which Lesser and Maloney (1993) used the term 'genetically engineered organisms (GEOs)'.

Gene manipulation in aquaculture is a fast growing field and a bright future is seen for the use of farmed transgenic aquatic organisms (Maclean and Penman, 1990). The consequent debate has focused mainly on transgenic organisms.

The American Fisheries Society has issued a position statement on transgenic fishes (Kapuscinski and Hallerman, 1990) which can be summarized as urging caution and controls in the use of transgenic fishes, with no introductions for production purposes and no stocking into natural waters 'unless and until a body of research indicates the merits . . . and ensures ecological safety'. This position statement is well-balanced with respect to pressing for biosafety but not constraining the further research that is required to develop transgenic aquatic organisms and to assess their possible impacts through studies in containment facilities.

Other forms of genetic manipulation, such as polyploidy and production of monosex progeny may also become commonplace in aquaculture. For example, there has already been commercialization of technologies for triploidy in molluscs and salmonids and the commercial production of genetically male, also known as 'YY supermale' tilapias is under investigation (e.g. Mair, 1993). The production of triploid fish for stocking has appeal because they would be sterile. They might still, however, outcompete other fish for food and try to occupy breeding sites.

Gene transfer, chromosome manipulation and other techniques for genetic management will probably become increasingly used in aquaculture breeding programmes, either using existing genetic material (whether close to wild-types or already substantially modified) or combined with new selection programmes. The best strategy to minimize the risk of such developments having adverse effects on wild or farmed stocks is to regard their products as exotic organisms and to apply the same biosafety precautions (Pullin, 1995).

Conserving Biodiversity for Aquaculture

Given the wealth of aquatic biodiversity still to be documented and evaluated for aquaculture potential, the emphasis should be on *in situ* conservation. Inland waters and coral reef systems are probably the most threatened habitats, through overexploitation of living resources and through physical and chemical damage. The threats to freshwater species are perhaps the most serious because of their more limited prospects for recolonization. The sites that merit special conservation efforts range in size and complexity from large inland lakes (such as Lake Malawi and its feeder streams, which hold probably about 1000 endemic species) to small lakes and islets that may hold only one endemic species or subspecies currently identifiable as being of interest for aquaculture or enhanced fisheries.

The connectivity of waterbodies poses special problems here (e.g. lakes shared between two countries, downstream effects in rivers, etc.). Conversely, the isolated nature of some waterbodies (for example, inland lakes on islands in an archipelagic nation) may mean that controls at the national level will be insufficient to protect endemic species. Exotic species movements within the Philippines completely changed the fish fauna of Lake Lanao, Mindanao (Villwock, 1972).

Again, documentation of wild genetic resources and of threats to their survival are essential for conservation efforts. The ICLARM-FAO database includes data on the degree of threat to finfish species. One approach to conserving aquatic biodiversity *in situ* is to piggyback this on to conservation efforts for charismatic species, such as large mammals in gameparks (Pullin, 1990) and on to recreational activities in tourist-valued habitats, such as coral reefs and sports fisheries. Economic reasons and greater public awareness of the importance of such conservation, principally through education, will be necessary to protect these and other habitats that have less public appeal (e.g. mangroves and swamps).

Ex situ gene banks are also being considered for farmed aquatic organisms, but have yet not been much developed. The two main approaches are: (i) keeping broodstocks of different species and strains in ponds, tanks or aquaria, and accepting the high cost of this, the genetic modifications that such captive breeding brings, and the need for very careful management to keep stocks separate and to maintain adequate N_e values; and (ii) cryopreservation of gametes or embryos (Rana, 1992; Pullin, 1993).

Broodstock collections of limited numbers of breeds are maintained for some farmed fishes, in support of breeding programmes: for example, at Szarvas, Hungary, for 25 common carp races and some hybrids, and at Muñoz, Nueva Ecija, Philippines for eight reference populations of *Oreochromis niloticus* and an increasing number of selectively bred strains. The Institute of Aquaculture, University of Stirling, Scotland, maintains a wide collection of tilapia species and hybrids in tanks and aquaria. Moreover, many experimental stations maintain one or a few well-documented breeds, particularly of the common carp, as resources for the station's own research and for sharing with others. The Fish Farming Experimental Station, Ilmatsabu, Estonia, maintains common carp strains: e.g. 'Estonian', origin Latvia, 1983 and a closed population since then; 'Ropsha', origin St Petersburg, Russia; and 'German', acquired from Lithuania, 1977 (Gross and Wohlfarth, 1994). Globally, this all adds up to an imperfectly documented and highly decentralized system.

Even for the few species for which breeds are well maintained with adequate records, nobody really knows what others have got. Attempts to capture and to disseminate the information are progressing (e.g. Agustin *et al.*, 1993) but up-to-date data are difficult to acquire. Somehow, the flow of information from decentralized *ex situ* broodstock collections to a centralized database, such as the ICLARM–FAO FishBase, and the dissemination of this to interested parties have to be better organized and funded.

Rana (1992) reviewed the current state-of-the-art for cryopreservation of the gametes and embryos of aquatic organisms. Cryopreservation is currently possible only for the spermatozoa of finfish, yielding a haploid gene bank, but cryopreservation of the eggs and early embryos of some marine invertebrates, including some farmed bivalves (*Mytilus* and *Crassostrea* spp.) is possible. Cryopreservation of finfish spermatozoa is likely to become much more widely used in breeding programmes and in fieldwork, collecting spermatozoa from wild individuals for breeding or conservation purposes. Clearly, centralization is more feasible for gene banks that are based on cryopreservation than for live broodstock collections. However, it is still not easy to envisage centralization on the scale that has been possible for crop genebanks, where literally hundreds of thousands of seed accessions can be centralized at relatively low cost. The most probable scenario is that cryopreservation units will become an integral feature of breeding programmes, to complement live broodstock collections and that international exchange of properly accredited cryopreserved gametes and embryos will develop; rather like the use of frozen semen and embryos in livestock breeding.

Learning from Agriculture

Despite the vastly different histories of aquaculture and agriculture, there is much that aquaculture research and development can learn from that for farmed plants and livestock. In aquaculture at present, as in agriculture, most farmers do not produce their own 'seed'. The hatchery, nursery and growout subsectors of aquaculture are often run by different people at different locations. A classic example is the organization of hatchery and nursery operations in West Java, Indonesia, for which Sukabumi is a 'seed' distribution centre. Here, hatching, rearing the fry, and nursing to various sizes the fingerlings of carps and other species are in the hands of different farmers and groups and are intimately connected with different farming systems, including integrated rice–fish systems.

Could more farmers raise their own aquatic 'seed', thereby achieving a widening diversity of farm breeds, as envisaged by Doyle et al. (1991)? or will the same forces that have shaped seed supply in agriculture inevitably apply as aquaculture develops? The history of crop and livestock breeding, particularly the latter, teaches that farmers will search for the most profitable breeds available (from anywhere), will swap germplasm among themselves and will probably require subsidies for, rather than make profits from, conservation of traditional breeds (e.g. of sheep, cattle), unless such breeds have market advantages. In aquaculture at present, with so few well-defined breeds available, farmers are making choices mainly at the species level, including exotics.

It seems unlikely that breeding programmes in aquaculture, even given its differences from agriculture (e.g. at present a relative abundance of wild relatives, very few gene banks, etc.) will evolve in substantially different ways from those in agriculture. The scope for genetic gain in aquaculture, through selective breeding and genetic manipulation, is great: at least for macroalgae, molluscs, and finfish; perhaps less so for crustaceans, on present evidence. Therefore, breeding programmes in aquaculture are likely to draw increasingly upon the experience and strategies of animal and plant breeders (e.g., see Schultz, 1986; Cunningham, 1990). Where non-protectable, non-patentable fish breeds are developed and supplied to provincial multiplier stations and thence to individual farmers, as in the Philippine National Tilapia Breeding Program (BFAR, 1993), seed supply in aquaculture might remain somewhat decentralized. However, when patentable material, such as GEOs, are adopted, aquatic seed supply will probably become concentrated among very few hands.

One lesson that aquaculture research and development can learn from agriculture is that commercial pressures can accelerate biodiversity loss. Thus, it is vital for the long-term future of aquatic breeding programmes that genetic diversity in nature and on-farms is treasured and conserved. Aquaculture has a great opportunity to avoid some of the biodiversity losses that have been associated with agricultural development if this is done. One danger here is that in applying genetic engineering to aquatic organisms, developing gene libraries etc., researchers and their donors and entrepreneurs may neglect the importance of documenting, evaluating and conserving wildtypes and farm breeds. Success in this conservation task, linked to efforts to increase the productivity and profitability of aquaculture, will require substantial investment and international collaboration, both for *in situ* and *ex situ* programmes. Aquatic habitats are as threatened as tropical rainforests.

The Consultative Group on International Agriculture Research (CGIAR) is now beginning a system-wide programme of research on genetic resources, involving all its various crop, livestock, natural resources centres (fisheries and forestry), rather than the separate programmes for rice, maize, livestock, etc., that the centres formerly undertook. This will help the CGIAR and its many collaborators in the UN-system, developing and developed country institutions, and non-governmental organizations (NGOs), to combine knowledge and experience from agriculture, forestry, aquaculture and fisheries and to formulate policies that will help to conserve aquatic biodiversity. For example, the categorization of genetic material proposed by Goodman (1990) for plants ('uncollected, unacquired, unadapted, unevaluated') could be adapted for aquatic organisms as well.

The key requirement for all of these endeavours is accurate and widely accessible information. In this context, an FAO (1993) consultation stated:

> The Consultation *recognized* that to conserve genetic resources requires documentation and evaluation, and therefore, *recommended* collaborative efforts between developing and developed countries to determine conservation values. Developing countries are the sites for collection, evaluation of material and breeding programmes. Developed-country institutions can assist by analyzing and interpreting data from tissue samples, computing, brokering information and assisting with certification and banking of material. The Consultation *noted* that well-equipped international laboratories in developing countries can also contribute – as do the CGIAR centres for crops and livestock. In this regard, the Consultation *recommended* that FAO should strive to facilitate such collaboration among developing and developed countries.
>
> The Consultation *recommended* that species databases should contain clear information on breeds, where they can be obtained, and all other genetic information available. The Consultation further *recommended* that FAO and other international agency projects should foster resources and activities which make genetic information on relevant species available at the national level.

It is to be hoped that the resources will be found to support these activities and the further evaluation and sustainable use of aquatic biodiversity in aquaculture.

Acknowledgements

The author acknowledges the help and support of Dr Rainer Froese and the FishBase project team at ICLARM HQ in providing data and many helpful suggestions for this paper.

Paper prepared for the XXVth General Assembly of the International Union of Biological Sciences and the International Forum on Biodiversity, Science and Development, 5-9 September 1994, UNESCO Headquarters, Paris. *ICLARM Contribution No. 1090.*

References

Acosta, B.O. and Pullin, R.S.V. (eds) (1991) Environmental impact of the golden snail (*Pomacea* sp.) on rice farming systems in the Philippines. *ICLARM Conference Proceedings 28.* 34 pp.

Agustin, L.Q., Froese, R., Eknath, A.E. and Pullin, R.S.V. (1993) Documentation of genetic resources for aquaculture – the role of FishBase. In: Penman, D., Roongratri, N. and McAndrew, B. (eds) *International Workshop on Genetics in Aquaculture and Fisheries Management*, ASEAN-EEC Aquaculture Development and Coordination Programme, Bangkok, Thailand, pp. 63-68.

Bardach, J.E., Ryther, J.H. and McLarney, W.O. (1972) *Aquaculture. The Farming and Husbandry of Freshwater and Marine Organisms.* Wiley-InterScience, New York.

Bartley, D.M. and Gall, G.A.E. (1991) Genetic identification of native cut-throat trout (*Oncorhynchus clarki*) and introgressive hybridization with introduced rainbow trout (*O. mykiss*) in streams associated with the Alford Basin, Oregon and Nevada. *Copeia*, 854-859.

Born, A.F., Verdegem, M.C.J. and Huisman, E.A. (1994) Macro-economic factors influencing world aquaculture production. *Aquaculture and Fisheries Management* 25, 519-536.

BFAR (1993) *Framework for Philippine National Tilapia Breeding Program.* Draft Working Document. Philippine Bureau of Fisheries and Aquatic Resources, Manila, Philippines.

Cataudella, S. and Crosetti, D. (1993) Aquaculture and conservation of genetic diversity. In: Pullin, R.S.V., Rosenthal, H. and Maclean, J.L. (eds) *Environment and Aquaculture in Developing Countries.* ICLARM Conference Proceedings 31, Manila, Philippines, pp. 60-73.

Coates, D. (1992) *Implementation of the EIFAC/ICES Code of Practice: Experience with the Evaluation of International Fish Transfers into the Sepik River Basin, Papua New Guinea.* Paper presented at the World Fisheries Congress, 3-8 May 1992, Athens, Greece.

Cowx, I.G. (1994) Stocking strategies. *Fisheries Management and Ecology* 1, 15-30.

Crucible Group (1994) *People, Plants and Patents: The Impact of Intellectual Property on Trade, Plant Biodiversity and Rural Society.* International Development Research Centre (IDRC), Ottawa, Canada.

Cunningham, E.P. (1990) *Quantitative Approaches to Animal Improvement.* Proceedings of the 4th World Conference on Genetics Applied to Livestock Production, Edinburgh, UK, pp. 4-14.

Doyle, R.W., Shackel, N.L., Basiao, Z., Uraiwan, S., Matricia, T. and Talbot, A.J. (1991) Selective diversification of aquaculture stocks: a proposal for economically sustainable genetic conservation. *Canadian Journal of Fisheries and Aquatic Sciences* 48 (Suppl. 1), 148-154.

Edwards, P. (1980) Food potential of aquatic macrophytes. *ICLARM Studies and Reviews* 5. ICLARM, Manila, Philippines. 51 pp.

Eknath, A.E. and Doyle, R.W. (1990) Effective population size and rate of inbreeding in aquaculture of Indian major carps. *Aquaculture* 85, 293-305.

Eknath, A.E., Tayamen, M.M., Palada-de Vera, M., Danting, J.C., Reyes, R.A., Dionisio, E.E., Capili, J.B., Bolivar, A.V., Bentsen, H.B., Gjerde, B., Gjedrem, T. and Pullin, R.S.V. (1993) Genetic improvement of farmed tilapias: the growth performance of eight strains of *Oreochromis niloticus* tested in different farm environments. *Aquaculture* 111, 171-188.

FAO (1993) Expert consultation on utilization and conservation of aquatic genetic resources. *FAO Fisheries Report No. 491. FIRI/R49.* FAO, Rome, Italy.

Froese, R. and Pauly, D. (eds) (1994a) FishBase users manual, A biological database on fish. *ICLARM Software* 7, pag. var. International Center for Living Aquatic Resources Management, Manila, Philippines.

Froese, R. and Pauly, D. (1994b) *A Strategy and a Structure a Database on Aquatic Biodiversity.* Paper presented at the 6th Meeting of the Committee on Data for Science and Technology, Data Sources in Asian-Oceanic Countries (CODATA/DSAO), 10-12 March 1994, Taipei, Taiwan.

Gausen, D. and Moen, V. (1991) Large-scale escapes of farmed Atlantic salmon (*Salmo salar*) into Norwegian rivers threaten national populations. *Canadian Journal Fisheries and Aquatic Sciences* 48, 426-428.

Gjedrem, T. (1985) Improvement of productivity through breeding schemes. *Geo Journal* 10(3), 233-241.

Goodman, M.M. (1990) Genetic and germplasm stocks worth conserving. *Journal of Heredity* 81, 11-16.

Gross, R. and Wohlfarth, G.W. (1994) Use of genetic markers in growth testing of common carp, *Cyprinus carpio* L., carried out over 2 or 3 year cycles. *Aquaculture and Fisheries Management* 25, 585-599.

Hilborn, R. (1992) Hatcheries and the future of salmon in the Northwest. *Fisheries* 17(1), 5-8.

Hindar, K., Ryman, N. and Utter, F. (1991) Genetic effects of cultured fish on natural fish populations. *Canadian Journal of Fisheries and Aquatic Sciences* 48, 945-947.

ICLARM (1992) *Recommendations of the Meeting on International Concerns in the Use of Aquatic Germplasm.* Report available from ICLARM, Manila, Philippines.

ICLARM (1994) A global database of coral reef systems and their resources. *Naga, the ICLARM Quarterly* 17(1), 16.

Kapuscinski, A.R. and Hallerman, E.M. (1990) AFS Position Statement. Transgenic fishes. *Newsletter of the Fisheries Genetics Section of the American Fisheries Society* 3, 5-12.

Lesser, W. and Maloney, A.P. (1993) *Biosafety: a Report on Regulatory Approaches for* the *Deliberate Release of Genetically Engineered Organisms. Issues and Options for Developing Countries*. Cornell International Institute for Food, Agriculture and Development, Cornell University, Ithaca, New York.

Macaranas, J.M., Taniguchi, N., Pante, M.J.R., Capili, J.B. and Pullin, R.S.V. (1986) Electrophoretic evidence for extensive hybrid gene introgression into commercial *Oreochromis niloticus* (L.) stocks in the Philippines. *Aquaculture and Fisheries Management* 17, 249–288.

Maclean, N. and Penman, D. (1990) The application of gene manipulation to aquaculture. *Aquaculture* 85, 1–20.

Mair, G.M. (1993) Chromosome-set manipulation in tilapia techniques, problems and prospects. *Aquaculture* 111, 227–244.

Martin, J., Webster, J. and Edwards, G. (1992) Hatcheries and wild stocks: are they compatible? *Fisheries* 17(1), 4.

Moriarty, D.J.W. and Pullin, R.S.V. (eds) (1985) Detritus and microbial ecology in aquaculture. *ICLARM Conference Proceedings* 14, 420 pp.

Moyle, P.B. and Leidy, R.D. (1992) Loss of biodiversity in aquatic ecosystems: evidence from fish farmers. In: Fielder, P.L. and Jain, S.K. (eds) *Conservation Biology: The Theory and Practice of Nature Conservation Preservation and Management*. Chapman and Hall, New York, pp. 128–168.

Msiska, O.V. and Costa-Pierce, B.A. (1993) History, status and future of common carp (*Cyprinus carpio* L.) as an exotic species in Malawi. *ICLARM Conference Proceedings* 40, Manila, Philippines.

Naevdal, G. (1994) Genetic aspects in connection with sea ranching of marine species. *Aquaculture and Fisheries Management* 25 (Suppl. 1), 93–100.

Nehlsen, W., Williams, J.E. and Lichatowich, J.A. (1991) Pacific salmon at the crossroads: stocks at risk from California, Oregon, Idaho, and Washington. *Fisheries* 16(2), 4–21.

Nelson, J.S. (1994) *Fishes of the World*, 3rd edn. John Wiley, New York.

New, M.B. and Crispoldi-Hotta, A. (1992) Problem in the application of the FAO definition of aquaculture. *FAO Aquaculture Newsletter* 1, 5–8.

Padhi, B.K. and Mandal, R.K. (1994) Improper fish breeding practices and their impact on aquaculture and fish biodiversity. *Current Science* 66(9), 624–626.

Pauly, D. and Froese, R. (1991) Fishbase: assembling information on fish. *Naga, the ICLARM Quarterly*. 14(4), 10–11.

Prescott-Allen, R. and Prescott-Allen, C. (1990) How many plants feed the world. *Conservation Biology* 4, 365–374.

Pullin, R.S.V. (ed.) (1988) Tilapia genetic resources for aquaculture. *ICLARM Conference Proceedings* 16, Manila, Philippines. (French edition available from 1989).

Pullin, R.S.V. (1989) Third-world aquaculture and the environment. *Naga, the ICLARM Quarterly* 12(1), 11–13.

Pullin, R.S.V. (1990) Down-to-earth thoughts on conserving aquatic genetic diversity. *Naga, the ICLARM Quarterly* 13(1), 5–8.

Pullin, R.S.V. (1993) *Ex-situ* conservation of the germplasm of aquatic organisms. *Naga, the ICLARM Quarterly* 16(2-3), 15–17.

Pullin, R.S.V. (1994) Exotic species and genetically modified organisms in aquaculture and enhanced fisheries: ICLARM's position. *Naga, the ICLARM Quarterly* 17(4), 19–24.

Pullin, R.S.V. (1995) Aquaculture and Biodiversity. In: Hartnoll, R.G. and Hawkins, S.J. (eds) *Marine Biology – a Port Erin Perspective*. Immel Publishing, London.

Pullin, R.S.V., Rosenthal, H. and Maclean, J.L. (eds) (1993) Environment and aquaculture in developing countries. *ICLARM Conference Proceedings* 31. 359 pp.

Pullin, R.S.V. and Prein, M. (1995) Fishponds facilitate natural resources management on small-scale farms in tropical developing countries. In: Symoens, J.J. and Micha, J.-C. (eds) *Proceedings of the Seminar on The Management of Integrated Freshwater Agro-Piscicultural Ecosystems in Tropical Areas*, 16–19 May 1994, Technical Centre for Agricultural and Rural Cooperation (CTA) and Royal Academy of Overseas Sciences, pp. 169–186, Brussels.

Pullin, R.S.V., Bimbao, M.A.P. and Bimbao, G.B. (1994) World outlook for tilapia farming. Paper presented at the First International Symposium on Aquaculture, 9–11 June 1994, Vera Cruz, Mexico.

Rana, J.K. (1992) Cryopreservation of aquatic gametes and embryos. International workshop on genetics in aquaculture. In: Penman, D., Roongratri, N. and McAndrew, B. (eds) *International Workshop on Genetics in Aquaculture and Fisheries Management*. AADCP Workshop Proceedings. ASEAN-EEC Aquaculture Development and Coordination Programme, Bangkok, Thailand, pp. 49–53.

Ray, G.C. (1991) Coastal zone biodiversity patterns. *Bioscience* 41(7), 490–498.

Ryman, N. and Laikre, L. (1991) Effects of supportive breeding on the genetically effective population size. *Conservation Biology* 5, 325–329.

Schultz, F. (1986) Developing a commercial breeding program. *Aquaculture* 57, 65–76.

Seshu, D.V., Eknath, A.E. and Pullin, R.S.V. (1994) *International Network on Genetics in Aquaculture*. International Center for Living Aquatic Resources Management, Manila, Philippines.

Smith, P.J., Francis, R.I.C.C. and McVeagh, M. (1991) Loss of genetic diversity due to fishing pressure. *Fisheries Research* 10, 309–316.

Smitherman, R.O. and Tave, D. (1988) Genetic considerations on acquisiton and maintenance of reference populations of tilapia. *Aquabyte* 1(1), 2.

Stickney, R.R. (1994) Use of hatchery fish in enhancement programs. *Fisheries* 19(5), 6–13.

Tave, D. (1986). *Genetics for Fish Hatchery Managers*. AVI Publishing Co., Westport, Connecticut.

Trono, G.C. Jr and Ganzon-Fortes, E.T. (1988) *Philippine Seaweeds*. National Book Store Inc., Manila, Philippines.

Turner, G.E. (1988) Codes of practice and manual of procedures for consideration of introductions and transfers of marine and freshwater organisms. *FAO, EIFAC/CECPI Occasional Paper* no. 23, 44 p.

Verspoor, E. and Hammer, J. (1991) Introgressive hybridization in fishes: the biochemical evidence. *Journal of Fisheries Biology* 39 (Suppl. A), 309–334.

Villwock, W. (1972) Gefahren für die endemische Fischfauna durch Einbürgerungsversuche und Akklimatisation von Fremdfischen am Beispiel des Titicaca-sees (Peru/Bolivien) und des Lanao-Sees (in German). [Danger for endemic fish fauna through attempts to introduce or acclimatize foreign fish – the examples of Lake Titicaca (Peru/Bolivia) and Lake Lana (Mindanao) Philippines]. *Verhandlung Internationaleu Vereinigung für Limnologie* 18, 1227–1234.

36 Conservation of Biological Diversity in Hatchery Enhancement Programmes

D.M. Bartley
Fisheries Department, Food and Agriculture Organization of the United Nations, Viale delle Terme di Caracalla, 00100 Rome, Italy

Introduction

As the need for protein from the aquatic environment increases due to an ever-increasing human population (Fig. 36.1), new management practices and innovative technologies must be applied to the world's aquatic resources and environments. One such management tool is the use of hatcheries to support, augment, or create new fisheries. This practice is often termed 'marine ranching', 'hatchery enhancement' or 'culture-based fisheries'. Although the terms have subtle differences in meaning, for the purpose of this presentation they are synonymous.

Pacific salmon (*Oncorhynchus* spp.) are perhaps the most widely ranched species of fish (Netboy, 1980; Isaksson, 1988a). However, other species including, among others, cod, *Gadus morhua* (Svåsand et al., 1990), Atlantic salmon, *Salmo salar* (Isaksson, 1988a,b), sturgeon (Acipenseridae) (Binkowski and Doroshov, 1985; FAO, 1992c), red drum, *Sciaenops ocellatus* (Rutlege, 1989), white seabass, *Atractoscion nobilis* (Kent et al., 1995), red seabream, *Pagrus major* (Smith, 1983; Sugama et al., 1988) and several invertebrate species are also currently being raised in hatcheries to augment natural fisheries (Lockwood, 1991). Japan is a leader in ranching technology conducting ranching and research on approximately 80 marine species and currently has several commercial fisheries based primarily on fish produced in hatcheries (Watanabe et al., 1989; Ikenoue and Kafuku, 1992).

However, the use of animals raised in captivity to replenish natural populations has been questioned and criticized because of three main factors: (i) lack of evidence that the hatchery enhancement benefited the fishery; (ii) lack of a cost effective use of resources, i.e. the hatcheries were too expensive to maintain (MacCall, 1989; Hilborn, 1992), and more recently (iii) the fear of the loss of natural genetic diversity through interaction of 'hatchery'

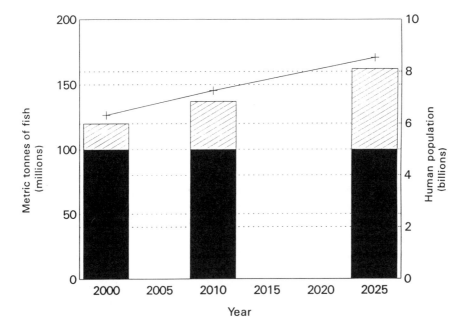

Fig. 36.1. Projected human population growth (solid line), production from capture fisheries, that is anticipated to plateau at 100 million metric tonnes (dark bars) and the need for increased production from the aquatic sector (hatched bars) in order to maintain current rate of aquatic food consumption (FAO, 1992a,b).

fish with native populations (Hindar *et al.*, 1991; Philipp *et al.*, 1993). These concerns must be addressed by hatchery enhancement programmes.

The history of ranching, especially marine ranching, has not been very bright. Billions of Atlantic cod were released into the Atlantic from both North America and Norway, yet the cod stocks continued to decline (MacCall, 1989; Griffin, 1993). Of the eggs produced in USA enhancement programmes of the 1940s, 98% were of the marine species cod, flounder species, and pollack, *Pollachius virens*; these programmes were ended when there was no increase in the associated fisheries (Larkin, 1991).

The reasons for the failure of many past ranching or enhancement programmes are mainly that the causes for the decline in the fishery were not adequately addressed or corrected and that a rather simplistic approach to restoring the population was adopted: simply dump millions of larvae into convenient water bodies and hope (Troadec, 1991). Furthermore, most hatchery enhancement programmes were not monitored and hatchery fish were not tagged. Therefore, it was usually impossible to differentiate 'hatchery' fish from 'native' fish (Parker *et al.*, 1990 and references).

However, there is now an increased understanding of the complexity of natural aquatic populations. Current hatchery enhancement programmes involve ecological and genetic considerations (Bartley et al., 1995). Tagging technology has improved to incorporate electronic tags (Kent et al., 1995) and genetic tags that can uniquely identify fish or groups of fish (Utter and Ryman, 1993; Wirgin and Waldman, 1994).

Recent reviews demonstrated that many anadromous salmonid populations are well adapted to their native habitat and that infusion of exotic genes from populations from different habitats, including hatcheries, may be detrimental to the native fish (Hindar et al., 1991; Waples, 1991a). Some of the reported effects of the interaction between hatchery fish and closely related or conspecific fish, are genetic erosion (loss of alleles), introduction of maladapted genes, introduction of disease organisms, changes in genetic subpopulation structure, changes in stock size, reductions in survival, changes in adaptive behaviour, reductions in homing accuracy, changes in migration patterns, reductions in recapture rate and increased disease susceptibility (Ryman et al., 1994). It was also demonstrated that effective population size of an enhanced population can theoretically be reduced by the addition of hatchery fish with variable reproductive success and reduced genetic variability (Ryman and Laikre, 1991). The above effects are due to direct competition, predation, and disease transmission between hatchery and wild stocks, and the introgression of 'hatchery' genes into the wild population.

Strategies to reduce or eliminate the interaction between hatchery and wild fish include (i) creating hatchery fish that are substantially different from wild stocks in migration patterns, spawning times, or food preference, so that hatchery and wild stocks would not come into contact, (ii) altering release or aquaculture sites to minimize contact (Isaksson, 1988b), and (iii) the production of sterilized fish so that reproduction between hatchery and native fish would be impossible (Turner, 1988; Seeb et al., 1993). Each option has its strengths and weaknesses in regards to conserving the native gene pools and each should be considered in a hatchery programme.

However, the goal of many enhancement projects is for the hatchery and wild fish to mix and form a new and larger mixed population (Bartley et al., 1995). Therefore, trying to create fish in the hatchery that are compatible with wild stocks and the environment will be an important option for certain enhancement programmes and for certain aquatic species. The creation of a 'wild' fish under culture conditions should be approached by minimizing artificial and inadvertent selection and by sound hatchery management to conserve genetic diversity. However, the ability to successfully create such a fish in an artificial environment such as a hatchery may be difficult (Durand et al., 1993; Howell, 1994).

Above all, the hatchery enhancement programme must have clearly defined objectives and goals (Barber and Taylor, 1990). A major objective of any aquatic development programme should be the conservation of aquatic resources (FAO, 1993). In light of the facts that the majority of these resources reside in wild populations and that fisheries make up the largest single source of animal protein (Table 36.1), every effort should be made to protect the wild resources.

Table 36.1. Global animal protein production in 1989. Data from the Food and Agriculture of the United Nations, Rome, Italy (FAO, 1992a,b).

Sector	Production (thousand metric tonnes)
Capture fisheries	88,514
Aquaculture	11,021
Swine farming	67,460
Beef and veal farming	49,436
Poultry farming	37,817
Mutton and lamb farming	6,473

The purpose of this chapter is to outline general principles and procedures that should be incorporated into a hatchery enhancement or ranching programme to conserve and utilize genetic resources (biodiversity). Hatchery enhancement must consider the hatchery fish to be released into the wild, the broodstock used to produce them, the wild populations in the area of release, and the mixture of the hatchery and wild fish.

Protocols for Conservation of Biodiversity in Aquatic Ranching

The following section outlines general considerations for aquatic ranching that will help conserve the genetic resources in both hatchery and wild populations. Items addressed include the understanding of natural levels and organization of genetic diversity, the objective of ranching programmes, the understanding of the reasons for decline in original fishery, special considerations for the establishment of a new species, assessment of hatchery technology and husbandry, assessment of ecological requirements of ranched species, and the monitoring of both the hatchery and wild populations.

Assess natural levels and organization of genetic diversity

The first step in conserving genetic resources is to know what genetic resources exist. This basic information is often lacking or incomplete for many aquatic species (Allendorf and Utter, 1979). Protein or isozyme analysis has provided a means to quickly and relatively cheaply examine the genetic variability of numerous species. Protein variation has been used to provide information on geneflow, population structure, species identification, taxonomy, hybridization, and evolutionary relationships in fishes (Allendorf and Utter, 1979; Utter and Seeb, 1990). Additional information is rapidly accumulating concerning DNA level variation in aquatic species (Wirgin and Waldman, 1994).

Because a prime desire is the conservation of natural diversity, a hatchery enhancement or ranching programme should strive to utilize the known information on natural diversity and complement or reproduce it in the hatchery product (Meffe, 1990). In cases where an exotic species is being ranched, knowledge of natural levels of diversity will help pick strains for evaluation in the new environment or will help identify strains for crossbreeding in hopes of establishing a well-adapted population (see discussion in Waples, 1991a).

Define objective of enhancement/ranching programme

This obvious step is often neglected or misinterpreted by managers and evaluators. An all too common goal of past enhancement projects has been simply the release of a number of eggs/fry or fish (Troadec, 1991). However, clearly defined goals are absolutely essential to design appropriate hatchery management strategies, as well as to evaluate realistic alternatives to hatchery enhancement (Allendorf and Ryman, 1987). It is given that a major objective is the conservation of genetic resources in wild and cultured stocks. Since the Earth Summit in Rio de Janeiro, development agencies as well as aquaculturists are emphasizing that sustainable production from natural systems must involve the conservation of genetic diversity. This requires the thorough documentation of the existing biological and genetic resources (Ryman *et al.*, 1994). Presently, nearly all the aquatic genetic resources are found in wild populations, as little domestication has taken place. Preserving the existing stock structure and migration patterns found in wild populations and ensuring that culture practices do not adversely affect the wild stocks will be necessary.

The conservation objectives that the hatchery should establish will be based on the results of the survey/documentation of the wild stocks. Components of genetic diversity to be conserved include more than simply alleles and genes, but may encompass numerous subpopulations or metapopulations. Fishery geneticists for the US Endangered Species Programme recommend the conservation of 'ecologically significant units' defined in the National Marine Fisheries Service statement on endangered species to be a genetic population/stock which is (i) genetically unique, (ii) exhibits unique adaptation, (iii) inhabits unique habitat, or (iv) whose extinction would represent loss of an evolutionarily significant legacy (Waples, 1991b). The preamble to the Convention on Biodiversity (UNEP, 1992) identifies intrinsically significant units as a genetic population/stock which has social, economic, scientific, educational, cultural, recreational, or aesthetic value, and thereby worthy of conservation.

Production goals in terms of real increases to a fishery and other measures of success of many enhancement projects are also necessary, but often omitted. Although such goals are essential to planning and evaluation and they must be compatible with conservation goals, they will not be discussed further here.

Assess reasons for decline in original fishery

This critical phase is necessary to evaluate the logic behind the enhancement programme and to discover means to optimize effort. For example, nearly 80% of salmon spawning habitat has been degraded in California (Netboy, 1980) and probably a similar loss of sturgeon spawning habitat has occurred in the tributaries to the Caspian Sea (FAO, 1992c). Therefore, hatchery production of juveniles was seen as a means to replace the function of these habitats. This is at least logical. The fishery for white sturgeon, *Acipenser transmontanus*, was collapsing in California due to overfishing. A ban on commercial fishing and a moratorium on sport fishing has been effective at partially reviving the stock in the San Francisco Bay and Sacramento River to the point where an active sport fishery is now allowed (Kohlhorst, 1980).

Hatchery enhancement of Black and Azov Sea fisheries, collapsing due to pollution and a exotic and voracious comb jellyfish, *Mnemiopsis leidyi*, has been suggested (Travis, 1993). Although the restocking of these seas would involve fish theoretically resistant to the comb jelly, enhancement would appear ill-advised until pollution and fishing controls are put in place.

Special considerations for use of 'exotic' species

The use of exotic species in ranching or enhancement projects is a special case that requires planning and evaluation of the ecological and socioeconomic risks associated with using a species in an area outside its natural range. The introduction of Kapenta, *Limnothrissa miodon*, a freshwater herring-like fish, into Lake Kariba, Zimbabwe, resulted in a fishery that produces 20,000 tonnes year^{-1} valued at over US$10,000,000 (Bartley, 1993a); farming of Nile tilapia, *Oreochromis niloticus*, as well as other tilapia species, provides fish farmers throughout the world with a highly desirable and marketable product (Pullin, 1988). In addition to additional animal protein introduced species may provide to an area, exotic species often command a higher market price than local species and may provide local economies means to acquire foreign currency from tourism or through export (Reynolds and Greboval, 1989).

Unfortunately, the results of an introduction are not always predictable (Mann, 1992) and not always positive. Introductions of aquatic species are now considered one of the main threats to natural aquatic populations (Williams *et al.*, 1989). European populations of crayfish, *Astacus astacus*, were decimated by a fungal disease inadvertently introduced when American crayfish, *Orconectes limosus*, was brought to Europe (Furst, 1984). The use of Atlantic salmon from the Baltic Sea in stock enhancement programmes in Norway introduced a parasite to the Norwegian Sea that now threatens native stocks (Bakke, 1991; Folsom *et al.*, 1992). The widespread movement of rainbow trout, *O. mykiss*, throughout the western United States, has reduced local aquatic invertebrate fauna and threatened other species and subspecies of salmonids through competition, predation, and hybridization (Nicola, 1976; Bartley and Gall, 1991). The introduction of the Nile perch, *Lates niloticus*, into Lake Victoria changed primarily small-scale artisanal fisheries into a multimillion dollar commercial fishery that supports industrialized processing and exportation ventures; the Nile perch may have also led to the extinction of several hundred of Lake Victoria's species of fish (Barel *et al.*, 1985; Reynolds and Greboval, 1989).

The purposeful movement of fish and shellfish from one area to another

430 D.M. Bartley

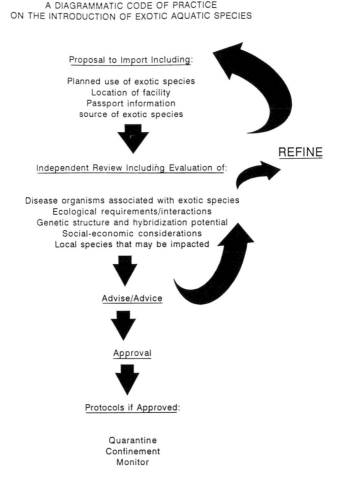

Fig. 36.2. A diagrammatic code of practice on the introduction of exotic aquatic species.

has been, and will continue to be, a viable method to increase aquatic food production (Welcomme, 1988). However, institutions responsible for introductions and transfers of aquatic organisms must be aware of and evaluate both the good and bad sides of this issue. In this light, professional fisheries societies and international organizations are promoting a responsible approach to the use of introduced or transferred species through the use of codes of practice and protocols (Fig. 36.2). Foremost in this regard are the International Commission for the Exploration of the Sea and the European Inland Fisheries Advisory Commission (ICES/EIFAC) Codes of Practice and Manual of Procedures for Consideration of Introductions and Transfers of Marine and Freshwater Organisms (Turner, 1988).

The code in Fig. 36.2 states that a proposal to import an exotic species should be made by the person, company or institute wishing to use an exotic species. An independent review of the proposal should be performed by local authorities or resource managers. Local authorities may wish to seek the advice of external authorities, such the ICES or EIFAC, on the risks/benefits of the proposal. The review could either lead to an approval to use an exotic, a refinement or redraft of the proposal, or a denial of the proposal. If approval is given, quarantine, monitoring, and evaluation will be essential to determine the success of the project and to benefit future uses of exotic species.

These codes and associated protocols provide a framework for decision-making that facilitates thorough consideration and evaluation of such things as, the need for the introduction, disease and quarantine safeguards, potential alternatives to the proposed introduction or transfer, and the protection of the local and neighbouring resources (Kohler and Stanley, 1984). 'These are similar considerations that any good financial analyst would look at to assure success of a new product or business enterprise: is there a need for the product? what other local product may be better developed? how will the new product fit in amongst established products? how will the new product affect the market in general? how will distribution and production be controlled?' (Bartley, 1993b).

Aquaculture is currently the number one cause of species introductions into freshwater habitats (Welcomme, 1988). To ensure that growth of this sector continues to be positive, it will be imperative to give due consideration to codes of practice promoting the responsible use of introduced and transferred aquatic species.

Assessment of established hatchery technology and husbandry

As it will be imperative to produce healthy offspring from a large base of healthy broodstock, controlled breeding, larval rearing and nutrition and disease control need to be optimized. An alternative scenario would involve harvesting naturally produced seed for grow out in the hatchery if controlled reproduction in the hatchery is not possible. Although this option may be cost effective in some marine cultures (Hoffman, 1991), this type of 'mining' of the natural population to support a hatchery may be detrimental to the wild populations and may not be sustainable. For a variety of aquatic species, such as salmon, tilapia, red seabream, carps, and oysters, hatchery technology is well established (Jhingran and Pullin, 1985; Pullin, 1988; Folsom et al., 1992). For others, such as many coral reef fish (excluding the popular clownfish, *Amphiprion* spp.), and many marine invertebrates, it is possible to raise only the early larval stages (Allen, 1991; Lee and Wickens, 1992) that would probably survive poorly in the wild. Often hatchery technology exists, but is overly expensive to produce animals of a given size, as in the case of homarid lobsters that are cannibalistic under culture conditions (Lee and Wickens, 1992).

Technological breakthroughs or advances may make previously difficult species more suitable for hatchery enhancement. For example the use of pituitary extract to induce spawning in Chinese carps greatly increased the use of these species in culture (Chapter 35). The development of techniques to mass-rear cod in seminatural ponds has led to renewed interest in cod-ranching (Svåsand *et al.*, 1990). Advances in larval and broodstock nutrition will also enable more species to be considered for hatchery enhancement (Watanabe *et al.*, 1989).

Assess proposed areas for reintroduction or enhancement

It will be of little benefit to stock a good fish into a degraded environment. Even in unpolluted areas, the site of release is an important component of survival of hatchery-produced fish. Research in Hawaii on striped mullet, *Mugil cephalus* (Leber *et al.*, 1995) and in California on white seabass (Kent *et al.*, 1995) demonstrated that survival was higher for fishes released in protected areas and bays than on the open coast. In anadromous species or other migratory species, timing of release, as well as location, will be important (Cresswell, 1981; Naiman *et al.*, 1987).

In the case of an enhancement project this assessment should also include the level of fishing activity likely to impinge on the developing enhanced population. Care should be taken not to increase the fishing activity or to relax regulations and limits on the enhanced fishery simply because fish are now being introduced into the wild. There is a real danger in optimistic fishers reducing the wild and enhanced stocks to a point below pre-enhancement levels (Larkin, 1977; Nelson and Soulé, 1987). Indeed, successful enhancement projects in Japan and off the Texas coast in the Gulf of Mexico were coupled with habitat renovation and more restrictive fishing regulations (Bhat, 1989; King *et al.*, 1993). Again, the objectives of the ranching programme will determine whether or not and to what extent fishing should be allowed on the hatchery product. It is likely that fishing regulations will need to be adapted to give the enhancement project time to work.

Monitoring

An often neglected step in ranching and enhancement programmes has been the monitoring of a hatcheries contribution to a fishery or to a wild population. This is due to the difficulty in accurately identifying hatchery fish without using expensive tags or unduly handling large amounts of fish. However, techniques for the mass marking of large amounts of fish are being developed (Parker *et al.*, 1990). Genetic analyses using allozyme and DNA markers now provide means to identify hatchery contributions to mixed fisheries and even to identify individual fish (Utter and Seeb, 1990; Wirgin and Waldman, 1994). An advantage of genetic markers over physical tags is that they can assess introgression and hybridization between hatchery and wild populations in subsequent generations (Compton and Johnston, 1985; Bartley and Gall, 1991).

The monitoring of the population in the hatchery is often overlooked in ranching programmes. It is necessary to maintain levels of genetic variability in the hatchery in order to produce viable and well-adapted fish for release.

Table 36.2. Toward a 'responsible approach' to marine ranching.

Dr Kenneth Leber of the Oceanic Institute in Hawaii stated, 'To ensure their successful use and avoid repeating past mistakes, we must take a responsible approach to developing, testing, and managing marine stock enhancement programmes'. Under the direction of Dr Leber, the following guidelines were established by an ad hoc committee of the World Aquaculture Society (1993/94). By adhering to these recommendations, the hope is to make marine ranching one viable option in efforts to increase fishery production.

1 Establish priorities and methodologies for selecting species to be ranched or enhanced.
2 Create a species management plan with long- and short-term goals, harvest regimes and genetic conservation objectives.
3 Incorporate life history and ecological attributes into enhancement strategies and tactics.
4 Create a genetic resource management plan to minimize inbreeding/outbreeding depression and to conserve genetic resources.
5 Create a disease and health management plan.
6 Define an empirical process for defining optimal release strategies.
7 Identify and implement means to identify hatchery-produced fish.
8 Assess the enhancement project in terms of stated objectives in management plan.
9 Include quantitative measures of success and socioeconomic evaluation.

Hatchery populations of broodfish will usually have reduced levels of variation compared with that found in the wild (Allendorf and Ryman, 1987). However, effective hatchery management can prevent this reduction from becoming problematic (Tave, 1986). Management would involve the use of large numbers of broodfish, avoidance of high levels of inbreeding, periodic infusion of new genes, and the maintenance of genetic lines or stocks (Bartley et al., 1995).

Allozyme studies on shrimp, *Penaeus japonicus*, cultured in Italy (Sbordoni et al., 1987) revealed that after seven generations in a hatchery, levels of heterozygosity were reduced and that coincident with this reduction was a reduction in the percentage hatch of shrimp larvae. An explanation for the decrease in hatchability was inbreeding depression, brought on by small effective population size of hatchery broodstock. The number of shrimp actually contributing genes to the next generation was much lower than the number of shrimp actually present in the hatchery. Analysis of allozyme genotypes and effective population size in a hatchery enhancement programme on white seabass revealed that only about half of the parents were contributing genetic resources to the progeny that were to be released (Bartley et al., 1992). Conservation goals were established that called for a specific number of adult white seabass spawners; monitoring revealed that the hatchery needed to keep twice that amount in captivity (Bartley et al., 1995).

Conclusion

Aquatic ranching is a complicated area involving technology, biology, ecology, socioeconomics and politics. Recently, a group of interested scientists from around the world got together and established a 'responsible approach' to hatchery enhancement that involves multidisciplinary assessment, monitoring and action (Table 36.2). Genetic considerations alone will not ensure a successful ranching programme, however, without the application of principles of genetic resource management, the long-term success of production and conservation efforts will be impossible.

References

Allen, G.R. (1991) *Damselfishes of the World*. Mergus, Melle, Germany.

Allendorf, F.W. and Ryman, N. (1987) Genetic management of hatchery stocks. In: Ryman, N. and Utter, F. (eds) *Population Genetics and Fishery Management*. University of Washington Press. Seattle, Washington, USA, pp. 141-160.

Allendorf, F.W. and Utter, F.M. (1979) Population genetics. In: Hoar, W.S., Randall, D.J. and Brett, J.R. (eds) *Fish Physiology*, Vol. 8. Academic Press, New York, pp. 407-454.

Bakke, R. (1991) A review of the inter- and intra-specific variability in salmonid host to laboratory infections with *Gyrdactylus salaris*. *Aquaculture* 98, 303-310.

Barber, W.E. and Taylor, J.N. (1990) The importance of goals, objectives, and values in the fisheries management process and organization: a review. *North American Journal of Fisheries Management* 10, 365-373.

Barel, C.D.N., Dorit, R., Greenwood, P.H., Fryer, G., Hughes, N., Jackson, P.B.N., Kawanabe, H., Lowe-McConnel, R.H., Nagoshi, M., Ribnbink, A.J., Trewavas, E., Witte, F. and Yamaoka, K. (1985) Destruction of fisheries in Africa's Lakes. *Nature* 315, 19-20.

Bartley, D.M. (1993a) *An Application of International Codes of Practice on Introductions of Aquatic Organisms: Assessment of a Project on the Use of Chinese Carps in Mozambique*. FAO Fisheries Circular no. 863. Food and Agriculture Organization of the United Nations, Rome, Italy.

Bartley, D.M. (1993b) Introduced species. Editorial. *FAO Aquaculture Newsletter* 5, 1-2.

Bartley, D.M., Bagely, M., Gall, G. and Bentley, B. (1992) Use of linkage disequilibrium data to estimate effective population size of hatchery and natural fish populations. *Conservation Biology* 6, 365-375.

Bartley, D.M. and Gall, G.A.E. (1991) Genetic identification of native cutthroat trout (*Oncorhynchus clarki*) and introgressive hybridization with introduced rainbow trout (*O. mykiss*) in streams associated with the Alvord Basin, Oregon and Nevada. *Copeia* 1991, 854-859.

Bartley, D.M., Kent, D.B. and Drawbridge, M.A. (1995) Conservation of genetic diversity in a white seabass (*Atractoscion nobilis*) hatchery enhancement programme in southern California. *Uses and Effects of Cultured Fish in Aquatic Systems*, Symposium of the American Fisheries Society (in press).

Bhat, V. (1989) Shrimp mariculture: sea ranching technology in Japan. *Seafood Export Journal* Sept-Oct, 29-43.

Binkowski, F.P. and Doroshov, S.I. (eds) (1985) *North American Sturgeons: Biology and Aquaculture Potential*. Dr W. Junk Publ. Boston.

Campton, D.E. and Johnston, J.M. (1985) Electrophoretic evidence of a genetic admixture of native and non-native rainbow trout in the Yakima River, Washington. *Transactions of the American Fisheries Society* 114, 782-793.

Cresswell, R.C. (1981) Post-stocking movements and recapture of hatchery-reared trout released into flowing waters - a review. *Journal of Fish Biology* 18, 429-442.

Durand, P., Wada, W.T. and Blanc, F. (1993) Genetic variation in wild and hatchery stocks of the black pearl oyster, *Pinctada margaritifera*, from Japan. *Aquaculture* 110, 27-40.

FAO (1992a) *Aquaculture Production (1984-1990). FAO Fisheries Circular no. 815*. Revision 4. Food and Agriculture Organization of the United Nations, Rome, Italy.

FAO (1992b) *Fishery Statistics. Catches and Landings. FAO Yearbook*. Food and Agriculture Organization of the United Nations, Rome, Italy.

FAO (1992c) *The Islamic Republic of Iran: Aquaculture Sector Fact Finding Mission*. FI:TCP/IRA/2251 (F). Food and Agriculture Organization of the United Nations, Rome, Italy.

FAO (1993) *Report of the Expert Consultation on Utilization and Conservation of Aquatic Genetic Resources*. FAO Fisheries Report No. 491 (FIRI/R49). Food and Agriculture Organiziation of the United Nations, Rome, Italy.

Folsom, W., Altman, D., Manuar, A., Nielsen, F., Revord, T., Sanborn, E. and Wildman, M. (1992) *World Salmon Culture. NOAA Technical Memo*. NMFS-F/SPO-3. Silver Spring, MD, USA.

Furst, M. (1984) Introduction of the North American crayfish (*Pacifastacus leniusculus*) into Sweden. In: *Documents Presented at the Symposium on Stock Enhancement in the Management of Freshwater Fish*. EIFAC Technical Paper 42/Suppl. 2, pp. 400-404.

Griffin, N. (1993) Fishermen hope to find future in cod ranching. *National Fisherman* (April), 22-23.

Hilborn, R. (1992) Hatcheries and the future of salmon in the Northwest. *Fisheries* 17, 5-8.

Hindar, K., Ryman, N. and Utter, F. (1991) Genetic effects of cultured fish on natural fish populations. *Canadian Journal of Fisheries and Aquatic Sciences* 48, 945-957.

Hoffman, E. (1991) A review of plaice (*Pleuronectes platessa*) transplantation trials in Denmark 1891-1990. *ICES Marine Science Symposium* 192, 120-126.

Howell, B.R. (1994) Fitness of hatchery-reared fish for survival in the sea. *Aquaculture and Fisheries Management* 25 (Suppl 1), 3-17.

Ikenoue, H. and Kafuku, T. (1992) *Modern Methods of Aquaculture in Japan*. Elsevier, Tokyo.

Isaksson, A. (1988a) Salmon ranching: a world review. *Aquaculture* 75, 1-33.

Isaksson, A. (1988b) Salmon ranching in Iceland. In: *Wild Salmon Present and Future*. Sherkin Island Marine Station, County Cork, Ireland, pp. 96-104.

Jhingran, V.G. and Pullin, R.S.V. (1985) *A Hatchery Manual for the Common, Chinese and Indian Major Carps. ICLARM Studies and Reviews*. Asian Development Bank and the International Center for Living Aquatic Resources Management. Manila, Philippines.

Kent, D.B., Drawbridge, M.A. and Ford R.F. (1995) The roadblocks and milestones to making marine stock enhancement work: perspectives for the middle of a twenty year programme in southern California. *Uses and Effects of Cultured Fish in Aquatic Systems*. American Fisheries Society Symposium, Vol. 15, pp. 492–498.

King, T.L., Ward, R. and Blandon, I.R. (1993) Gene marking: a viable assessment method. *Fisheries* 18, 4–5.

Kohler, C.C. and Stanley, J.G. (1984) A suggested protocol for evaluating proposed exotic fish introduction in the United States. In: Courtenay, W.R. and Stauffer, J.R. (eds) *Distribution, Biology, and Management of Exotic Fishes*. Johns Hopkins University Press, Baltimore, Maryland, pp. 387–406.

Kohlhorst, D.W. (1980) Recent trends in the white sturgeon population in California's Sacramento–San Joaquin estuary. *California Fish and Game* 66, 210–219.

Larkin, P.A. (1977) An epitaph for the concept of maximum sustained yield. *Transactions of the American Fisheries Society* 106, 1–11.

Larkin, P.A. (1991) Mariculture and fisheries: future prospects and partnerships. In: Lockwood, S.J. (ed.) *The Ecology and Management Aspects of Extensive Mariculture*. ICES Marine Science Symposium, Vol. 192. International Council for the Exploration of the Sea, pp. 6–14.

Leber, K.M., Brennan, N.P. and Arce, S.M. (1995) Marine enhancement with striped mullet: are we replenishing or displacing wild stocks? In: *Uses and Effects of Cultured Fish in Aquatic Systems*. American Fisheries Society Symposium, Vol. 15, pp. 376–387.

Lee, D.O'C. and Wickins, J.F. (1992) *Crustacean Farming*. Blackwell Scientific Publications, London.

Lockwood, S.J. (ed.) (1991) *The Ecology and Management Aspects of Extensive Mariculture*. ICES Marine Science Symposium, Vol. 192. International Council for the Exploration of the Sea.

MacCall, A.D. (1989) Against marine fish hatcheries: ironies of fishery politics in the technological era. *California Cooperative Oceanic Fisheries Investigations Report* 30, pp. 46–48.

Mann, R. (1992) Management of introductions and transfers: a commentary on the changing role of the biologist. In: Rosenfield, A. and Mann, R. (eds) *Dispersal of Living Organisms into Aquatic Ecosystems*. Maryland Sea Grant, College Park, Maryland, pp. 291–293.

Meffe, G.K. (1990) Genetic approaches to conservation of rare fishes: examples from North American desert species. *Journal of Fish Biology* 37 (Suppl. A), 105–112.

Naiman, R.J., McCormick, S.D., Montgomery, W.L. and Morin, R. (1987) Anadromous brook charr, *Salvelinus fontinalis*: opportunities and constraints for population enhancement. *Marine Fisheries Review* 49, 1–13.

Nelson, K. and Soulé, M. (1987) Genetical conservation of exploited fishes. In: Ryman, N. and Utter, F. (eds) *Population Genetics and Fishery Management*. University of Washington Press. Seattle, Washington, pp. 345–368.

Netboy, A. (1980) *Salmon: the World's Most Harassed Fish*. Andre Deutsch Limited, London.

Nicola, S.J. (1976) Fishing in western national parks – a tradition in jeopardy? *Fisheries* 1, 18–21.

Parker, N.C., Giorgi, A.E., Heidinger, R.C., Jester, D.B., Prince, E.D. and Winans, G.A. (1990) *Fish Marking Techniques*. American Fisheries Society Symposium 7. Bethesda, Maryland.

Philipp, D.P., Epifano, J.M. and Jennings, M.J. (1993) Point/counterpoint: conservation genetics and current stocking practices – are they compatible? *Fisheries* 18, 14–16.

Pullin, R.S.V. (ed.) (1988) *Genetic Resources for Tilapia Aquaculture.* International Center for Living Aquatic Resources Management, Manila, Philippines.

Reynolds, J.E. and Greboval, D.F. (1989) *Socio-economic Effects of the Evolution of Nile Perch Fisheries in Lake Victoria: a Review. Commission on Inland Fisheries of Africa Technical Paper 17.* Food and Agriculture Organization of the United Nations, Rome, Italy.

Rutledge, W.P. (1989) The Texas marine hatchery program – it works! *California Cooperative Oceanic Fisheries Investigations Report* 30, 49–52.

Ryman, N. and Laikre, L. (1991) Effects of supportive breeding on the genetically effective population size. *Conservation Biology* 5, 325–329.

Ryman, N., Utter, F. and Laikre, L. (1994) Protection of aquatic biodiversity. In: Voigtlander, C.W. (ed.) *The State of the World's Fishery Resources. Proceedings of the World Fisheries Congress, Plenary Sessions.* Oxford and IBH Publishing Co. Pvt. Ltd, New Delhi, pp. 92–115.

Sbordoni, V., La Rosa, G., Mattoccia, M., Cobolli Sbordoni, M. and De Matthaeis, E. (1987) Genetic changes in seven generations of hatchery stocks of the Kuruma prawn, *Penaeus japonicus.* In: Tiews, K. (ed.) *Selection, Hybridization, and Genetic Engineering in Aquaculture.* Vol. I. Heenemann, Berlin, pp. 143–155.

Seeb, J.E., Thorgaard, G.H. and Tynan, T. (1993) Triploid hybrids between chum salmon female × chinook salmon male have early sea-water tolerance. *Aquaculture* 117, 37–45.

Smith, P.J. (1983) Hatchery rerearing and reseeding in Japan. In: Prospects for snapper farming and reseeding in New Zealand. *Fisheries Research Division Occasional Publication* No. 37, 14–17.

Sugama, K., Taniguchi, N. and Umeda, S. (1988) An experimental study on genetic drift in hatchery populations of red seabream. *Nippon Suisan Gakkaishi* 54, 739–744.

Svåsand, T., Jfrstad, K.E. and Kristiansen, T.S. (1990) Enhancement studies of coastal cod in western Norway. Part I. Recruitment of wild and reared cod to a local spawning stock. *Journal Conseil International Exploration de la Mer* 47, 5–12.

Tave, D. (1986) *Genetics for Fish Hatchery Managers.* AVI Publishing Co., Westport, Connecticut.

Travis, J. (1993) Invader threatens Black, Azov Seas. *Science* 262, 1366–1367.

Troadec, J.P. (1991) Extensive aquaculture: a future opportunity for increasing fish production and a new field for fishery investigations. In: Lockwood, S.J. (ed.) *The Ecology and Management Aspects of Extensive Mariculture. ICES Marine Science Symposium,* Vol. 192. International Council for the Exploration of the Sea, pp. 12–15.

Turner, G.E. (ed.) (1988) *Codes of Practice and Manual of Procedures for Consideration of Introductions and Transfers of Marine and Freshwater Organisms. EIFAC Occasional Paper No. 23.* European Inland Fisheries Advisory Commission. Food and Agriculture Organization of the United Nations, Rome.

Utter, F. and Ryman, N. (1993) Genetic markers and mixed stock fisheries. *Fisheries* 18, 11–21.

Utter, F.M. and Seeb, J.E. (1990) Genetic marking of fishes: overview focusing on protein variation. In: Parker, N.C., Giorgi, A.E., Heidinger, R.C., Jester, D.B., Prince, E.D. and Winans, G.A. (eds) *Fish Marking Techniques.* American Fisheries Society Symposium 7, Bethesda, Maryland, pp. 426–438.

UNEP (1992) *Convention on Biological Diversity, June 1992.* No. 92–8314. United Nations Environment Programme, Nairobi, Kenya.

Waples, R.S. (1991a) Genetic interactions between hatchery and wild salmonids: lessons form the Pacific Northwest. *Canadian Journal of Fisheries and Aquatic Sciences* 48, 124–133.

Waples, R.S. (1991b) Pacific salmon, *Oncorhynchus* spp., and the definition of 'species' under the Endangered Species Act. *Marine Fisheries Review* 53, 11–22.

Watanabe, T., Davy, F.B. and Nose, T. (1989) Aquaculture in Japan. In: Takeda, M. and Watanabe, T. (eds) *The Current Status of Fish Nutrition in Aquaculture. Proceedings of the Third International Symposium on Feeding and Nutrition in Fish.* Toba, Japan, pp. 115–129.

Welcomme, R.L. (1988) *International Introductions of Inland Aquatic Species. Fisheries Technical Paper 294.* Food and Agriculture Organization of the United Nations, Rome.

Williams, J.E., Johnson, J.E., Hendrickson, D.A., Contreras-Balderas, S., Williams, J.D., Navarro-Mendoza, M., McAllister, D.E. and Deacon, J.E. (1989) Fishes of North America endangered, threatened, or of special concern. *Fisheries* 14, 2–20.

Wirgin, I.I. and Waldman, J.R. (1994) What can DNA do for you? *Fisheries* 19, 16–27.

Banking Fish Genetic Resources: The Art of The Possible

B. Harvey
*International Fisheries Gene Bank, c/o World Fisheries Trust,
202–205 Fisgard St, Victoria, BC, Canada V8W 1R3*

Introduction

Concerning the disappearance of plant genetic resources, it has been said that ten thousand years of agricultural development can meet extinction in a single bowl of porridge. Fish genetic resources are equally priceless and threatened – with the sobering difference that, unlike agriculture, the rational development of fish as a crop has scarcely begun. By the time aquaculturists and hatchery managers progress very far in developing genetically improved broodstocks, there may be very little wild raw material left to work with. But when considering the loss of biodiversity in general it is useful to recall the story of the boiled frog: if you put a frog in cool water and raise the temperature slowly enough, it doesn't realize it's being cooked. Human societies are a lot like that frog where many things are concerned – including fish stocks.

FAO recognized the cultural and biological impact of dwindling aquatic biodiversity in its *Expert Consultation on the Conservation and Utilization of Aquatic Genetic Resources* in 1992. The International Fisheries Gene Bank was established to implement some of the recommendations of that consultation, and has as its mission 'to assist in the preservation of aquatic biodiversity by collecting, storing and disseminating fish genetic resources through an international network of fish gene banks'.

Gene banking is an important stone in the 'conservation arch', the centrepiece of which is habitat protection (*in situ* conservation). For threatened stocks and species, gene banking ensures that genetic variability will not be lost while efforts to preserve habitat continue. For species with economic value in aquaculture, gene banking safeguards biodiversity for later use in selective breeding.

Over the past four decades, the international effort to conserve plant genetic resources has had to confront issues of technology and policy. Fish

© 1996 CAB INTERNATIONAL. *Biodiversity, Science and Development: Towards a New Partnership* (eds F. di Castri and T. Younès)

gene banking is proving no different, as this chapter will detail. First, though, a few definitions and clarifications.

First, what is a gene bank? It is a representative collection of the genes – the genetic blueprint – of an organism. It can be the living plants or animals themselves, conserved *in situ* (in place) or *ex situ* (at a remote location). Another option is conservation of reproductive elements like seeds or frozen gametes, usually *ex situ* and considerably less expensive. Either way, a gene bank is raw material for food security: without gene banks we cannot breed better crops, farm better fish, or repopulate depleted stocks.

Second, a gene bank is only part of the picture, and it makes no sense without habitat. Those who feel gene banks are intended to somehow replace habitat conservation – and this is a recurring theme, particularly among enviromentalists – simply have not been told clearly enough that the two are interdependent. A gene bank is a tool – albeit one that, when mentioned in the context of dwindling fish stocks, serves at the very least to 'concentrate the mind' (to quote Mark Twain).

Third, a bank of frozen fish sperm does indeed lack the female genetic contribution. In practical terms, though, that is not much of a problem. First, because even in the worst case scenario where there are absolutely no females left from a particular stock, two generations of backcrossing with a female from a related stock serves to reestablish the genome from the desired stock. Second, 50 sperm samples from 50 different males is all that is needed to overcome genetic bottlenecking – the biggest danger small stocks face – and it only needs the eggs from one female to make the process work.

Lastly, the question of ownership – who has legal access to genetic material, who can benefit from later use, how can it be transferred?. Ownership issues are complicated, and for fish genetic resources the policies are only beginning to evolve. Fortunately, there is precedent to draw on from decades of agricultural gene banking. We need to work on ownership issues, but the water around us frogs is already pretty warm, and if we think we can get all the policy issues sorted out first we will most assuredly get cooked.

Practical Valuation of Genetic Resources

In the mid-1970s in Indonesia, a virus caused a yearly loss of three million tons of rice – enough to feed nine million people. A gene for resistance to this disease was found in one sample of a variety of rice collected in 1963 in India. In fact, only three resistant individual plants were found, but they provided the gene that is now in every high yield rice variety grown anywhere. That variety sat for ten years in a gene bank, unused; if it had not, it is unlikely that a gene for disease resistance would ever have been found (Hoyt, 1992).

Another example: the California barley crop, worth $160 million annually, is protected from dwarf yellow virus by a single gene from Ethiopia – also found in a gene bank. In all, the USDA estimates that genetic resources held in gene banks account for an increase in US crop production worth $1 billion per year.

The point here is that the genes used in these examples – and there are literally thousands more – were available in gene banks when disaster struck a crop. The genes were often in another country, because crops and their genes do not recognize national boundaries, but they were there for breeders because people had collected them. Rebuilding depleted fish stocks and farming fish is breeding too, but where is the raw material? Where can breeders go to find fish genetic resources? At present, only to the wild, and the genetic diversity of wild stocks is finding its way faster and faster into commercial fishing nets.

Fish Gene Banking is Just Beginning

There are over 460 plant gene banks in the world: about a dozen major ones (part of the CGIAR system (the Consultative Group for International Agriculture Research)) and many smaller national banks (IBPGR, 1992). These banks all communicate with each other, and they all exchange material – although that exchange is becoming less free than in the past, as the coming into force of the Convention on Biological Diversity makes the old principle of 'germplasm held in trust for all mankind' less tenable for many developing nations. In contrast, how many fish gene banks are there? There are a few in Norway, India, and a few collections in Europe, the Philippines and in Canada and there is virtually no coordination among them: no sharing of technology and experiences, let alone collections.

Why is this? Gene banks are started for two reasons only. First, when there is an immediate need for the genes, in selective breeding or crop improvement. Second, when enough people sound the alarm about disappearance of the resource. Plant gene banks thus had their origins not only in the needs of breeders for genes, but also in Vavilov's insistence that genetic resources were vanishing, and needed to be saved for future use. The classic 'Catch-22' of fisheries is that, just as we have begun to figure out how to do captive breeding of many fish species, for enhancement and aquaculture (this 'state of infancy' in aquaculture is well described in Chapter 35), the raw material is vanishing before we can use it. Nevertheless, if fisheries resource managers and aquaculturists want an appropriate stock or strain in their rivers and pens, they need the right starting material. Salmon farms in British Columbia are finally figuring this out and have started to draw on private gene banks in their broodstock programmes.

Biodiversity and Food Security

Most people know the surface of the earth is roughly 80% water. Few know that fresh water accounts for only 1% of this. How many know that this tiny 1% produces as many species of fish as the other 49% of salt water? This is an example of biodiversity, defined, vaguely, as the 'total variability among living organisms'. What biodiversity represents – and this is not vague at all – is food security, construction material, and the raw material for industry and medicine.

There are believed to be between ten and 100 million species on earth, but only 5% of these have been discovered and classified. We have a lot of work to do just to make an inventory – pharmaceutical companies certainly know this, and they spend a lot of money prospecting for new drugs from plants and marine animals. The deal between Merck and the government of Costa Rica, in which Merck will pay a royalty on commercialized pharmaceutical discoveries, is frequently cited. But biodiversity – all kinds of biodiversity, not just fisheries diversity – is being lost daily, which makes all our food production systems less reliable at a time when world population is ballooning.

By 2025, food production must double, but at the same time we're losing the biodiversity that makes this production secure. Evidence for the increasing uniformity of our crop systems can be striking. For example, 60% of all of humankind's calories and protein comes from just three plants. Conserving biodiversity, in gene banks or in protected areas or best of all in combinations of the two, is simply an investment in the future.

A Crisis in Fisheries Resources

What does this have to do with fish? For much of the world, fish are the most important source of protein. Fish farmers certainly know this. In the next decade we are going to need 150 million tons of fish to help feed humankind, but the wild fishery can not produce more than 100 million, and in fact has started on the downturn after hitting 90 million. The remaining 50 million tons have to come from aquaculture, a sector that has managed only 14 million so far.

The world's capture fisheries are showing the strain. Consider:

1. The collapse of Canada's Atlantic cod fishery (40,000 jobs lost, with $400 million per year in government payouts);
2. Overfishing of New England groundfish has already cost 14,000 jobs and $350 million annual loss (American Fisheries Society, 1994);
3. Overall, the US marine harvest has dropped 40%, with an annual loss of $3 billion;
4. Atlantic bluefin tuna have been fished to 10% of the levels in 1970 (American Fisheries Society, 1994);
5. Overall, 28% of North American freshwater fish species are extinct or threatened;
6. The US salmon fishery in Washington and Oregon was shut down in 1994 to allow stocks to rebuild;
7. In the US Pacific Northwest, 106 salmon extinctions have been documented, and over 200 stocks are listed as threatened (Nehlsen et al., 1991).

In British Columbia, estimates for the number of unique salmon stocks range from 5000 to 20,000; there are a great many rivers, tributaries, streams and creeks in the province. We cannot save them all, and we should not even try. Most frightening is the fact that not only do we not know what we have, we also have a very incomplete picture of how much we are losing. A study in progress, under the auspices of the American Fisheries Society, will document the state of British Columbia's salmon stocks, but the database for this study is scattered and much of the information is informal. While we sort the data out and fund new studies to fill in the gaps, stocks erode further, and with them goes our fisheries security.

We already know hundreds of stocks have been lost – dozens of chinook salmon stocks in the Vancouver area, for example – and many more are threatened. High profile stocks like the Puntledge and Coquihalla steelhead are well known; as small stocks they are particularly vulnerable to incidental interception in a mixed-stock fishery. In the Shuswap Traditional Territory in Southeastern British Columbia – a very small portion of the province – there are at least 30 threatened stocks, and that is in just a few tributaries of the North Thompson River – not the mainstem Fraser, not the Nass, the Stikine, or the Skeena (Shuswap Nation Fisheries Commission, 1993, personal communication).

I use the general term 'security' to emphasize how much we all depend on fish stocks. In British Columbia the average person thinks about commercial fisheries and salmon farms, but in fact the recreational fishery in British Columbia outperforms the commercial fishery in terms of dollars generated.

Conserving Biodiversity is an Investment in the Future

Internationally, conserving and using biodiversity are beginning to be seen as an investment in the future: if we want sustainable development, business has to invest. Government alone cannot foot the bill, nor should it. Putting a dollar value on biodiversity is critical, however, for industry will only support environmental initiatives that make economic sense. A pharmaceutical company wishing to manufacture in Brazil, for example, has an absolute requirement for clean water.

An interesting lesson in the application of new conservation strategies is provided by the official acceptance of gene banking as a tool to protect salmon stocks in British Columbia. To insure a resource that maintains an industry worth almost $1 billion annually for the recreational fishery alone, the government's outlay on gene banking is currently less than $5000. And even this expenditure is only reluctantly acknowledged, due to concerns that once the public is aware genes are being banked the alarm on stocks will truly be sounded. This is a natural bureaucratic response, and it can only be countered by education of both the government and the public. Both must come to understand that conservation of genetic resources must make use of all the available tools, often employed in creative ways. Another problem is lack of exposure to international undertakings: many fisheries managers are unfamiliar with the Convention on Biological Diversity and its enormous implications for conservation and utilization of genetic resources. By the nature of their jobs their concerns are local, but it is on the international stage that innovative strategies marrying technology and policy are starting to emerge.

Practical Contributions of the International Fisheries Gene Bank

The contribution of the International Fisheries Gene Bank (IFGB) is straightforward and practical. The gene banking technology was developed in Canada by MTL Biotech Ltd, and comprises a field kit that simplifies the cryopreservation of fish sperm, making it possible for anyone to do it, in the field. In the last two years, IFGB has started projects in British Columbia (with the Shuswap Nation), Venezuela, Colombia and Brazil, and is developing one in China to help preserve germplasm from Yangtze carps, the basis for much of that country's fish culture. IFGB trains local people to do the freezing – in many cases people with no high school education – and acts as the coordinating hub for a growing network of regional gene banks. Projects in South America focus at present on cachama or pacu, members of the genera *Colossoma* and *Piaractus* indigenous to the Amazon and Orinoco basins and now farmed widely throughout tropical areas of the world (Mace and Harvey, 1995).

The readily transferable technology package – equipment and training – is going to allow a wide variety of organizations to collect, store and exchange fish genetic resources. Users include national governments, the aquaculture sector and private industry involved in projects that have an impact on fish stocks. One thing is certain: as the Convention on Biological Diversity begins to dictate environmental statutes – as it already is in developing countries, so that industry has to pay attention at the very least to compensation to national governments for removing potentially valuable genetic resources – there will be a rapidly increasing need for exportable technologies like that of IFGB.

In conclusion, I want to provide a comparison that has to do with the casualties of war, but not the human casualties. It provides the perfect illustration of why we need gene banks, even of healthy stocks. In 1989, the Somali government sent out of the country 300 samples of sorghum and maize seeds. They were hand-carried by staff member of the International Plant Genetic Resources Institute and deposited in national gene banks in Kenya, India, and Nigeria. In 1992, fighting in Somalia claimed 400,000 lives. Research stations at Afgoi and Baidoa, which held gene banks for these crops, are now derelict, and the seeds looted for food. Farmers' underground seed collections were destroyed; others were eaten to prevent starvation. These seeds were adapted to Somali conditions; they were unique. Thanks to the foresight of someone in the Somali government, they still exist, and now, they are beginning to be reintroduced (IBPGR, 1993).

The same thing has happened in Rwanda – this time the programme, called *Seeds of Hope*, is better publicized, but it is accomplishing the same thing: banking priceless, locally adapted genetic resources as an insurance policy, to be repatriated when the hostilities are over.

In British Columbia and all of the other areas where IFGB is working there is no war on. But genetic resources are the loser in hostilities of another kind: dams, bulldozers, and above all nets have taken their toll on stocks. And it is still going on: in 1994 there was a 'fish war' between Canada and the United States, and recreational, aboriginal and commercial interests continue to argue about allocations. In the face of these hostilities, we need a backup. We need a 'Seeds of Hope' programme for threatened fish stock. That is what I would tell any local fisheries manager: restore the habitat, regulate the fishing, improve the allocations, but be sure you have something to put back when you have finished. By doing so, a model for fish genetic resource conservation for the world will begin to develop.

References

American Fisheries Society (1994), *Fisheries* 19(7), 10.

FAO (1992) *Report of the Expert Consultation on Utilization and Conversation of Aquatic Genetic Resources*, Grottaferrata, Italy, 9–13 November 1992. FAO Fisheries Report No. 491. FAO, Rome.

Hoyt, E. (1992) *Conserving the Wild Relatives of Crops*. International Board for Crop Genetic Resources, the International Union for the Conservation of Nature, and the World Wildlife Fund.

IBPGR (1993) *Geneflow*. International Board for Plant Genetic Resources, Rome.

IBPGR (1992) *Partners in Development* 2nd edn. International Board for Plant Genetic Resources, Rome.

Mace, T. and Harvey, B. (1994) Establishing a gene bank for aquaculture in Colombia. *World Aquaculture* 25(2), 26–28.

Mace, T. and Harvey, B. (1995) Gene banking in Aquaculture in South America. *Pacific Coast Aquaculture*, February 1995, 7–9.

Nehlsen, W., Williams, J.E. and Lichatowich, J.A. (1991) Pacific salmon at the crossroads: stocks at risk from California, Oregon, Idaho and Washington. *Fisheries* 16(2), 45–21.

38 Introduction of Fish Species in Freshwaters: A Major Threat to Aquatic Biodiversity?

C. Lévêque

ORSTOM, 213 rue La Fayette, 75480 Paris Cedex 10, France

Introduction

Freshwater ecosystems occupying less than 1% of the surface of the Earth, are inhabited by more than 25% of the vertebrates. Fish which consitute the bulk of this fauna, are currently exposed to the consequences of major impacts resulting from anthropogenic activities: management of water systems for hydropower or agriculture purposes, pollutions, overexploitation of natural resources including living resources, introduction of alien species. The result is that a great number of endemic fish species are presently endangered.

Among major threats, the question of the introduction of alien fish species became a matter of debate among freshwater ecologists and managers during the last decades. The main goals of voluntary introductions, most usually decided by fishery officers, were initially to improve sport fisheries, artisanal fisheries and aquaculture, or to develop biological control of aquatic diseases, insects and plants. More recently, transfers of ornamental fish for the aquarium trade have also sharply increased, as well as 'accidental' introductions of these species into freshwater ecosystems.

The future promises a continuing spread of exotic species, and managers of aquatic ecosystems will be confronted increasingly with a shifting mix of native and non-native species. However, although the introduction of species has been encouraged all around the world both by managers and scientists for centuries, many ecologists consider it today to be like Pandora's box. The consequence is that any planned introduction is viewed as a potential ecological catastrophe, and that both its ecological and economical potential values are not objectively considered.

Are introductions of fish species a way to encourage in order to improve fisheries, or a major threat to freshwater biodiversity, or a game of chance as suggested by number of ecologists? Whereas many scientists, more or less intuitively, claim that the introduction of exotic species is risky, there are also examples of assumed success when the goal was to improve artisanal fisheries. We need therefore a very careful assessment of past experiences in order to develop general guidelines and policies about fish introductions which are scientifically based.

Purposes of Introductions

The first attempts to introduce true exotic species are not exactly known, but the modern era of introductions started during the second half of the nineteenth century. In Europe, these introductions of exotics involved mainly North American species: salmonids (the brook trout was introduced to the United Kingdom in 1869), catfishes (*Ictalurus* spp.) and centrarchids.

There are different motives for performing introductions (Welcomme, 1988; Moreau *et al.*, 1988). During the nineteenth century and until about World War II, fish introductions were largely an outgrowth of colonialism and nostalgia by settlers in their newly adopted homelands for the familiar species and surroundings left behind (Welcomme, 1984). Acclimatization societies proliferated for instance in New Zealand in the late 1800s to a point that of the 46 fish species now found, 20 are exotic (McDowell, 1990). Several species, usually salmonids or top predators, have been introduced in different parts of the world as sport fish. The most spectacular example is that of the brown trout (*Salmo trutta*), native from Europe and now occurring worldwide (Baglinière and Maisse, 1991).

More than a hundred fish species have been introduced for fish culture, but less than ten now have a worldwide distribution. The common carp (*Cyprinus carpio*) and the herbivorous carp (*Ctenopharyngodon idella*), now occur in many temperate countries. In the intertropical area, several species of tilapiines (family Cichlidae) native to Africa, such as *Oreochromis niloticus* and to a lesser extent *Oreochromis mossambicus*, make up the bulk of the fish culture production in Southeast Asia. They have also been introduced in South America, as well as Australia.

One of the purposes of fish introductions has been the control of the aquatic vegetation. The grass carp, *Ctenopharyngodon idella*, has been successfully used in many countries for the control of aquatic macrophytes. *Tilapia rendalli* as well has been introduced into Sudanese irrigation channels and some artificial lakes of Shaba to control aquatic vegetation (Philippart and Ruwet, 1982).

Fish introduction has also been suggested for disease control. That is the case for instance for *Astatoreochromis alluaudi*, a 'mollusc crushing' cichlid under threat of extinction in East Africa, whose introduction to Cameroon has been proposed for control of snails involved in schistosomiasis transmission (Slootweg, 1989). The mosquitofish (*Gambusia affinis*) native from southern United States, and guppy (*Poecilia reticulata*) native from north-eastern South America, were introduced worldwide after the World War II to control larvae of mosquito vectors of malaria.

A current major cause of introduction is the trade of ornamental fish. For the past few decades, several thousands of small, usually tropical species, have been disseminated around the world by the aquarium fish trade. A famous and ancient example is the popular gold fish, *Carassius auratus auratus*, which has achieved a wide distribution for rearing in ornamental ponds. Most of these fish are rarely recorded as introduced, but can accidentally escape or be introduced in natural water bodies. It is generally assumed that they cannot survive in temperate climates, although populations may survive in artificial warm waters.

Many introductions of alien fish into natural environments appear to have been more or less accidental, and occurred during the deliberate introduction of another species, because of confusions in identification. That is the case, for instance, with batches of stock-fish contaminated with unnoticed juveniles of another species.

What Have We Learned about the Consequences of Fish Introductions on Biodiversity and the Aquatic Environment?

From the different experiences in fish introduction, some general reflections should be derived.

The introduction of large predators can be almost catastrophic in complex fish communities which evolved in tropical lakes with considerable specialization in feeding and reproduction. A well-documented example was the accidental introduction of the cichlid piscivore, *Cichla ocellaris*, to Gatun lake, Panama, around 1967 (Zaret and Paine, 1973). In less than five years, the introduced species which has voracious predatory habits, led to the elimination of six of the eight common fish and decimated a seventh.

The large predator *Lates niloticus* was introduced into Lake Victoria during the 1950s and early 1960s to boost fisheries. During the first 25 year after its introduction, catches of the Nile perch were insignificant. However, in the early 1980s an explosive increase in the catches took place and simultaneously stocks of endemic cichlids have been decimated (Ogutu-Ohwayo, 1990). It can be assumed that the establishment of the Nile perch was initially supported by the large quantities of native haplochromines which were their main prey. As a result, about two-thirds of the flock of 300 haplochromine cichlid species have disappeared or are threatened with extinction (Witte *et al.*, 1992).

One of the major problems in fish species introductions, is their irreversibility, at least on the scale of a human's lifetime. Once introduced and established, it is impossible, given current technology, to eradicate a fish species from a large natural water body.

Most information about the effects of introduced species on native fish is often anecdotal and fragmentary, not only because so little is known about native fish, but because the effects of the introduced fish have been overshadowed by the effects of environmental changes which have occurred simultaneously. For instance, although it has been claimed that the introduced

Nile perch was responsible for the decline of the native cichlids, the impact of human activities in the vicinity of Lake Victoria could also be involved. The lake has experienced profound changes over the last 30 years (Witte et al., 1992). Eutrophication, as well as the introduction of new fishing techniques (trawling) certainly contributed significantly to the decline of the *Haplochromis* fauna. It has been shown that the haplochromine stock was already affected by fisheries before the establishment of *Lates* (Ogutu-Ohwayo, 1990).

In newly created habitats (geologically young lakes, reservoirs) only a few species of riverine fish are adapted to take advantage of the deep or open waters and to occupy adequately the newly created niches. The creation of reservoirs provides the habitat but not the preadapted inhabitants (Fernando, 1991). From the vast number of tropical freshwater fish only Cichlidae and Clupeidae, and to a limited extent Cyprinidae, have proved to be preadapted to colonize lacustrine habitats and to increase the total fish yield substantially. However, it seems reasonable to limit their introduction into reservoirs or natural lakes of recent origin, while introduction into ancient lakes must be very carefully considered (Fernando and Holcik, 1982).

In aquatic ecosystems containing a very depauperate indigenous freshwater fauna as a result of historical and biogeographical events, the introduction of appropriate fish species that would occupy presently vacant or underutilized niches, could significantly improve the fishery. For example, *O. mossambicus* introduced in the Sepik River (Papua New Guinea) makes up about 50% of the present catch in inland fisheries (Coates, 1987). It should also be recalled that Ireland's current freshwater fish fauna consists of 20 species, whereas after the last glaciation, the fish fauna probably consisted of only about eight anadromous species. At least seven species were introduced into Ireland within the last 400 years or so (Fitzmaurice, 1984).

Introductions for the purposes of fish conservation is a new option, for conservationists. A captive breeding programme may be considered as a way to save taxa threatened with extinction in the wild and to provide propagules for the repopulation of natural habitats. However, the restocking of captive bred fish to their native habitats is not always realistic and cannot necessarily be considered as the ultimate goal of breeding programmes. Indeed, reintroduction should take place only in localities where the original causes of extinction have been removed, and where habitat requirements of the species are satisfied. In the case of Lake Victoria for instance, reintroduction would be futile as long as the *Lates* threat persists, and the eutrophication of the lake is not controlled. To save endangered fish species, conservationists, therefore suggested introducing them into other ecologically suitable habitats, similar or identical to the original one, in order to initiate new safety net populations in appropriate waters. In other words, fish introductions may receive particular attention in that case, as a way of saving endemic species from extinction, providing similar habitats may be found for these species.

Conclusions

Are introductions a game of chance? Most scientists more or less agree that the introduction of exotic species is risky, and many examples have been given that provide evidence of the potential threats. But there are also examples of apparent success which demonstrate the need for very careful impact assessments prior to any introduction. The assessment of the impact of fish introduced in Asia (De Silva, 1989), is that this continent has suffered few losses of native species to introductions.

The need for better communication between the scientific community and managers or administrators has been raised many times (Balon and Bruton, 1986) as well as the need for establishing formal protocols for the evaluation of risks, both for indigenous species and for ecological balance, prior to any introduction. Balon and Bruton (1986) also clearly pointed out the difficulties of such procedures: are decision-makers willing to accept advice which does not conform to their schemes? Do scientists have sufficient knowledge to give clear and unequivocal advice? In fact, biologists have a limited ability to predict the outcome of introductions because of an inability to answer some fundamental biological questions.

Applied research programmes should be devised urgently to improve the documentation of the consequences of fish introductions, starting with the more common species. Both field and controlled experiments should be conducted using the already available information to assess direct and indirect effects of introductions. But decision-makers should also realize that the lack of regulation and control makes it difficult nowadays to avoid possible major ecological threats in the near future.

References

Allan, J.D. and Flecker, A.S. (1993) Biodiversity conservation in runnning waters. *BioScience* 43, 32–43.

Baglinière, J.L. and Maisse, G. (1991) *La Truite: Biologie et Écologie*. INRA Editions, Paris.

Balon, E.K. and Bruton, M.N. (1986) Introduction of alien species or why scientific advice is not heeded. *Environmental Biology of Fishes* 16, 225–230.

Coates, D. (1987) Consideration on fish introductions into the Sepik River, Papua New Guinea. *Aquaculture and Fisheries Management* 18, 231–241.

De Silva, S.S. (1989) Exotic aquatic organisms in Asia. *Special Publication of the Asian Fisheries Society* 3, 156 pp.

Fernando, C.H. (1991) Impacts of fish introductions in tropical Asia and America. *Canadian Journal of Fisheries and Aquatic Science* 48 (Suppl. 1), 24–32.

Fernando, C.H. and Holcik, J. (1982) The nature of fish community: a factor influencing the fishery potential and yields of tropical lakes and reservoirs. *Hydrobiologia* 97, 127–140.

Fitzmaurice, P. (1984) The effects of freshwater fish introductions into Ireland EIFAC Technical paper (suppl. vol) 449–457.

McDowall, R.M. (1990) *New Zealand Freshwater Fishes: a Natural History and Guide.* Heinnemann, Auckland, New Zealand.

Moreau, J., Arrignon, J. and Jubb, R.A. (1988) Les introductions d'espèces étrangères dans les eaux continentales africaines. In: Lévêque, C., Bruton, M.N. and Ssentongo, G.W. (eds) *Biology and Ecology of African Freshwater Fishes.* ORSTOM, Paris, pp. 395-425.

Ogutu-Ohwayo, R. (1990) The reduction in fish species diversity in lakes Victoria and Kyoga (East Africa) following human exploitation and introduction of non-native fishes. *Journal of Fish Biology* 37, 207-208.

Philippart, J.C. and Ruwet, J.C. (1982) Ecology and distribution of Tilapias. In: Pullin, R.V.S. and Lowe-McConnell, R. (eds) *The Biology and Culture of Tilapias.* ICLARM Conference Proceedings 7. ICLARM, Manila, Philippines, pp. 15-59.

Slooteweg, R. (1989) Proposed introduction of *Astatoreochromis alluaudi*, an East African mollusc-crushing cichlid, as a means of snail control. *Annales du Musée Royal d'Afrique Centrale, Sciences Zoologiques* 257, 61-64.

Welcomme, R.L. (1984) International transfers of inland fish species. In: Courtenay, W.R. and Stauffer, J.R. (eds) *Distribution, Biology, and Management of Exotic Fishes.* The Johns Hopkins University Press, Baltimore, Maryland, pp. 22-40.

Welcomme, R.L. (1988) International introductions of inland aquatic species. *FAO Fisheries Technical Paper* 294, Rome, 318 pp.

Witte, F., Goldschmidt, T., Wanink, J., van Oijen, M., Goudswaard, K., Witte-Maas, E. and Bouton, N. (1992) The destruction of an endemic species flock: quantitative data on the decline of the haplochromine cichlids of Lake Victoria. *Environmental Biology of Fishes* 34, 1-28.

Zaret, T.M. and Paine, R.T. (1973) Species introductions in a tropical lake. *Science* 182, 449-455.

39 Biodiversity and Fisheries and Aquaculture Development in Taiwan

I. Chiu Liao[1] and Yew-Hu Chien[2]

[1]Taiwan Fisheries Research Institute, 199 Hou-Ih Road, Keelung, Taiwan 202; [2]Department of Aquaculture, National Taiwan Ocean University, 2 Pei-Ning Road, Keelung, Taiwan 202

Introduction

Biodiversity has been defined in various ways, ranging from the diversity in genes, species, ecosystems, to biological resources in general (e.g. Solbrig, 1991; Flint, 1991; Solbrig and Nicolis, 1991; WRI/IUCH/UNEP, 1992; CGIAR 1992). Since aquaculture – the husbandry of aquatic organism – and fisheries (capture) – the exploitation of aquatic resources – are both practised mainly for economic purposes and are considered primary industries that deal closely with living organisms and livelihood, biodiversity in aquaculture and fisheries will be elaborated in this chapter from both bio- and socioeconomic viewpoints so that a broader understanding between biodiversity and fisheries and aquaculture development in Taiwan can be made.

It is natural for Taiwan, an island nation surrounded by seas and with a deeply rooted seafood-loving custom, to rely heavily on seafood for animal protein. This tendency results in a vast and increasing demand for seafood. Such demand has established both fisheries and aquaculture as the basic purpose of a primary industry. Depletion of fishery resources due to overfishing and deteriorated aquatic environment further powered the development of aquaculture and brought to light the importance of sea ranching. Large-scale expansion of the aquaculture industry started when aquaculture was used to produce commodities for export and earning foreign exchange to cope with the development of an export-oriented and industry-based economy (Chien, 1990) – the culture species being used is thus focused on export-oriented ones. The adverse impacts on the natural environment brought about by a rapidly growing aquaculture industry through the overutilization of natural resources, such as land and water, have raised the issue of conservation and have constrained the direction and pace of aquaculture development toward a more environment friendly one. This chapter aims to see what possible roles biodiversity has played in Taiwan during the shift from industry development to conservation.

Overview

Coastal fisheries production in Taiwan has declined from around 57,000 tons in 1986 to around 45,000 tons in 1992. Improvements and diversification of fishing gears and methods have placed tremendous pressure on the fishery resources. Illegal fishing including poisoning, dynamite fishing, electric fishing, undersize fishing, and fishing during offseason, and pollution from land are believed to have depleted and damaged the fishery resources and aquatic ecosystem. The decline in production is most significant in the mullet and anchovy fisheries. Land pollution has resulted in the shrinkage of fry populations and has jeopardized the livelihood of fry collectors. The major fry collected for aquaculture use are eel, milkfish, mullet, grouper, and other high-value finfish and juveniles of lobster. Partial replacement of coastal commercial fisheries with leisure fishery should to some extent relieve the fishing pressure and allow the recovery of the biodiversity of the coastal ecosystem.

The influence of aquaculture on biodiversity is much more significant than fisheries. The species considered for aquaculture by farmers are demand-oriented, either popular family food animals, although relatively low priced, such as tilapia, milkfish, Cyprinidae, hard clam, and oyster, or high-value restaurant animals, such as prawn, grouper, bream, perch, and porgy, or high-value export fishes, such as prawn and eel. Although the demand and price fluctuate, the number of commercial culture species in Taiwan has increased from 44 in 1978 (Li and Liao, 1979), 55 in 1987 (Liao and Shyu, 1992), to 70 in 1991 (Liao, 1991).

Significant diversification of culture species occurred when the grass prawn and kuruma prawn farming industries collapsed in 1988 and 1992, respectively, and farmers switched from monoculture to polyculture or completely to marine finfish culture.

Another recent diversification is that eel farmers have started to culture American and European eels when Japanese eel fry is not available. The consequences of land subsidies due to overtaxed ground freshwater have encouraged farmers to culture species that rely on less freshwater.

The 38 aquatic species known to have been introduced to Taiwan up until 1989 (Liao and Liu, 1989), include 25 finfishes, five crustaceans, two molluscs, three reptiles, and three seaweeds. Among these, 20 species have been commercially cultured, and at the other extreme, a few of them have been later regarded as pests and thus discarded. The purpose of introducing these exotic species for culture were at first, to enhance the animal protein supply in Taiwan, but later on, to export and to fulfil consumers' preferences, or to improve breeds. The most and probably the only prominent success in genetic expansion of aquatic animals in Taiwan is the hybridization of tilapias. This success has made tilapia the most common food fish. In 1988, tilapia production in quantity ranked first, or 18% of total aquaculture production. Introduction of exotic species increases the biodiversity for aquaculture but at the same time may decrease the biodiversity of the indigenous biota because some exotic species may be predators, competitors, and even disease carriers.

Using the cold effluent from OTEC (Ocean Thermal Energy Conversion) in the future to culture cold water species will further add biodiversity in aquaculture (Chien, 1989). The adverse impact of exotic species on the natural environment has not been fully studied. Different strains of the indigenous species have been imported to Taiwan to increase the seed supply for aquaculture. One is the grass prawn female broodstock from Southeast Asia, the tropical strain and the other is Japanese eel fry from China, Japan, and Korea, the temperate strain. The Japanese strain of ayu was imported in the 1960s in order to repopulate local waters of the extinct Taiwanese strain of ayu. The consequences of replacing the indigenous strain with the exotic strain, or to mix these two strains, need to be assessed. Biotechnology application on aquaculture, especially the feasibility of creating super transgenic fish, has been widely studied.

Sea ranching started in the 1970s. The species raised are: Japanese eel, grass prawn, small abalone, and black porgy. Since sea ranching is conducted on a limited scale, the enhancement of fishery resources has not been significant and the effects on the gene pool of the wild population have not been assessed. However, release of surviving animals, usually the disease-contaminated ones after culture failure, may have introduced inferior genes into the wild population. Only the screened culture animals being released into the open ecosystem for sea ranching has gained increasing attention. An average of around 3000 artificial reefs per year have been placed around the coast for the past 15 years to improve the habitat in the hope of enchancing fishery resources.

When Taiwan becomes a member of General Agreement on Tariff and Trade (GATT), the industries which are not competitive, either in aquaculture or fishery, will be phased out, and the biodiversity of these industries will certainly decrease.

References

CGIAR (1992) Biodiversity and plant genetic resources. In: *Contribution to the Debate on Biodiversity for the United Nations Conference on Environment and Development, Brazil.* 20 pp. Consultative Group on International Agricultural Research, Washington, DC.

Chien, Y.H. (1989) The feasibility of using artificial upwelling water in aquaculture in Taiwan. In: Liu, C.K. and Sun, L.C. (eds) *Artificial Upwelling and Mixing. Proceedings of the International Workshop on Artificial Upwelling and Mixing in Coastal Water.* National Taiwan Ocean University, Keelung, Taiwan, pp. 153-166.

Chien, Y.H. (1990) Aquaculture in the Republic of China: A biosocioeconomic analysis of the aquaculture industry in Taiwan. In: Liao, I.C., Shyu, C.Z. and Chao, N.H.

(eds) *Aquaculture in Asia: Proceedings of the 1990 APO Symposium on Aquaculture.* TFRI Conference Proceedings 1. pp. 31-50.

Flint, M. (1991) *Biological Diversity and Developing Countries: Issues and Options.* Overseas Development Administration, London, UK. 50 pp.

Li, Y. and Liao, I.C. (1979) Introduction to aquaculture in Taiwan. *Proceedings of the World Mariculture Society* 10, 229-237.

Liao, I.C. (1991) Aquaculture: The Taiwanese experience. *Bulletin of the Institute of Zoology, Academia Sinica, Monograph* 16, 1-36.

Liao, I.C. and Liu, H.S. (1989) Exotic aquatic species in Taiwan. In: de Silva, S.S. (ed.) *Exotic Aquatic Organisms in Asia: Proceedings of a Workshop on Introduction of Exotic Aquatic Organisms in Asia.* Asian Fisheries Society Special Publication No. 3. Asian Fisheries Society, Manila, Philippines, pp. 101-117.

Liao, I.C. and Shyu, C.Z. (1992) Evaluation of aquaculture in Taiwan: Status and constraints. In: Marsh, J.B. (ed.) *Resources and Environment in Asia's Marine Sector.* Taylor & Francis, Washington, DC, pp. 185-197.

Solbrig, O. T. (1991) Biodiversity. *MAB Digest 9*, May. UNESCO, Paris, France.

Solbrig, O.T. and Nicolis, G. (1991) Biology and complexity. In: Solbrig, O.T. and Nicolis, G. (eds) *Perspectives in Biological Complexity.* IUBS, Paris, France, pp. 1-6.

40 Biodiversity and Marine Biotechnology: a New Partnership of Academia, Government and Industry

R.R. Colwell
*University of Maryland Biotechnology Institute,
4321 Hartwick Road, Suite 550, College Park,
MD 20740, USA*

Introduction

Marine biotechnology has emerged as a major component of the biotechnology revolution and is rooted in the traditions of marine biology. Marine biotechnology was first defined only a decade ago (Colwell, 1983). The marine environment offers a source of new drugs, cures, and chemicals, and a rich source of protein and nutrients. Presented here is the use of marine biotechnology as an example of potential, and actual, interaction of industry, government, and the university to develop this new economic sector.

In a strict sense, marine biotechnology may be defined as 'a set of scientific techniques that use living marine organisms, or parts of marine organisms, such as cells, to make or modify products, to improve plants or animals, or to develop organisms for specific applications'. Recently, a study was done to identify companies in the US that could be identified as being dedicated to marine biotechnology. In the summer of 1991, from information provided by the companies and from interviews of scientists and managers employed in industry, corporate strategies for marine biotechnology were explored, as well as strengths and weaknesses in corporate R&D programmes, and links between industry and universities (Zilinskas *et al.*, 1994).

Marine Aquaculture and Biotechnology

Total world fish production yields from aquaculture reached 14 million metric tons (mmt) in 1991. Expansion of marine aquaculture will contribute greatly to meet the growing world demand for fresh seafood. In addition, aquaculture of seaweed and phytoplankton yields high-value products, especially algal-derived polysaccharides and chemicals. A good example of the progress being made in finfish aquaculture is the work being done on the regulation of the hormonal control of reproduction of finfish and shellfish in fish farming. During the past decade, direct genetic manipulation of fish, using recombinant DNA techniques, has begun to revolutionize the aquaculture industry and make it possible to develop an entirely new approach to fish farming. Fish are highly amenable to genetic manipulation, because fish eggs are characteristically large and, therefore, can be micro-injected with DNA constructs, after which external fertilization and, subsequently, development will take place. Major research efforts in genetic manipulation have been directed at enhancement of growth and production of fish with superior resistance to cold temperatures. Also, disease-resistant fish is an increasingly important objective.

Of more practical significance has been the achievement of growth enhancement using fish growth hormone. T. Chen, at the Center of Marine Biotechnology, University of Maryland Biotechnology Institute (UMBI/COMB) and D. Powers, then at the Johns Hopkins University and a joint faculty member of UMBI/COMB, demonstrated that growth hormone in rainbow trout is encoded by two separate genes (Agellon et al., 1988a). A large amount of biologically active rainbow trout growth hormone was prepared by expressing one of the rainbow trout GH genes in the bacterium *Escherichia coli*. This hormone was administered to rainbow trout by injection or dipping, yielding enhanced growth of the trout (Agellon et al., 1988b). Obviously, exogenous GH application by injection is not very practical for large-scale aquaculture, since it is labour-intensive and requires individual treatment of each fish. However, implants and the use of ultrasound offer mechanisms for automation of application of the hormones. Another approach is generation of transgenic fish. This was first achieved by transfer of a rainbow trout GH gene to common carp and channel catfish (Zhang et al., 1990).

'All-fish' gene constructs have also been used for growth enhancement of fish. Hackett and colleagues at the University of Minnesota developed expression vectors containing the proximal promoter and enhancer regulatory elements of the carp-B actin gene and the polyadenylation signal from the salmon growth hormone gene (Liu et al., 1990). Growth enhancement was subsequently obtained in Atlantic salmon using an all-fish gene construct. The construct was an antifreeze protein gene promoter linked to a chinook salmon GH gene. These transgenic Atlantic salmon demonstrated enhanced growth (Du et al., 1992).

Many marine fish inhabiting cold waters produce proteins which act as an 'antifreeze', i.e. protect fish by inhibiting the formation of ice crystals in their serum. These proteins are termed antifreeze glycoproteins and antifreeze polyproteins/polypeptides. Atlantic salmon lack genes coding for these proteins and, therefore, cannot survive in icy waters (Hew et al., 1991). The mechanism whereby antifreeze proteins bind ice crystals and inhibit ice formation has been described (Raymond et al., 1989). Genes coding for antifreeze proteins in Arctic flounder have been transferred and expressed in Atlantic salmon and inheritance of the genes was demonstrated (Shears et al., 1991).

Expression of an adequate concentration of antifreeze proteins in salmon blood can extend the geographical range within which this fish can be cultured. From the view of industrial application, antifreeze proteins from fish may also prove valuable for hypothermic preservation of mammalian organs, especially for transplant operations (Lee et al., 1992).

Not only finfish, but also shellfish, are amenable to genetic manipulation, especially for enhancing both the rate of growth and size of the adult. The exogenous application of bovine GH enhanced the growth rates of California red abalone (Morse, 1984). Similar results were reported when biosynthetic rainbow trout GH was applied to juvenile oysters (Paynter and Chen, 1991).

Many fish species demonstrate poor ovulation and spawning in captivity. Manipulation of water temperature and photoperiod has been used with some success in attempts to improve spawning.

Spawning in fish is initiated by a surge of gonadotropin (GtH) secretion from the pituitary. There is growing evidence that this surge is frequently absent in fish raised in captivity (Zohar, 1988a). An important factor controlling induction of a GtH ovulatory surge is gonadotropin releasing hormone (GnRH). Administration of GnRH, or its analogues, offers, therefore, an efficient method for control of ovulation and spawning. Analogues of GnRH have been synthesized, which have been found to possess increased resistance to degradation (Zohar et al., 1990b), as well as possessing higher affinity to pituitary receptors (Pagelson and Zohar, 1992). These analogues have subsequently proven to be superactive in the induction of spawning.

Controlled release delivery systems for the hormones show potential for inducing and synchronizing spawning in several fish species important in aquaculture, including Atlantic salmon (Crim and Glebe, 1984), trout (Breton et al., 1990) and seabream (Zohar, 1988b). Novel approaches have been developed by utilizing advanced polymer technology to produce implants that slowly dissolve and release hormones at a steady rate. Manipulation of spawning in farmed fish has been achieved by sustained administration of GnRH analogues via polymer-based delivery systems (Zohar et al., 1990b). Low intensity ultrasound has been shown to enhance dramatically the uptake of test peptides into the circulation of fish blood and offers an alternative approach to the use of polymer-based delivery systems. The use of low intensity ultrasound has tremendous potential for improved hormone and drug delivery in aquaculture (Zohar et al., 1990a) because it is non labour-intensive and minimizes handling of the treated fish.

Aquaculture of marine macroalgae, i.e., seaweeds, has been practised for several centuries in Asian countries, particularly Japan, and products from these algae have been widely used as medicines and food. Microalgae culture is practiced in Australia, Israel and the US. Macro- and microalgae yield a wide range of products, including food additives and supplements, culture media, pesticides, plant growth regulators, and antibacterial, anticancer and antiviral agents (Harvey, 1988).

Microalgae have proven useful for large-scale production of the long-chain fatty acids, eicosapentaenoic acid (EPA) and docosahexaenoic acid

(DHA) (Kyle et al., 1991). Diets rich in these omega-3 oils have been suggested to reduce the risk of coronary vascular disease. The green microalga, *Dunaliella salina*, is grown in large-scale, intensive culture in California and Australia to produce beta-carotene (Kranzfelder, 1991), a vitamin A precursor associated with the prevention of cancer. *Dunaliella* cells can accumulate β-carotene to up to 10% of their dry weight and are, therefore, an excellent source of this substance (Avron and Ben-Amotz, 1992). It has even been suggested that 'oceanic farming' of marine algae can reduce global carbon dioxide levels (North, 1991) by increasing rates of fixation of carbon dioxide into organic material.

Application of biotechnology to cultivation of marine algae presents an opportunity for riparian countries, especially developing countries, with extensive coastal areas. This potential is most likely to be realized by the formation of partnerships with industrialized countries. Although molecular techniques have not yet been widely applied to achieve strain enhancement or production of transgenic plants and algae of commercial importance, this approach is being taken in several laboratories in the USA, Asia, and Europe. Most of this work employs protoplast fusion. Protoplast fusion has been somewhat limited in application to seaweeds because of difficulties in obtaining plant regeneration from protoplasts of complex algae. Successful genetic manipulation, using protoplast fusion, has been achieved. For example, protoplast fusion has proven to be a useful tool in production of red algae hybrids, specifically the commercially valuable, agar-producing seaweed *Gracilaria* (Cheney, 1990). Successful protoplast fusions and regeneration have also been reported for *Porphyra perforata* (Polne-Fuller and Gibor, 1984) and *Porphyra nereocystis* (Waaland et al., 1990); important achievements since the edible product, nori, is derived from *Porphyra* species and recently in red microalgae (S. Arad, 1994, Israel, personal communication). Direct DNA manipulation, using vectors for gene transfer or techniques such as electroporation or biolistics are just beginning in genetic studies of macroalgae.

Natural fish and shellfish populations, as well as marine mammals, are susceptible to viral, bacterial, fungal, and protozoan infections. Animals raised in intensive aquaculture are especially vulnerable to disease. Since massive use of antibacterials and/or antibiotics in aquaculture can be counterproductive, molecular techniques employed in marine biotechnology to counteract disease are important.

Important diseases of fish include those caused by infectious pancreatic necrosis virus (IPNV) and infectious haematopoietic necrosis virus (IHNV), that infect salmonids. Recombinant DNA technology has been used to construct viral subunit vaccines for IPNV (Manning and Leong, 1990) and IHNV (Gilmore et al., 1988). These vaccines have induced protective, long-lasting immunity in laboratory trials (Leong et al., 1991). Because there are no effective antiviral treatments available, unlike bacterial diseases where antibiotic treatment is frequently used with good effect in aquaculture, vaccines protective against viral disease can be highly effective.

Enzymes

Enzymes resulting from discoveries of novel marine microorganisms have made possible some important new techniques in biotechnology, e.g. high-temperature-resistant polymerases, from hydrothermal vents, which are employed in the polymerase chain reaction. The polymerase chain reaction makes possible the selective amplification of DNA sequences of interest and this important new technique has many applications in molecular biology. Other such enzymes are the proteases that are being developed for commercial application.

Bioactive Compounds from Marine Invertebrates

The production of bioactive chemicals is a common means of defence in marine invertebrates, especially sessile organisms and vulnerable soft-bodied organisms. Groups of organisms that have been found to produce bioactive natural products include marine bacteria, dinoflagellates, algae, coelenterates (namely the corals), echinoderms (such as sea cucumbers and starfish), bryozoans, sponges, soft-bodied molluscs (such as sea hares and nudibranchs), and tunicates. The chemical basis of some of the marine ecological interactions among invertebrates has been discussed by Scheuer (1990), who has also edited two comprehensive review volumes on organic chemicals of biological marine origin (Scheuer, 1987, 1988). Bioactive substances from marine organisms have been studied for several decades and thousands of these chemicals have been described. Recent discoveries of marine natural products with interesting biological and pharmaceutical properties have been the subject of a series of comprehensive reviews (Faulkner, 1992).

Discovery of bioactive compounds have been made from a wide range of organisms, including sponges (James et al., 1991; Kushlan and Faulkner, 1991; Stierle and Faulkner, 1991) and algae (Trimurtulu et al., 1992), anti-inflammatory and antiviral agents, mainly from corals (Groweiss et al., 1988; Roussis et al., 1990), and antifungal disulphides from ascidians (Lindquist and Fenical, 1990).

Sponges, in particular, have proven to be an important source of bioactive compounds. Researchers in Scheuer's laboratory at the University of Hawaii in Manoa have isolated a number of cytotoxic compounds from marine sponges (Akee et al., 1990; Carroll and Scheuer, 1990). The Harbor Branch Oceanographic Institution, Ft Pierce, Florida, team of researchers has isolated many bioactive compounds from sponges, including an antitumour compound (Sakai et al., 1986).

The wide range of bioactive compounds produced by marine organisms emphasizes the great potential of compounds for biomedical applications, which has encouraged large-scale systematic screening of marine organisms. For example, the National Cancer Institute (NCI), US Public Health Service, has established a screening system consisting of 60 *in vitro* cell lines representing seven cancer sites: blood cells; brain; colon; kidney; lung; ovary; and skin (Ansley, 1990). Extracts from many marine organisms are tested for their cytotoxic activity and additional tests are performed to detect anti-HIV activity, using a human lymphoblastic cell line infected with the AIDS virus. Some pharmaceutical companies also screen marine isolates for anti-inflammatory, insecticidal, and herbicidal activities, in addition to cytotoxic and antiviral screening (Cardellina, 1986).

Many compounds isolated from marine macro-organisms, such as sponges, are produced by bacteria associated with those sponges. For example, several diketopiperazines previously ascribed to the sponge, *Tedania ignis*, have been shown to be produced by a marine *Micrococcus* sp. associated with this sponge (Stierle et al., 1988). It has also been observed that secondary metabolites from certain molluscs, sponges, and tunicates closely resemble natural products from cyanobacteria, formerly taxonomically described as 'blue green algae'. In molluscs, these metabolites are generally derived from ingestion of cyanobacteria, whereas in sponges and tunicates these products are apparently produced by symbiotic cyanobacteria (Moore, 1991).

Marine invertebrates need to be collected from natural ecosystems where they may be inaccessible or present only in low numbers. Large-scale collection of invertebrates for natural product production may threaten endangered populations. Alternatively, specialized conditions can be established to grow invertebrates in captivity. Thus, the possibility exists for cloning genes from invertebrates into bacteria for production of the described natural product. In cases where true symbiotic relationships exist between host and bacterium, it may be extremely difficult or impossible to isolate and maintain the bacterium in pure culture. Molecular approaches are very useful in such cases. For example, luminescent symbionts of some marine fish have not yet been isolated into pure culture but have, instead, been characterized by 16S rRNA (Haygood and Distel, 1993).

Marine toxins are marine natural products that have specific pharmacological activities resulting in adverse effects in animals, generally at very low concentrations. Many marine toxins are produced by dinoflagellates and may be retained or concentrated through several trophic levels before exerting adverse effects on predators higher in the food chain (including man). Examples of toxins from dinoflagellates capable of causing fatal poisoning in man are ciguatoxin and saxotoxin. Ciguatoxin is a sodium channel agonist and is generally considered to be produced by dinoflagellates associated with coral reefs. There are, however, indications that ciguatoxin may be produced by bacteria, including the cyanobacterium *Oscillatoria erythraea* (Hahn and Capra, 1992).

An important marine toxin found in many marine animals is the potent sodium channel blocker, tetrodotoxin, also known as puffer fish toxin. However, this toxin has also been found in a wide range of marine bacteria (Simidu et al., 1987). The presence of this neurotoxin in many distantly related animal genera, therefore, may indicate production of the toxin by bacteria associated with these animals. A variety of bacteria, including *Vibrio* species, have been shown to produce tetrodotoxin (Simidu *et al.*, 1987). The production of tetrodotoxin by *Vibrio cholerae* was reported by Tamplin *et al.* (1987).

Toxins are of interest in the context of marine natural products because they may have useful medical applications, if appropriate dosages and delivery systems can be devised. They also have application as research tools, particularly in studies on neuromuscular systems (Colwell, 1983).

The Marine Biotechnology Industry

To understand the dimensions of the emerging marine biotechnology industry, a list of 59 companies was prepared and included in a study, but the actual number of US companies involved in some type of marine biotechnology activity may be as high as 110. About 72% are US corporations and 21% multinational corporations.

From the survey, it was found that the focus of research in private industry was on aquaculture (19%) and natural products chemistry (15%), with the next most frequent being bioremediation and microbiology, each 8.2%.

The major areas of interest were: aquaculture (21.4%); pharmaceuticals/fine chemicals (18.6%); fermentation processes (14.3%; and environment/bioremediation (10%).

Interestingly, 52% of the academic scientists interviewed indicated that they had some form of collaboration with private industry. Researchers in private industry (77%) indicated that they did, indeed, have a collaboration underway with an academic unit.

More than 60% of the academic scientists indicated that they are carrying out some type of collaboration with foreign scientists in the area of marine biotechnology. Work of the Pacific Institute of Bioorganic Chemistry (Far East Division of the Russian Academy of Sciences, Vladivostok) provides a good example. Results of chemical investigation of marine invertebrates, especially echinoderms, have demonstrated a great variety of steroids to be present in these animals. Starfish were also shown to be sources of steroid polyols, polyol sulphates, and glycosides with a very high level of oxidation in the steroid moiety. Recently, new steroid biosides have been isolated from the Far-Eastern starfish *Crossater papposus* (G. Ilyakov, 1994, Vladivostok, personal communication).

Investigation of marine microorganisms associated with marine macroorganisms has been investigated by the Russian scientists and have been found to be producers of many well-known marine toxins, including tetrodotoxin, saxitoxin, and prosurogatoxin.

The Russian team has isolated new cyclic depsipeptides with pH-dependent cytotoxic activity against tumour cells. These were found in the culture of *B. pumilus* (KMM 150), associated with the Australian sponge *Ircinia* sp.

Actinomycetes of marine sediments from various depths and various geographical zones of the World Ocean have yielded more than 150 strains of marine actinomycetes of the genera *Nocardia, Streptomyces, Micromonospora* and *Promicromonospora*. A new protein antibiotic – palmycromycin was reported, isolated from the actinomycete *Streptomyces pluricolorescens*. This compound did not show cytotoxic or proteolytic activities.

Antiviral activity of bacteria on animals in sea water and in sediments of the Great Barrier Reef was shown for 72 strains of bacteria from sponges (13.3%), bottom sediments (26.5%), ascidians (17.6%), coelenterates (14.6%), sea water (10.6%). Metabolites of 16 strains from 224 appeared to have reverse transcriptase activity. The largest number of producers of such inhibitors (33.3%) was found in microorganisms from sea water. Of 224 strains 12 inhibited RNA-polymerase. In particular, many such strains were isolated from coelenterates. Thus, the Russian group working in Vladivostok has obtained some interesting findings.

Australia, another country highly involved in marine biotechnology, is exceedingly well endowed with marine biological resources. Its marine territory is impressively large, i.e. almost nine million square kilometres encompassing a larger area than its land area. The world's largest coral reef system, the Great Barrier Reef, lies off the eastern coast of Australia, stretching for more than 2200 km. Australia's marine environs include tropical, temperate, and cold-water seas. It is likely that Australia possesses a greater variety of marine life than any other nation. The Great Barrier Reef is populated by approximately 2000 fish species (twice as many as the second richest habitat, located in New Caledonia) and 500 coral species (compared with 300 in New Caledonia) (Groombridge, 1992). Other regions of Australia, particularly western and southern Australia, may have the most diverse algal flora in the world, with an unusually large proportion of endemic species and genera (L. Borowitzka, Australia, 1994, personal communication). The rich biodiversity of animals and plants present in Australian waters forms a treasure trove of raw material, amenable to sustainable economic exploitation via marine biotechnology.

Several small biotechnology companies in Australia show signs of success, including Bresatec Ltd, Bioclone Australia Pty Ltd, Memtec Ltd, Calgene Pacific Pty Ltd, Australian Medical Research and Development Corporation Ltd (AMRAD Corporation), Progen Industries Ltd, and Peptide Technology Pty Ltd. The Australian biotechnology industry has moved from being dominated by a few large institutions and several small specialized firms (often associated with an educational and/or research institution) to being much more diversified, with a wide ranging institution base and more large multi-interest firms entering the field at both the state and Commonwealth levels (F. Roseby, Australia, 1994, personal communication). The future, indeed, appears bright; a recent report, *Biotechnology in Australia*, predicts that biotechnology will grow into a multibillion dollar industry by the year 2000.

The marine natural products industry in Australia has, until recently, been limited to only a few firms, e.g. Betatene Ltd and Western Biotechnology Ltd, culturing microalgal species, mainly *Dunaliella salina*, to produce beta-carotene. Total sales of beta-carotene, used for animal feed, human dietary supplements, and food colouring, were A$ 2 million in 1987–1988 (Review

Committee on Marine Industries, 1989). Both companies are profitable. Sales increased to A$ 5 million by 1992–1993 and continue to rise. Reflecting the increasing size of the market for beta-carotene, Western Biotechnology expanded production facilities at Hutt Lagoon in Western Australia from 50 to 75 h of ponds (L. Borowitzka, Australia, 1994, personal communication).

A successful example of international collaboration is a research project between the James Cook University, Australia, and the International Centre for Living Aquatic Resources Management (ICLARM), that is focused on the giant clam (*T. gigas*). ICLARM, which became a member of the Consultative Group on International Agricultural Research network in 1992, includes six countries in the Asia–Pacific region, with headquarters in the Philippines and a Coastal Aquaculture Centre in the Solomon Islands. The giant clam grows to a 50 kg in weight within nine to ten years, producing a delicious meat highly prized in countries throughout the Pacific and Asia. In addition, giant clam shells are sought after by tourists. Because they are so popular, fishermen, mainly from Taiwan, have decimated the giant clam population in the Pacific Islands, almost to the point of extinction. However, they have survived in large numbers on the Great Barrier Reef, because they are one of Australia's protected natural resources. Work done at James Cook University Orpheus Island facility on the Great Barrier Reef has resulted in an effective flow-through culture technique, including early-stage nutrient supply and micro-encapsulated foods, allowing control of the settling of giant clam larvae and rate of metamorphosis of larvae to juveniles. Operations have expanded to include other sites in Australia, as well as the Philippines, Cook Islands, Fiji, and Tonga (B. Smith, Australia, 1994, personal communication).

Thus, the last decade has witnessed a striking escalating of interest in the secondary metabolites of marine animals, plants and bacteria. At the University of Maryland Center of Marine Biotechnology, one of four research centres of the University of Maryland Biotechnology Institute, a programme of study of marine microbial ecology and natural products has included, to date, investigations of the chemistry of bacterial cultures derived from algae (Hanzawa *et al.*, 1993) and tissues of sponges (Stierle *et al.*, 1988; Stierle and Faulkner, 1991).

Marine invertebrates, primarily tunicates, collected from the Great Barrier Reef of Australia at the University of Queensland Heron Island Marine Station, included invertebrate hosts: tunicates *Ascidia aperta, Cystodytes dellachiajei, Diplosoma similis, Diplosoma virens, Eudistoma glaucus, Herdmania momus, Leptoclinides lissus, Lissoclinum bistratum, Lissoclinum patella, Phallusia julinea, Trididemnum cyclops* (or *mineatus*), *Trididemnum paracyclops, Trididemnum tegulum,* and an unidentified tunicate (tentatively identified as either *Aplidium* or *Eudistoma*); echinoderms (*Holothuria atra* and *Holothuria leucopebia*); and the zoanthid, *Palythoa* sp.

Bacteria associated with *Ascidia aperta, Cystodytes dellachiajei, Diplosoma similis, Diplosoma virens, Eudistoma glaucus, Herdmania momus, Leptoclinides lissus, Lissoclinum bistratum, Lissoclinum patella, Phallusia julinea, Trididemnum cyclops* (or *mineatus*), *Trididemnum paracyclops,*

Trididemnum tegulum and an unidentified tunicate were examined in this study, along with cultures from *Palythoa* spp. (Anthozoa), and *Holothuria leucopebia* (Echinodermata). About 9% of the culture extracts exhibited detectable antimicrobial activity and eight of the strains exhibited varying degrees of toxicity for brine shrimp larvae at concentrations of 200 μg ml^{-1}. *Palythoa* isolates comprised a quarter of these strains. Screening of microbial extracts in a primary plant growth regulatory screen, i.e. germination and growth of rye grass seeds, revealed significant growth promotion (26% of the extracts) and inhibition (10%). Bioactive metabolites were produced by some cultures isolated from all hosts, except for *Ascidia aperta*. Cultures were taken only from the blood of *A. aperta* and none of these were active in the plant growth assays. Five of the culture extracts showed significant inhibition of seed germination. In the microtitre assays, the seven extracts exhibited very strong inhibition of seed germination at a concentration of 1.0 mg; the extracts previously identified as growth inhibitor completely inhibited seed germination (Hanzawa et al., 1993). Thus, bacteria associated with marine invertebrates, indeed, can yield intriguing results.

In conclusion, the marine environment offers a very wide diversity of species, with significant potential for commercial application.

In the two examples offered, aquaculture and marine natural products, utilization of marine resources, especially the rich diversity of the marine environment, in industrial applications is proving to be highly profitable. Both segments of the rapidly developing marine biotechnology industry are environmentally protective. Both utilize marine genetic resources, without depleting them by application of molecular biology and bioengineering tools. The marine biotechnology partnership of academia, government, and industry is proving both productive and rewarding.

Acknowledgement

I wish to acknowledge the collaboration of my colleagues, Drs R. Zilinskas, D. Lipton, and R.T. Hill, for their contributions to a major study on the marine biotechnology industry, which we have published together, and which was drawn upon in preparing this chapter.

References

Agellon, L.B., Davies, S.L., Lin, C.M. and Chen, T.T. (1988a) Growth hormone in rainbow trout is encoded by two separate genes. *Molecular Reproduction and Development* 1, 11-17.

Agellon, L.B., Emery, C.J., Jones, J.M., Davies, S.L., Dingle, A.D. and Chen, T.T. (1988b) Promotion of rapid growth of rainbow trout (*Salmo gairdneri*) by a recombinant fish growth hormone. *Canadian Journal of Fisheries and Aquatic Science* 45, 146-151.

Akee, R.K., Carroll, T.R., Yoshida, W.Y. and Scheuer, P.J. (1990) Two imidazole alkaloids from a sponge. *Journal of Organic Chemistry* 55, 1944–1946.

Ansley, D. (1990) Cancer institute turns to cell line screening. *The Scientist* 4, 3.

Avron, M. and Ben-Amotz, A. (1992) *Dunaliella: Physiology, Biochemistry and Biotechnology*. CRC Press, Boca Raton, Florida.

Breton, B., Weil, C., Sambroni, E. and Zohar, Y. (1990) Effects of acute versus sustained administration of GnRHa on GtH release and ovulation in the rainbow trout, *Oncorhynchus mykiss*. *Aquaculture* 91, 373–383.

Cardellina, J.H. II. (1986) Marine natural products leads to new pharmaceutical and agrochemical agents. *Pure and Applied Chemistry* 58, 365–374.

Carroll, A.R. and Scheuer, P.J. (1990) Four beta-alkylpyridines from a sponge. *Tetrahedron* 46, 6637–6644.

Cheney, D.P. (1990) Genetic improvement of seaweeds through protoplast fusion. Penniman, C.A. and Van Patten, P. (eds) *Economically Important Marine Plants of the Atlantic*. Connecticut Sea Grant College Program, Groton, Connecticut, pp. 15–26.

Colwell, R.R. (1983) Biotechnology in the marine sciences. *Science* 222, 19–24.

Crim, L.W. and Glebe, B.D. (1984) Advancement and synchrony of ovulation in Atlantic salmon with pelleted LHRH analogue. *Aquaculture* 43, 47–56.

Du, S.J., Gong, Z.Y., Fletcher, G.L., Shears, M.A., King, M.J., Idler, D.R. and Hew, C.L. (1992) Growth enhancement in transgenic Atlantic salmon by the use of an all-fish chimeric growth hormone gene construct. *Bio/Technology* 10, 176–181.

Faulkner, D.J. (1992) Marine natural products. *Natural Products Reports* 9, 323–364.

Gilmore, R.D., Engelking, H.M., Manning, D.S. and Leong, J.C. (1988) Expression in *Escherichia coli* of an epitope of the glycoprotein of infectious hematopoietic necrosis virus against viral challenge. *Bio/Technology* 6, 295–300.

Groombridge, B. (ed.) (1992) *Global Diversity: Status of the Earth's Living Resources*. Chapman & Hall, New York.

Groweiss, A., Look, S.A. and Fenical, W. (1988) Solenolides, new anti-inflammatory and antiviral diterpenoids from a marine octocoral of the genus *Solenopodium*. *Journal of Organic Chemistry* 53, 2401–2406.

Hahn, S.T. and Capra, M.F. (1992) The cyanobacterium *Oscillatoria erythraea* – a potential source of toxin in the ciguatera food chain. *Food Additive Contamination* 9, 351–355.

Hanzawa, N., Singleton, F.L., Hamada, E., Colwell, R.R. and Miyachi, S. (1988) *Journal of Marine Biotechnology* 1, 105–108.

Harvey, W. (1988) Cracking open marine algae's biological treasure chest. *Bio/Technology* 6, 486–492.

Haygood, M.G. and Distel, D.L. (1993) Bioluminescent symbionts of flashlight fishes and deep-sea anglerfishes form unique lineages related to the genus *Vibrio*. *Nature* 363, 154–156.

Hew, C.L., Davies, P.L., Shears, M. and Fletcher, G. (1991) Antifreeze protein gene transfer in Atlantic salmon. International Marine Biotechnology Conference. Baltimore, Maryland, Society of Industrial Microbiology, Washington, DC.

James, D.M., Kunze, H.B. and Faulkner, D.J. (1991) Two new brominated tyrosine derivatives from the sponge *Druinella* (=*Psammaplysilla*) *purpurea*. *Journal of National Products* 54, 1137–1140.

Kranzfelder, J.A. (1991) β-carotene production in large-scale intensive cultures of *Dunaliella salina*. International Marine Biotechnology Conference, Baltimore, Maryland.

Kushlan, D.M. and Faulkner, D.J. (1991) A novel perlactone from the Caribbean sponge *Plakortis angulospiculatus*. *Journal of Natural Products* 54, 1451–1454.

Kyle, D.J., Gladue, R., Reeb, S. and Boswell, K. (1991) Production and use of omega-3 designer oils from microalgae. International Marine Biotechnology Conference, Baltimore, Maryland.

Lee, C.Y., Rubinsky, B. and Fletcher, G.L. (1992) Hypothermic preservation of whole mammalian organs with antifreeze proteins. *Cryo-letters* 13, 59–66.

Leong, J.C., Anderson, E., Bootland, E., Drolet, B., Chen, L., Mason, C., Mourich, D. and Trobridge, G. (1991) Biotechnologic advances in fish disease research. International Marine Biotechnology Conference, Baltimore, Maryland.

Lindquist, N. and Fenical, W. (1990) Polycarpamines A–E, antifungal disulfides from the marine ascidian *Polycarpa auzata*. *Tetrahedron Letters* 31, 2389–2392.

Liu, Z., Moav, B., Faras, A.J., Guise, K.S., Kapuscinski, A.R. and Hackett, P.B. (1990) Development of expression vectors for transgenic fish. *Bio/Technology* 8, 1268–1272.

Manning, D.S. and Leong, J.C. (1990) Expression in *Escherichia coli* of the large genomic segment of infectious pancreatic necrosis virus. *Virology* 179, 16–25.

Moore, R.E. (1991) Role of blue-green algae (cyanobacteria) in biosynthesis of natural products from marine animals. 1991 International Marine Biotechnology Conference, Baltimore, Maryland.

Morse, D.E. (1984) Biochemical and genetic engineering for improved production of abalone and other valuable mollusks. *Aquaculture* 39, 263–282.

North, W.J. (1991) Oceanic farming, a new dimension in cultivating marine macroalgae. UNESCO International Conference on Global Impact of Applied Microbiology and Biotechnology, IX, Valleta, Malta.

Pagelson, G. and Zohar, Y. (1992) Characterization of gonadotropin-releasing hormone (GnRH) receptors in the gilthead seabream (*Sparus aurata*). *Biology of Reproduction* 47, 1004–1008.

Paynter, K.T. and Chen, T.T. (1991) Biological activity of biosynthetic rainbow trout growth hormone in the eastern oyster, *Crassostrea virginica*. *Biological Bulletin* 181, 459–462.

Polne-Fuller M. and Gibor, A. (1984) Developmental studies in *Porphyra*. I. Blade differentiation in *Porphyra perforata* as expressed by morphology, enzymatic digestion, and protoplast regeneration. *Journal of Phycology* 20, 607–618.

Raymond, J.A., Radding, W. and DeVries, A.L. (1989) Inhibition of growth of nonbasal planes in ice by fish antifreezes. *Proceedings of the National Academy of Sciences USA* 86, 881.

Review Committee on Marine Industries, Science and Technology (1989) Oceans of Wealth? Department of Industry, Technology and Commerce, Canberra.

Roussis, V., Wu, Z. and Fenical, W. (1990) New antiinflammatory pseudopterosins from the marine octocoral *Pseudopterogorgia elisabethae*. *Journal of Organic Chemistry* 55, 4916–4922.

Sakai, R., Higa, T. and Y. Kashman, (1986) Misakinolide-A, an antitumor macrolide from the marine sponge *Theonalla* sp. *Chemical Letters* 1, 1499–1502.

Scheuer, P.J. (1987) *Bio-organic Marine Chemistry*. Springer-Verlag. Berlin.

Scheuer, P.J. (1988) *Bio-organic Marine Chemistry*. Springer-Verlag. Berlin.

Scheuer, P.J. (1990) Some marine ecological phenomena: chemical basis and biomedical potential. *Science* 248, 173–177.

Shears, M.A., Fletcher, G.L., Hew, C.L., Gauthier, S. and Davies, P.L. (1991) Transfer, expression and stable inheritance of antifreeze protein genes in Atlantic salmon (*Salmo salar*). *Molecular Marine Biology and Biotechnology* 1, 58–63.

Simidu, U., Noguchi, T., Hwang, D.F., Shida, Y. and Hashimoto, K. (1987) Marine bacteria which produce tetrodotoxin. *Applied Environmental Microbiology* 53, 1714–1715.

Stierle, A.C., Cardellina, J.H. II, and Singleton, F.L. (1988) A marine *Micrococcus* produces metabolites ascribed to the sponge *Tedania ignis. Experientia* 44, 1021.

Stierle, D.B. and Faulkner, D.J. (1991) Antimicrobial N-methylpyridinium salts related to the xestamines from the Caribbean sponge *Calyx podatypa. Journal of National Products* 54, 1134–1136.

Tamplin, M.L., Colwell, R.R., Hall, S., Kogure, K. and Strichartz, G.R. (1987) Production of sodium channel inhibitors by *Vibrio cholerae* and *Aeromonas hydrophila. Lancet* i, 975.

Trimurtulu, G., Kushlan, D.M., Faulkner, D.J. and Rao, C.B. (1992) Divarinone, a novel diterpene from the brown alga *Dictyota divaricata* of the Indian Ocean. *Tetrahedron Letters* 33, 729–732.

Waaland, J.R., Dickson, L.G. and Watson, B.A. (1990) Protoplast isolation and regeneration in the marine red alga *Porphyra nereocystis. Planta* 181, 522–528,

Zhang, P., Hayat, M., Joyce, C., Gonzalez-Villasenor, L.I., Lin, C.M., Dunham, R.A., Chen, T.T. and Powers, D.A. (1990) Gene transfer, expression and inheritance of pRSV-rainbow trout-GHcDNA in the carp, *Cyprinus carpio* (Linnaeus). *Molecular Reproduction and Development* 25, 3–13.

Zilinskas, R.A., Colwell, R.R., Lipton, D.W. and Hill, R. (1994) The global challenge of marine biotechnology: a status report of marine biotechnology in the United States, Japan, and other countries. Report prepared for the National Sea Grant Program and the Maryland Sea Grant College.

Zohar, Y. (1988a) Fish reproduction: its physiology and artificial manipulation. In: Shilo, M. and Sarig, S. (eds) *Fish Culture in Warm Water Systems*. CRC Press, Boca Raton, Florida, pp. 65–119.

Zohar, Y. (1988b) Gonadotropin-releasing hormone in spawning induction in teleosts: basic and applied considerations. In: Zohar, Y. and Breton, B. (eds) *Reproduction in Fish – Basic and Applied Aspects in Endocrinology and Genetics*. INRA Press, Paris, pp. 47–62.

Zohar, Y., D'Emanuele, A., Kost, J. and Langer, R. (1990a) Ultrasound-mediated administration of compounds into aquatic animals. International Patent Number 07/583, 573.

Zohar, Y., Goren, A., Fridkin, M., Elhanati, E. and Koch, Y. (1990b) Degradation of gonadotropin releasing hormone in the gilthead seabream, *Sparus auratus*: II. Cleavage of native salmon GnRH, mammalian LHRH and their analogues in the pituitary, kidney, and liver. *General and Comparative Endocrinology* 79, 306–319.

Biotechnologies and the Use of Plant Genetic Resources for Industrial Purposes: Benefits and Constraints for Developing Countries

A. Sasson
United Nations Educational, Scientific and Cultural Organization (UNESCO), 7 Place de Fontenoy, 75352 Paris 07 SP, France

Introduction

The major centres of plant biological diversity are found in tropical regions, where two-thirds (250,000–300,000 species) of the world's plant genetic resources lie. Developing countries – especially those situated in the tropics – are, therefore, directly or indirectly, the suppliers of plant genetic resources used in improving agricultural and foodstuff production, increasing yields of fibre and woody species, and in finding new plant raw materials, food additives, pigments, or pharmaceutical substances.

In the tropics, plant genetic resources are threatened by destruction of the natural environment. The wild relatives of crop species or the semi-domesticated crop varieties, if not preserved *in situ* or *ex situ*, are also indirectly threatened by the spread of monocultures; higher-yielding varieties, resulting from conventional crop breeding or biotechnological processes could replace the sturdy traditional cultivars, less yielding but more resistant to, or tolerant of, environmental stresses, pathogens, and pests.

However, plant biotechnologies can play a key role in the massive production of improved crop varieties (through *in vitro* tissue culture followed by clonal propagation), as well as in their genetic improvement. They can also help in propagating plant species which contain useful and biologically active substances. Organ tissue and cell cultures could be more efficient than conventional extraction. Developing countries can benefit from these applications; an increasing number of them are screening their plant genetic resources for this purpose, sometimes with the help of corporations from the industrialized countries or through joint ventures.

© 1996 CAB INTERNATIONAL. *Biodiversity, Science and Development: Towards a New Partnership* (eds F. di Castri and T. Younès)

Utilization of Plant Genetic Resources in Conventional Breeding: a Few Examples

The Japanese primitive dwarf wheat variety Norin 10, which was introduced into America in 1946, played an important role in efforts to improve wheat varieties: the selected dwarf cultivars reacted well to fertilizers, giving higher yields. Semidwarf wheat varieties occupied 60% of wheat acreage in the USA (Brady, 1985). A Turkish wheat variety collected by Harlan in 1948, whose agronomic traits initially excluded utilization, in fact proved very useful because it contained the genes for resistance to *Puccinia striiformis*, 35 strains of *Tilletia caries*, *T. foetida*, 10 strains of *T. controversa*; this variety was also tolerant to some species of *Urocystis*, *Fusarium* and *Typhula*, and was therefore used in cross-breeding schemes aimed at increasing the resistance of wheat to these various parasites (in Esquinas-Alcázar, 1987).

The primitive rice cultivars in north-western India have been particularly useful in breeding rice varieties resistant to several diseases and parasites. In the case of corn, six hardy varieties identified in Mexico and Guatemala contributed to introducing a much higher resistance to viruses and to parasitic insects into varieties currently cultivated. These viruses and insects had caused annual losses of at least $500 million around the world.

Furthermore, a Mexican variety appeared to be the only genetic resource available to help increase resistance to two major virus diseases in corn cultivated in the USA (B mosaic and a chlorosis). During the 1970 corn leaf blight epidemic, American breeders produced large quantities of seeds of resistant varieties in greenhouses in Hawaii and Florida, thus enabling cereal growers to obtain good harvests the following year.

In using wild groundnuts found in the Amazonian forest for crossbreeding purposes, resistance to a leaf disease of this legume species has been improved, the resulting increase in harvest being evaluated at some $500 million by the International Crop Research Institute for the Semi-Arid Tropics.

Similar examples concern fodder species. Varieties of *Lolium multiflorum*, collected in Uruguay in the 1940s, have been used in selection schemes aimed at increasing the resistance of this species to crown rust. A variety of *Bromus biebersteinii*, collected in Turkey in 1949, has been used for the transfer of good agronomic traits and increased vigour in the Regar variety grown in the USA. The American commercial alfalfa ecotype AWPX3 was the result of cross-breeding between 13 ecotypes collected in nine countries at different periods. An alfalfa ecotype, collected in Iran in 1940, has been used to transfer to this fodder legume species resistance to nematodes, which attacked the stem (in Esquinas-Alcázar, 1987).

The wild species of the genus *Lycopersicon* which cross-breed well with the cultivated tomato, *Lycopersicon esculentum*, have been used as gene pools for the selection of varieties resistant to parasitic fungi (*L. hirsutum*, *L. peruvianum* and *L. pimpinellifolium*), viruses (*L. chilense* and *L. peruvianum*), nematodes

(*L. peruvianum*), insects (*L. hirsutum*), for the improvement of fruit quality (*L. chmielewskii*) and adaptation to unfavourable environmental conditions (*L. cheesmanii*) (see Esquinas-Alcázar, 1987). Two wild tomato species, discovered in the early 1960s on the highlands of Peru, helped increase the pigmentation of tomatoes and their solid matter content, which resulted in an annual gain of some $5 million at that time for the canned tomato industry in the USA.

One-third of the rubber used throughout the world was of natural origin and 99% of this was extracted from plantations of a few selected varieties of *Hevea*. The *Hevea* plantations of Southeast Asia were threatened by the South American leaf blight. The only hope of fighting it lay in the selection of resistant trees indigenous to the Amazonian forest. However, deforestation of the latter increased the risks of losing these precious trees, the seeds of which could not be stored in cold conditions (as was true for many other equatorial forest species) and therefore had to be protected in their natural environment.

Regarding resistance to pathogens and pests, the genetic sources of resistance to the golden nematode in potatoes came from Peru. Strains resistant to Southern corn leaf blight, corn rust, and maize dwarf mosaic virus were the result of collaboration with scientists in developing countries, as was resistance to the soybean mosaic virus (Brady, 1985).

The conservation of sufficient genetic diversity within the same species was indispensable to the future utilization of its full genetic potential. For instance, one single population of *Oryza nivara*, and not the species itself, has been used to select rice varieties resistant to the grassy stunt virus (see Esquinas-Alcázar, 1987).

Access to abundant plant genetic resources has therefore been an important factor for any work on improving and selecting crop varieties, either through conventional cross-breeding and hybridization techniques, or cultures of plant cells, tissue and organs, or gene transfer.

Biotechnologies and Production of Plant-derived Drugs and Phytochemicals

According to a 1988 report by the consultancy firm McAlpine, Thorpe and Warrier (United Kingdom), 'the market potential for herbal drugs in the "Western" world could range from $4.9 billion in the next ten years to $47 billion by the year 2000 . . .'. This expected growth is due to shifting consumer preferences away from chemicals to plant-derived drugs and the fact that pharmaceutical companies are continuously seeking new compounds for medicines to improve their competitiveness (Komen, 1991).

The World Health Organization (WHO) estimates that about 80% of people in developing countries still rely on traditional plant-derived drugs, the main reason being their low price. For modern pharmaceuticals, they depend largely on imported drugs and technology. Thus, plant-derived drugs offer an interesting potential as resources for local industry and a substitute for costly pharmaceutical imports (Komen, 1991).

Most of the world's medicinal plants are located in the tropics, which store about 250,000 to 300,000 (two-thirds) of all plant species. Although at least 35,000 are estimated to be of medicinal value, up until now only 5000 have been exhaustively studied for possible medical applications. The development of highly specific bioassays to detect picogramme quantities of potentially useful compounds and automated screening technology make it possible to screen thousands of plant samples daily and efficiently select those with pharmaceutical, agrochemical or other industry usefulness (Komen, 1991).

Screening operations for medicinal substances

In the USA, the largest tropical plant collecting effort was sponsored by the Federal Government's National Cancer Institute (NCI). In 1986, the NCI launched a five-year $2.8 million programme to screen thousands of exotic plants from tropical forests of Latin America, Southeast Asia and southern Africa. Drug development from plants would be left to private companies. In addition, the NCI intended to develop agreements that would benefit the countries of origin (Komen, 1991). Since 1986, Biotics Ltd, UK, has undertaken a programme supported by the European Commission aimed at promoting cooperative exploitation of the indigenous genetic resources of developing countries.

It should be recalled that ethnobotanical evidence shows that many plants have already proved useful to humankind as sources of medicines, insecticides, pigments, gums, flavours, and other useful products; at least 24,000 different plants are thought to have been used worldwide in traditional medicine. The cultivation of some of these tropical and subtropical species is of some considerable industrial importance. Biotics Ltd's Phytochemical Screening Programme was designed to improve the situation prevailing between developing and industrialized countries with regard to access to plant genetic resources and the compensation owed to the country of origin of these resources. It therefore has aimed to implement cooperative screening projects based on formally negotiated agreements. In this Programme, plant materials are screened for useful natural products by industrial and other specialist research organizations using the latest bioassay technology.

Companies participating in screening projects have access to low-cost sustainable supplies of plant materials and receive appropriate commercial exclusivity to develop promising compounds. In return, those countries supplying plant materials are awarded royalty or equivalent payments linked to the commercialization of any indigenous natural product. The Programme is based on a series of individually tailored bilateral projects, each building on the experience gained at an initial pilot stage and backed by a formal agreement. It was specifically intended to complement rather than compete with existing similar national or regional programmes.

Screening projects were launched in 1989 in cooperation between British pharmaceutical groups and Ghana, Cameroon, Mauritius, and Malaysia. In 1991, Biotics' programme involved the screening of more than 1000 plant samples by bioindustrial organizations in the USA and UK. Biotics Ltd set up a commercial laboratory facility, BioEx, for the preparation of plant extracts, which was also expected to train plant chemists from developing countries. It was also hoped that BioEx would lead, in developing countries, to the setting up of a series of small-scale extraction laboratories operating in association with Biotics' screening programme (Komen, 1991).

There is also considerable potential for initiating collaborative screening of flora from Arab countries. These countries span a wide range of climatic and soil conditions, and have a broad plant genetic diversity; the Arabian Peninsula alone hosts 4500–5000 different plant species. This potential was highlighted in 1982 by a UNIDO Report (Medicinal and Aromatic Plants for Industrial Development) which underlined that phytochemicals and plant extracts were widely used as therapeutic agents in many Arab countries. A detailed survey by leading companies in Egypt, Sudan, Syria, and Iraq revealed that more than 25% of the trade items marketed by them contained one or more plant products.

Biotics Ltd is considering the potential for extending the above strategy to include the in-depth phytochemical screening of economically important agricultural species. The aim is to identify new sources of marketable chemicals at considerably reduced costs. Using existing technologies, chemical inventories could be complied of otherwise discarded agricultural residues, such as sugar-cane bagasse or, in the Arab region, date-palm residues; the same analytical procedure might equally be applied to marginal crops of importance to farmers in poorer rural areas.

Production and processing of plant-derived medicinal substances in developing countries

Research on traditional drugs in the Philippines has led to the identification of more than 300 medicinal plants. Researchers interviewed over 1000 village medicine men throughout the country. In 1989, a herbal medicine processing factory was set up in the Cotabato Province (Komen, 1991).

The first of its kind in Asia when it opened in 1993, the Centre for Natural Product Research (CNPR) in Singapore studies plants and other natural products as potential cures for certain diseases, including cancer. It also studies traditional medicines, such as Chinese and Malay herbal remedies, to understand why these have been effective. Marine, microbial (equatorial fungi) and plant organism extracts have been screened to isolate small molecules of therapeutic value.

The CNPR owes its existence to a joint venture between the British pharmaceutical group, Glaxo Holdings plc, Singapore's Economic Development Board (EDB) and the Institute for Molecular and Cell Biology (IMCB), at the National University of Singapore (NUS).

As far as funding is concerned, Glaxo provided an initial S$20 million and the EDB a total of S$10 million, the IMCB contributing the equivalent of S$10 million in research and intrastructural support; the three partners guarantee funding up until 2003.

The CNPR approach and its partnership with Glaxo Holdings plc (comparable to the partnership between the INBIO in Costa Rica and the multinational pharmaceutical group, Merck & Co. Inc.) show clearly that a venture of this kind is not within the reach of all research institutions in a developing country. It is true that a number of developing countries wish to embark on screening their genetic resources to isolate new medicinal products; it is also true that building up an 'extract library' is not that difficult to achieve and that it could even be traded. Nevertheless, a screening venture could not be successful without strong research backstopping to provide the sophisticated screens (i.e. cells, enzymes, physiological signals, etc.) needed for identifying the biologically active compounds. At a later stage, once a compound is discovered, its development into a commercial product is another complex and costly venture.

Established by the National University of Singapore and the Economic Development Board (EDB) in 1994, the Bioscience Centre (BSC) was conducting and coordinating multidisciplinary research from its facilities in the University's Zoology Department. An initial grant of about S$6 million was approved by the EDB for setting up the Centre. Enjoying core facilities and specialized equipment, the BSC facilitates and promotes research by staff and students of the biological sciences departments of the NUS, on an individual department or collaborative basis. The Centre's main objectives are to:

1. Provide a multiuser central facility for research in biotechnologies relating to agrotechnology, aquaculture, horticulture, environmental management, and marine living resources and pharmaceutical industries;

2. Conduct research on bioactive compounds from animals, plants and microorganisms;

3. Serve as a vehicle for collaborating with industries and other institutions in Singapore or abroad in biotechnology-related programmes; provide a mechanism for establishing linkages and collaborations with overseas institutions and companies, as well as for initiating exchange programmes with institutions abroad;

4. Provide training for both undergraduate and postgraduate students through practical or project work undertaken at the Centre. The hands-on training and research experience would support the development of bio-industries in Singapore.

In China, more than 40,000 different kinds of traditional plant drugs are produced. In 1986, 57 factories were producing plant drugs, providing the country with a significant source of foreign exchange that is expected to grow in coming years (Komen, 1991).

On 16 April 1994, the WHO made public the results of a study on treatment of acute forms of malaria. Carried out in Eastern Thailand over two years, the study was supervised by Dr J. Karbwang of Mahidol University,

Bangkok. It concerned 97 people affected by acute forms of malaria. The study showed that injections of artemether, an artemisinin derivative, were effective in treating these patients. Artemisinin or qinghaosu, extracted from *Artemisia annua*, had previously been identified as a possible antimalaria drug, following a study made in China in 1972 on traditional medicine. The WHO's results demonstrated that, when treated with quinine, 18 patients out of 50 died whereas when treated with artemether, only six out of 47 died. The WHO specialists concluded that, despite the restricted test group, the results are encouraging (Nau, 1994).

Other clinical trials with the same drug are being carried out in China, Vietnam, Brazil and in several African countries where malaria was present in an endemic form. The artemether is produced by a Chinese laboratory and is being marketed at world level (e.g. in Ivory Coast, Madagascar, Kenya and Nigeria) under the name of Paluther by the French chemical and pharmaceutical group Rhône-Poulenc-Rorer. The Government of Vietnam decided to develop a cheap process of cultivating the plant locally as well as a simple extraction procedure of artemisinin from the leaves (Nau, 1994).

The main risk relates to the uncontrolled use of artemisinin and its derivatives, to which the malaria parasites could become resistant. That is why the WHO recommended that the substances of this family be used in association with other drugs, such as mefloquine or doxycycline, only in areas where the parasites show multiresistance to antimalaria drugs. For its part, Rhône-Poulenc-Rorer stated that, for ethical reasons, this injectable drug is only recommended in the treatment of acute forms of malaria (Nau, 1994).

In November 1994, the Swiss chemical and pharmaceutical multinational Ciba-Geigy AG announced an agreement signed between the multinational and three Chinese partners (Institute of Microbiology and Epidemiology in Beijing, Kunming Pharmaceutical Factory in Kunming and CITIC (China International Trust Investments Corp.) Technology Inc. of Beijing) to develop a new antimalaria drug. The latter would combine benflumetol, an anti-malaria substance developed at the Institute of Microbiology and Epidemiology, and artemether. The synergistic combination has proven effective against the disease.

Research on medicinal plants in Madagascar has resulted in the discovery of valuable drugs. The most important medicinal plant in Madagascar is *Catharanthus roseus* (rosy periwinkle), which is used throughout the world to treat a variety of cancers and specifically for treating leukaemia in children. Two chemical compounds extracted from *C. roseus*, vinblastine and vincristine, possess tumour-inhibiting properties. Approximately two tonnes of crushed leaves yield one gramme of the two compounds, enough to treat a child for six weeks. Vincristine and vinblastine were discovered in the research laboratories of the American company, Eli Lilly & Co., which immediately recovered exclusive marketing rights through patent protection. However, with the exception of *C. roseus*, all medicinal plants in commerce are collected from wild sources, resulting in the gradual disappearance of some species such as *Rauwolfia confertiflora* which contains large quantities of reserpine, used in

Table 41.1. Plant cell cultures with high amounts of metabolites.

Product	Plant species	Tissue culture (% dry matter)	Plant (% dry matter)
Ajmalicine	*Catharanthus roseus*	1.00	0.30
Anthraquinones	*Cassia tora*	0.33	0.21
	Morinda citrifolia	18.00	2.20
Diosgenin	*Dioscorea deltoidea*	2.00	2.00
Ginsenoside	*Panax ginseng*	0.38 (fresh matter)	0.3–3.3

treating hypertension. *C. roseus* is widely grown on a commercial scale because of the increasing demand; the plant is easily cultivated.

Utilization of biotechnologies for the production of pharmaceuticals and phytochemicals

Plant tissue cultures could be used for the production of pharmaceuticals (alkaloids, steroids, terpenoids, flavonoids, enzymes), food additives (carotenoids, anthocyanins, betalains, vanilla), perfumes (rose, lavender, sandalwood oil, agarwood oil), and biopesticides. The utility of plant tissue cultures was recognized in the early 1940s, but it was only in 1956 that the first patent on plant cell cultures was awarded to Routin and Nickel. It would take another decade to demonstrate that plant cells in culture could produce metabolites in reasonable amounts (Table 41.1).

It has been observed that molecular, cellular and organ differentiation influences product biosynthesis. For instance, Heble *et al.* (cited in Heble, 1993) reported that plumbagin (2-methyl, 5-hydroxy 1,4-naphthaquinone) synthesis occurs in the highly specialized pigmented cells of tissue cultures of *Plumbago zeylanica*; the cells could be selected from a heterogenous mass and culture as a pure strain. Selecting cell lines with high productivity can be simply done on the basis of colour as in the case of *Plumbago*, *Nicotiana tabacum* and *Lithospermum*, by plating the cell suspensions and/or protoplasts. Working with *Coptis japonica*, Yamada *et al.* (cited in Heble, 1993) observed that selection of aggregates was more effective than single cells for the production of berberine.

A large number of compounds are the products of organogenesis. In tissue cultures of *Atropa belladonna*, the synthesis of tropane alkaloids is associated with complete differentiation and development of plantlets. Similar observations were made with tissue cultures of *Tylophora indica* and *Artemisia annua*. Partially differentiated structures, such as multiple shoots and roots in culture, also produce substantial amounts of useful compounds: working with shoot cultures of *Rauwolfia serpentina*, Roja *et al.* (cited in Heble, 1993) observed that they produced more ajmaline than the roots of the parent plant and the cultures exhibited biochemical stability for long periods; shoot cultures of *Withania somnifera* produce significant amounts of withanolides; multiple shoot cultures of *Catharanthus roseus* produce vinblastine; root cultures of *Hyoscyamus muticus* synthesize high amounts of scopolamine (0.3%), a high-value drug. Organ cultures are generally stable, but are difficult to cultivate in big bioreactors (cited in Heble, 1993).

However, cultures could produce secondary metabolites even in the absence of organogenesis, e.g. those of *Catharanthus roseus, Dioscorea deltoidea* and *Rauwolfia serpentina*. Also, some of the important cultures belonging to this category are those of *Morinda citrifolia, Panax ginseng, Linum flavum* and *Glycyrrhiza glabra* (Heble, 1993).

The regulatory influence of auxins and cytokinins or the production of secondary metabolites by tissue culture were reported in the 1970s in *Solanum khasianum* and *Nicotiana tabacum*. Working with *Catharanthus roseus*, Zenk et al. (cited in Heble, 1993) observed that growth hormones added to the medium strongly influenced the production of ajmalicine and serpentine. Similar observations have been made in *Dioscorea* cultures where diosgenin synthesis was influenced by the type of auxin added to the culture medium (cited in Heble, 1993).

Plant cells and organs have been successfully grown in 2-litre to 75,000-litre capacity bioreactors under precise nutritional and phytohormonal regimes. Industrial processes have been established for shikonin from *Lithospermum erythrorhizon*, ginsenosides from *Panax ginseng*, anhydrovinblastine from *Catharanthus roseus* and polysaccharides from *Echinacea pallida* (Heble, 1993).

The biggest bioreactor for plant cells was developed in Germany (75,000-litre capacity) by the Diversa Company for the cultivation of *Echinacea* cells for the production of polysaccharides. A 20,000-litre bioreactor has been used by Nitto Electrical Industrial Co. Ltd for the production of *Panax ginseng* cells. Bioreactors of 200-800-litre capacity have been used for the growth of *Coleus blumei, Coptis japonica* and *Catharanthus roseus* cell cultures. There is considerable interest in developing bioreactors for organ cultures like multiple shoots and hairy roots, the latter of which have been grown in 500-litre bioreactors (Heble, 1993).

By using immobilized plant cells, the working volume for a given product formation could be effectively reduced; cell efficiency could be increased. Different types of bioreactors have been used with success for immobilized cells: Morris et al. (cited in Heble, 1993) cultured immobilized cells of *Catharanthus* in a fluidized bed chemostat; Kabayashi et al. (cited in Heble, 1993) developed a bioreactor for producing berberine by immobilized cells of *Thalictrum minus*. In these systems, products are leached out into the culture media and downstream processing would be even easier (cited in Heble, 1993).

Cell cultures have been used for the biotransformation of various exogenous substrates. These transformations include stereospecific reactions, such as hydroxylation, oxidation, reduction, glycosylation, esterification, epoxidation, hydrolysis, isomerization, and dehydration. Alkaloids, such as codeinone, cathenamine, papaverine, tetrahydroberberine, and anhydrovinblastine, have been used as substrates. Immobilized cells of *Digitalis lanata* perform efficient biotransformation of beta-methyldigitoxin to beta-methyldigoxin. A large number of monoterpenes have been transformed by plant cell cultures (Heble, 1993).

Certain enzymes involved in the biosynthesis of phytochemicals have been purified and the gene coding for them identified for transfer. For instance, the gene for a key enzyme in the indole alkaloid biosynthesis, strictosidine synthase, has been transferred to *Escherichia coli*; the gene for the sweet protein monellin has been transferred to plants. Gene cloning and transfer for viable production of phytochemicals is still in its infancy since the biosynthetic pathways for most of the chemicals are not clearly understood (Heble, 1993).

The transformation of roots by *Agrobacterium rhizogenes* into hairy roots and their subsequent cultivation in large-capacity bioreactors could be advantageous for the production of some useful compounds. Fast growth – from a single root tip, hairy root culture could grow from 2500 to 5000 times its size in three weeks – gives root cultures several other advantages, including a natural defence against infection. Japanese firms using this technique to scale up ginseng root cultures have applied for protection of their results (Komen, 1991; Sasson, 1992a,b).

Escagenetics Corp. (California, USA) reported the successful production of taxol using the hairy root technique. Escagenetics' commercial production of tissue culture-derived vanillin has led the company into taxol tissue culture. Currently derived from the bark and needles of the Pacific yew (*Taxus brevifolia*), taxol is effective in treating several forms of cancer.

Supplies of taxol are limited, both as a result of the scarcity of Pacific yew trees and because of the minimal concentration of the compound found in the tree. Escagenetics Corp. has been able to produce taxol at concentrations higher than those found in the bark and needles of the yew. The company's proprietary technique is also said to outyield the callus culture technique that Phyton Catalytic (New York, USA) is licensing from the US Department of Agriculture to produce 'taxol or taxol-like compounds' on a pilot scale (Komen, 1991).

During the past three decades, several phytochemicals with biological activity have been identified from higher plants. Vincristine, vinblastine, artemisinin, Navelbine, taxol, and podophyllotoxin are on the list of approved drugs. Some products, such as forskolin, castanospermine, ginkgolides and hypericin, are undergoing chemical trials. These products are of very high value, ranging from $0.5 million to $9 million per kg; they are also excellent candidates for biotechnological production or mixed biotechnological and chemical synthesis (Table 41.2).

Production of food additives, such as pigments, flavours and fragrances, is also the subject of active research in different laboratories worldwide for the synthesis of crocin, betaxanthines, bixin, and anthocyanins. Among the flavours and fragrances, capsaicin and vanilla are commercially important candidates. There are global vanilla requirements of about 2000 tonnes, of which only about 20 tonnes come from the original source. The consumer preference for natural vanilla makes it a candidate for tissue culture;

Table 41.2. High value products, candidates for biotechnological production.

Product	Plant species	Price (DM g^{-1})	Culture type	Yield
Ajmalicine	Catharanthus roseus	56.00	SC	0.2 g l^{-1}
Vinblastine	C. roseus	15,800.00	ShC	Trace
Vincristine	C. roseus	37,800.00	ShC	Trace
Ajmaline	Rauwolfia	15.50	SC	0.04 g l^{-1}
Vincamine	Vinca minor	29.50	SC	3.3 g l^{-1}
Ellipticine	Ochrosia elliptica	3940.00	SC	0.005% dry matter
Camptothecin	Camptotheca acuminata	720.00	SC	0.00025% dry matter
Emetine	Cephaelis ipecacuanha	39.50	RC	0.3-0.5% dry matter
Berberine	Coptis japonica	16.50	SC	7 g l^{-1}
Sanguinarine	Papaver	72.00	SC	0.25 g l^{-1}
Colchicine	Colchicum autumnale	75.00	CC	1.5% dry matter
Aconitine	Aconitum	1120.00	Nil	
Diosgenin	Dioscorea deltoidea	0.10	SC	2.0% dry matter
Gingenosides	Panax ginseng	-	SC	0.4% dry matter
Podophyllotoxin	Podophyllum peltatum	-	SC	0.2 mg l^{-1}

SC, suspension culture; ShC, shoot culture; CC, callus culture; RC, root culture.
From Verpoorte and Heijden (1991), in Heble (1993).

Escagenetics Corp., USA, has a patent for production of vanilla by plant cell cultures. The aromas of coffee and cocoa have also been produced by cell cultures (Heble, 1993) (Table 41.3).

With regard to perfumes, agar wood oil, a group of sesquiterpenes, is an important and expensive product; the oil is produced from the wood of the tree *Aquilaria agallocha*, when it is infected by fungi; the tree is highly localized in northeastern parts of India and Myanmar (ex-Burma). Other interesting candidates are patchouli, vetiver, geranium, mentha, and sandalwood (Heble, 1993).

Therapeutic and digestive enzymes are also good candidates for biotechnological production. Thus, the production of papain, the major plant proteolytic enzyme, could be effectively increased by cloning elite papaya plants. Peroxidase is commercially produced by horseradish cells. Sweet proteins of commercial importance, monellin and thaumatin (3000 times sweeter than cane sugar) are also good candidates (Heble, 1993).

Clonal propagation of identified source plants would give uniform plants which could be used in herbal formulations such as therapeutics, tonics, and cosmetics. Potential candidates for cloning are *Withania somnifera, Rauwolfia serpentina, Centenella asiatica, Aloe, Phyllanthus embelica, Terminalia*

Table 41.3. Food additives.

Product	Plant species	Culture type	Yield (% dry matter)
Betaxanthine	*Beta vulgaris*	SC	0.5
Bixin	*Bixa orellana*	CC	3.4
Anthocyanins	*Vitis vinifera*	SC	3.4
Crocin	*Gardenia jasminoides*	CC	8 times less than *in vivo*
Vanillin	*Vanilla*	SC	100 mg l^{-1}
Mentha terpenes	*Mentha*	ShC	–

SC, suspension culture; ShC, shoot culture; CC, callus culture.
Source: Verpoorte and Heijden (1991), in Heble (1993).

arjuna, T. belerica, T. chebula, Commiphora mukul, Azadirachta indica, etc. (Heble, 1993).

The first commercial success with plant cell cultures has been for the production of shikonin, a pigment used in cosmetics by Mitsui Petrochemical Co. Ltd, Japan. This was followed by several ventures, which are today in the spotlight (Table 41.4).

More than 50 major industries worldwide are working on plant tissue cultures, with a wide range of target molecules, like pharmaceuticals, food additives, aromas, biopesticides and enzymes. Technology Centre, by Kyowa Hakko Kogyo Co., Mitsui Petrochemical Co. Ltd, Mitsui Toatsu Chemical Inc., Hitachi Ltd, Suntory Ltd, Toa Nenryo Kogyo Co. and Kirin Breweries Co. Ltd. Japan is followed by Germany and the USA (Heble, 1993).

An important consideration is the cost/profit ratio. The first biotechnology product, shikonin, was produced at a cost of $4500 kg^{-1} in a 200-litre bioreactor. A study carried out in Japan showed that any substance of plant origin costing more than $80 kg^{-1} could be produced by plant tissue culture techniques; this range would include a high number of phytochemicals. In the case of micropropagated elite plants, the production cost would be around 50 cents–$1 per plant and the product yields could be effectively increased by two- to tenfold. Techniques of cell immobilization, hairy roots and multiple shoot cultures could effectively lower the final production cost. Candidates for plant tissue culture costing between $0.5 million and $9 million are given in Table 41.5.

Among the constraints and risk factors associated with plant cell and tissue cultures are the following.

1. They have slow growth compared to microorganisms, with a doubling time in the range 20–60 hours; as a result, the cultures are vulnerable to microbial contamination and a great deal of expenditure goes toward maintaining aseptic conditions.

2. Mutation frequency caused by the culture conditions, stress, and phytohormones poses problems for the maintenance of stability of cultures.

Table 41.4. Major commercial phytochemicals.

Product	Plant species	Annual needs	Cost ($ kg^{-1})	Biological activity
Vinblastine	*Catharanthus roseus*	5-10 kg	5 million	Anticancer
Vincristine	*C. roseus*	-	-	-
Ajmalicine	*C. roseus*	3-5 tonnes	1500	Anti-hypertension
Podophyllotoxin	*P. hexandrum*	-	-	Anticancer
Codeine	*Papaver somniferum*	80-150 tonnes	650-900	Expectorant
Digoxin	*Digitalis lanata*	6 tonnes	3000	Cardiotonic
Diosgenin	*Dioscorea deltoidea*	200 tonnes	20-40	Steroid hormones
Jasmine oil	*Jasminum*	100 kg	5000	Fragrance
Mint oil	*Mentha*	3000 tonnes	30	Fragrance
Vanillin (natural)	*Vanilla*	30 tonnes	2500	Fragrance
Taxol	*Taxus brevifolia*	-	-	Anticancer

Table 41.5. High value medicinal substances to be produced by plant tissue culture.

Product	Plant species	Biological activity
Taxol	*Taxus brevifolia*	Anticancer (ovarian)
Camptothecin	*Camptotheca acuminata* *Nothopodytes foetida*	Anticancer
Forskolin	*Coleus forskohlii*	Cardiotonic
Artemisinin	*Artemisia annua*	Antimalaria drugs
Castanospermine	*Castanospermum australe*	Anti-AIDS
Hypericin	*Hypericum esculentum*	Anti-AIDS

3. Highly productive cell lines only occur rarely, generally obtained through selection.

4. There is difficulty in inducing the cells to produce the desired compound (the genes coding for plant chemical products are turned on only under special conditions).

5. There is laborious excretion of secondary products by plant cells.

6. The technology development costs are high, the gestation period being generally three to four years (Komen, 1991; Sasson, 1992a,b).

Strong research-and-development support is a prerequisite. In fact, all commercial ventures are backed by the team leaders in medicinal plant biotechnologies (e.g. Staba in the USA, Zenk and Reinhard in Germany, Yamada, Furuya and Tabata in Japan) (Heble, 1993).

Plant Germplasm Appropriation: Impact on Developing Countries

Intellectual property rights on plant germplasm

Appropriation may concern:

1. Plant germplasm essential to breeding work and plant tissue cultures;
2. Breeding lines possessing desired characteristics and used for developing new varieties;
3. New commercial varieties or new cell lines, new products and new uses or applications of these products;
4. Innovative technical processes involved in the creation of new varieties or products.

In promulgating the Plant Patent Act in 1930, the USA became the first country to protect, through the award of a patent, plants which were asexually or vegetatively reproduced, mainly ornamentals. In Europe during the 1940s and 1950s, there were so many seeds of allegedly improved varieties on the market that the goverments of several countries regulated the commercialization of seeds; a system of certification was established to limit abuse; only the sale of seed varieties listed on national registers was authorized.

The rules applying to the award of patents do not apply, in most cases, to cultivated plant varieties. In fact, breeders' 'inventions' are not generally protected in the initial national regulations. Under the 1973 European Patent Convention, plant varieties and biological inventions were excluded from the patent domain (article 53b). When a breeder was awarded a patent for a variety, the patent was rendered almost useless because of extension of the granted right: due to the application of the very general principle concerning the 'exhaustion of right', buyers might use freely any seeds or plants bought from patentees or from their licensees. Buyers might propagate the plant material for their own needs or for marketing purposes. Patent legislation was not therefore adapted to living things, but concerned mainly inanimate objects (UPOV, 1985; Le Buanec, 1987).

The legal protection of new plant varieties and the ensuing legislation were designed by governments as a means of making research in plant genetics and breeding more attractive to private firms which up until then had preferred to invest in other sectors of research and development. An awareness of the need to protect plant obtentions led to the adoption, on 2 December 1961 in Paris, of the Convention of the International Union for the Protection of New Varieties of Plants (the UPOV Convention), under the aegis of the World Intellectual Property Organization (WIPO). Revised in 1978, the Convention was considered an appropriate legal framework for the protection of plant species, which Member States could enforce through legislation. In 1978, membership of the UPOV was extended to non-European countries. Member countries, at that time, represented more than 95% of the seed market of market-economy industrialized countries and over 70% of the seed market of all countries with a market economy (UPOV, 1985; Le Buanec, 1987).

The UPOV Convention ruled that in a member country there could not be double protection, i.e. it was not possible to request the protection of a variety through a patent when a certificate of protection (breeder's right) could be obtained (Le Buanec, 1987).

The American Plant Variety Protection Act of 1970 and its subsequent amendments in 1980 protect sexually reproduced varieties and concern the seeds and propagation organs of these varieties and of more than 350 plant species. The law, however, excludes bacteria, fungi, tuber-propagated plants, uncultivated plants and first-generation (F1) hybrids (Williams, 1984). The period of plant protection was also extended from 17 to 20 years, in conformity with the UPOV Convention.

The majority of breeders seem satisfied with the UPOV Convention and the breeders' rights granted to protect their new varieties. The Convention has been revised, but more drastic changes regarding the protection of plant varieties could be brought about by the development of plant biotechnologies and the pressure exerted by the industrial companies or groups involved in the propagation of plants and production of selected seeds.

The totipotency of plant cells, i.e. their aptitude to regenerate a whole plant, enables the *in vitro* micropropagation of plants from tissue fragments, or the utilization of embryoids or somatic embryos obtained from tissue cultures for the same purpose. The obtention of haploids, somaclonal variation, protoplast fusion, as well as genetic recombinations, can contribute to the selection of new plant varieties which would be protected efficiently. On the other hand, the major groups of agricultural food, chemical, petrochemical, and pharmaceutical industries, which make substantial investments in the breeding of plant varieties and in plant biotechnologies, want to strengthen the protection of plant obtentions and believe that the patent system (to which these industrial groups are more accustomed) is more appropriate.

The Supreme Court of the United States of America ruled in 1980 that genetically engineered bacteria were patentable. By so authorizing the award of a patent to a bacterial strain transformed by Chakrabarty, who had filed an application in 1972, the Court ruled that 'the relevant distinction was not between living and inanimate things, but between products of nature, whether living or not, and man-made inventions'. Following this decision, the number of patents in genetic engineering increased twice as fast as those in other areas of technology (National Research Council, 1984). Furthermore, the first protection by an industrial patent in 1985, in the USA, of a maize variety with a content of tryptophan (Hibbert case), changed attitudes to extending patents to include all biotechnological inventions, even when these concerned genetically modified living beings.

Experts were of the opinion that the inventor of a gene which might be transferred into a plant with a view to breeding a new variety, should be rewarded for his/her work. But is it justified to extend this right to include the whole transformed plant? Is it right that the inventor of the gene should own the variety to which a certificate of protection was granted? Access to new varieties should remain free to enable them to become the initial source of variation. According to Le Buanec (1987), it is advisable to keep the present protection status of plant varieties, as described in the UPOV Convention, but to introduce technical improvements and an important modification: a patented gene remains protected in the plant; the right of access to genetic variability should be maintained even while becoming liable to payment. According to this expert, the system of double protection of plant varieties, through patents and breeders' rights, should be avoided (Le Buanec, 1987).

Impact on developing countries

Measures taken by technologically advanced countries to protect the products of increasingly expensive plant genetic research and to make this research more profitable entail, for developing countries, the payment of high fees for biotechnology-derived plant products. But, as 'the North is rich in seeds and the South in genes', developing countries today conscious of the economic importance of their plant genetic resources intend to protect these by, for instance, prohibiting export of any reproductive plant part, i.e. its germplasm. Developing countries claim that the price for re-acquiring the varieties selected and improved from their own plant genetic resources is too high and that it is unfair to be thus condemned indirectly to purchase back part of their own plant genetic heritage (Hobbelink, 1987). Developing countries intend also to promote the national production of improved seeds.

Search for an international agreement

One extreme position would be to deny any right of intellectual property. This could result in the isolation of a country taking such a position, in research by foreign corporations or institutions existing in that country being hindered and in the failure to transfer or acquire technology. Another extreme position would be to accept all the imposed rules of the game, which might lead a country to become merely a focus for the interests of multinational corporations.

As scientific, technological, economic and financial interdependence is increasing throughout the world, it is ever more difficult for a country to isolate itself. On the other hand, the disparities in the bargaining power between rich and poor countries, between developing and underdeveloped ones, make it increasingly difficult to arrive at balanced agreements offering reciprocal benefits.

That is why the problem of protecting biotechnological innovation is important and should be approached in a pragmatic way. There are two distinct approaches to an international system of protecting biotechnological inventions. The first is that of the Committee of Experts for Biotechnological Inventions and Industrial Property of the World Intellectual Property Organization (WIPO), which operates a system of patents similar to that existing for conventional industrial products and processes, and protects processes or products designed for a specific use through the award of patents.

Therefore, transformed plants, animals and microorganisms could be protected by patents.

The other approach is that of the UPOV Convention, i.e. that protecting breeders' rights adopted by all the industrialized member countries of this Union. In this case, the right of the inventor is ensured through the payment of royalties by all those who utilize the product, in agreement with the standards and conditions defined in the regulations of the UPOV Convention and enforced in the laws of member countries.

There is a difference between the two positions. In the first approach, the patentee or inventor has an unlimited right, this right being either exclusive or restricted to a certain area. Defended by the scientifically and technologically advanced countries and by the large multinationals, this approach leads to a monopoly or oligopoly. In the case of the UPOV Convention, the inventor is guaranteed a right to compensation for use of the invention, in addition to the right to transfer the 'invention' to anyone wishing to use it or start commercial production. This is due to the fact that, in breeding a new plant variety, the innovative role largely belongs to nature; that is why breeders' rights are not treated in the same way as patents covering an invention.

The alarming decrease in genetic diversity or 'genetic erosion', together with the restrictions imposed on the distribution of plant material, have led to the search for an international agreement on conserving plant genetic resources, considered to be part of the common heritage of humankind, and on ensuring their equitable utilization, rather than leaving their exploitation under national jurisdiction alone. A total of 156 countries represented at the 22nd session (November 1983) of the FAO Conference recognized that 'plant resources were part of the common heritage of mankind and should be accessible without any restriction'. These resources include wild species or species closely related to cultivated varieties which need to be inventoried and protected, because they are threatened with extinction, but also crop varieties and the most recent cultivars, which permit the production of higher-yielding hybrid seeds.

It is important to find a compromise between the legitimate wish to reward human ingenuity by awarding rights and the necessity for developing countries to acquire improved plant germplasm at a cost compatible with their limited means and their agricultural development needs. Furthermore, the toil of generations of farmers in developing countries to improve their crop varieties should also be taken into consideration when selling back to them improved varieties of their major crop species (Mooney, 1984; Hermitte, 1985, 1986; Alaux, 1987; Hobbelink, 1987).

In a November 1991 resolution adopted by the FAO Conference, the Member States came as close as possible to acknowledging the need for a mandatory international fund devoted to implementing farmers' rights. To this end, it was agreed that 'an international fund on plant genetic resources would support (...) conservation and utilization programmes, particularly, but not exclusively, in the Third World [and that] the resources for the international fund (...) should be substantial, sustainable and based on the principles of equity and transparency.' This resolution was incorporated as the third annex to the International Undertaking on Plant Genetic Resources.

The FAO's Commission on Plant Genetic Resources, at its fourth session, in April 1991, agreed that a Code of Conduct for international collectors of germplasm would have the following objectives: to promote the sustainable use of biotechnologies for the conservation and use of plant genetic resources; to ensure free access to plant genetic resources; to mitigate the risks biotechnologies represent for the environment; to distribute advantages fairly between those developing, and those supplying, the genetic material used by biotechnologies.

The Commission reiterated that intellectual property rights were not designed to obstruct the free exchange of genetic material, information and technologies for scientific purposes. A system aimed at protecting intellectual property in the field of plant genetic resources would need to take into consideration both the rights of innovators and the interests of farmers, contributors for centuries to the adaptation and improvement of crop species. The Commission highlighted the need to train scientists and technicians in biotechnologies in developing countries, in order to facilitate the transfer of knowledge and avoid deepening the gap between developing and industrialized countries.

A draft of the Code of Conduct was presented to the 5th session of the Commission on Plant Genetic Resources in April 1993.

The Keystone Center, an American non-profit making organization, initiated a dialogue on plant genetic resources in 1988, which brought together more than 90 participants from 30 countries. They represented seed corporations, non-governmental organizations, national and international organizations and research institutions. The group's third and final plenary session was held in Oslo, in June 1991.

Preventing 'genetic erosion' is, according to the group, an urgent task requiring international commitment. A Global Plant Genetic Resources Initiative, as proposed by the group, would require $300 million annually to strengthen existing programmes and institutions. For the 1993-2000 period, an estimated $1.5 billion would be required. In the Final Consensus Report of the Keystone Oslo meeting, suggestions were provided for the governance of the fund needed to support the initiative, ensuring national sovereignty and commitment, as well as equity between suppliers of plant genetic resources, technology and information at global, regional, national and community level. The Dialogue group also dealt with intellectual property protection and tried to reconcile the positions of industrialized and developing countries, particularly between private industry and community activities.

References

Alaux, J.P. (ed.) (1987) *Les ressources génétiques végétales, atouts du développement?* Paris, Institut français de recherche scientifique pour le développement en coopération (ORSTOM, Direction de l'information, de la formation et de la valorisation – DIVA), 201 pp.

Brady, N.C. (1985) Agricultural research and U.S. trade. *Science* 218, 847–853.

Esquinas-Alcázar, J.T. (1987) Plant genetic resources: a base for food security. *Ceres, The FAO Review* 118, 39–45.

Heble, M.R. (1993) High value chemicals by tissue culture. *Chemical Industry Digest* September, 113–118.

Hermitte, M.A. (ed.) (1985) *La protection de la création végétale: le critère de nouveauté*. Paris, Librairie Techniques.

Hermitte, M.A. (ed.) (1986) *Le droit du génie génétique végétal*. Paris, Librairie Techniques, 256 pp.

Hobbelink, H. (1987) *New Hope or False Promise? Biotechnology and Third World Agriculture*. Brussels, Interational Coalition for Development Action (ICDA), 73 pp.

Kloppenburg, J. and Rodriguez, S. (1992) Conservationists or corsairs? *Seedling* 9(2,3), 12–17.

Komen, J. (1991) Screening plants for new drugs. *Biotechnology and Development Monitor* 9, 4–6.

Le Buanec, B. (1987) La protection des obtentions végétales. *Biofutur* 61, 49–53.

Mooney, P.R. (1984) The law of the seed: an introduction. International Foundation for Development Alternatives (Nyon, Switzerland). *IFDA Dossier* 39, 77–78.

National Research Council (1984) *Genetic Engineering of Plants. Agricultural Research Opportunities and Policy Concerns*. Board on Agriculture, National Research Council, US Academy of Science Press, Washington, DC, 83 pp.

Nau, J.-Y. (1994) Selon l'OMS, un remède traditionnel chinois permet de lutter efficacement contre les formes graves de paludisme. *Le Monde* 19 April, p. 22.

Sasson, A. (1992a) Production of useful biochemicals by higher-plant cell cultures: biotechnological and economic aspects. In: DaSilva, E.J., Ratledge, C. and Sasson, A. (eds) *Biotechnology: Economic and Social Aspects; Issues for Developing Countries*. Cambridge University Press, Cambridge UNESCO, pp. 81–109.

Sasson, A. (1992b) *Biotechnology and Natural Products. Prospects for Commercial Production*. African Centre for Technology Studies (ACTS), Nairobi, Kenya, 97 pp.

UPOV (International Union for the Protection of New Varieties of Plants) (1985) *Les brevets industriels et les certificats d'obtention végétale; leurs domaines d'application et les possibilités de démarcation*. UPOV, Geneva.

Williams, S.B. Jr (1984) Protection of plant varieties and parts as intellectual property. *Science* 225, 18–23.

42 A Partnership: Biotechnology, Bio-pharmaceuticals and Biodiversity

M. Comer and E. Debus
*Boehringer Mannheim GmbH, Nonnenwald 2,
82377 Penzberg, Germany*

Introduction

The use of plants or extracts derived from them in the healing of wounds or the treatment of disease is almost as old as human history (Mothes, 1984).

Nearly half of today's medicines, and one quarter of all drugs prescribed in the US have originated from around only 100 species of plants. Also, from an estimated 250,000–500,000 species of plants less than 1% has been thoroughly investigated (Wallace, 1993). Other sources of potential therapeutically active substances are tropical insects and related species. However, although insects are known to metabolize a wide variety of secondary metabolites as defensive agents, e.g. venoms and pheromones, they have received even less attention. It can be expected that the myriad of this still largely unexplored species of animals and plants, may provide additional sources of powerful new medicines in the future. The potential which exists in marine organisms is still only a matter of guesswork. For instance, the deadly venom of the marine cone shell snail is composed of hundreds of peptides each of which can be targeted to affect a different organ (Bovsun, 1994).

With the rapid development of modern technology in the 1980s, the interest in traditional medicine and nature as a source of pharmaceuticals was quite low. But this has changed dramatically (Cox and Balick, 1994). Today huge programmes for natural compound screening are under-way; these have been initiated by international organizations, scientific institutions, and transnational pharmaceutical companies. The diversity in humans is also receiving a much more intensive scrutiny in order to search for cures to life-threatening diseases (Wallace, 1993).

Aspects of Biodiversity for Human Health Care

Even with the sophisticated methods of modern biotechnology, like drug design and genetic engineering, the biodiversity of Nature herself is still the best source of therapeutics, since she is still the best drug designer. In the search for new products from natural resources pharmaceutical companies, like Boehringer Mannheim (BM) invest large sums of money.

Today with increasing insight into biological and biochemical interactions, there is a growing understanding of the importance of biodiversity in fauna and flora, its conservation and application in health care and the life sciences.

This includes the study of the diversity in the human genome (Gillis, 1994), the extensive genetic variation within and among different human populations around the globe and particularly those special groups which have been called 'Isolates of Historical Interest' (Roberts, 1993). There are already two huge projects concerned with the genetic examination of humans under way. The 'Human Genome Project' is designed to sequence the entire genetic code, but will not necessarily evaluate population-level variation. However, the 'Human Genome Diversity Project' aims to examine differences in the genetic make-up and will provide a greater knowledge about susceptibility or resistance to disease (Lewin, 1993; Roberts, 1993; Cavalli-Sforza et al., 1991). It will also highlight whether variations are the result of adaptation to local conditions or merely random changes in the DNA molecule. Areas of particular interest under investigation at the moment are diabetes, thalassaemia, hypertension and sickle-cell anaemia. Perhaps when all the data are collected and analysed it may also supply the answer to the burning question with regard to human evolution, did Eve originate in Africa? (Bertranpetit and Cavalli-Sforza, 1991; Vigilant et al., 1991; Moutain et al., 1992; Roberts, 1992; Gibbons, 1993).

Examples of Plants and Microorganisms as a Source of Drugs

Drugs derived from plants

Well-known examples of plant-derived drugs with a long and successful story are aspirin and quinine.

Already Hippocrates recommended the bark of the willow tree against the suffering of pain, this then fell into oblivion but was later to be rediscovered several times by people who were skilled in the art of plant-derived medicines. The bark's active ingredient, acetyl salicylic acid (aspirin), is today produced synthetically and is probably still the widest known drug in the world. Currently, this preparation is not only recommended for pain relief but also for the treatment of fever, the prophylactic prevention of thrombosis and indeed several other applications (Mann and Plummer, 1993).

Quinine is extracted from the bark of the cinchona tree, it was a heritage discovery of the Indians of South America (Galeffi, 1993). From the beginning of the nineteenth century, the purified substance has been used against intermittent fever and malaria which was also prevalent in Southern Europe. When the supply of the cinchona bark proved more difficult during the war, chemists tried to synthesize the active substance and discovered instead chloroquine. This was also found to be effective in the treatment of malaria. However, the negative news is the steadily increasing resistance of some varieties of the malaria pathogens to these drugs. This illustrates, in a smaller dimension the dynamic change of nature and the progression in biodiversity.

To the present day, the natural product quinine remains the most effective agent against malaria and it is still cheaper to produce than when compared to synthetic substitutes. But quinine is not the only active substance found in the bark, there are indeed other alkaloids like quinidine which is used in the treatment of cardiac dystony.

Towards the end of the nineteenth century, Boehringer started its own production of quinine. Since then, the development of our company for over 100 years has been closely associated with quinine and the supply of the natural product. Boehringer Mannheim is still worldwide the principal producer, with large plantations for the raw material in Africa (Fischer, 1991). However, despite the major importance of quinine it comprises only a minor component of our total business today.

Pharmaceutics from microorganisms

Famous examples of drugs derived from microorganisms, like bacteria and fungi, are the penicillin antibiotics. However, there are many other well-known and important classes of compounds such as enzymes which can be manufactured from microbial sources.

Through modern biotechnology it is possible to grow these organisms *in vitro* on a large scale and subsequently, as highly purified proteins to play an important role as life-saving medicines or as valuable diagnostic reagents.

Table 42.1 gives an overview to show the potential discovery still existing with respect to microorganisms, i.e. algae, bacteria, fungi and viruses. From 1.7 million estimated species not even one tenth have been described (some will never be described, because of environmental destruction or other influences caused by civilization, many species will simply have vanished before they have been discovered).

Biotechnological methods have made it possible to preserve some of these described organisms, at least *in vitro*, thus making them available for future research.

Worldwide, various culture collections already exist, often either run by companies or research institutions. Also BM has a substantial collection, which has been established for the company's special need for working in the field of pharmaceutical biotechnology. The function of the culture collection of microorganisms established by BM is summarized in Table 42.2.

The establishment of a culture collection requires a significant effort since

Table 42.1. Microbial diversity.

	Number of species		Species in culture collections	
Group	Described	Estimate of Total	Number	% of total
Algae	40,000	60,000	1,600	2.5
Bacteria	3,000	30,000	2,300	7.0
Fungi	69,000	1.500,000	11,500	0.8
Viruses	5,000	130,000	2,200	2.0
Total	117,000	1.720,000	17,600	1.0

From Nisbet, (1992).

Table 42.2. Function of the Boehringer Mannheim culture collection of microorganism.

Conservation of incoming strains
Documentation using a highly sophisticated data base system
Internal distribution of microbial strains for screening and production
Classification and characterization of unidentified microorganisms
Examination and control of strains for purity and/or genotypic variability
Classification with regard to safety (risk grouping)
Comprehensive data files covering every known microorganism
Communication with culture collections worldwide
Consultancy for all internal microbiological problems

it can also mean a wide variety of different methods of conservation. For instance, the medium required for the propagation of one organism may be entirely different from those requirements of another. It is therefore essential that all the information known is compiled and documented, so that it remains available for posterity and in order that exchange of data can take place. However, it will be obvious that such a volume of information creates its new problems, not only in the communication of that information to those who may need it but often involving a large administrative effort. This is why in recent times one has enlisted the aid of computers with the resulting so-called 'Data Banks' (Canhos et al., 1993).

To prevent misuse of the collected data and to protect the rights of all partners, international conventions are required; this must be established under the aspects that biodiversity is of international importance and for the common good.

Pharmaceutical Industry and Environment

The protection of the natural environment must have a priority, if alone or as well as an obligation to preserve biodiversity for future generations.

At Boehringer Mannheim, a pharmaceutical company working for health care, we recognized early the importance of a healthy environment and its influence on the welfare of people. Many years ago we implemented environmental management as one of our highest corporate priorities. We had already set high standards in research, development and production for all of the business units in order to maximize the quality requirements of our products. These were then to conform to many of the proposals which were formulated by Agenda 21 from the summit in Rio.

Example of a policy statement

Many companies worldwide have recognized the importance of environmental protection and possibilities to maintain biodiversity. They have accepted the responsibilities and have written it into their company policy.

The Executive Committee (EC) of our worldwide operating Corange Group, of which Boehringer is part, has presented a strategy for environmental protection and safety. A summary is given as an example of many similar statements made by other pharmaceutical companies who have recognized that safety, health, and environment (SHE-concept) are tightly linked together.

1. A clear aim for the Corange Group is defined: The firms of the Corange Group are established on the markets as environment-orientated and safe companies. Environmental protection, safety, and high ethical standards are Corporate aims.

2. Guidelines to reach this aim: The Corange EC passes its own set of guidelines based on internationally introduced programmes such as 'Responsible Care' and 'Sustainable Development'

3. Organizational measure: The Group organization is built on already existing structures. It uses synergisms and makes know-how and information group-wide available. Health services are integrated.

4. A set of instruments have been named:
(a) Regular pooling of experience
(b) International audit programme for production locations
(c) Intensive use of information technology
(d) 'Ethics Committee'

It has been categorically stated that this is the basis for a long-term strategy.

Environmental protection in the past

We believe that in the past BM has made some major contributions to environmental protection, especially with biotechnological methods. Although the degree of impact for environmental protection may not have been fully foreseen by our developers at the time, an increasing awareness of the environmental burden caused by hazardous waste disposal, drives the development in our R & D departments nowadays. Here, more and more research is conducted under the awareness and with the aim of harmony with the environment.

Table 42.3. The determination of triglyceride: a comparison of substances used in a chemical method and the enzymatic method.

Chemical method	Enzymatic method
Methanol	Glycerokinase
Chloroform	Pyruvate kinase
Petroleum ether	Lactate dehydrogenase
Sodium arsenate ($NaAsO_2$)	Nicotinamide-adenine dinucleotide
Sodium periodate ($NaIO_4$)	Adenosine triphosphate
4,5-dihydroxydisulphonic acid	Adenosine diphosphate
	Phosphoenolpyruvate

Replacement of hazardous chemicals

The biochemists at BM were surely pioneers in their field, when in 1954 the first 'enzymatic' test kit for the detection of alcohol in blood was developed. This was at a time when 'Environmental Protection' was still a relatively unusual topic. With the introduction of enzymatic methods for diagnostics, it was possible to replace traditional procedures, many of which involved potentially hazardous chemicals, like inorganic acids, heavy metals and large amounts of organic solvents. The enzymatic test kits or biological reagents do not normally present hazardous waste. They are therefore safer for ground water and the environment since they are, without exception, biodegradable.

A quite impressive comparison are the components of the test for the analysis of triglycerides in blood (Table 42.3).

For 35 million tests (which are performed in Germany alone per year) with the chemical method the following amounts are used: approx. 2,000,000 litres of solutions and 400 kg of salts.

The same amounts of tests performed with the enzymatic method uses: approx. 70,000 litres of enzymes and buffers.

The above enzymes and related biochemicals are so-called biologicals and represent a fraction of the quantities of the chemical counterpart. Also, they are of biological origin with no harm resulting to the environment if they should be accidentally released or improperly disposed of. The buffers used are all biodegradable or biologically utilized (Waessle et al., 1990).

Today in the physician's laboratory and clinic, enzymatic tests have thankfully replaced many chemical methods. They can run either automatically on suitable instruments or are performed with so-called 'dipsticks'. Besides the advantage of being fast and clean, the environmental burden has been reduced enormously when compared to that of the chemicals used in traditional methods.

Also in the case where a strain, cell, or organ expresses the desired product but only insufficiently or is a pathogenic species, the DNA can be transfected into a strain which is not pathogenic and therefore, it is easier to handle the product which can be purified more efficiently and usually in higher yields.

Table 42.4. Environmental relief by the use of recombinant *E. coli* for the production of cholesterine oxidase instead of using the original strain, *Rhodococcus* sp. for the annual production.

(a) Consumption of energy and waste burden

Energy and disposal	*Rhodococcus erythropolis*	rec. *E. coli*
Steam	330 tons	32 tons
Cooling water	22,000 m^3	240 m^3
Electricity	92,200 kWh	2208 kWh
Process/air supply	81,400 m^3	4560 m^3
Decarbonated water	1100 m^3	./.
Ice water	4000 m^3	400 m^3
Drinking water	120 m^3	24 m^3
Demineralized water	400 m^3	16 m^3
Liquid waste	176,000 population equivalent	2400 population equivalent
Disposal biomass	19 tons	800 kg

(b) Ingredients and quantities used for the production process

Ingredients	*Rhodococcus erythropolis*	rec. *E. coli*
Butanol	88,000 l	./.
Siliceous earth	12,000 kg	./.
Triton X-100	60 l	8 l
Yeast extract	1650 kg	224 kg
Potassium phosphate	550 kg	12 kg
Potassium acetate	5456 kg	./.
Cholesterin	4400 kg	./.
Acetic acid	1650 kg	./.
Phosphoric acid	1430 kg	./.
Lecithin	550 kg	./.

Introduction of gene technology

With the advent and subsequent application of gene technology it has been possible to obtain microorganisms as high-producer strains. With genetic alteration, the yield of one fermentation batch could be raised by a factor of a hundred or even more. In this way it is possible to reduce significantly the energy consumption, water, and waste by increasing production yields considerably.

Table 42.4 demonstrates the saving of various chemicals in a process for the manufacture of cholesterine oxidase, produced by a recombinant *Escherichia coli* and not from the original strain. Purification is from the fermentation broth. This illustrates a further advantage obtained from modern DNA techniques.

Partial replacement of potentially hazardous chemical methods

With the introduction of biocatalysts for the manufacture of, for example, modified antibiotics, harsh and complex chemistry can often be replaced by the use of enzymes. This not only makes available possible new antibiotics for use in health care but contributes considerably to the welfare of the environment.

Partial replacement of radioactivity

Radioactive substances are often used as markers in biochemical and diagnostic test systems. These can cause serious difficulties in disposal as laboratory waste, and at the same time present a potential hazard to the operator. Boehringer Mannheim has invented the so-called DIG-System (Digoxigenin). It is a most convenient and effective system for the labelling and the detection of DNA, RNA, or oligonucleotides. It uses the steroid hapten digoxigenin as a label for hybridization and subsequent detection of nucleic acids instead of a radioactive one.

Biological waste water plant

Our waste water plant, which was extended in 1978, had then the capacity of 58,000 population equivalence. With the development of recombinant gene technology in microbiological systems, the reduction in waste disposal was so effective, that until today and for the next 16 years, the plant will have adequate capacity despite a steady increase in production of approximately 10% per year.

At the time we had thoroughly investigated the best system for our waste water plant under real conditions. Indeed we simulated potential accidents where in this unlikely event rec-microorganisms would be released to the outside world. This was performed by using a contained laboratory model, scaled down to the exact dimensions of the sewage plant, in order to test the efficiency of inactivation and thereby the safety of use of transfected organisms.

The results showed that the recombinant organisms could only survive a short time and were quickly outgrown by their natural counterparts. Thus, our relatively extensive experience with biological waste has also of use to the general good since we have often been consulted about improvements of other sewage systems.

The Biotechnological Advantage

The developments in the field of biotechnology have opened new opportunities for a rationalized growth in the pharmaceutical industry.
  1. It has provided:
  (a) new insight into life sciences, biological processes, and biodiversity;

(b) new techniques to gather data and exchange information;
(c) application of this knowledge to human health care.

2. In relation to environmental protection, successes have been achieved through:
   (a) the avoidance of hazardous chemicals, or techniques which may be harmful to the environment;
   (b) a production which makes a better use of the sustainable resources and thus maintains them;
   (c) reducing volumes of waste and the conservation of the most precious of our resources, water;
   (d) possibilities to remedy already contaminated areas, land, or ocean.

The fast development in biological science will provide ways which offer further possibilities to minimize the rapid depletion of important and often irreplaceable natural resources. It will reduce the burden to nature, thus civilization can live in better harmony with our unique planet to achieve a more ecologically balanced environment. Ultimately the best protection for biodiversity and human health care is to remember that prevention is better than cure.

Conservation of Biodiversity

Time is getting short; if we do not continue efforts to preserve biodiversity now the many forms of life will be lost irretrievably. The increasing urbanization threatens the integrity of small isolated populations; wars, famine, and diseases can accelerate the extinction of unique forms of diversity expressed by the human genome. Also worldwide deforestation causes the yearly loss of several plant species which may have vanished forever and, at the same time, eradicates the natural animal habitat (Ehrlich and Wilson, 1990).

For a health care company there are three approaches for attacking the problems of human disease, i.e. prevention, treatment, and cure. These principles can of course also be applied to the maintenance of biodiversity. Prevention of the loss may refer to the *in situ* preservation of genetic resources within the dynamic evolutionary ecosystem of their original or natural environment.

Treatment, i.e. precautionary measures, have to be taken where species are endangered in their natural habitat. A realistic approach to this immense task, for both the *in situ* as well for the *ex situ* preservation, can only be achieved through the sophisticated methods of modern technologies, like informatics and biotechnology.

The *ex situ* conservation requires the removal of germplasm resources like seed, pollen, sperm and individual organisms from their original habitats and preserving them in botanical gardens, zoos or aquaria, or with the use of biotechnological methods like collecting specimens for gene-or seed-banks, gene mapping, and cryopreservation of the germplasm (Benford, 1992; Coghlan, 1994; Sirica, 1994).

But also, breeding endangered species in zoos or botanical gardens can depend on the use of gene technology, for example to avoid inbreeding to preserve the genetic diversity. A collected data base with inventories of genetic diversity will have to be established. Apart from the problems involved in maintaining long-term integrity of the germplasm, one has, however, to realize that no further evolution can take place in an artificial *ex situ* conservation situation (Ford-Lloyd and Maxted, 1993; Kumar, 1993).

In situ conservation is therefore more favourable than *ex situ*, but at least *ex situ* concentration is better than total loss.

The cure for the situation is the development of biodiversity, which is a dynamic process, with species disappearing and new ones appearing. Therefore the aim of programmes for conservation *in situ* should not be to establish a natural history museum but resorts, where an undisturbed nature, can exist in a progressive harmony or at least not too adversely influenced. An extension to the ideas of the National Park Reserves could be a solution to fulfil this.

Impact of the Pharmaceutical Industry

Therefore, what role can the biopharmaceutical industry play?

In annual communiqués many chemical and pharmaceutical companies report of their progress and successes in environmental protection, especially through the application of modern technologies. They have included such aims of 'responsible care' and 'sustainable development', according to the proposal made by the summit in Rio, which our company has also adopted. Although the principle that biodiversity needs to be preserved has been recognized, many companies and institutions (indeed countries) have still failed to take adequate measures. However, in our opinion, the degree of responsibility has been accepted and it can only be a question of time before more universal policies proposed from community and countries will have been implemented (Burk *et al.*, 1993).

International pharmaceutical companies in the developed world have the knowledge and the technologies to utilize sensibly the resources of biodiversity. The majority of genetic resources, especially centres of taxonomic diversity, lie mainly in developing countries. However, the technology to use the abundance of their biodiversity without depletion is not readily available (Mannion, 1993). Here, there is a lack of finance which would be required to establish such technologies. However, the numbers of suitably trained people with the capability to apply their knowledge within these countries is gradually increasing. For the transnational high-tech companies and institutions it is in the interest of all those concerned to form partnerships with their developing counterparts in the so-called 'third world'.

Ambitious initiatives have been translated into many projects between international institutions (e.g. UNESCO, WHO, IUBS, WWF, WRI), governmental agencies (e.g. NIH), universities (e.g. Cornell University) and companies together with their developing counterparts (Colwell, 1993).

These programmes will lay the groundwork to pave natural ways towards new partnerships in science and development and create a transcription from words to meaningful actions. I believe there is still a great deal to be accomplished, there is an immense amount of work to be done. The way to success is like a great orchestra, we all need to be in tune and come in at the right time in order to achieve the perfect melody but above all we have to play in harmony and together.

References

Benford, G. (1992) Saving the 'library of life'. *Proceedings of the National Academy of Sciences USA* 89, 11098-11101.

Bertranpetit, J. and Cavalli-Sforza, L.L. (1991) A genetic reconstruction of the history of the population of the Iberian Peninsula. *Annals of Human Genetics* 55, 51-67.

Bovsun, M. (1994) Aussies dig in pharmaco 'gold mine' - toxic venom from deadly snails. *Biotechnology* (Newswatch), 1-4.

Burk, D.L., Barovsky, K. and Monroy, G.H. (1993) Biodiversity and biotechnology. *Science* 260, 1900-1901.

Canhos, V.P., Canhos, D.L. and de Souza, S. (1993) Establishment of a computerized biodiversity/biotechnology network. The Brazilian effort. *Journal of Biotechnology* 31, 67-73.

Cavalli-Sforza, L.L., Wilson, A.C., Cantor, C.R., Cook-Deegan, R.M. and King, M.C. (1991) Call for a worldwide survey of human genetic diversity: a vanishing opportunity for the human genome project. *Genomics* 11, 490-491.

Coghlan, A. (1994) Seed bank builds on frozen assets. *Technology/New Scientist* 24 pp.

Colwell, R.R. (1993) Biodiversity - an international challenge. *Science* 259, 1107.

Cox, P.A. and Balick, M.J. (1994) Neue Medikamente durch ethnobotanische Forschung. *Spektrum der Wissenschaft* 40-46.

Ehrlich, P.R. and Wilson, E.O. (1990) Biodiversity studies: science and policy. *Science* 253, 758.

Fischer, E.P. (1991) *Wissenschaft für den Markt.* Piper Verlag, München, Zürich.

Ford-Lloyd, B. and Maxted, N. (1993) Preserving diversity. *Nature* 361, 579.

Galeffi, C. (1993) The contribution of American plants to pharmaceutical sciences. *FARMACO* 48, 1175-1195.

Gibbons, A. (1993) An array of science from mitochondrial EVE to EUVE. *Science* 259, 1249-1250.

Gillis, A.M. (1994) Getting a picture of human diversity. *Bioscience* 44, 8-11.

Kumar, P.V.S. (1993) Biotechnology and biodiversity - a dialectical relationship. *Journal of Scientific and Industrial Research* 52, 523-532.

Lewin, R. (1993) Genes from a disappearing world. *New Scientist* 138, 25-30.

Mann, C.C. and Plummer, M.L. (1993) *Aspirin.* Drölmer/Knaur Verlag, München.

Mannion, A.M. (1993) Biotechnology and global change. *Global Environmental Change Human and Policy Dimensions* 3, 320-329.

Mothes, K. (1984) Zur Wissenschaftsgeschichte der biogenen Arzneistoffe. In: Czygan, F.C. (ed.) *Biogene Arzneistoffe,* Verlag Friedrich Vieweg 8 Sohn, Braunschweig/Wiesbaden, p. 5.

Moutain, J.L., Lin, A.A., Bowcock, A.M. and Cavalli-Sforza, L.L. (1992) Evolution of modern humans: evidence from nuclear DNA polymorphisms. *Philosophical Transactions of the Royal Society London Biology* 337, 159–165.

Nisbet, L.J. (1992) Useful functions of microbial metabolites. *Ciba-Foundation Symposium* 171, 215–225.

Roberts, J. (1993) Global project under way to sample genetic diversity. *Nature* 361, 675.

Roberts, L. (1992) Anthropologists climb (gingerly) on board. *Science* 258. 1300–1301.

Sirica, C. (1994) Taking stock of tropical biodiversity. *Science* 265, 690.

Vigilant, L., Stoneking, M., Harpending, H., Hawkes, K. and Wilson, A.C. (1991) African populations and the evolution of human mitochondrial DNA. *Science* 253, 1503–1507.

Waessle, W. (1990) Umweltschutz durch moderne Bio-Diagnostica. In: Form, M., Näher, G., Otte, H., Schmidbauer, B., Schumacher, G., Seydler, B., Tischer, W. and Wassle, W. (eds) *Symposium Umweltschutz durch Biotechnik*. Boehringer Mannheim GmbH, Penzberg, p. 32.

Wallace, J. (1993) Back to nature (renewed attention to plants as sources of medicine). *Across the Board* 30. 32–37.

43 In Defence of Biotechnology*

M.F. Cantley
Biotechnology Unit in the Directorate, for Science, Technology and Industry, Organisation for Economic Cooperation and Development, 2, rue André-Pascal, 75775 Paris Cedex 16, France

Introduction

But why should biotechnology be on the defensive? Unfortunately, because all too often, it is being unreasonably attacked; even in the context of discussions about biodiversity, to whose conservation it can greatly contribute. Why is this, and what can be done about it?

Consider the Editorial by IUBS President Francesco di Castri, writing in July 1994:

> the niche of public visibility in these fields (ecology, biodiversity, hydrobiology, environmental biology, etc.) is taken over – much too often – by 'instant scientists', with unknown university backgrounds, almost devoid of research experience, with no associated students, and not belonging to the so-called scientific community. These individuals however, tend to answer to the whims and demands of the media, that is to say, sensationalism, catastrophicism, or a romantic attachment to a wilderness past.

The same phenomenon is confirmed by systematic public opinion measurement: the sources of information about new technologies with (by far) the greatest public credibility are the environmental and consumer organizations (e.g. see European Commission, 1993). Government has little credibility, industry practically none. The risks of biotechnology or genetic engineering are seen as high, and almost all respondents feel the technology should be controlled by governments.

---

*Not a statement of policy of the OECD or its Member governments; the opinions expressed are those only of the author.

© 1996 CAB INTERNATIONAL. *Biodiversity, Science and Development: Towards a New Partnership* (eds F. di Castri and T. Younès)

The effects of this stigmatization on policy for biotechnology are numerous; many examples can be cited. Of particular importance in the present context is the Convention on Biological Diversity. On matters of intellectual property, access to germplasm and its economic value, there is a significant gap between expert advisers and policy-making circles. A similar gap prevails in the area of biosafety, and this is reflected in the discussions relating to Article 19.3 of the Convention.

This is the background against which we review some of the contemporary policy debates at the interfaces between biotechnology and biodiversity; starting with the experience of the OECD (Organisation for Economic Cooperation and Development).

OECD Activities in Biotechnology

The OECD forms part of the intergovernmental learning machinery of the developed world. Apart from its general objectives of promoting economic growth, employment, social welfare, and world trade, it enables governments to exchange ideas on best practices in many specific areas.

The interest of the OECD Committee for Scientific and Technological Policy (CSTP) in biotechnology dates from the early 1980s, when there was established a Group of National Experts on Safety in Biotechnology (GNE). By ad hoc studies and through the work of the GNE, a series of general reports was produced on various, policy-related aspects of biotechnology, as follows:

1982: Biotechnology: International Trends and Perspectives (Bull, Holt, and Lilly);
1985: Biotechnology and Patent Protection: an international review (Beier, Crespi, and Straus);
1988: Biotechnology and the Changing Role of Government;
1989: Biotechnology: Economic and Wider Impacts.

These were general policy studies. Specifically on safety, the GNE was responsible for a series of reports of which the first was and remains the best known and most influential; the 1986 'Blue Book', 'Recombinant DNA Safety Considerations' (OECD, 1986). The main points of this report were as follows.

> Any risks raised by recombinant DNA organisms are expected to be of the same nature as those associated with conventional organisms. Such risks may, furthermore, be assessed in generally the same way as non-recombinant DNA organisms;

> Although recombinant DNA techniques may result in organisms with a combination of traits not observed in nature, they will often have inherently greater predictability than conventional methods of modifying organisms.

> There is no scientific basis to justify specific legislation for recombinant DNA organisms.

The last point has been of particular importance in emphasizing the need to address biotechnology-related safety issues on a sectoral basis. Thus, there are evident risks in handling pathogens, e.g. in vaccine production; but the risks derive from the pathogenicity of the organisms, not from the techniques used to modify them. The many subsequent safety reports from the OECD have followed this approach, in such fields as the scale-up of field trials of agricultural crop plants; live vaccines; biofertilizers; biomining agents, and others.

These reports have been widely influential. The World Bank, for example, has described them as 'the most authoritative set of internationally agreed-upon guidelines presently available', which 'provide a sound basis for national policymakers in all countries'.

How Do Societies Learn?

The new knowledge and techniques summarized by the much-defined word 'biotechnology' represent a massive surge of knowledge and basic understanding of the structure and functioning of living organisms. The consequences will be permanent, and widespread across all areas of application of the life sciences and technologies; in agriculture and food, in health care, and in managing man's interfaces with the natural environment. The surge of new knowledge is hard to digest. The answer to the question above is: 'slowly, and repetitively, with long time-lags'. In democracies, for better or for worse, politicians depend upon public opinion, and move too far ahead of it at their peril.

On the other hand, failures to learn will also be penalized; as is emphasized by such clichés as: 'the one thing that man learns from history, is that man does not learn from history'; or 'he who does not learn from history is condemned to repeat it'. Both quotations could be illustrated by specific references from the history of biotechnology regulations. The ultimate penalties for societies unable to digest or assimilate effectively and rapidly the new surge of biological knowledge will be economic.

In the two decades since the Berg letter of June 1974 (proposing the famous voluntary moratorium on certain rDNA experiments) and the Asilomar conference of February 1975, much has been learned, much written, and much ignored. In a massive compendium published in 1981 and entitled 'The DNA Story' James Watson (Nobel Prizewinner for the discovery of the DNA double-helix) and John Tooze (Secretary of the European Molecular Biology Organisation) collected in 605 pages many of the primary documents from the DNA wars of the late 1970s, and added incisive commentary (Watson and Tooze, 1981). In their final chapter, 'Epilogue', they summarize the period as follows: 'Politics and politicking preoccupied the first years of the recombinant DNA story, but that phase, fortunately, is fast becoming history. This book is our epitaph to that extraordinary episode in the story of modern biology.'

Their triumphant expression of hope was to be defeated by experience over the following decade, particularly in Europe. Scientific advice was ignored; antiscience movements and political opportunism influenced legislation at national and European levels. Initially, this merely inconvenienced researchers; but by the early 1990s its effect upon investment and upon the economic prospects for Europe in the new, knowledge-based sectors seen as harbingers of the 21st-century economy, was too grave to be longer ignored. In the December 1993 'Delors White Paper' on 'Competitiveness, Growth, and Employment' the European Commission identified the need to revise and review their technology-based regulatory framework for biotechnology. Measures were put forward by the Commission in 1994 with a view to implementing these aims; but the prospect of Environment Ministries abandoning their recently-acquired regulatory territory seems at best uncertain. And uncertainty deters investment.

The continued public disquiet and misunderstanding about genetic engineering supports the continuing stigmatization of these precise new methods. In spite of its 20-year safe track record, regulations specific to biotechnology are now defended less on the grounds of 'defending the public from biotechnology', than on the grounds of 'defending biotechnology from the public'. The existence of such regulations conveys to the public – in spite (or because) of government assurances – the implicit, subliminal message that the technology is inherently dangerous.

The resulting and continuing stigmatization of biotechnology has many unfortunate effects; not least, the misunderstanding of the constructive role which it can and should play in the conservation of biological diversity.

The Convention on Biological Diversity

The Convention on Biological Diversity should have been one of the success stories of the Earth Summit held at Rio in June 1992. Overall, it represents an important political statement about the global need to conserve our vast inheritance of diverse natural species. But 'the devil is in the details'. The implications of some of the central political concepts do not appear to have been thought through in detail, for example, the concept of national sovereignty over national 'genetic resources'. Several illusions underlie the 'bargain' summarized by Article 16, which appears to trade access to the biotechnology of the developed world for access to the germplasm of developing countries, of supposed economic value.

This economic value has in fact been undercut by at least two developments.

1. The existing availability in *ex situ* collections, and consequently also in private breeders' collections, of practically all the relevant germplasm providing the genetic base of the world's 30 or so major crop species;

2. The provision, through combinatorial chemistry, of a vastly greater repertoire of diverse molecules (oligonucleotides and oligopeptides) than can

be provided by 'bio-prospecting' in the tropical forest. Interest in such prospecting had justifiably been stimulated during the 1980s by the availability through gene cloning of a growing number of cell surface receptor molecules (the key targets of the pharmaceutical, agrochemical and other industries), and the laboratory automation enabling such companies to screen many thousands of samples per day, of extracts of materials of natural origin, plant or microbial. But these facilities are now more effectively deployed to screen the synthetic diversity available from combinatorial chemistry.

This erosion of the already uncertain economic value of germplasm should have been irrelevant to the aims of the Convention. There was never any realistic prospect that the revenues available could approach the amounts required for a serious global conservation effort, an inevitably vast scientific and logistic enterprise to which the techniques of biotechnology are of great relevance (e.g. as outlined by Giddings and Persley, 1990). What is illustrated by the misconception is the extent to which the original aims of the Convention have been 'hijacked' by other contemporary political arguments.

A similar irrelevance dominates the debate on biosafety, which was translated into the uncertain language of article 19.3:

> The Parties shall consider the need for and modalities of a protocol setting out appropriate procedures, including, in particular, advance informed agreement, in the field of safe transfer, handling and use of any living modified organisms resulting from biotechnology that may have adverse effect on the sustainable conservation and use of biological diversity.

An apt reply to the question of need has been furnished by Montesquieu: 'When it is not necessary to make a law, it is necessary not to make a law'; and such has been the position of several governments including those of Japan and the United States; but not those of a noisy majority of countries having much less experience of biotechnology, nor of certain European Environment Ministries.

These examples illustrate the continuing problems caused by the stigmatization and misperception of biotechnology. It is a challenge to the scientific community, and particularly to serious ecologists and conservation movements to overcome this. It was Rachel Carson (1962) in *Silent Spring* who called for:

> biological solutions based on understanding of the living organisms they seek to control, and of the whole fabric of life to which these organisms belong. Specialists representing various areas of the vast field of biology are contributing – entomologists, pathologists, geneticists, physiologists, biochemists, ecologists – all pouring their knowledge and their creative inspirations into the formation of a new science of biotic control.

Similarly Edward Wolf (1985), writing from the WorldWatch Institute more than 20 years later, emphasized that: 'A biotechnology attentive to natural history may provide some of the most powerful tools to reduce the pressures on genetic resources and enhance the value and conservation of wild species'.

The implications are clear. There is a range of challenges: to scientists, industrialists, and policy-maker alike: to make best use of the new knowledge and techniques in the interest of conserving our heritage; and through open communication with the wider public and policymaking communities to overcome the ignorance and prejudice which inhibit such use.

References

Carson, R. (1962) *Silent Spring*, Houghton Mifflin, New York.
di Castri, F. (1994) Matching rigor with openness in biology. *Biology International* 29, 1–2.
European Commission (1993) Biotechnology and Genetic Engineering: What Europeans think about it in 1993. Survey conducted in the context of Eurobarometer 39.1.
Giddings, V.L. and Persley, G. (1990) *Biotechnology and Biodiversity*. UNEP/Bio.Div/SWGB, 1/3 October 1990.
OECD (1986) *Recombinant DNA Safety Considerations*. OECD, Paris.
Watson, J.D. and Tooze, J. (1981) *The DNA Story: A Documentary History of Gene Cloning*. W.H. Freeman, San Francisco.
Wolf, E. (1985) Conserving biological diversity. In: Brown L.R., Chandler, W.U., Flavin, C., Pollock, C., Postel, S., Starke, L. and Wolf, E.C. (eds) *State of the World, 1985*, Worldwatch Institute, Norton, Washington, DC.

44 Tropical Biodiversity and the Development of Pharmaceutical Industries

E.J. Adjanohoun
38, rue de Bois Grammont, 33320 Eysines, France

General Considerations

The present study fits into the framework of the action plans prepared by the United Nations Conference on Environment and Development (UNCED, 1992). Key words likely to result in positive directives for implementation in the short, medium and long term are: sustainable and ecologically rational development; growth in economic activities; biodiversity and environment; conservation and protection of biosphere resources; improvement of the quality of life and of human health, and the six documents and instruments issued by this world conference:

1. Agenda 21 on the preservation of the future of humanity.
2. Declaration on Environment and Development.
3. United Nations Framework Convention on Climate Change.
4. Convention on Biological Diversity.
5. Statement of principles for a Global Consensus on the Management, Conservation and Sustainable Development of All Types of Forests.
6. Creation of the World Environment Fund.

The concerns of the organizers of our meetings have been focusing the debate on the exploitation of biodiversity, no doubt taking as accepted the measures recommended to protect it against the degradation which is currently taking place, and to conserve it. Conservation has two complementary aspects: a static aspect of strict protection, and a dynamic aspect of rational development, which should make it possible to make better use of natural resources and the habitats of the biosphere. One can easily see why discussions are proposed in focal areas such as agriculture, animal husbandry, forestry, aquaculture, the urban and peri-urban environment, culture, ethics and of course the pharmaceutical industry and biotechnology.

© 1996 CAB INTERNATIONAL. *Biodiversity, Science and Development: Towards a New Partnership* (eds F. di Castri and T. Younès)

The tropical environment is one in which the plant and animal kingdoms are currently displaying remarkable development, and we are justified in feeling that the future of the planet will depend on its survival. Its survival, however, is threatened by catastrophic onslaughts caused by climatic changes (artificial or natural in origin), and by human activity, intentional or not. It therefore seems to us more urgent to discuss and examine the impact of industry on tropical biodiversity.

The pharmaceutical industry, like so many others, basically depends on research and the trial, standardization and marketing of medicines needed for our own health and that of livestock. The raw material for the industry is principally plants, and animal and mineral elements in smaller quantities. The plant kingdom comprises some 350,000 identified species; amongst these are the *Spermatophyta* or flowering plants estimated at some 250,000 species, followed by Conifers in the broad sense of the term with 700 species; the *Pteridophyta* or ferns comprise some 12,000 species; the *Bryophyta* or mosses of which there are some 25,000 species; and finally the *Thallophyta*, which can be subdivided into algae and mushrooms with respective inventories of some 30,000 species of algae and 31,000 species of mushrooms, all taxonomic subdivisions included. This vast quantity of raw pharmaceutical material presently known to botanists does not include microflora, microfauna or plankton, whose real availability we have not yet begun to evaluate; indeed, we firmly believe their number may reach several million taxa.

We should bear in mind the fact that at least 80% of this biodiversity is concentrated and flourishes in the tropical regions of Africa, Asia, Madagascar, Oceania, and America, in multiple terrestrial or aquatic fresh- and sea-water ecosystems.

A brief historical analysis shows that the most ancient pharmacopoeias, known as 'Pent Sao' (China, 3000 BC), 'Vedas' (India, 2000 BC), 'Papyrus' (Egypt, 1500 BC), did not generate an industry of any significant scale. The national modern pharmacopoeias which have fostered current and future pharmaceutical industries have grown in scale and importance during the twentieth century and are still only in their infancy.

It all began in the developed countries with the observation of the spectacular results of traditional medicine used to treat a wide range of illnesses, resulting in the introduction of a whole process of pharmacognostic, phytochemical, pharmacological, pharmacotechnical, and clinical research. The resulting drug has to surmount various legal hurdles before authorization for its sale is obtained. However, the massive supply of the plant or animal material which has demonstrated active medicinal properties that can be used worldwide, poses serious problems that will have to be overcome by the search for synthetic equivalents and appropriate biotechnological procedures that are, and will continue to be, not only very expensive, but a long way from producing expected results. Cultivation of medicinal plants is supervised to a greater or lesser degree in the developed countries, but is much less so in tropical developing countries whose flora is still very poorly exploited. The use of synthetic molecules in many cases produces allergies and side effects which decrease either their effectiveness or their credibility. Potential users revert to phytotherapy, herb medicine, homeopathy and folk remedies. Traditional tropical medicine, never having had the means to rise to the level of modern medicine, is condemned to establishing ethnopharmacopoeias which are nothing more than the compilation of ethnobotanical studies.[1] The first African pharmacopoeia in the strict sense of the term was drawn up and published in 1985 by the Organization of African Unity (OAU), but it only comprises excerpts from the pharmacopoeias of industrialized countries. It concerns in fact about only 100 African medicinal plants which have already been studied and put to use by the modern pharmaceutical industry. The purpose of the publication was to provide African industrialists with a condensed technical instrument enabling them to organize a parallel industry capable of meeting the supply and demand of the world trade, bearing in mind the level of raw materials available on the spot both in terms of quality and quantity, and significant financial and economic revenue they generate.

Contributions of Tropical Biodiversity

Tropical plants adequately studied and industrialized

Among species whose natural molecules have survived competition with chemosynthesis, there still exist today a few hundred extremely well-known tropical plants that are marketed or exported to pharmaceutical industries of the developed countries. The following are a few examples.

Highly specialized products

  1. *Catharanthus* spp. (*Apocynaceae*) or periwinkle: various Afro-Asian and Madagascan species of herbaceous plants including the Madagascar periwinkle, a self-sown or cultivated plant which is the source of vinca-leucoblastine, an alkaloid successfully used to treat certain forms of leukaemia.
  2. *Carica papaya* (*Caricaceae*) or papaya: a shrub of Mexican origin which is widely cultivated in all tropical regions, the latex of which is the source of papain, an extremely powerful proteolytic complex with two stable peptidase enzymes used in pharmaceuticals and medicine to treat a wide number of illnesses.
  3. *Cinchona* spp. (*Rubiaceae*): various shrubs of Latin American origin, later introduced and cultivated in many tropical countries, the bark of which contains quinine, a well-known alkaloid for the treatment of malaria.
  4. *Rauwolfia* spp. (*Apocynaceae*): various low-lying Afro-Asian trees whose roots contain reserpine and other derivatives used in the treatment of arterial hypertension.
  5. *Centella asiatica* (*Umbelliferae*): a trailing herbaceous plant of Asian

origin which has become pantropical since it is to be found in all humid areas; it is the source of madecassol, a substance prized for its use in retinopathies, the healing of incurable wounds and leprosy.

6. *Strophanthus* spp. (*Apocynaceae*): various bushes and lianas from intertropical Africa whose seeds are the source of various compounds including ouabain, used in treating heart disease.

7. *Tabernathe iboga* (*Apocynaceae*): a shrub from central Africa known as iboga whose roots are the source of ibogain, an alkaloid used as a stimulant for nervous and muscular systems.

8. *Voacanga* spp. (*Apocynaceae*): Afro-Madagascan trees, the bark and roots of which are rich in voacangine, a cardiotonic alkaloid, and whose seeds contain tabersonine which can be transformed into vincamine to improve oxygenation of the brain.

9. *Coffea* spp. (*Rubiaceae*) or coffee-trees: shrubs of African and Arabian origin, cultivated in Africa and tropical America, the seeds of which contain caffeine, an alkaloid widely used in medicine.

10. *Thea* spp. (*Theaceae*) or tea-plants: shrubs of Chinese origin which are cultivated in intertropical mountain areas and whose leaves also contain caffeine.

11. *Azadirachta indica* (*Meliaceae*), also known as neem, Indian lilac, or Margosa: a small tree of Indian origin but widely introduced and cultivated throughout tropical regions for various uses, including medical applications based on the multiple chemical substances extracted from its vegetative system (sterols, terpenes, flavonoids).

12. *Cassia angustifolia* (*Caesalpiniaceae*), known as Indian senna: an Afro-Asian shrub which is the source of the sennas widely used in pharmacy as laxatives.

13. *Cassia alata* (*Caesalpiniaceae*), or 'scabious': a shrub of American origin widely cultivated in the tropics; it is the source of chrysophanic acid used in the treatment of mycosis and skin diseases.

14. *Cassia occidentalis* (*Caesalpiniaceae*), or false 'kinkeliba': a shrub of Brazilian origin which spread very rapidly and has now become pantropical either by cultivation or self-seeding; it produces alkaloids, flavonoids, fatty acids, etc., that are widely used.

15. *Pausinystalia yohimbe* (*Rubiaceae*), commonly known as yohimbé: a central African tree whose bark contains the alkaloid yohimbine, used as an aphrodisiac and a stimulant.

16. *Pygeum africanum* (*Rosaceae*), an Afro-Madagascan tree whose bark is the source of tadenan, a well-known drug for the treatment of prostatic tumours.

Products with a wide range of uses

To the short, purely demonstrative compendium above should be added an equally significant list, which will be limited here to a brief presentation of commercial genera of specifically tropical plants used by industry for a wide range of applications in medicine, food processing, hygiene and cosmetics, etc.

(The examples chosen are designated by their common names or those used by industry.) These include:

1. Spices (ginger, pimentos, cloves, black and grey pepper, malaguetta pepper, cinnamon, grains of paradise, cardamom, caraway, basil, Indian saffron, nutmeg, etc.);
2. Aromatic plants (ylang-ylang, bay laurel, lemon balm, vanilla, orange blossom, vetiver, star or Chinese anise, geranium, eucalyptus);
3. Dyes (sundew, annatto, teak, algae, indigo, mangrove, lawsonia, acacia), colorants – tannins (Campeachy wood, umbrella wood, tropical almond wood, filao, etc.);
4. Essential oils (oils or essences of paper-bark, Indian hardwood, soothing oil, cineole, niaouli, eucalyptus, camphor, chaulmoogra oil, etc.);
5. Oilseed and vegetable oils (palm, palmetto, coconut, sesame, shea, peanut, castor and cotton oil, balsam of Tolu, etc.);
6. Resins (incense, gum arabic, gutta percha, tragacanth, etc.);
7. Hallucinogens (the higher fungi, iboga, narcotics, etc.);
8. Stimulant plants (cola nut, tobacco, guar, Paraguay tea, cocoa-tree, coffee-tree; tea-plants, coca, ipecacuanha, etc.);
9. Toxic or poisonous plants (Calabar beans, nettles, Barbados nuts, erythrina, jimsonweed, derris, euphorbia, thevetin, strychnos, phytolacca, ricin, tephrosia, cubeb, amanita, jequerity, quassia, rose bay, etc.);
10. Antimicrobial, antifungal, antiviral, antibiotic plants (scabious, centaurea, cassia, *Euphorbiaceae*, garlic, *Chenopodiaceae*, kalanchoe, various lower fungi, etc.);
11. Various infusions (mint, vervain, lemon balm, harpagophyton, etc.);
12. Wines and fermented drinks (palm, Palmyra and raphia wines, millet, sorgho, tamarind, ginger and lemon beers, etc.);
13. Medicinal fruit (bitter orange, lemon, tamarind, gourd, avocado, mombin, anacardium, custard-apple, *Annonaceae*, mango, banana, pomegranate, grenadillo, guava, etc.).

Tonnage

The culling and marketing of these tropical plants, picked or grown for pharmaceutical companies in developed countries, are organized either in an orderly way or in an illicit fashion, in most cases by non-authorized and unsupervised private companies to the detriment of the domestic economy of the producer country. The following are a few examples of the tonnage of raw medicinal materials exported each year:

- Madagascar periwinkle: 400 tons of dried roots;
- *Rauwolfia vomitoria*: 500 tons of dried root bark;
- *Strophanus gratus*: 300 tons of seeds;
- *Pygeum africanum*: 300 tons of dried stem bark;
- *Carica papaya*: 300 tons of latex produced by the extraction of 600 million unripe fruit for export to the United States alone:
- *Voacanga* spp.: 200 tons of seeds.

Tropical plants in use but inadequately or incompletely studied

Definition

Here we refer to plants whose analysis, although not completed, is sufficiently advanced to permit their use by the pharmaceutical industry. They have produced a number of worthwhile discoveries, but research on them is continuing since they contain other active elements about which little is known. Estimates put the number of these plants at several hundred species, a large number of which have been the subject of monographs published in the pharmacopoeias of the industrialized countries.

Examples

Combretum micranthum (Combretaceae), contains certain active elements which are complex and have yet to be specified; as a result, it is difficult to explain all the therapeutic effects attributed to the plant. (cf. African Pharmacopoeia, OAU, 1985).

Euphorbia hirta (Euphorbiaceae), a pantropical plant widely used in folk medicine, is well-known for its efficiency in the treatment of amoebae. The whole extract continues to be generally used pending a precise study on its pharmaceutical preparations. (cf. African Pharmacopoeia, OAU, 1985).

Aframomum melegueta (Zingiberaceae), or malaguetta pepper, seeds are widely used as an aromatic or stimulant. There has been no recent pharmacological research on the species, etc.

Conclusion

The analyses of these plants are generally performed in universities and technology institutes of the Third World, and in those of industrialized countries which welcome foreign students and trainees. The ideal situation would be to equip the specialized laboratories of these countries with up-to-date instruments and to create very high-level regional centres of excellence to conclude the analyses and draw up product patents for their industrial development. The United Nations Industrial Development Organization (UNIDO) is following a similar line, the results of which should be taken into consideration here.

These plants are used for a large number of simple, well-known, very old and highly popular preparations, which have proved their worth. It should be possible to cultivate them in their native sites, and package and distribute them to countries where they are not found.

Plants in use but not scientifically studied

Definition

In this category we refer to a large number of medicinal plants which are used in traditional pharmacopoeias but which have not been the subject of any scientific study; estimates put their numbers at several thousand species. Together with the preceding category, they comprise the major part of the material which is immediately available for a wide range of research and new discoveries with scientific, medical, industrial, and economic benefits. Protection and conservation of these plants encounter problems of mutilation, destruction and extinction.

Examples

The African Pharmacopoeia (OAU, 1985) published a list of some 100 species which are to be the subject of monographs for its second edition.

In the *Encyclopaedia of Natural Medicines* Adjanohoun (1991) mentions some 60 plants on which precise pharmacological research is needed.

In the ethnopharmacopoeias published by the French Cultural and Technical Co-operation Agency (ACCT), most of the 3000 species recorded should be taken into account.

Conclusion

Increasing numbers of Third World scientists are carrying out research on the multiple possibilities offered by the use of the plants listed in ethnopharmacopoeias. Ethnobotanical studies, the methodology for which has been designed by tropical experts under the auspices of the ACCT, have already been carried out in 12 French-speaking countries that are members of the Agency, and are currently being extended to the English-speaking members of OAU and to most of the other developing countries. The studies take into account the entire natural environment, the ecosystems involved and their evolution, and the arrangements needed to arrive at a better knowledge of the taxons which have been catalogued.

Specialized data banks are increasingly being set up to record the inventories and the specific details needed for the rational exploitation of plant resources. Worth noting is the existence of PHARMEL, a data bank on medical and traditional pharmacopoeias set up in 1986 by the ACCT. The headquarters of the data bank is in Brussels, led by Professor Jean Lejoly, Director of the Systematic and Phytosociology Laboratory of Brussels Free University. The bank comprises a data input guide and a software user guide. At present it has recorded 16,399 pharmacopoeic formulae from 22 countries, mostly African, drawn from 30 large-scale bibliographical references, using nearly 3541 medicinal plants with 27,741 therapeutic applications.

Also in operation is the NEPRALERT data bank in the United States, and the NAPRECA data bank in West Africa; that of the CICIBA (Gabon) is currently being set up, as is the one for the islands east of Africa (by Professor Petitjean).

Fundamental Problems

Despite the above, tropical plants used by the pharmaceutical industry represent less than 0.1% of the world's genetic heritage currently available or catalogued (this estimate takes into account only 300 taxons used to produce manufactured medicines out of the 350,000 plants identified around the world). Considerable potential therefore remains for future use. A number of fundamental problems also remain, however, which must be neither avoided nor minimized. The most important of these will necessarily address the consequences of culling these plants, the political will of all the potential partners, and the transfer of appropriate new technology.

Culling

All parts of plants are culled, with or without any special precaution. In traditional, popular pharmacopoeias, the whole plant is used (when it is an annual herbaceous plant), as well as the seedlings, roots and tubers, the bark of the roots and the stem, including the leaves, inflorescences, flowers, fruit, and seeds, which frequently contain secretions such as sap, latex, juices, gums, resins, oils, essences, etc., the constituents specific to the phanerophytes and the standing or trailing ligneous chamaephytes. In addition, the epiphytes, parasites, fungi, algae, bryophytes and pteridophytes complete the arsenal of medical tools supplied by tropical forests.

Over-culling of herbaceous plants disturbs their growth cycles and ecotopes, resulting in their rarefaction and ultimate disappearance. The same fate awaits the seedlings normally used for the necessary restocking of forests. *Catharanthus roseus*. which has been over-culled and sold on the black market, has practically disappeared from the Ivory Coast, Ghana, Togo, and Benin.

The intensive culling of the bark of roots and stems results in the suppression of the physiological functions of the plant, and the death of widely used trees such as *Pygeum africanum* and *Pausinystalia yohimbe*. Useable specimens of these trees are found at increasingly large distances from conurbations or have taken refuge in inaccessible sites.

Massive culling of leaves, buds and other parts of plants' vegetative system by pruning, pollarding or other methods of eliminating apical meristems, results in the death of the seed-bearers and the degeneration of the species, a self-defensive reaction which is fatal for forests. There are certain forests in which germination has totally ceased and seedlings no longer exist, such as those of *Okoubaka aubrevillei* (Octoknemataceae), a species widespread from the Ivory Coast to Zaire.

To get an idea of the scale on which medicinal plants are used by people, without forgetting the fact that samples culled in forested and savannah areas are frequently found at markets, one need only estimate the constituent parts of plants sold in all the urban and rural market-places of countries in the tropics. This very old form of trading, which has survived colonization, has developed to quite an extraordinary degree due to the lack of processed, manufactured medicines, and to the fact that it is impossible for 80% of poor people to have access to very expensive pharmacies located in cities.

According to demographic statistics, in the year 2025 the population of developing countries will have reached 6.8 billion. If we assume that each individual has an average annual consumption of 10 kg of medicinal plant material, this means 68 million tons of plants will be culled each year from tropical vegetation. Although the figure is only an estimate, it is significant as a new figure to be added to those usually given for forestry exploitation.

Political will

Maintenance and development of the floristically very rich tropical vegetation basically depends on the will and determination of people at the end of the twentieth century and in coming centuries. Every aspect of the degradation and destruction of the fragile natural habitat in the tropics has been widely condemned and published in the minutes of symposia and seminars of UNESCO, IUBS, ICSU, IUCN, ACCT, UNEP, WHO, CAMES, CSTR-OAU, FAO, UNDP, and most recently UNCED. Recommendations, resolutions, actions and agreements drafted on the subject are piling up without being effectively or widely applied. Currently, the forests of tropical Asia are the most exploited, whereas serious inroads have been made into those of Africa. Only those of tropical America remain relatively untouched. We are in the midst of a regressive systemic situation.

UNCED wishes to obtain resources to put an end to worldwide environmental degradation. In order to achieve that aim it must change the current mentality of the inhabitants of our planet, it must free creative initiatives, identify all potential partners, coordinate their work and ensure global management of the resources of the biosphere. But is this not simply yet another recommendation that will be put on a shelf, like so many others?

Technology transfer

New technology for the rational exploitation of tropical plant resources
The watchword since the beginning of the 1960s has been technology transfer to benefit North–South development. Industrialized countries have made available vast resources to transform that aim into a practical reality, but few positive results have been recorded. Over the last 20 years, it is biotechnology which has oriented research towards the transformation of living organisms and their particles in order to develop the food processing and pharmaceutical industries. Albert Sasson (1988) in his book *Biotechnologies and Development* described and analysed numerous aspects of foreseeable or possible contributions for developing nations, the basic components of which are tropical countries. The modern pharmaceutical industry performs highly advanced pharmacological research using biotechnological methods in order to increase and optimize production. Much work has begun or is under way on plant aromas, sea-based pharmaceutical products, the cosmetics industry, the production of alkaloids, antiviral plants, the food industry, modern pharmacology, etc., carried out by specialists, ultraspecialists and research teams working in ultramodern research facilities. The extremely high cost of material and equipment used is beyond the means of tropical countries.

There is a deep gap between the industrialized temperate nations and the tropical countries. The same gap can be observed between the different regions of the tropics, namely the humid tropics, the semihumid tropics, semiarid tropics, the arid tropics, the desert tropics, the tropical mountains and the subtropics. The level of development is highly unequal between the climatic and ethnic areas corresponding to these different types of tropics. The example of the African continent shows that Mediterranean Africa and temperate southern Africa can hope to keep up with current biotechnological progress, but that is not true of Sahelian Africa or of tropical forested Africa, whose low level of development contrasts with their extraordinary wealth of biological diversity.

What kind of technology transfer is appropriate for the rational exploitation of tropical plant resources? 'Wisdom' calls for taking the time to use conventional models that have demonstrated their worth, backed by powerful financial means.

Outlook for the Future

The pharmaceutical industry based on the rational exploitation of the resources drawn from tropical biodiversity must correspond to the completion of a whole multidisciplinary system, organically linked and complementary, which is why the scheme in Fig. 44.1 is entitled an 'integrated general organization'.

Conditions for the development and success of this industry are based on the way in which the various disciplines and sections that make up the diagram are dealt with. The scientific study and exploitation of medicinal plants involves *ipso facto* on the one hand, botanists, geneticists, forestry technicians, agronomists, biotechnologists, pharmacognosists, physiologists, ethnosociologists, environmentalists and phytogeographers, etc., and on the other, doctors, pharmacologists, pharmacotechnicians, agroforestry technicians, industrialists, economists, etc. These specialists work in six individual, interconnected modules, which are not self-contained, combining the following disciplines:

Module No. 1: Botany–Pharmacognosy–Genetics–Microbiology–Virology––Biotechnology–Agroforestry;
Module No. 2: Phytochemistry–Biochemistry–Molecular biology;
Module No. 3: Pharmacology–Toxicology;
Module No. 4: Pharmacotechnology–Legislation;
Module No. 5: Clinic–Health Centre;
Module No. 6: Industry–Marketing.

The objectives for each module are indicated. The structure, which may be termed an 'Industrial Research Centre', would be managed and supplied by two main technical departments:

1. A commission for applied pharmacopoeia research, corresponding to

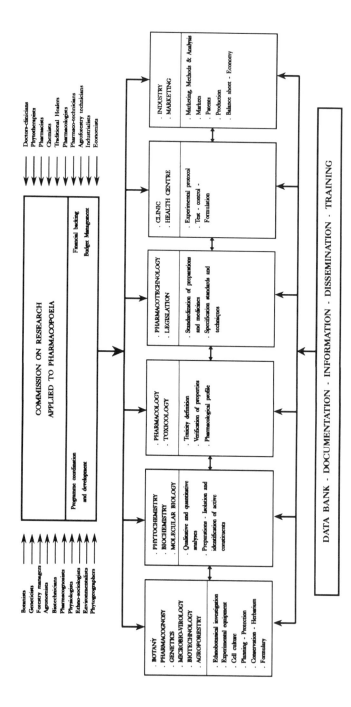

Fig. 44.1. An integrated general organization.

the central administration responsible for programming, management and funding.

2. A documentation department, more like a database comprising information, dissemination and training divisions.

No such centre of excellence as complete and as functional as this is known to exist in tropical regions, despite their wealth of underanaylsed and hence poorly exploited biological diversity.

General Conclusion

The first section of this chapter drew attention to the scale of biological diversity in the tropical regions of the world, where few factors intervene to limit the life-span of plants and animals with the exception of humans, who must fight against disease from the moment we are born. A brief historical analysis outlined the existence of natural resources whose exploitation and even industrialization is still in its infancy, which forces us to think about the various models possible for research and rational, sustainable economic development.

The second section briefly indicated the contribution made by tropical biological diversity, with examples of well-known, industrialized medicinal plants, examples of medicinal plants currently under study, and plants pharmacologically unknown. A third section was devoted to examining the fundamental problems of culling, political will and technology transfer, all of which force us to be increasingly concerned with conventional technological models.

To apply the preceding analyses concretely, the fourth section outlined future prospects in order to illustrate our proposals (condensed into an integrated general organization chart) (Fig. 44.1).

Pharmaceutical industries: why? who for? and to do what? One must admit that, on the one hand, in cold or temperate climates, there are the already industrialized, rich countries which are running out of new ideas or changing rapidly. On the other hand, in hot tropical climates, there are the poor underdeveloped nations, poorly organized and poorly supported, which are lagging behind. In view of the opulence and profusion of tropical biological diversity, an integral part of humanity's world heritage, UNCED forces us to take on our responsibilities and promises to provide us with the means to continue in a better way the technological and scientific research currently being carried out.

To give an impetus to the aptitudes of their researchers, who are competent, motivated and available, the underdeveloped countries need large-scale financial and technical backing focused on organizing and developing a pharmaceutical industry tailored to their real needs and practices, and drawing its resources from a scientifically controlled biodiversity. In order to achieve that aim, we need to provide these countries with complete, developed and functional, multidisciplinary research centres on site in the tropical environment,[2] based on Adjanohoun's model of organization. Economic and industrial research centres such as these do not exist anywhere in tropical Africa, and we should like to see them taken into consideration in the recommendations of this forum on biodiversity.

The centres of excellence in question would be regional and interregional, independent, different and not superimposable on the current university system, which is held hostage almost everywhere by the agitation and revolt of their users, anxious about their future.

The establishment of these centres also fits in perfectly with the 'Priority: Africa' programme launched by UNESCO and with the 'actions and conventions' of UNCED, held in Rio in 1992.

Notes

1. From 1977 to 1987 the French-language Agency (ACCT) funded ethnobotanical studies in six countries in Black Africa, one country in North Africa, one country in the Caribbean, five countries in the Indian Ocean, one study in Vietnam and one in Madagascar.
   The OAU recently began drafting ethnopharmacopoeia for its English-language members (Nigeria, 1988; Uganda, 1993).
2. 'In place' (on site) and not 'in their place', with reference to the current deployment of numerous specialized research institutes collecting ethnobotanical data from Africa and elsewhere for processing in the United States and United Kingdom, etc. (News Letter OAU/STRC/1994).

Further Reading

Adjanohoun, J.E. (1991) La médecine traditionnelle africaine. *Encyclopédie des Médecines Naturelles*. Editions Techniques, Paris.

Adjanohoun, J.E. (1991) Forêts tropicales et pharmacopées africaines. *Proceedings of the N'sélé regional symposium* (Zaire).

Bourdoux, J.-L. (1983) *Plantes Médicinales de la Flore Amazonienne*.

CTFT Ministère de la Coopération et du Développement (1989) *Mémento du Forestier*, 3rd edn. Paris.

Kabala, M.D. (1992) Modification de l'environnement et des milieux naturels africains subsahariens. Unpublished PhD, Thesis, Université de Bordeaux III, France.

OAU – CSTR (1985) *Pharmacopée Africaine*. Vol. 1, Lagos (Nigeria).

Penso, G. (1980) *Inventaire des Plantes Médicinales Employées dans les Différents Pays*. WHO, DPM 8.3, Geneva.

Ramaut, J.L. (1978) *Plantes Médicinales et Condimentaires*. Société Botanique de Liège, Université de Liège, Belgium.

Sasson, A. (1988) *Biotechnologies and Development*. UNESCO, Paris.

Biodiversity Prospecting: Opportunities and Challenges for African Countries

C. Juma and J. Mugabe
*African Centre for Technology Studies
PO Box 45917, Nairobi, Kenya*

Introduction

Recent years have seen a major resurgence in screening plants for their medicinal properties. Plant extracts provide important leads for the industrial development of new drugs, some of which are already on the market. The screening of wild plants for various compounds by scientists from the industrialized countries as well as the development, by industrialized country-based companies, of new produce from extracts from the plants has been the focus of international debate over the past ten years. The main concern has been that the industrialized country scientists and firms have had free access to plant germplasm of the developing countries. However, some of these concerns have partly been resolved, and the relevant compromises have been deposited in the Convention on Biological Diversity which was signed in Rio de Janeiro in June 1992 and came into force on 29 December 1993.

One of the key aspects of the Convention is emphasis on the need to enable the developing countries – which supply most of the genetic material used in industry and agriculture in the industrialized countries – to benefit from their contributions to the global economy. It places emphasis on the equitable sharing of the benefits derived from the conservation of genetic resources. Its Article 1 states:

> The objectives of this Convention, to be pursued in accordance with its relevant provisions, are the conservation of biological diversity, the sustainable use of its components and the fair and equitable sharing of the benefits arising out of the utilization of genetic resources, including by appropriate access to genetic resources and by appropriate transfer of relevant technologies, taking into account all rights over those resources and to technologies, and by appropriate funding.

To achieve the objectives related to the 'fair and equitable sharing of the benefits arising out of the utilization of genetic resources', countries need to create the requisite institutional structures and accumulate technological capabilities at the national level that will enable them to enhance the value of genetic resources. One of these measures is biodiversity prospecting, which is defined as the search for wild species, genes and their products with actual or potential use to humans (Eisner, 1992; Sittenfeld and Gámez, 1993).

This chapter argues that for African countries to benefit from their biological diversity they need to create programmes and undertake institutional innovations that enhance their technological capabilities in biodiversity prospecting. The African countries need to organize their institutions and build capabilities in research and development (R&D) to take advantage of the growing market interest in genetic resources for the biotechnology industry. They should be to diversify their R&D activities and invest in new areas of medical biotechnology which will also enable them to add value to their biodiversity. Much of the R&D work on biotechnology in Africa has concentrated on agriculture, and that orientation has been influenced by Africa's food production needs and the strong tradition of research in agriculture that the countries have (Juma et al., 1994). This bias however, may undermine developments and opportunities in other fields such as medical biotechnology that are equally important. This chapter does not cover all the major issues related to biodiversity prospecting. For example, the legal issues related to this subject have been extensively discussed in other studies (Reid et al., 1993). It also deals with plants, although biodiversity prospecting also covers animals and microorganisms.[1] In addition, attention is now starting to shift to marine biological resources.

Biodiversity Prospecting

Biodiversity prospecting (also referred to as bioprospecting) is the exploration of biodiversity for commercially valuable genetic and biochemical material. Activities of bioprospecting depend on there being a wide variety and diversity of living organisms. Such diversity is manifested at the genetic, species, ecosystem and cultural levels. Biological diversity is the main source of raw materials for agricultural, medical and industrial innovations and thus forms the foundation of sustainable development.

Interest in plant-based genetic and biochemical resources has increased in recent years.[3] The growth in interest in biodiversity prospecting is accounted for by a number of factors. These include declines in innovativeness in the chemical and pharmaceutical industry, the rise of biotechnology as a new avenue for corporate competitiveness, concern over the loss of biological diversity, the search for new economic activities in the developing countries and advances in techniques for biodiversity prospecting. These factors have also been associated with major growth in awareness over the importance of genetic resources, especially during the negotiations for the Convention on Biological Diversity. (UNEP, 1992).

Biodiversity prospecting can contribute to conservation and development. It can generate revenue and provide benefits to local communities, who are the custodians of genetic resources, in developing countries. But biodiversity prospecting can also have potentially harmful effects on biodiversity conservation (Reid, 1994). In order for biodiversity prospecting activities to contribute to conservation as well as generate returns to local communities appropriate public policies, laws and institutions are needed. The policies, laws and institutions should explicity ensure that 'the commercial value obtained from genetic and biochemical resources is a positive force for development and conservation' (Reid, 1994).

It should be noted that commercialization of genetic resources will not necessarily lead to the conservation of the resources. As Reid (1994) has noted 'unregulated biodiversity prospecting and drug development could speed the destruction of the resource. In one particularly egregious example, the entire adult population of *Maytenus buchananni* – source of the anti-cancer compound, maytansine – was harvested when a mission sponsored by the US National Cancer Institute collected 27,215 kg in Kenya for testing in its drug development programme.

In order to guarantee that most of the biodiversity prospecting activities contribute to conservation, it is crucial that institutions established to commercialize genetic resources explicitly incorporate conservation in their activities. Over the past years there has been no link between the commercialization of genetic resources mainly in the industrialized countries and the conservation of the resources in the developing countries. The industrialized countries have been the main beneficiaries of genetic resources obtained from developing countries. However, new opportunities for developing countries, particularly those of Africa, can be obtained through the implementation of the Convention. The Convention provides a legal basis that the developing countries can use to negotiate for sharing the benefits arising from the use of their genetic resources. But the countries need to further strengthen their abilities to commercialize genetic resources by themselves instead of leaving it all to companies of the industrialized world. Unless the developing countries build up technological and institutional competencies to add value and realize benefits from genetic resources, invoking the Convention and their sovereign rights to benefit from the commercialized resources will be difficult.

There are new technological opportunities for enhancing the value of genetic resources. Developments in the field of biotechnology and increasing interest as well as demand for biological compounds have stimulated considerable interest in a search for plant materials that have chemical compounds for pharmaceutical products. Biodiversity prospecting has become an established enterprise with a wide range of actors and procedures. It is an enterprise where there is a considerable convergence of disciplines: for example botany, ethnobiology, parasitology, genetic engineering, and biochemistry. It demands interdisciplinary expertise and involves a series of procedures including search, collection, scientific identification of plant material, preparation of extracts and chromatographic analysis, pharmaceutical screening of crude extracts, bioassay of each fraction of the extracts, verification of purity of isolated compounds, eludication of the chemical structures, and large sample isolation of compounds for pharmacological and toxicological test (World Wide Fund for Nature, 1993). It is only those countries that build the requisite expertise in the various disciplines, and more particularly in biotechnology, that will be able to engage in biodiversity prospecting.

Biodiversity and Biotechnology: Vital Links

The link between biodiversity and biotechnology has continued to receive international attention. The most significant articulation of this link is the formulation of key articles in the Convention on Biological Diversity. The development and use of genetic material as well as its conservation through *in vitro* methods is the stuff of biotechnology. The Convention recognizes the vital links between biodiversity and biotechnology. Article 16 (on Access to and Transfer of Technology) of the Convention explicitly recognizes the role of biotechnology in conservation and sustainable use. It provides that:

> Each Contracting Party, recognizing that technology includes biotechnology, and that both access to and transfer of technology among Contracting Parties are essential elements for the attainment of the objectives of this Convention, undertakes subject to the provisions of this Article to provide and/or facilitate access for and transfer to other Contracting Parties of technologies that are relevant to the conservation and sustainable use of biological diversity or make use of genetic resources and do not cause significant damage to the environment.

The Convention under Article 16(2) suggests that technology should be made available to developing countries on concessional terms but with due recognition of property rights over the technology: 'Access to and transfer of technology ... to developing countries shall be provided and/or facilitated under fair and most favourable terms, including on concessional and preferential terms where mutually agreed, and, ..., such access and transfer shall be provided on terms which recognize and are consistent with the adequate and effective protection of intellectual property rights.' The Convention fails to provide specific measures that may allow developing countries to have access to 'protected' technologies. The Convention addresses in a general way intellectual property rights (IPRs) on genetic resources in Article 16. It also recognizes that IPRs may act in some cases as a barrier to technology transfer and calls for mechanisms, which it does not specify, to be developed to ensure that IPRs do not act as a barrier to technology transfer. It further calls for cooperation to ensure that IPRs are supportive of the objectives of the Convention.

It should be noted that technological developments in biotechnology, particularly in the area of genetic engineering and new screening techniques, have stimulated interest among pharmaceutical companies which see the potentials of developing and commercializing new products from plants. It is on the basis of this that developing countries were negotiating for the inclusion of transfer of biotechnology in the Convention. Their concerns have been partly responded to in the Article 16 of the Convention. However, there are still a number of issues that remain untackled. First, the issues of distribution of economic benefits (of local communities in developing countries who are the custodians of plant genetic resources) from plants is still not effectively tackled. Second, it has been recognized that biodiversity prospecting activities utilize traditional knowledge of local communities. Most developing countries and some social scientists have called for a regime that recognizes the intellectual contribution of local farmers and herbalists to the identification and development of new plant varieties as well as new drugs. The rights and contribution of local and indigenous people are yet to be provided adequate recognition in both national and international laws.

On the whole, genetic resources are among the few strategic resources that developing countries still control, and ignoring their legitimate interest in obtaining some benefits from these resources would be a mistake. While the links between biodiversity and biotechnology cannot be denied, several institutional limitations continue to rob developing countries of opportunities for deriving benefits from biodiversity prospecting. In order for developing countries in general, and Africa in particular, to benefit from their resources it is crucial that they create the requisite institutional structures and build scientific and technological capabilities in various areas of biotechnology.

The Rise of the Biotechnology Industry

Slump in chemical innovations

The economic slump of the 1970s and 1980s forced the major corporations of the world to start looking for new ways of enhancing their market competitiveness. Technological innovation, the main source of such competitiveness, became increasingly important in corporate planning. Since the 1960s, the chemical and pharmaceutical industries have experienced a decline in the discovery and market application of new inventions. From the 1930s to the early 1980s, there were 63 major innovations in the chemical industry of which 40 were introduced in the 1930s and 1940s, 20 more in the 1950s and 1960s and only three in the 1970s and 1980s (Juma, 1989a).

It was becoming very costly to put a new chemical product on the market. A number of responses were proposed to deal with this situation. The first was to tighten intellectual property regimes so that investments in inventive activity could be recovered. The pharmaceutical industry has been very active in this process and relies heavily on patented products (Cook et al., 1991). The other was to shift from chemical inventions to biological inventions. This shift increased interest in biodiversity prospecting. The growing concern over environmental pollution and the related environmental regulations also stimulated research in biological innovations.

Growth of the pharmaceutical sector

The rise of the biotechnology industry as an investment sector has stimulated more interest in biodiversity prospecting. Most of the major industrialized countries have already formulated policies whose aim is to use biotechnology as a way of enhancing national competitiveness (OECD, 1988; OTA, 1991). A number of subsectors of the industry are emerging, and various countries are carving out niches using new techniques. Japan, for example. is using the techniques to produce new perfumes and thereby displace France from some of its traditional markets. The market for aromas and fragrances is attracting a number of biotechnology firms because of its size and prospects for growth. In 1980, the value of the market was US$4 billion and it is currently estimated at $US6.5 billion. Since access to biological material is essential to the performance of the biotechnology industry, biodiversity prospecting has been given prominence at the policy level.

A new field of biotechnological interest. referred to as 'aromachology' is starting to emerge. This science aims at studying 'effects of scents and fragrances on the physical and mental conditions of human beings, and at using these substances to modify these conditions (Sasson, 1992). Car devices which release a jasmine fragrance to keep drivers awake were tested in Japan in 1989. Leading perfume firms have been keeping their eyes open for any new developments in the field of aromachology.

Kanebo of Japan sells a fragrance containing lavender, camomile and anise which has sedative and sleeping effects. Shiseido, also of Japan, sells an 'antidote': an automatic alarm clock which, before ringing, releases a perfume made of pine and eucalyptus fragrances to stimulate the body. Research is underway on fragrances which increase vigilance, affect violence and modify other human behaviour.

Advances in new techniques, especially in the pharmaceutical industry, have increased the chances of obtaining a new drug from biological material. The use of new screening techniques has increased the chances of identifying a potentially useful drug a hundredfold. These techniques have not only reduced the costs of screening, but they have also increased the number of firms interested in screening. The effect of this has been a renewed interest in biodiversity prospecting, which had faltered.

The future of the biotechnology industry is currently being shaped by the high growth in R&D expenditure. In the US, for example, R&D expenditure on biotechnology-based pharmaceuticals in 109 major firms increased by 71% of the 1991–92 period to a staggering US$11.9 billion (or an average of US$17.4 million per firm).[3] The increase in the US R&D expenditure is associated with the expected new biotechnology products. In 1991, for example, there were 135 biotechnology-derived products in clinical trials,

Table 45.1. Total US biopharmaceutical R&D spending, 1991 (US$ millions).

Genentech	270,147
Amgen	182,297
Chiron	142,927
Centocor	94,762
Biogen	60,399
Synergen	60,264
Alza	52,089
Genetics Institute	47,000
Gensia Pharmaceuticals	40,353
Immunex	32,745

Source: Spalding (1993).

23% more than the previous year. In the same year, biotechnology firms filed 290 investigational-new-drug applications, a 26% increase from the previous year (Spalding, 1993; Table 45.1).

The pharmaceutical sector is one of the most profitable industries in the world, with many of the leading firms achieving profit margins of more than 20%. Five of the most profitable firms in the world are pharmaceutical companies. Total sales were estimated at US$180 billion in 1990, of which Western Europe accounted for 30%, the US 25% and Japan 30%. The rest of the world accounted for less than 20%. Of the sales, patented prescription drugs, or 'ethical pharmaceuticals', were valued at US$130 billion. The total sales for over-the-counter medicines were US$20 billion. In the 1990s, however, these margins have been under pressure, as it has become technically more difficult to develop new drugs (UNIDO, 1991).

World sales in this industry grew by 10% per year in the 1980s, but they are expected to grow by 5% per year in the 1990s. The project growth in the developing countries is higher than 5% per year, but other economic factors may inhibit such growth. Cost-awareness is forcing state health agencies and insurance groups to limit any major increases in pharmaceutical prices. Firms, especially in the US, see health cover for employees as a significant factor in reducing their competitiveness. Measures that would reduce pharmaceutical prices, especially through R&D in new products, become important aspects of corporate strategies. This has added to the attractiveness of the biotechnology.

The pharmaceutical industry is dominated by large multinational firms and is characterized by major entry barriers, especially in relation to the development of new products. It depends on the development of high value-added products and is highly regulated. The major firms are vertically integrated and control all aspects of the business (including R&D, production and marketing). Their major drug types include cardiovascular, anti-infective, internal medicine, pain control, and respiratory medicines.

Merck, the largest drug firm in the world, accounted for about one-quarter of the share of the profits of the top ten drug firms in the world

Table 45.2. World's most profitable drug firms, 1989/90 ($ million).

Firm	Profit	% share	Sales	% share	Margin[a]	Country
Merck	2,155.8[b]	28.40	6,058.1	24.45	35.60	USA
Glaxo	1,313.3[c]	17.30	4,679.5	18.88	28.10	UK
Johnson & Johnson	883.0[c]	11.63	2,652.0	10.70	33.30	USA
Abbott	797.0[d]	10.50	2,785.0	11.24	28.60	USA
Warner-Lambert	776.0[c]	10.22	2,694.0	10.87	28.80	USA
ICI	654.2[c]	8.62	2,187.2	8.83	29.90	UK
Schering-Plough	626.9[e]	8.26	2,432.3	9.82	25.80	Germany
Fisons	209.4[e]	2.76	775.5	3.13	27.00	UK
Ares-Sereno	148.7[f]	1.96	420.8	1.70	35.30	France
Mylan	26.2[c]	0.35	95.4	0.38	27.50	US
Total	7,590.5	100.00	24,779.8	100.00		

[a] Profit as percentage of sales.
[b] Pre-tax operating income. Figures for Merck relate to human and animal products.
[c] Operating income.
[d] Trading profit.
[e] Activity profit.
[f] Net profit.
Source: UNIDO (1991).

(Table 45.2). It accounts for 5% of the world's market for prescription drugs and is responsible for six of the 50 best-selling drugs in the world. Merck's best-selling drug is Vasotec, a heart disease medicine, which fetches some US$1.5 billion a year. It is the world's second largest selling drug after Zantac (an ulcer drug), made by Glaxo of the UK and which earns US$2.8 billion a year. Merck invests heavily in R&D and marketing. Its annual R&D budget is over US$600 million, and this activity employs over 6000 people.

Of critical importance to the pharmaceutical industry has been the design of rational drugs to complement conventional drug screening (OTA, 1991). This approach to drug design requires a detailed understanding of the physiological basis of diseases and deals with enzyme activity, hormones and hormone receptors, cell replication, protein synthesis and other molecular aspects of disease and drug treatment (Ganellin, 1989).

This knowledge allows researchers to identify the specific molecular structures that would form the basis for drug design. This targeted approach requires access to a wide range of information on the molecular structures already available in nature. In this respect, biodiversity prospecting provides the bioinformation needed for the design of rational drugs. This process also involves extensive computer modelling to match the physiological conditions with the proposed rational drug. Entry into this kind of research requires national or corporate competence in computer modelling, which is still weak in most African countries.

Economics of drug development

Financial projections on the potential value of biotechnology products are influenced, to a certain extent, by the current value of plant-based products. The pharmaceutical sector has been the source of such information. It is estimated that every fourth drug prescribed in the US is derived from plant extracts. The value of such sales was estimated at US$4.5 billion in 1980 and stood at an estimated US$16 billion in 1992. In the mid-1980s, the annual over-the-counter and prescription drug value in Europe, Japan, Canada, and the US is estimated at US$43 billion (Principle, 1989). The annual worldwide sales of the pharmaceutical industry was estimated to be over US$200 billion and will reach US$500 billion by the year 2000, of which the industrialized countries will account for nearly 75%. This is the equivalent of the annual GDP of the UK or France.

The plant-derived prescription drugs in the US originate from 40 species, 20 of which are from the tropics.[4] Given the sales value of these drugs, the average value per species utilized is US$200 million, bringing the value of the contributions of the tropics to US$4 billion for the species utilized. This, of course, is not the value of the species in their natural form because it covers all the costs of R&D, manufacturing and marketing. In addition to the drugs, more than 60 plant species yield prescription drugs which are used in the US but which have not been recognized by the US Food and Drug Administration (USFDA).[5] The 1990 estimate for the US market for 'unconventional therapy', most of which is based on plant-derived medicines, was US$13.7 billion (Gupta, 1993). Over 300 species are sold as herbal teas in the US alone.

It should be noted that in any given drug-development process, the chances of discovering a valuable compound for the pharmaceutical industry is considerably low. It is estimated that, only about one in 10,000 chemicals yields a promising lead into clinical tests. Of the chemicals going to clinical test stage less than one-fourth get developed and approved as a new drug. The uncertainty in drug development is illustrated by the work of the US National Cancer Institute (NCI) which has signed specimen supply contracts with over 25 countries worldwide. NCI had, by 1991, screened over 50,000 extracts for HIV of which only three were likely to reach clinical trials. Only five of the 33,000 extracts screened for cancer are undergoing further study (Sears, 1992). Tropical candidates with unique anticancer compounds include *Tabebuia serratifolia, Jacaranda caucana* and *Croton tiglium*. Over the 1960–81 period, NCI mounted the largest screening programme under which 114,045 individual extracts from up to 35,000 plant species were tested. Of these, some 4897 extracts from 3394 species exhibited biological activity. About 40 of the active compounds were considered for further development (Principle, 1991). Much of the screening attention has turned to anti-HIV extracts. There are already some promising indications, although it may take up to a decade before a drug is developed. An extract from a species of the vine genus *Ancistrocladus* found in Cameroon shows strong anti-HIV activity in the test-tube (McKenna, 1993).

The prospects of generating large incomes from the sale of extracts from plants are limited by the high rate of uncertainty and the long periods involved in drug development. The value of supplying extracts to pharmaceutical firms is relatively low given the low royalty rates (1-5% of net sales) and the long periods of product development. A royalty payment of 3% for an extract that leads to a drug with an annual market value of US$10 million would yield only US$52,500 in royalty payments. This figure assumes one 'lead' in 10,000 chemicals and a discount rate of 5%.

There are ways of increasing the probability rate of getting a new drug, mainly by enhancing the local scientific capacity (Davis et al., 1993). These include raising the local capacity in screening, greater knowledge of the physiological aspects of the disease, cumulative knowledge of the molecular structures of different plant extracts and competence in bio-informatics. Such knowledge-intensive approaches could increase the chances of putting a new drug on the market tenfold and thereby raise the royalty income to US$461,000 per year. A drug with an annual net value of US$1 billion – which is quite unlikely – will earn the supplier of biological material US$461,000 annually.

When a viable product has been developed, the turn-over can be high, and investments in research can be recovered quickly. This is illustrated by the case of shikonin, the antibacterial and anti-inflammatory extract from the roots of *Lithospermum erythrorhizon* used in lipsticks. Shikonin, which is also a dye, is extracted from roots. It takes at least seven years before the tree reaches commercial value, and therefore an alternative way of producing shikonin was needed. The annual world market for shikonin was US$600,000 in 1988, but by then some US$50 million had been invested in biotechnology R&D.

Although this figure looked excessive, it was subsequently justified by the sale of biotechnology-based shikonin lipsticks. When a biotechnology product was developed (and costed at US$4000 per kg compared with US$4500 per kg using conventional extraction), its manufacturers, Kanebo, realized a turnover of US$65 million in two years in Japan alone. The five million lipsticks sold for US$13 each (Sasson, 1992). This case also illustrates how a technology can drastically alter the size of the market and create conditions for further expansion.

The approaches adopted in biodiversity prospecting rely on isolating the active substances in plants. This poses problems for formulations which rely on a combination of different plant extracts as in the case of much of Chinese herbal medicine. However, some movement has been noted in bringing such knowledge into conventional medicine. The NCI, for example, introduced a US$20.5 million programme to identify naturally occuring foods which can be concentrated into preventive anticancer concoctions (Emmett, 1992).

Technological and Economic Challenges for Africa

Africa has added considerably to the collection of clinically useful plants. Its contributions include *Catharanthus roseus* (antitumor), *Centella asiatica* (vulnerary). *Gossypium* (male contraceptive), *Pausinystalia yohimbe* (aphrodisiac), *Ricinus communis* (laxative) and *Stophanthus gratus* (cardiotonic). African countries, however, are not major sources of cultivated medicinal plants. Notable exceptions include South Africa, Namibia, Lesotho, and Botswana which produce *Harpagophytum procumbens* as a crude drug; Sudan and Egypt which produce *Hibiscus sabdariffa* as a crude drug; Cameroon, Nigeria and Rwanda which export *Pausinystalia yohimbe*; Zaïre which exports *Peumus boldus* and Zaïre, Madagascar, and Mozambique which produce *Rauwolfia vomitoria*. African material is also being used in a wide range of health-related research activities. One of the most notable is *Phytolacca dodecandra* (endod) whose extract is an effective molluscicide and is now used in a number of countries to control schistosomiasis (bilharzia) (Lemma, 1991). This option is cheaper than using Bayluscide (niclosamide), which is marketed, at US$30,000 a tonne by Bayer Company (Aliro, 1993). Further research has shown that endod also controls the zebra mussels (*Dreisena polymorpha*) which cause damage to water utilities in the US and Canada.

Though some African countries have contributed useful plants for drug development, there is no indication that they have benefited from their contributions. It should be noted that the countries continue to register declining economic status and a large portion of their populations face serious health problems. Returns to traditional exports of African countries continue to decline as a result of global technological and economic changes. Major technological innovations in the industrialized countries are displacing traditional exports – mainly raw materials – from African countries. Already, genetic material from pyrethrum (*Chrysanthemum cinerariafolium*) is being used to produce pyrethrins through cell culture. In 1992, AgriDyne Technologies (Utah, USA) received a US$1.2 million grant from the US Department of Commerce for research on a new process to develop a plant-based insecticide. The firm will spend a total of US$3 million on the project with the aim of putting on the market a wide range of pyrethrin, a move that would undercut such exports from countries such as Kenya, Tanzania, Equador, and Rwanda (Juma *et al.*, 1994).

These technological changes are likely to affect African countries in various ways. As already noted above, the countries are going to lose their market niches and experience declines in their economies. It should be noted that most of the countries depend on a narrow range of economic activities. The displacement of the exports in essence means the erosion of economic prospects for the countries. Can Africa stem the tide? Yes only if they create measures that promote the accumulation of technological capabilities to undertake major technological innovations that will enable them to diversify their economies. Unlike previous technological developments, biotechnology offers a wide range of new opportunities for African economies. The fact that biotechnology is science-intensive reduces the traditional barriers associated with capital-intensive technologies. The success of African countries depends on the nature and levels of technological capabilities in biotechnology as well as the kinds of policies and institutions that are put in place to promote scientific and technological development.

African countries are not going to deal with the technological changes and the associated battle over access to their genetic resources by merely appealing to the Convention on Biological Diversity. Their ability to invoke the Convention and benefit from it will depend largely on the capacity of the African countries to strengthen their scientific and technological capabilities. One of the prerequisites for this is to have an effective science and technology policy. Biotechnology cannot be discussed outside the broader context of scientific and technological development. Effective science and technology policies should lead to the enhancement of the capabilities of the African countries to generate technologies. This is a learning process which does not emerge without investment in resources and time. Morever, it requires strategies that are carefully worked out and reflect the long-term national and regional development needs.

During the negotiations for the Convention on Biological Diversity African countries joined other developing countries in arguing for linking questions of technology transfer to the issue of access to genetic resources. Their position was that they will facilitate access to genetic resources by industrialized nations only if, in return, they allowed access to technologies particularly biotechnology developed in the North. It should be noted that excessive restrictions on the flow of genetic resources will slow technological development in pharmaceutical and agricultural biotechnology, as well as conventional agricultural research.

There are a number of institutional issues that deny African countries opportunities for deriving benefits from biodiversity prospecting. First, institutions that safeguard intellectual property in the form of technological innovations are larger and stronger than those that protect the interests of local communities involved in conservation efforts (Juma and Ojwang, 1989). Indeed, most possess scientific, technical and legal expertise available to few conservation groups. More generally, African countries are not likely to benefit fully from the provisions of the Convention on Biological Diversity or other transactions involving genetic resources until new legal and institutional regimes are established. Accordingly, African countries should establish public agencies to address intellectual property concerns within the framework of national legislation and use them as a basis for international negotiation (Reid, 1992).

Second, the countries have limited national institutional and legal capacity for regulating access to genetic resources (Juma, 1989b). Currently, only the assertion of national sovereignty over the resource can be used in negotiations over access to biodiversity and biotechnology. During the negotiations of the Convention on Biological Diversity, for example, the question of sovereignty arose during discussions of a proposed global list of habitats and species threatened with loss. By arguing that the publication of such a list would impinge upon national sovereignty, developing countries were able to keep it from being published. Unless legal and institutional regimes are established to assert and enforce that sovereignty, how can developing countries in general, and Africa in particular, benefit fully from the provisions of the Convention on Biological Diversity covering genetic rrsource transactions?

Third, African countries have tended to separate institutions involved in biodiversity management from institutions involved in biotechnology. In fact, ratification of the Convention on Biological Diversity could stimulate the development of effective biodiversity prospecting institutions by providing the policy, legal and institutional bases for linking biodiversity and biotechnology. Today, few of the key principles of this Convention are reflected in the character and organization of national and international biodiversity institutions.

The development of biotechnology represents a convergence of a wider range of skills and knowledge than any single institution or country is likely to possess (Clark and Juma, 1991). Thus, institutional cooperation will be critical to the success of biodiversity prospecting. Clearly, the coordination of biotechnology activities, as part of the larger enterprise of science and technology, needs to be given legitimacy and impetus at government's highest levels.

Most African countries have set up stand-alone institutions to promote biotechnology or biodiversity conservation. But such units – often established in anticipation of funding rather than out of genuine interest to promote the use and conservation of biodiversity – are unlikely to yield long-term benefits unless they play a coordinating role. New institutional arrangements aimed at promoting biotechnology development should be constructed with national goals in mind, not donor politics.

Institutional cooperation should also be extended beyond national boundaries – difficult for most African countries since their institutional arrangements for technological cooperation are limited. There are two main ways of dealing with this issue. One is through multilateral arrangements. For instance, African countries can use the facilities of international institutions such as the International Centre for Genetic Engineering and Biotechnology (ICGEB) to acquire technological skills. Another is through bilateral arrangements such as those of the US Agency for International Development (USAID) and the Dutch government (Directorate General International Cooperation, 1992; Komen, 1993).

Building Capabilities for Prospecting

Human resource development

As noted above, biodiversity prospecting demands a wide range of skills. In order for the African countries to effectively engage in prospecting – to add value and draw benefits from their genetic resources – they need to institute training programmes aimed at creating the requisite human skills and manpower. Biodiversity prospecting is a science intensive process. A country cannot effectively engage in the enterprise without availability of manpower trained in specialized scientific disciplines such as taxonomy, genetics, botany, zoology, pharmacology and others. Therefore, if an African country is to participate in biodiversity prospecting, it must devote a significant portion of its resources to investment in building human capital in core scientific areas.

Another important facet of human resources development involves the mobilization of the already existing expertise and directing these to biodiversity prospecting activities. It should be noted that in some African countries with scientists in areas such as botany, it will be crucial that such resources be efficiently used and programmes aimed at enhancing the qualities of the existing scientists be put in place. The training and skill enhancement processes should also focus on new technological areas such as biotechnology. Biotechnology is a very dynamic field and operates on the basis of knowledge generated in other disciplines which are themselves rapidly moving. The complementary growth in product and process innovation makes it necessary for researchers to keep abreast with a wide range of disciplines to be able to move effectively into the biotechnology field. In this regard, it is important for countries to invest in certain areas of knowledge and develop international competitiveness in those areas.

It should be noted that even small countries with limited industrial capacity can move to the frontiers of biotechnology in specific fields by enhancing their human resource capacity. By investing in training, establishing systems that provide ready access to information about both biodiversity and new technologies and seeking ways to add value to genetic resources through screening and characterization. African countries can turn short-term economic benefits into a long-term development strategy.

Improved coordination and institutional links

Biodiversity prospecting requires the bringing together of a wide range of expertise ranging from systematics to molecular biology. Biotechnology development is already leading to the emergence of new forms of expertise such as biomolecular engineering. The required expertise, as well as the related infrastructure, tends to be located in institutions established to carry out research in more narrowly defined fields. The outcome has often been the underutilization of expertise and equipment, lack of knowledge on the national requirements for effective biodiversity prospecting programmes and failure to attract support for these activities.

Changing this situation will require major reforms in the way research institutions are organized and relate to each other. A strategic focus on biodiversity prospecting in Kenya, for example, would require the involvement of all the national research institutions and the relevant university departments. This would also require new ways of organizing different institutions to focus on such a programme.

The need to improve institutional coordination under such conditions is obvious. Normally, countries have established national biotechnology centres to undertake such coordination. Such centres would bring together outstanding experts from relevant institutions to form *ad hoc* committees, subcommittees, or working groups to promote policy coordination and planning. In order to facilitate such coordination, it is important for countries to train personnel in the basic principles of R&D management, product development and marketing.

Institutional reforms

Government has an important role to play in promoting biodiversity prospecting, especially in promoting the institutional reforms necessary to promote biotechnology development. The ability of the government to create the necessary institutional environment will depend largely on how well the characteristics of biotechnology are understood. The interdisciplinary nature of biotechnology requires that national institutional arrangements maximize the use of locally available resources while at the same time forging international alliances for technology development.

Evidence from a number of newly industrializing countries has shown that there are specific institutional arrangements that need to be introduced to promote biotechnology development. These reforms must be guided by a national science and technology policy which places emphasis on biotechnology in general, and specified aspects such as biomedical research, in particular. Without such comprehensive context and sharp focus, it is impossible for African countries (with their meagre resources) to make substantial contributions to the field of biodiversity prospecting.

Public policy considerations

Governments play a fundamental part in the promotion of technological innovation and biodiversity prospecting. The role of government is crucial particularly in African countries with weak economies and where the abilities of the sector to promote technological innovation, and biodiversity prospecting are constrained by the fragmented nature of the markets. In these conditions governments face problems of scarce resources for allocation to technological activities in general and biodiversity prospecting in particular.

Some governments have set national goals and priorities for biodiversity conservation and sustainable use. Public policy discourse is also marked by frequent references to the role of genetic resources to sustainable development in general and to national economic progress in particular. On the whole, there seems to be a broad acceptance that biodiversity is one of the major strategic resources for sustainable development. We do not have grounds to demise national goals and interests of biodiversity conservation and prosepecting. But in order for the countries to both conserve and engage in prospecting they require considerable technological capabilities as well as policies that promote conservation and prospecting. There are many cases where policies of African countries nurture trade-offs between short-term economic growth and long-term sustainability and management of biodiversity (WRI, IUCN,

UNEP, 1992). Furthermore, the policies fail to articulate the role of biodiversity prospecting in economic growth. As noted above, the emphasis of policies in the region is on export of raw materials. The value premises of such policies must he confronted when addressing issues of biodiversity prospecting and sustainable development.

Government policies that promote biodiversity prospecting should support institutional and technological capability building. Such policies first and foremost recognize the strategic role of genetic resources in economic change and renewal. It should also be noted that the area of biodiversity prospecting requires specific policy measures. The biodiversity-prospecting policies should foster equitable sharing of the benefits of genetic resources as well as promote the development of national capabilities in biotechnology. Efforts that do not will fall victim to the historical mistakes of other export industries based on raw materials in developing countries.

A narrow focus on the sharing of returns on the sale of products derived from biological material is misguided. This approach can give African countries financial incentives to conserve biodiversity, but even longer-term benefits will stem from technological cooperation and capacity building in science and technology. For this reason, biodiversity prospecting should be considered as a public policy issue in its own right.

Another important facet of policy pertains to promoting technological innovation and the creation of specific incentive schemes for it. The incentives necessary to promote technological innovation are lacking in most African countries. Governments, especially their finance ministries, still see the provision of certain tax incentives to the scientific community as loss of revenue to the treasury. Governments also create disincentives to research. For example, some governments will reduce their allocation of funds to research institutions if those institutions start to generate revenue from their innovations. Such measures, although intended to save the treasury revenue, have the effect of discouraging research institutions from moving into the commercialization of their products. Unless these kinds of measures are reformed, it will be difficult for African countries to get into new areas of biotechnology and biodiversity prospecting.

In addition to incentives to research institutions, it is equally important to provide market incentives to allow for the participation of the private sector in the commercialization of biotechnology products. The provision of a suitable entrepreneurial environment would encourage innovators from various institutions to set up private enterprises. It will also encourage public enterprises to set up commercial facilities that operate as private enterprises.

Improving the regulatory environment to accommodate new technologies is an important aspect of economic change. It is not common to see African countries formulate new laws and regulations to promote particular technologies. The field of biodiversity prospecting may require the introduction of enabling regulatory measures. These could cover issues such as sample collection, intellectual property rights, biosafety, product standards and biodiversity conservation.

Conclusion

This chapter has argued that success in biodiversity prospecting will depend largely on how the African countries organize their institutions to take advantage of the growing market interest in genetic resources for the biotechnology industry. This is mainly because identifying a potentially useful product is highly uncertain and requires a high degree of institutional organization to reduce the associated risks. It is also science-intensive and requires effective policy guidance. The case of biotechnology has been used to illustrate the point. It placed emphasis on issues related to national institutional reform and international cooperation in technology.

Notes

1. Fungi, for example, have been an important source of medicinal ideas. For example, a fungus *Taxomyces andreanae*, which is found in the bark of the Pacific yew that has given the world the anticancer drug, Taxol, also makes taxol and a related substance in small amounts (Anon., 1993).
2. Interest in the use of local medicinal plants is not new. For a comprehensive survey of such uses in East Africa, see Kokwaro (1976). For a historical study of the links between plants and pharmacy, see Stockwell (1989).
3. Biotechnology-based pharmaceuticals are defined as recombinant protein drugs, recombinant vaccines and monoclonal antibodies (including therapeutic MAbs and imaging agents). For details, see Bienz-Tadmor (1993).
4. Medicinal plants contribute to the pharmaceutical industry in at least three ways: through constitutions isolated and used directly; by providing the base materials for the synthesis of drugs: and by providing natural products which serve as models for the synthesis of pharmacologically active compounds.
5. More than 50 drugs which exhibit anticancer properties are not marketed because of adverse effects but continue to be widely used in research.

References and Further Reading

Aliro, O.K. (1993) 'Natural soap washes away a deadly bug. *Panoscope* April, pp. 5–6.
Alper, J. (1993) R&D alliances give biggest bang for buck. *Bio/Technology* 11, 150.
Anon. (1993) Taxol-making fungus found. *Science News* 143, 230.
Aubert, J.-E. (1992? 'What evolution for science and technology policies? *OECD Observer*. February/March, p. 5.
Barton J.H. (1993) *Genetic Resource Negotiations: Relating Scientific and Commercial Worlds*. Biopolicy International. Acts Press, Nairobi.
Bienz-Tadmor, B. (1993) Biopharmaceuticals go to market: patterns of worldwide development. *Bio/Technology* 11, 168–172.
Blum, E. (1993) Making biodiversity conservation profitable: a case study of the Merck/INBio Agreement. *Environment* 35, 39.

Brown, M. (1992) Science, technology and the environment. *OECD Observer*, February/March, p. 12.

Clark, N. and Juma, C. (1987) *Long-Run Economics: An Evolutionary Approach to Economic Growth*. Pinter Publishers, London.

Clark, N. and Juma, C. (1991) *Biotechnology for Sustainable Development: Policy Options for Developing Countries*. Acts Press, Nairobi.

Cook, T. *et al.* (1991) *Pharmaceutical Biotechnology and the Law*. Stockton Press, New York.

Davis, C. *et al.* (1993) *Biotechnology in Thailand: Scientific Capacity and Technological Change*. Biopolicy International 10. Act Press, Nairobi.

DiMasi, J. *et al.* (1991) Cost of innovation in the pharmaceutical industry. *Journal of Health Economics* 10, 107–142.

Directorate General International Cooperation (1992) *Biotechnology and Development Cooperation: Priorities and Organization of the Special Programme*. Directorate General International Cooperation, Ministry of Foreign Affairs, The Hague.

Downes, D., Laird, S.A., Klein, C. and Kramer Carney, B. (1993) Biodiversity prospecting contract. In: Reid, W., Laird, S.A., Meyer, C.A., Gámez, R., Sittenfeld, A., Janzen, D.H., Gollin, M.A. and Juma, C. (eds) *Biodiversity Prospecting: Using Genetic Resources for Sustainable Development* World Resources Institute, pp. 254–266.

Eisner, T. (1992) Chemical prospecting: a proposal for action. In: Bormann, F.H. and Kellert, S.R. (eds) *Ecology Economics, and Ethics: The Broken Circle*. Yale University Press, New Haven, Connecticut.

Elisabetsky, E. and Nunes, D.S. (1990) Ethnopharmacology and its role in Third World countries. *Ambio* 19, 419–421.

Emmett, A. (1992) Where East does not meet West. *Technology Review* 95, 50–56.

Gámez, R., Piva, A., Sittenfeld, A., Léon, E., Jiménez, J. and Mirabelli, G. (1993) Costa Rica's conservation programme and National Biodiversity Institute (INBio). In; Reid, W., Laird, S.A., Meyer, C.A., Gámez, R., Sittenfeld, A., Janzen, D.H., Gollin, M.A. and Juma, C. (eds) *Biodiversity Prospecting: Using Genetic Resources for Sustainable Development*. Washington, DC, World Resources Institute, pp. 53–67.

Ganellin, C.R. (1989) Discovering new medicines. *Chemistry and Industry* 1, 9–15.

de Groot, C. (1992) China and Europe together in biotechnology. *Biotechnology and Development Monitor* 13, 20.

Gupta, A. (1993) Losing the edge. *Down to Earth* 2,(5), 29.

IFC (1992) *Investment and Environment: Business Opportunities in Development Countries*.World Bank and International Finance Corporation, Washington DC.

Juma, C. (1989a) *The Gene Hunters: Biotechnology and the Scramble for Seeds*. Zed Books and Princeton University Press, London, Princeton, NJ. pp. 81–82.

Juma, C. (1989b) *Biological Diversity and Innovation: Conserving and Utilizing Genetic Resources in Kenya*. Acts Press, Nairobi.

Juma, C. (1991) Difusion de la Biotecnología en África. Revisión de Avances en este Terano Oriental y del Sur. *Revista Mexicana de Sociología* 2, 91–123.

Juma, C., Mugabe, J. and Rainneri-Mbote, P. (1994) (eds) (1989) *Coming to Life: Biotechnology in African Economic Recovery*. Nairobi: Acts Press, Nairobi and Zed Books, London.

Juma, C. and Ojwang. J.B. (eds) (1989) *Innovation and Sovereignty: The Patent Debate and African Development*. African Centre for Technology Studies, Nairobi.

Juma, C. and Sagoff, M. (1992) Policies for technology. In: Dooge, J.C. et al. (eds) *An Agenda of Science for Environment and Development into the 21st Century*. Cambridge University Press, Cambridge.

Juma, C. and Sihanya, B. (1993) Policy options for scientific and technological capacity-building. In: Reid, W., Laird, S.A., Meyer, C.A., Gámez, R., Sittenfeld, A., Janzen, D.H., Gollin, M.A. and Juma, C. (eds) *Biodiversity Prospecting: Using Genetic Resources of Sustainable Development*, World Resources Institute, Washington, DC, pp. 216–218.

Kokwaro, J. (1976) *Medicinal Plants of East Africa*. East African Literature Bureau, Nairobi.

Komen, J. (1993) New initiative links: US universities and companies to developing country partners. *Biotechnology and Development Monitor* 15, June, 22.

van Latum, E.B.J. and Gerrits, R. (1992) *Bio-pesticides in Developing Countries*. Biopolicy International 1. Acts Press, Nairobi.

Lemma, A. (1991) The potentials and challenges of endod, the Ethiopian soapberry plant, for control of schistosomiasis. In: *Science in Africa: Achievements and Prospects*. American Association for the Advancement of the Sciences, Washington DC.

Lewington, A. (1990) *Plants for People*, Natural History Museum, London.

McKenna, N. (1993) Third World plants may offer AIDS treatment. *Panos Features*, March 5.

Mody, A. (1989) *Staying in the Loop: International Alliance for Sharing Technology*. World Bank discussion paper no. 16. World Bank, Washington, DC.

OECD (1988) *Biotechnology and the Changing Role of Government*. Organization for Economic Cooperation and Development, Paris.

OECD (1992) *The OECD Environment Industry: Situation, Prospects and Government Policies*. Organization for Economic Cooperation and Development, Paris.

OTA (1993) *Development Assistance, Export Promotion, and Environmental Technology*. Office of Technology Assessment. Congress of the United States, Washington, DC.

OTA (1991) *Biotechnology in a Global Economy*. Office of Technology Assessment, Congress of the United States, Washington, DC.

Porter. G. (1993) *The United States and the Biodiversity Convention*. Biopolicy International 12. Acts Press, Nairobi.

Principle, P. (1989) The economic significance of plants and their constituents as drugs. In: Wagner H. et al. (eds) *Economic and Medicinal Research*, Vol. 3, Academic Press, London, pp. 1–17.

Principle, P. (1991) Valuing the biodiversity of medicinal plants. In: Akerele, O. (eds) *Conservation of Medicinal Plants*, Cambridge University Press, Cambridge.

Reid, W. (eds) (1993) *Biodiversity Prospecting: Using Genetic Resources for Sustainable Development*. World Resources Institute, Washington, DC.

Reid. W. (1992) *Genetic Resources and Sustainable Agriculture*. Biopolicy International 2. Acts Press, Nairobi.

Reid, W. (1994) Biodiversity prospecting: strategies for sharing benefits. In: Sanchez, V. and Juma, C. (eds) *Biodiplomacy: Genetic Resources and International Relations*. Acts Press, Nairobi, p. 242.

Sánchez, V. and Juma, C. (eds) (1994) *Biodiplomacy: Genetic Resources and International Relations*. Acts Press, Nairobi.

Sasson, A. (1992) *Biotechnology and Natural Products: Prospects for Commercial Application*. Acts Press, Nairobi.

Sears, C. (1992) Jungle potion. *American Health*, October.

Simpson, D. and Sedjo, R.A. (1922) Contracts for transferring right to indigenous genetic resources. *Resources* 109, 1-6.

Sittenfeld, A. and Gámez, R.(1993) Biodiversity prospecting by INBio. In: Reid, W., Laird, S.A., Meyer, C.A., Gámez, R., Sittenfeld, A., Janzen, D.H., Gollin, M.A. and Juma, C. (eds) *Biodiversity Prospecting: Using Genetic Resources for Sustainable Development*, World Resources Institute, Washington, DC.

Spalding, B.J. (1993) Biopharmaceutical firms up R&D spending 71%. *Bio/Technology* 11, 768.

Stockwell, C. (1989) *Nature's Pharmacy: A History of Plants and Healing*. Arrow Books, London.

UNEP (1992) *Convention on Biological Diversity*. United Nations Environment Programme, Nairobi.

UNIDO (1991) *Industry and Development: Global Report*. United Nations Industrial Development Organization, Vienna.

United Nations (1992) *Biotechnology and Development: Expanding the Capacity to Produce Food*. Advanced Technology Assessment System Issue 9. United Nations, New York.

Violetta, D.M. and Chestnut, L.G. (1986) *Valuing Risks: New Information on Willingness for Pay for Changes in Fatal Risks*. United States Environmental Protection Agency, Washington, DC.

Walsh, V. (1993) Demand, public markets and innovation in biotechnology. *Science and Public Policy* 20, 138-156.

World Wide Fund for Nature (1993) *Ethics, Ethnobiological Research, and Biodiversity*. WWF, Gland Switzerland.

WRI, IUCN, UNEP, (1992) *Global Biodiversity Strategy*. World Resources Institute, Washington, DC.

Biodiversity in Urban and Peri-urban Zones

R. Folch
Fundació Enciclopèdia Catalana, Diputació, 250, 08007 Barcelona, Spain

Following a large number of chapters dedicated to fundamental aspects of biodiversity in natural environments, the time has come to focus our discussions on biodiversity in urban and peri-urban environments – a very interesting, yet often neglected, topic. In fact, the traditional ecological approach is not very fond of cities. We are accustomed to thinking of cities as anti-nature, home to all types of environmental disasters. At present, however, we can say that the problem with this approach is not that it is so traditional, rather that it is outmoded, out of date. Cities are not so much centres for disaster as they are centres for human beings (although some of them, you must admit, act in disastrous way. . .).

There are two very different aspects of biodiversity to take into consideration. On the one hand, there is the issue of biodiversity in cities, meaning the presence of biological elements linked to urban or peri-urban areas. On the other hand, there is the question of the role which cities should play as centres for the creation and maintenance of biodiversity for the planet.

The City: a Natural Artifice

The city is a natural habitational invention of the human species. It is, therefore, an artifice, but constructed in the midst of nature using transformed natural elements – natural elements that we still find everywhere in any city.

The city is a natural system. A natural system made artificial, the functioning of which requires the presence of a certain number of different species. The human species is not the only one living in cities: there are always either remnants of nature which preceded the city or biotic exotic elements introduced by humans. Controlling and conserving these elements, improving and increasing their numbers, is a primary objective of cities, for scientific as well as social reasons. Urban biodiversity, in the end, is the only biodiversity which plays a part in the everyday life of a majority of people.

Cities and Biodiversity Management

Diversity, including biodiversity, is thus a fundamental component of urban systems. This is why Chapter 53 will discuss 'The urban dimension of biodiversity', bringing to light the experience of an environmental architect, an urban planner if you like, in the most applied sense.

This diversity begins with individuals of the human species itself. Ultimately, a real city only deserves this name if it functions as a place of interaction for different cultures and groups. However, this diversity is the product of many other species, some of which find the city to be a very adequate environment. In this respect, Chapter 48 presents the views of a horticultural and botanical engineer concerning 'The importance of biodiversity in urban environments and the role of cities in biodiversity conservation'. Chapter 50 by Maciej Luniak, a zoologist expert in urban ecology, discusses 'Synurbanization of animal populations, factor increasing diversity of urban fauna'.

Whatever the case may be, a great deal needs to be done in order to maintain and increase urban and peri-urban biodiversity. Chapters 52 and 59 by a geneticist, and an ecologist, will consider, respectively, 'General considerations on biodiversity changes in urban and peri-urban environments with native forests' and 'Restoration of biodiversity in urban and peri-urban environments with native forests'. Chapter 51 by an expert in the biology of populations, will present the concrete case of 'Biodiversity management in peri-urban environments in Switzerland'.

Two Examples: Buenos Aires and Barcelona

Natural colonization of environments spontaneously works in favour of biodiversity. A particularly remarkable case of increase in local biodiversity indexes in urban and peri-urban zones is that of the Reserva Ecológica Costanera Sur, a large humid zone along the right bank of the Rio de la Plata, and perfectly integrated into a large metropolitan area, that of Buenos Aires in Argentina. This example is worth commenting on.

In 1918, the city of Buenos Aires resolved to build a riverside promenade downstream of the Rio de la Plata. This area is known as the Costanera Sur and it was conceived by the French architect Jean-Claude Forestier as a large boulevard full of magnificent gardens and protected by the ramparts on the river. This urban dynamic was considered adverse for such a beautiful promenade, and in 1972 it was decided to fill in part of it in order to free up new land for development. With the debris and other such material, Dutch style polders were built – in other words – dikes capable of holding large quantities of water which thus became lakes that were to be drained and then refilled. Desiccation was completed, but not the refilling, as the operation was abandoned at the beginning of the 1980s.

The remaining basins thus gradually filled up with ground water and rain water, and vegetation quickly gained its place and created a favourable habitat for the fauna. In addition, colonization by fauna was enormously encouraged by the continuous arrival of animals travelling on 'camalotes', or small wooden rafts and floating plants brought in by the Uruguay and Paraná Rivers. The abandoned polders, in a very short time, became a marsh overflowing with life, despite its location right next to a port zone belonging to a huge conurbanization of 12 million inhabitants. This created a magnificent humid zone, even though human-made, 350 ha long.

In 1986, the marshes were classified as a natural park and ecological reserve. Currently, forests of alders and weeping willows (*Tessaria integrifolia*, *Salix humboldtiana*) can be found there, as well as grasslands of the 'cortadera' Gramineae (*Cortaderia selloana*), 'chilca' bushes (*Baccharis punctulata*) and of course marshes with reeds, water hyacinths, etc. (*Scirpus*, *Eichhornia*, etc.) This vegetation has become both a refuge and a source of food for an abundant and diversified fauna: 235 species of birds, 23 species of reptiles, nine species of amphibians and four species of mammals. This entire landscape, adjacent to an urban and industrial zone, formed spontaneously, in less than 20 years, out of debris. It is difficult to imagine an ecosystem more linked to urban activity. This ecosystem demonstrates, without a doubt, not only the colonizing power of life in humid zones lacking nutritional elements, but also the importance of peri-urban biota.

A second example, of quite a different nature, is that of the Serra de Collserola. This chain of mountains reaching an altitude of 512 m and adjoining the city of Barcelona, covers approximately 6500 ha and was classified as a metropolitan natural park in 1976. This park, entirely surrounded by the conurbanization of Barcelona (3 million inhabitants), houses a truly remarkable and imposing pine (*Pinus halepensis*) and oak (*Quercus ilex*, *Q. cerrioides*) forest. It is a monumental forest as far as the entire Mediterranean area is concerned, and yet it is a metropolitan park at the same time! And surprising as it may seem, there is also an abundant population of wild boars (*Sus scrofa*) in the forest. In other words, in the suburbs of the Olympic city Barcelona, that fashionable city housing some of the most futuristic designs, wild boar still roam. Boars, along with 23 other species of mammals, 15 species of reptiles, 11 amphibians, 140 species of birds (79 of them nest forming), and even six species of freshwater fish. A study, completed in autumn 1989, of prey-feeding birds which use Collserola as a migratory stop, identified up to 21 different species: 902 individuals of honey buzzards (*Pernis apivore*), 252 kestrel hawks (*Falco tinnunculus*), 168 marsh harriers (*Circus aeruginosus*) and 135 sparrow hawks (*Accipiter nisus*). In total, 1612 migrating birds were detected.

Conclusions

In summary, it is very important that we be concerned with urban and peri-urban biodiversity, not simply because the majority of humans live in cities or urban areas, but also because its scientific importance. This will be further considered in Chapter 47.

Studying urban biodiversity goes much further than just looking at bioornamental curiosities in the city. And like everywhere, whether it be in cities, forests or in the country, our messages concerning biodiversity should help people to understand the advantages and the need for diversity. We must not forget the very important educational aspect of our role. As scientists, we should aim to understand and to help others to understand.

The Importance of Urban Environments in Maintaining Biodiversity

V.H. Heywood
*School of Plant Sciences, University of Reading, Whiteknights,
PO Box 221, Reading RG6 2AS, UK*

Introduction

In most considerations of biodiversity, scant consideration is given to urban or peri-urban biodiversity. Indeed some authors would deliberately exclude it, or at least those parts of it that are human-created, such as urban parks, street and roadside plantings, and domestic gardens, on the grounds that it is artificial and not therefore properly a part of biodiversity as such which should concern itself with ecological integrity (Angermeier, 1994). At the other extreme there are those who adopt a restorationist approach and would argue against taking human beings out of nature and making wilderness of it, and propose a new ecological paradigm of ecological restoration, construction and invention (Jordan, 1994; Turner, 1994). The debate, which centres on our understanding of the relationship between nature and humans, is analysed by Kane (1994).

As has been pointed out frequently, there is little left of nature and our landscapes that has not been extensively altered by the actions or activities of human beings. Moreover, it is difficult to draw any clear line between the remaining so-called natural ecosystems and those that are modified by humans. Urban and peri-urban biodiversity can be considered as a special case: as we shall see below, it covers both natural and artificial situations, although it deals largely with the latter, i.e. habitats or ecosystems that are invented or restored. The role of humans in such habitats is by definition dominant, cities being created by and for them.

Urban nature conservation is one of the most remarkable developments of the past two decades. As Goode (1993) observes, what started as a minority interest in the mid-1970s has now become accepted as a major component of nature conservation in the 1990s. Although some thought was given by a few individuals to the problems of wildlife and habitats in cities in the 1930s and

1940s, it was not until the 1960s that much effort was devoted to the subject, at which time, in much of Europe, there was a movement of people away from the cities while wildlife was migrating more and more into the cities (Nicholson-Lord, 1987). There were several factors involved in the evolution of the urban conservation movement, such as:

1. The increasing amount of urban wasteland, including bomb-sites and industrial relicts such as disused factories, railway stations, etc.;
2. The development of environmentalism;
3. The recognition of the discipline of urban ecology;
4. The creation of urban wildlife organizations;
5. The reduction in some forms of pollution (e.g. smoke-free zones);
6. A growing recognition of the fact that the urban environment is a mosaic of ecological niches that house a diversity of species ('the landscape does not end where cities begin' Nicholson-Lord, 1987);
7. A responsive attitude on the part of local authorities.

Today, there is now a substantial literature on urban landscapes and their wildlife in many parts of the world (e.g. López-Moreno, 1993; Adams, 1994); on the ecology and maintenance of natural and human-created greenways (e.g. Smith and Hellmund, 1993); and on preserving and restoring urban biodiversity (Platt *et al.*, 1994). Various symposia have been held on these topics (e.g. Barker *et al.*, 1994). Several journals deal with issues of urban biodiversity, such as *Urban Forests*, which has devoted two recent issues to urban ecosystems, and *Urban Wildlife News* published by the UK inter-Agency Habitat network, which is also used to network information on UNESCO's Man and the Biosphere project No. 11 (urban systems) and serves as a newsletter for the members of the Urban Ecology Group of INTECOL. Hundreds of articles and reports on various aspects of urban wildlife have been published. Detailed listings of various taxonomic groups such as birds, plants, lepidoptera, and soil arthropods have been published, and monitoring of urban biodiversity is being carried out in many cities and urban areas. Some cities, such as Edinburgh, Scotland, and Moscow, Russia, have prepared an Urban Nature Conservation Strategy.

The Nature of Urban Biodiversity

A large percentage of the human population lives in or around cities: in terms of cities of over 750,000 inhabitants, it ranges from virtually 100% of the countrys's total population in the case of Singapore to 0% in the case of Sri Lanka (Table 47.1). By the year 2025 it is estimated that 80% of the residents of developed countries will live in cities, and that throughout the world only 40% will live in rural areas (UN, 1989). A considerable part of urban and peri-urban areas is devoted to 'green space'. In the United States, urban parkland ranges from 42.2% in Honolulu to 0.3% in Baton Rouge. It has been estimated that about a fifth of towns and cities is open space

Table 47.1. Percentage of total national population in cities of over 750,000 inhabitants.

Singapore	100
Belgium	97
Kuwait	96
United Kingdom	92
Venezuela	90
Mexico	73
Australia	60
USA	41
Bangladesh	18
China	9.9
Burundi	6
Uganda	4.3
Malawi	0
Sri Lanka	0

and a study in Brussels suggested that up to 50% is actually green and photosynthesizing.

Urban 'green space' comprises a wide range of habitats. These are of three main sorts: natural, seminatural (although recognizing that no clear distinction exists between these two), and artificial (invented) or completely human-made (Table 47.2).

Natural habitats

Natural habitats include remnants of native vegetation as can be found, for example, in Rio de Janeiro, Brazil where the celebrated Mata Atlântica forests, among the richest in the world, (Monteiro and Kaz, 1991-92) extend into the city (although it may be noted in passing that the Tijuca area was reforested in the nineteenth century on the orders of the Emperor Peter II). Another celebrated example is the 4 ha of original evergreen forest preserved in Singapore Botanic Garden, in the centre of the world's second largest port city (Tinsley, 1983). Or again, in Venezuela, the National Park El Avila which rises out of the north of the city of Caracas, forming a majestic wall of mountains, and home to over 1700 species of flowering plants (Steyermark and Huber, 1978). There are many other less dramatic examples in various parts of the world, ranging from remnants of natural forests, such as the four natural forest stands left on the York University Campus, Toronto, Canada, or of wetlands to individual trees.

Seminatural habitats

Seminatural habitats abound in cities and most are of recent origin, having developed on derelict land or disused industrial land or on other open land, such as old quarries, sewage works, gravel pits, cemeteries, railway

Table 47.2. Types of urban 'green spaces'.

National parks
Municipal parks and gardens
Botanic gardens
Zoological parks
Wildlife corridors
Urban commons
Garden centres
Urban forests
Business parks
Canals
Nature centres
Vacant lots
Nature reserves
Domestic gardens
Arboreta
Roadside verges
Greenways
Allotments
Nurseries
Streams and ponds
Industrial parks
Cemeteries
Golf courses
Street trees

embankments and hospital and school grounds, Such habitats which form part of what has been called the unofficial countryside, a term coined by Richard Mabey (Mabey, 1973), often contain a greater diversity of plant and animal life than the officially created municipal parks and open spaces (which have been described as sterile and monotonous by comparison.)

But as Goode (1993) has pointed out, there is more to urban conservation than maintenance or protection of existing habitats: on the contrary is also a creative process including enhancement of existing habitats and creation of new ones. It has more to do with habitat restoration and rehabilitation than conservation. Just as experiments are in progress to create plant communities of both sown and naturally occurring weeds suitable for insects in agricultural landscapes, so similar efforts are being made to provide habitats for various animals in urban and peri-urban habitats.

Artificial habitats

Artificial, invented or human-made habitats are enormously diverse: in towns and cities and their immediate surrounds we find everything from vineyards (such as the celebrated one on the butte de Montmartre, Paris), crop fields, banana plantations, sugar plantations, intensive horticulture (such as the

Huerta of Valencia), to parks, recreated wetlands in Utrecht, woodlands, ecological parks, botanic gardens, private gardens, and urban landscaping. In such habitats, most of the plant diversity consists of cultivated plants and weeds although native species are sometimes found.

Organismal Diversity in Urban Habitats

It is no more possible to estimate how much biodiversity occurs in urban and peri-urban environments than to assess how much biodiversity occurs globally. As inventories or environmental audits are increasingly being commissioned by urban authorities and other bodies, evidence of the enormous array of diversity such habitats contain is coming to light. In Italy, for example, over 600 papers on urban avifauna had been published by 1988 and advanced ornithological studies had been carried out in 27 towns or cities. A total of 103 breeding bird species have been recorded in Italian towns and cities (Dinetti, 1994).

Of course, much of the species diversity is of introduced species while native species are being lost, and this raises a whole series of issues that cannot be explored here. Yet the number of native or spontaneous species that are found in cities can be remarkably high: in the case of higher plants, more than 50% of the species recorded in cities as different as Mexico and Rome are native and in a study of roadside verges in the Netherlands, it was found that, about 50% of the country's flora were identified as occurring, including 16.8% of rare to rather rare species (Sykora *et al.*, 1994).

In terms of exotic species, many thousands are cultivated in parks and gardens, both public and domestic. Botanic gardens are a special case. There are over 1600 botanic gardens and arboreta worldwide (cf. Heywood *et al.*, 1991); about 540 of these occur in Europe, and 451 are listed for the United States (Watson *et al.*, 1993). At least 80,000 species are in cultivation in botanic gardens and represent the greatest array of plant diversity outside nature. Private gardens, especially those of specialist collectors, also house many thousands of plant species. It is, therefore, possible to see very large samples of the world's native flora in cities and the scientific, cultural and educational significance of this should not be underestimated.

Most of the cultivated or introduced biodiversity, apart from naturalized aliens, does not form breeding populations although it should be noted that artificial habitats in public and private gardens do often provide a home for animal populations. Another point that should be stressed is the relatively poor range of sampling of natural populations represented by introduced or cultivated plants in the majority of cases. Their genetic diversity is not usually very great. In some cases, however, as a result of intensive breeding or selection, large amounts of 'new' variation is produced and cultivated, for example in agricultural and horticultural crops (cultivars) and in ornamental groups such as orchids, narcissus, tulips, roses, begonias, delphiniums, polyanthus, violas and pansies.

Significance of Urban Biodiversity

Urbanized societies, as already noted, are increasingly becoming a predominant way of life for our species, and the extent and speed of this urbanization is a relatively new phenomenon. It must be evident, therefore, that urban and peri-urban biodiversity, its maintenance and use, have already become major considerations in conservation and biodiversity strategies. Of course, to advocate the maintenance of urban and peri-urban biodiversity is in no sense to suggest that it is a substitute for wildlife in nature but anyone who travels the world soon realizes how little genuinely 'natural' native, relatively untouched habitats are left: everywhere one finds destruction, conversion, fragmentation, with strictly protected areas rapidly becoming islands of conservation in a rapidly changing landscape.

The bioregional approach to biodiversity conservation, developed in the Global Biodiversity Strategy (WRI/IUCN/UNEP, 1992), recognizes this situation and provides us with a context and framework within which the urban and peri-urban milieu can be developed. It is based on the perception that biodiversity conservation is no longer a question of establishing national parks and other protected areas as islands of protection in a sea of unregulated agriculture, forestry, fisheries and urban development, but calls for strategies and actions that integrate the needs and perceptions of local communities. Bioregions are not delimited by political boundaries but by the geographical limits of ecological systems and human communities. Also, since protected areas will probably never reach more than 10% of our planet's surface, most biodiversity will occur outside such protection and therefore the conservation of biodiversity has to be planned and managed in such a way as to coordinate protected areas with the total landscapes in which they occur, and this of necessity includes urban and peri-urban environments.

Moreover, urban landscapes provide their inhabitants with social values and benefits that derive from contact with nature, albeit a nature often artificially created or developed. As Box and Harrison (1994) have noted,

> if the contribution of urban green spaces to future generations is to be justified solely in terms of their contribution to the stock of environment assets, then urban environmental assets will always be deemed to be poor substitutes for their rural counterparts. On the other hand, if urban green space policies acknowledge the social and educational assets of accessible natural greenspace, then the inheritance value of these areas is seen to be unrivalled.

Here then we have another important reason for giving urban biodiversity due weight in our planning: its inheritance value for our children.

Finally, it should be noted that in many parts of the tropics, away from the major conurbations, in small towns and villages, the landscape is essentially green and includes a great variety of habitats which combine natural, seminatural and artifical elements – cultivated fields, community forests, sacred groves, home gardens, spice gardens, semidomesticated fruit trees, paddies (sometimes the habitats of extremely rare native species), plantations, roadside verges that merge almost imperceptibly into the vegetated landscape. Although perhaps less structured and managed in many instances than the urban and peri-urban habitats found especially, although by no means exclusively, in temperate countries, they are sometimes under strong local social control and management. Their contribution to the conservation of biodiversity is major and much more attention should be paid to them in overall conservation strategies.

References

Adams, L.W. (1994) *Urban Wildlife Habitats*. University of Minnesota Press, Minneapolis and London.
Angermeier, P.L. (1994) Does biodiversity include artificial biodiversity? *Conservation Biology* 8, 600–602.
Barker, G.M., Luniak, M., Trojan, P. and Zimny, H. (eds) (1994) Proceedings of the II European Meeting of the International Network for Urban Ecology. *Memorabilia Zoologica* 49 (Warszawa).
Box, J. and Harrison, C. (1994) Minimum targets for accessible natural greenspace in urban areas. *Urban Wildlife News* 11, 10–11.
Dinetti, M. (1994) The urban ornithology in Italy. *Memorabilia Zoologica* 49, 269–281.
Goode, D. (1993) Local authorities and urban conservation. In: Goldsmith, F.B. and Warren, A. (eds) *Conservation in Progress*. John Wiley, Chichester, pp. 335–345.
Heywood, C.A., Heywood, V.H. and Jackson, P.S. (1991) *International Directory of Botanical Gardens*, 5th edn. Koeltz Scientific Books, Koenigstein.
Jordan, W.R. (1994) 'Sunflower forest': ecological restoration as the basis for a new environmental paradigm. In: Baldwin, A.D., de Luce, J. and Pletsch, C. (eds) *Beyond Reservation. Restoring and Inventing Landscapes*. University of Minnesota Press, Minneapolis, pp. 17–36.
Kane, G.S. (1994) Restoration or preservation? Reflections on a clash of environmental philosophies. In: Baldwin, A.D., de Luce, J. and Pletsch, C. (eds) *Beyond Preservation. Restoring and Inventing Landscapes*. University of Minnesota Press, Minneapolis, pp. 69–84.
López-Moreno, I. (ed.) (1993) *Ecología Urbana Aplicada a la Ciudad de Xalapa*. Instituto de Ecología, Xalapa.
Mabey, R. (1973) *The Unofficial Countryside*. Collins, London.
Monteiro, S. and Kaz, L. (eds) (1991–92) *Atlantic Rain Forest*. Edições Alumbramento, Rio de Janeiro.
Nicholson-Lord, D. (1987) *The Greening of the Cities*. Routledge and Kegan Paul, London.
Platt, R.H., Rowntree, R.A. and Muick, P.C. (eds) (1994) *The Ecological City. Preserving and Restoring Urban Biodiversity*. University of Massachusetts Press, Amherst.
Smith, D.S. and Hellmund, P.C. (eds) (1993) *Ecology of Greenways. Design and Function of Linear Conservation Areas*. University of Minnesota Press, Minneapolis.
Steyermark, J.A. and Huber, O. (1978) *Flora del Avila*. Sociedad Venezolana de Ciencias Naturales y Ministerio del Ambiente y de los Recursos Naturales Renovables, Caracas.

Sykora, K., de Nijs, L. and Pelsma, T. (1994) Plant communities in road verges and their importance for the conservation of plant communities. Abstract TU/2A. *Symposium on Community Ecology and Conservation Biology, Bern, Switzerland.* 14–18 August.

Tinsley, B. (1983) *Singapore Green. A History and Guide to the Botanic Gardens.* Times Books International, Singapore.

Turner, F. (1994) The invented landscape. In: Baldwin, A.D., de Luce, J. and Pletsch, C. (eds) *Beyond Preservation. Restoring and Inventing Landscapes.* University of Minnesota Press, Minneapolis, pp. 35–66.

UN (1989) *World Populations Prospects.* United Nations, New York.

Watson, G.W., Heywood, V. and Crowley, W. (1993) North American Botanic Gardens. *Horticultural Reviews* 15, 1–62.

WRI/IUCN/UNEP (1992) *Global Biodiversity Strategy.* World Resources Institute, International Union for Conservation of Nature and Natural Resources, United Nations Environment Programme, Washington, DC.

The Role of Biodiversity in Urban Areas and the Role of Cities in Biodiversity Conservation

J.-P. Reduron

Mission Interservices pour le Respect de l'Environnement (MIRE), Association Française pour la Conservation des Espèces Végétales (AFCEV), Mairie de Mulhouse, 2, rue Pierre Curie, 68200 Mulhouse, France

Introduction

With the development of cities, the natural environment has been profoundly transformed. Nature has remained present and can be seen at varying degrees depending on several factors.

1. *The conceptual approach behind urban construction.* The density of constructed areas and infrastructures has left more or less room for nature. In some cities, zoning regulations call for numerous open spaces, most often made into parks. In other cities, chains of buildings, refuse dumps and rest areas have left little room for nature.

2. *The presence or absence of major natural sites* (rivers, geological variations, land unfit for construction, etc.). These sites have permitted a certain type of wildlife to survive in urban areas.

3. *The distance of the modern city from the old city centre.* City centres were built during a time when people had to fight against natural elements and when cities were a place to find protection. In these cases, there is little room for nature as the old city centres were heavily built. A few cities, however, pushed their walls back quite far allowing for agricultural activity inside city limits (for example, in Sienna, Italy).

Thus depending on the city, there is an equilibrium or a disequilibrium between the presence of natural elements and artificial or constructed elements. There are two types of natural elements remaining in cities.

1. Natural biological areas, where there has been little human intervention and where there is a high presence of indigenous species. These are the

'islets of nature' found in the heart of cities: the plain of Maine in Angers, the hills of Besançon, the bed of the Loire river in Orléans, rocky hills in Marseille, etc.

2. Cultivated biological areas which for most city dwellers are part of 'nature' in the city. These areas have been constructed by humans in order to make the city more pleasant. They often require considerable human intervention, and consist of species which were introduced or products of horticulture. There are a number of horticultural cities in France known for the quality of their gardens and flower beds: Nantes, Orléans, Bourges, Caen, Metz, Besançon, Dijon, etc.

The distinction between these natural areas is only schematic because, today, the limit is often unclear. Areas developed by humans use indigenous species (concept of 'natural' garden) and similarly, species which have been introduced adapt in the city and propagate without intervention.

All of these areas bring plant and animal biodiversity together. They may also bring about biohomogeneity. The following sections will discuss the importance of maintaining biodiversity in cities and the need to establish new partnerships between the scientific community and city representatives if we are to reconcile such historical entities such as cities and biodiversity.

The Importance of Biodiversity in Cities

Aesthetic and social importance

City residents generally have different tastes and hardly appreciate uniformity in the vegetation which can be valid in a localized area for a particular aesthetic composition. Using a large range of flora and fauna allows for them to be adapted in an aesthetic way according to the various sectors of the city (city centre, city limits, communication routes, natural zones, parks, new neighbourhoods, etc.). This variation can also correspond to natural annual rhythms (spring awakenings, blooms, colours of leaves) which may be aesthetic events in neighbourhoods. It is thus important to increase the variation of species used during the year in order to maintain flowers or colourful atmospheres, for as long as possible.

Horticultural technical importance

Diversification in the types of land development and plant species used has many positive aspects. It provides photosanitary protection and thus an epidemic will not decimate the majority of trees in a city. Treating zones naturally and planting well-adapted species reduces the cost of maintenance, because the more artificial the landscaping, the costlier it is. A larger range of species enables areas with several strata of vegetation to develop (herbaceous, low shrubs, high shrubs, arboreous). This is true not only for the fauna, but for the health of trees as well. Planting ground covering shrubs at the base of trees encourages soil structuring and leads to better growth and resistance of trees.

Ecological interest

A high level of biodiversity in cities must be correlated with the various types of development planning. In natural spaces, made up of several strata of vegetation, a 'natural' urban ecosystem may function, permitting the development of diversified fauna. More cultivated spaces are not devoid of interest. Very tall trees have a positive impact on the local climate by improving air quality. Many horticultural species (conifers, woody shrubs) are often well equipped for the special climate in cities and protect a significant part of the fauna.

We need, therefore, to use a variety of methods when conceiving and planting the 'untouched' areas in cities. Each method should correspond to a well-affirmed desire to develop that area in a particular way. These spaces can then welcome a greater number of urban biotopes when they are developed intentionally using diverse materials and species.

Educational importance

It is clear that biodiversity in cities provides a very important educational tool for schools: fauna observations, school plantings, observing different species of trees, distinguishing between wild and cultivated flora, studying the role of insects for pollination, etc. These activities can diminish the negative effects of urban life and sensitize future generations to the pleasantness of living in a city with an equilibrium between inert structures and more or less organized and managed nature zones. In Mulhouse, a botanical instructorship has been running for more than twelve years to explain the importance of vegetation in urban areas.

The Role of Cities in Preserving Biodiversity

Environment is a transversal theme: it requires different disciplines to interact and it is necessary for it to be relevant at various levels. In fact, the solution to the problem lies at both the individual (personal behaviour) and collective (public groups and private businesses) levels. In order to preserve biodiversity, cities must also pass through these levels of intervention and invest in this topic. There are several types of action which they may take in this direction.

First, an inventory of the natural and cultivated biodiversity in the city should be carried out. With this information, the most interesting and useful spaces and species can be selected. The Swiss city, Basel, has established a map of the natural spaces and species in all of the communes – an extremely detailed and priceless tool for integrating the theme of biodiversity into urban planning. The appropriate city services should now adopt a different method of development for these spaces and give them adequate legal status in order to preserve them over time.

Cities can also go further by keeping a range of forgotten or local varieties in their parks. For example, particular ornamental plants can be used to mark certain developed areas, fruit varieties (urban orchards, vineyards as a remembrance of past cultures). Cities thus contribute to genetic conservation if they use the appropriate technologies.

Animal or plant collections might be oriented towards endangered species and varieties. Zoological parks and botanical gardens also have a conservation mission as regards biodiversity. The Zoological and Botanical Park in Mulhouse has, for more than 15 years, oriented its animal collections towards endangered fauna (felines, birds, monkeys). In Mulhouse there is now a botanical conservatory working on local endangered species and others of particular interest (e.g. chemical).

Other groups in France have become involved in the National Botanical Conservatories. They have formed associations in order to develop conservatories in Bailleul, Brest, Gap, Nancy, Porquerolles and the Mascarin on Reunion Island.

Scientific partnership: an indispensable condition for urban biodiversity

Being concerned with biodiversity in cities is a pertinent and appealing idea, but conditions and modalities of its application must not be forgotten. How should applied research in this domain be carried out? The difficulty comes from the division between the various services which must interact in order for this to happen. The city's technicians, especially the developers and managers of urban spaces, take part in landuse planning, thus transforming and humanizing environments. Landuse planning is, in theory, in conflict with the protection of nature – a domain grounded in a naturalistic conception of space against the development and transformation of environments. Different services frequently have different rationales, often irreconcilable, because they communicate so little.

Few developers call upon natural scientists, and few scientific naturalists are interested in urban environments, a sector where wild species are lost. Furthermore, urban technicians belong primarily to a body of engineers, responsible for putting technology into action for specific projects. They are not there to perform experiments or scientific applications. Few university professionals work within these groups, thus the applied research dimension cannot develop from contacts between people working together on a daily basis.

In reality, integrating nature into cities must be a truly collective and well-grounded project and it requires mobilizing specialists found outside of internal structures. A well-argumented project (e.g. quality of life in cities), innovative and motivating, and especially one that gives meaning to both those in charge of the project and to the collectivity, could bring about the necessary synergy between city developers and mangers and the scientific community involved in life sciences.

Phytosociology: the cornerstone

In order to take biodiversity into consideration in urban environments, an inventory of biodiversity must first be completed. Some cities have already moved in this direction: in Basel they have elaborated a map of nature; in Geneva and Mulhouse they have published an inventory of trees in the city; in Paris and Grenoble they have inventoried, in some sectors, their urban nature. Other cities have put together projects to take detailed inventories of their natural zones; for example in Besançon where natural penetrations are found right in the centre of town. In Mulhouse they have hired a botanist to inventory plant species in the commune.

My feeling, however, is that we need to do more and establish a true methodology for analysing natural environments in cities. This should be in line with phytosociology, a technique for describing plant coverage, but which extends much further because all plant communities are in fact dependent upon the type of soil and climate, and comprise different fauna depending on the community. In other words, plant groups are excellent ecosystem indicators and locating them is a good way to go about mapping the biotopes of a given territory.

Furthermore, the concept of vegetation series translates well the dynamics of the succession of plant types with a given soil, from its virgin state (rock, bare soil, abandoning of crops) to stable vegetation (climactic), often in forests, passing through stages of fallow lands and shrublands. This approach enables stages to be defined that can be used in planning or management projects according to, for example, biological diversity or landscape interests.

Furthermore, the vegetation 'transect' method shows biotope succession in space from one point to another in the city (for example, from a riverbank to the top of a hill, etc.). It complements the definition of vegetation series mentioned above. Developed in this manner, phytosociology gives a qualitative image of biotopes, situates them in space and imagines their potential evolution. It is the primary tool for defining policies for natural areas in cities and for biodiversity. The National Botanical Conservatory and the International Centre of Phytosociology in Bailleul constitute a precious and particularly promising tool for this endeavour.

Applying phytosociology in urban environments seems promising to me if an effort is made to elaborate a less esoteric and better-adapted language so that partners from other disciplines and the public at large may increase their understanding. It would be useful to try this approach in developed spaces as well, mainly in parks, where we could imagine a typology based on plants used there and the conditions for life that they demand or obtain. We can imagine without irony the 'urban land of junipers' or even the 'urban maple plantations'. All of these urban environments represent different living conditions which need to be made apparent in order to understand urban ecosystems and the assets and drawbacks of the natural palette of cities.

Scientific cooperation for biodiversity conservation

As mentioned earlier, cities may have two roles: to highlight and reinforce the value of biodiversity of its own territory, or to work for the general interest, orienting efforts of certain structures, such as zoos, botanical gardens or botanical conservatories, towards endangered species.

In both cases, there needs to be cooperation between city services or groups and the scientific community. Conservation *in situ*, or in a specialized environment (*ex situ* = *in horto*), cannot be improvised, and is not at all similar to perfected horticulture techniques applied in cities. It is necessary to experiment, because management of natural areas for biodiversity has only just begun. In fact, little is known about the biological reproduction of wild plants (more is known about fauna). How is it possible to establish effective management techniques when we don't know the biological behaviour of the species concerned? It is therefore important to innovate and to look for competent partners.

This is why in Mulhouse, in the context of its Botanical Conservatory, scientific cooperation is being encouraged, calling upon several bodies for their expertise. The National Natural History Museum of Paris is perfecting a 'conservatory profile' of the species concerned; the National School of Advanced Studies in Chemistry of Mulhouse is studying the fine details of the chemical make-up of plants – their toxicity, their possible application; the National Botanical Conservatory in Nancy is collaborating on projects concerning flora and seed germination. The Botanical Conservatory of Mulhouse is also working with site managers such as the Conservatory of Alsacian Sites (CSA). This kind of cooperation results in technological exchanges between sectors, often isolated, on the basis of a common project.

At the national level, the French Association for the Conservation of Plant Species (AFCEV) is trying to promote these kinds of technological and methodological exchanges. It encompasses more than 100 organizations, including institutions (Ministries, Research Institutes, Bureau of Genetic Resources, National Botanical Conservatories, Botanical Gardens, etc.) and associations (orchard conservatories, managers of protected areas, botanists, collectors). In its activities it helps promote dialogue about conservation among its various members (work groups, colloquia, liaison bulletins) often resulting in common methodologies (prescription books), national inventories, training, recommendations concerning regulations. It therefore encourages exchanges among professionals as regards the latest technological advances and large-scale conservation problems with structures involved at the local level and having at hand both knowledge of the area and the necessary methods. The AFCEV has not been very involved in urban environments, but a similar work procedure would be useful to bring together city technicians and biologists familiar with the environment. This is where methods are lacking, books of techniques, address books of various specialists, all of which would allow decision-makers and leaders to consider the feasibility of giving more importance to biodiversity in urban settings.

Conclusion

Cities benefit from biodiversity which improves the quality of city life. In return, they should give more funds to the general conservation of biodiversity. After a period of awareness-raising about the advantages of maintaining nature in cities (through education, communication, activities of associations), it is indispensable to go one step further by integrating environmental concerns into projects (whether for development or management) and by professionalizing procedures like those of other domains in city activities.

In order for these two objectives to be carried out successfully, new fields must be summoned to collaborate. As regards nature in cities, from now on biologists must be integrated into teams responsible for projects. In the executive phase, it should be species and biotope specialists who deal with biodiversity in natural areas. It is clear that new partnerships must be formed in order for cities to benefit from advances in the biological sciences, and similarly to incite members of the scientific community to think about the applications of their research. Development and urban planning should now take respect for biodiversity into account in an operational and not just in an emotional way.

The topic of environment is becoming increasingly integrated into cultures. This is necessary for the greatest number of people to value it. In order for this evolution to be relevant, it must result in new practices and methodologies, calling upon the numerous specialists interested in the future of the environment and human beings.

49 Restoration of Biodiversity in Urban and Peri-Urban Environments with Native Forests

A. Miyawaki

Japanese Centre for International Studies in Ecology (JISE), Shonan Village Centre, Hayama-machi, Kanagawa, 240-01, Japan

Summary

Monotonous biota and standardized artificial structures invested with non-biological materials and petrochemical fuels, characterize the typical urban sceneries today. The most desirable and reliable method to restore biodiversity of the urban and peri-urban environments is to create native forests using native trees based on the study of the potential natural vegetation.

The Miyawaki team has succeeded in restoring biodiversity by extending human-made forests to cities, industrial and housing complexes over 340 locations throughout Japan since 1973. The revegetation technique is the integration of the knowledge of Japan's indigenous Chinju-no-mori forest and modern vegetation ecology. The technique has also been successfully applied to commercial sites in Malaysia and Thailand.

Introduction

The glories of the human civilizations over the past 6000 years are great assets of humankind. The studies of ancient cities in Mesopotamia, Egypt, Greece, and the Roman Empire reveal, however, that impoverishment and loss of diversity, and eventual extinction of their native forests triggered the fall of the civilizations. The areas surrounding these historical sites still remain dry and barren today.

In Japan, forests have also been cleared over the past 2000 years to make villages, towns and cities; however, the sites sensitive to human impact, such as ridges, steep slopes, reservoirs and water fronts, were kept intact. Whenever new villages and cities were made, native forests with native trees were enshrined and preserved by evoking reverence for the sites in people's mind.

The native forests were sanctuaries. After World War II, rapid industrialization and urbanization and a loss of religious awe of nature accelerated the extinction of these native forests.

The Kanagawa Prefecture is a good example. It is adjacent to Tokyo, occupying 1/200 of Japan's total land area with a population of 8.3 million. If population density were an indication of urban advancement, it is highly advanced. However, most of the 2850 Chinju-no-mori forests identified in a previous survey have disappeared; only 40 sites retain the natural level equivalent to natural monuments (Miyawaki et al., 1979b). We integrated the theories of vegetation science and plant ecology of Europe, and the knowledge of Japanese Chinju-no-mori forests, and have applied the technique to the creation of native forests since the early 1970s (Miyawaki, 1975).

The efforts have been successful, with the cooperation of industries, local municipalities and some Ministries of the Japanese Government, in most of the vegetation zones from the northern Island of Hokkaido to the subtropical Ishigaki Island of Okinawa Prefecture, covering a distance of 3000 km (Miyawaki, 1989; Miyawaki et al., 1983, 1993; Miyawaki and Golley, 1993). The conventional succession theory has it that it takes 300 years to restore original native forests once they are destroyed. Our restoration technique based on the study of potential natural vegetation, however, has successfully created native forests in about a tenth of that time span (Tüxen, 1956). The seedlings planted 22 years ago now form a well-developed crown of 20 m. The soil fauna, an important element of ecosystems, steadily develops together with the trees (Aoki and Harada, 1985, 1989). In about 15–20 years, a diverse forest ecosystem which is approximately equivalent to natural native forests may be restored (Fig. 49.1).

This process was claimed to take over 200 years by Clements (1916) in his Succession Theory. This technique has been successfully extended to commercial areas in Malaysia, to the rain forest in Borneo, Belem in Amazon/Brazil and to the *Nothofagus* forest in Concepcion in Chile (Miyawaki, 1993a,b).

Methodology

The initial stage of the vegetation-ecological process of restoration of native forests consists in a phytosociological field study (Braun-Blanquet, 1928, 1951, 1964) to obtain information on the actual vegetation of the site. Remnants of house forests and Chinju-no-mori forests, land use, topography and soil profiles in the vicinity are also surveyed. Comprehensive study of these elements is made to detect the potential natural vegetation, a concept developed by Tüxen (1956). Then maps of the actual vegetation and of the potential natural vegetation are prepared. Our team has conducted intensive field studies and completed the maps of actual vegetation and potential natural vegetation, though at a small scale, for the total land area of Japan (Miyawaki, 1980–1989).

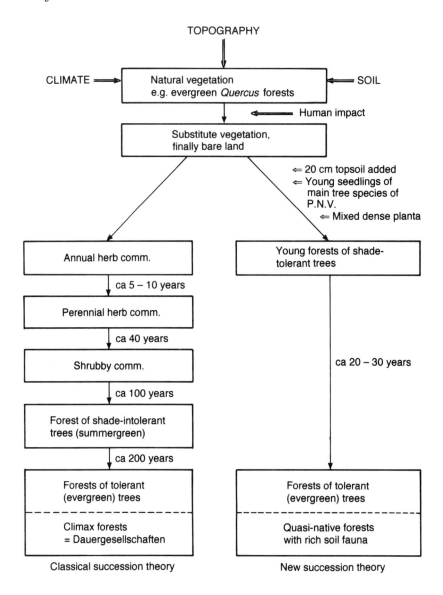

Fig. 49.1. Comparison between new succession and classical theory (Miyawaki, 1992).

The process of the phytosociological field study is illustrated in Fig. 49.2.

It is most important to select, for planting, the main tree species, or at least one of the main tree species with a potential to form a forest. In Japan's laurel forest zone, for example, the main tree species in lowland and alluvial areas are *Persea thunbergii* and *Castanopsis cuspidata* var. *sieboldii* in shallow soil, and evergreen *Quercus* species i.e. *Q. myrsinaefolia*, *Q. glauca* etc. in

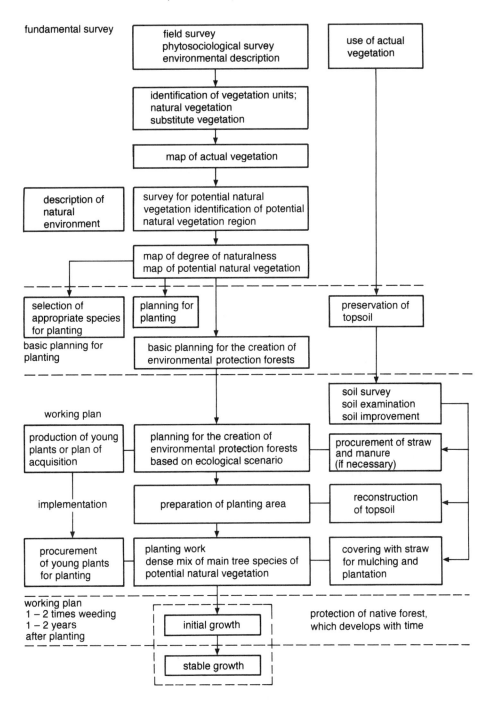

Fig. 49.2. Flow chart for the restoration and creation of native forests.

inland areas (Miyawaki, 1976, 1979, 1980-1989). Since the deep, straight roots of these species make transplanting difficult, the seeds are grown in containers for about 2 years until the seedlings are 30-50 cm high and the containers are filled with the root systems.

Drainage is important in areas where the annual rainfall exceeds 1000-1500 mm such as on the Japanese Archipelago. Low mounds of loose soil are created, therefore, along streets, boundaries between public facilities and industrial sites and railways to improve air and water drainage even if the width of the mound is only 1 m. Steep slopes of 30-45° along highways, for example, must be reinforced with bamboo or wood to make beds for seedlings. Mixed and dense planting of different species is the best way to restore natural diversity in urban areas. The number of seedlings is 2-3 m^{-2} in normal stands and 5-6 m^{-2} in beachfront and windy sites, to encourage a density effect. After the first winter, the seedlings planted in this manner are most likely to grow 1 m per year. In the first 2-3 years, weeds must be eliminated and then placed on the mound as extra mulch. After planting, mulching is provided, in Japan, by placing 4 $kg\,m^{-2}$ of rice straw around the seedlings. The mulching is effective in preventing weeds, and protecting the seedlings from cold and dryness. With mulching, erosion does not take place on slopes even under heavy rainfall. Decomposed straw contributes to the improvement of the soil. In three years the canopy develops and shields the forest bed, preventing the growth of weeds. After this point, it may be safely said that 'no management is the best management'.

Objectives and Results

Our method and technique have been successful throughout Japan from the summergreen broad-leaved forest zone, partly subalpine zone to the subtropical zone, and from the beachfront lowland to the summergreen broad-leaved forest zone with *Fagus crenata*, *Quercus mongolica* var. *grosseserrata*, *Q. dentata*, etc. at an altitude of 1000 m. On human-made islands and reclaimed land along urban and industrial complexes, railways, highways and around public and educational facilities, efforts were made to create native forests using native trees. The northernmost human-made native forest is at Sempaku industrial park in Hokkaido and the southernmost stand is at the second thermal power plant in Ishigakijima Island in the Okinawa Prefecture. It is always the citizens and school children who play the important role of planting seedlings according to the ecological scenarios (Miyawaki *et al.*, 1979a, 1993; Miyawaki and Golley, 1993). Efforts have also been successful overseas, in Kuala Lumpur and Malacca in Malaysia and in Bangkok. In Bangkok, for example, seedlings planted around a new Jusco shopping centre have grown to be 4 m in 20 months (Miyawaki, 1994).

Restoration of biodiversity has been successfully achieved in native forests by native trees throughout Japan since 1970. The achievements are highly valued in Japan as well as abroad (Miyawaki, 1993a,b; Miyawaki and Golley, 1963).

In subtropical Malaysia, an indigenous green environment and biodiversity has been restored around a shopping centre. Low native flowering shrubs are planted as an edge community along roads and alleys so that citizens can enjoy the blossoms as well as the greenery. In case of emergency, the native forests function as shields against typhoons, fires and earthquakes.

Conclusions and Further Development

Since 1990, more than 300,000 potted seedlings of 33 species of *Dipterocarpaceae*, the main tree species of the potential natural vegetation of Bintulu, Sarawak (Borneo) have been grown there. Currently, under the supervision of our group and the University of Agriculture in Malaysia, the seedlings are steadily growing and covering the part of the university campus (800 ha), which were once burned for farming purposes.

In Belem, in the lowland around the estuary of the Amazon River, the Miyawaki team and the University of Agriculture in Para formed a joint study group and created 7-m high native forests in 20 months. About 200,000 seedlings of *Virola* and 92 other species of main native trees have been grown in a year with the cooperation of the Mitsubishi Trading Company and EDB. The planting has passed the experimental stage. The native trees are growing steadily to recover the biodiversity of that part of the globe. Simultaneously, in Concepción in Chile, *Nothofagus* forests have replaced the conventional species of Australian *Eucalyptus* and American *Pinus taeda* with the support of the Chilean Government. *Nothofagus* trees were planted in areas which were burnt for the Purpose of farming or by accident. Recovery of biodiversity in native forests by native trees has started and so far been successful in the southern hemisphere as well.

In all of these instances, the government, local municipalities and the private sector are the producers and stay behind the scene. It is the citizens who play the main role of planting seedlings. The grass-roots movement for the recovery of biodiversity and green environments has started throughout the world (Miyawaki *et al.*, 1987).

References

Aoki, J. and Harada, H. (1985) Formation of environmental protection forests and changes in soil fauna, especially oribatid mites. *Bulletin of the Institute of Environmental Science and Technology Yokohama National University* 12, 125–135.

Aoki, J. and Harada. H. (1989) A Soil-Zoological Study of Urban Ecosystem around Tokyo. Research Report of the Special Expenditure on Education and Research, 1988. Ecological and Bio-technological Studies for the Conservation, Restoration and Creation of the Environment for Human Life, Yokohama National University, pp. 23–28.

Braun-Blanquet, J. (1928) *Pflanzensoziologie: Grundzüge der Vegetationskunde.* Berlin. 3. Auflage 1951 and 1964, Wien. New York.

Clements, F. (1916) *Plant Succession: An Analysis of the Development of Vegetation.* Carnegie Institution of Washington, Washington, DC.

Miyawaki. A. (1975) *Entwicklung der Umweltschütz-Pflanzungen und Ansaaten in Japan.* Sukzessionsforschung. Bericht über das Internationale Symposium der Internationalen Vereinigung für Vegetationskunde, Vaduz, Cramer, pp. 237-254.

Miyawaki, A. (1976) *Die Umwandlung immergrüner in sommergrüne Laubwälder in Japan.* Werden und Vergehen von Pflanzengesellschaften. Bericht über das Internationale Symposium der Internationalen Vereinigung für Vegetationskunde. Vaduz, Cramer, pp. 367-381.

Miyawaki, A. (1979) Vegetation und Vegetationskarten auf den Japanischen Inseln. *Bulletin of the Yokohama Phytosociology Society Japan* (Festschrift for Prof. R. Tüxen), Yokohama, 16, 49-70.

Miyawaki, A. (1981) Energy policy and green environment on the basis of ecology. Beyond the energy crisis opportunity and challenge. In: Fazzolage, R.A. and Smith, C.B. (eds) Oxford, New York, pp. 581-587.

Miyawaki, A. (1982) Umweltschutz in Japan auf vegetationsökologischer Grundlage. *Bulletin of the Institute of Environmental Science and Technology* Yokohama, Yokohama National University, 11, 107-120 (in German with English Synopsis).

Miyawaki, A. (1989) Restoration of evergreen broad-leaved forest (laurel forest) in Japan. In: Academy, Ch. (ed.), *The World Community in Post-Industrial Society,* 5. The human encounter with nature: destruction and reconstruction. pp. 130-147.

Miyawaki, A. (ed.) (1980-1989) Vegetation of Japan vols. 1-10 (each vol. 400-700pp with color vegetation maps and tables). Tokyo Shibundo. (In Japanese with German and/or English summary).

Miyawaki, A. (1993a) Restoration of native forests from Japan to Malaysia. In: Lieth, H. and Lobmann, M. (eds) *Restoration of Tropical Forest Ecosystems.* Kluwer Academic Publishers, Dordrecht, pp. 5-24.

Miyawaki, A. (1993b) Global perspective of green environments – restoration of native forests from Japan to Malaysia and South America, based on an ecological scenario. IGARSS '93, Better Understanding of Earth Environment, vol. 1, p. 6-8.

Miyawaki, A. (1994) Toward restoration of global green environments – from Japan to SE Asia and South America. *Japanese Scientific Monthly* 47(5), 6-17 (in Japanese).

Miyawaki, A., Fujiwara, K. and Box, E.O. (1987) Toward harmonious green urban environments in Japan and other countries. *Bulletin of the Institute of Environmental Science and Technology Yokohama National University* 14, 67-82.

Miyawaki, A., Fujiwara, K., Nakamura, Y. and Kimura, M. (1983) Ökologische und Vegetationskundliche Untersuchungen zur Schaffung von Umweltschutzwäldern in den Industrie-Gebieten Japans. *Bulletin of the Yokohama Phytosociology Society* 22, I. 84pp, II. 151pp, (In Japanese with German and English summaries).

Miyawaki, A., Fujiwara, K. and Ozawa, M. (1993) Native forest by native trees – restoration of indigenous forest ecosystem – (Reconstruction of environmental protection forest by Prof. Miyawaki's method). *Bulletin of the Institute of Environmental Science and Technology Yokohama National University* 15, 95-102 (Japanese and English).

Miyawaki, A. and Golley, F.B. (1993) Forest reconstruction as ecological engineering. *Ecological Engineering* 2, 333-345.

Miyawaki, A., Harada, H. and Ude, H. (1979) Vegetationskundliche Untersuchung zur Schaffung von Umweltschützwäldern um Industrie-Anlagen, erläutert am Beispiel der 11 Fabriken der Tore-Industrie-AG, *Bulletin of Yokohama Phytosociology Society* 8, 50pp, (In Japanese with German summary).

Miyawaki, A., Tohma, H. and Suzuki, K. (1979b) Pflanzensoziologische Untersuchungen der Shinto-Schrein- und Buddistischen Tempelwälder in der Präfektur Kanagawa (Hauptstadt Yokohama), (In Japanese with German summary), The Board of Education of Kanagawa Prefecture. Yokohama, 180pp.

Tüxen, R. (1956) Die heutige potentielle natürliche Vegetation als Gegenstand der Vegetationskartierung. *Angewandte Pflanzensoziologie, Stolzenau/Wes.* 13, 5–42.

50 Synurbization of Animals as a Factor Increasing Diversity of Urban Fauna

M. Luniak

Museum and Institute of Zoology (Polish Academy of Sciences), Wilcza 64, 00-679 Warszawa, Poland

Introduction: The Term and The Subject

The term 'synurbization' has been introduced by zoologists/ecologists (Andrzejewski *et al.*, 1978) as an analogy to the commonly used and more general term, 'synanthropization', which denotes coexistence of wild animals (or plants) with man. Synurbization (in respect to animals) means colonization of urban areas and adjustment to specific conditions of urban enviroment. This term is applied to whole local groups (populations) of a given animal species, which regularly occur in towns or cities, but not to individual animals which have come there (or have been brought by humans) accidentally. The synurbization phenomenon has been described mainly for birds and mammals.

Human settlements were always inhabited by various wild animals. Some of them are so well adapted to 'urban life' that they exist there more successfully than in their primary natural habitats. Examples are house sparrow (*Passer domesticus*), feral pigeon (*Columba livia domestica*), some species of leafhoppers (Auchenorryncha, Homoptera) or aphids (Aphididae). Some of these animals are, for humans, unwanted neighbours, e.g. rats, mice, and all kinds of insects in our houses. Nevertheless, urban fauna consists of a low number of species as compared to that found in natural or rural habitats. But recently (during the last few decades) more and more animal species, represented by an increasing number of populations, have colonized cities. This phenomenon is a topic of growing interest for scientists and specialists in applied urban ecology, because it suggests the opportunity for coexistence of fauna with global expansion of urban development – and that means the hope for enriching nature in our cities. Thus, the need has arisen to name the phenomenon by creating a new term: synurbization.

City as a Habitat for Wild Animals

Intensive urban development (urbanization) has created, in all parts of the world, an entirely new type of landscape. Cities are sites of extreme human impact on nature. Wildlife, particularly animals, face very specific conditions of existence in urban environments. General characteristics of urban ecosystems have been given by several authors (e.g. Numata, 1976; Andrzejewski et al., 1978; Klausnitzer, 1987; Philipson, 1989; Sukopp, 1981, 1990; Zimny, 1994). The most distinctive features of the city as a habitat of wild animals, in comparison to non-urban habitats, are as follows:

1. Changes in mesoclimatic and hydrological conditions: higher temperatures, reduced humidity, lower groundwater levels, reduced wind speed and insolation, artificial lights at night. In general, the urban mesoclimate is milder and drier. Furthermore, microclimates inside buildings, pipes, tunnels and other technical structures where animals can live, are often highly independent of outdoor atmospheric conditions. This creates favourable conditions for wintering of animals and for the existence of exotic species from warmer and drier regions, but many local species cannot exist in such modified conditions.

2. Rapid and radical degradation of biotopes as a result of urban development and management disturbs, or completely destroys, historical continuity and survival of animal communities. It also elminates many elements of original fauna.

3. Concentration of technical structures and traffic increases fragmentation and isolation of animal habitats and reduces the space favourable for existence of wildlife. This factor strongly restrains distribution of fauna and also causes direct destruction of animals (e.g. by various traps, thermal and electric shocks, and animals are hit by vehicles). Instead, new artificial sites are being formed. They are occupied by some local and alien species. For example, several birds nest on buildings: cockroaches, mice and rats live inside houses and spread through sanitary installations.

4. High levels of pollution in air, water and soil, and of chemicals, dust and garbage. This factor affects animals directly (poisoning) or indirectly (e.g. by changes in vegetation). It determines existence conditions particularly for the soil and water fauna.

5. Large amount of human food accessible for wild animals in storage, rubbish and sewage, as well as given directly (mainly to birds) by people. This creates favourable feeding conditions for some species, which become independent from natural resources.

6. Intensive gardening and other human impacts upon soil and vegetation (e.g. trampling) causes impoverishment of its natural components and increases the portion of cultivated and exotic plants. Horticultural practices remove leaf litter and old trees, and also reduce understorey. These factors change the habitats of animals towards ecological degradation and decreasing of their natural food resources. They also enable settling of exotic species.

7. Permanent human presence and domestic animals have direct impacts on the fauna in various ways. Humans may frighten or even kill wild animals, but may also protect them or improve their natural conditions. Dogs and cats are raptors to birds and mammals, but they do not depend on their prey, which is contrary to the essential principle of natural zoocoenosis functioning.

Intensity and mode of impact of the above factors vary significantly within the wide range of urban biotopes. For instance, in peripheral forest parks or in large city parks with old treestands, habitat conditions for wild animals are, of course, more natural than in the densely built-up areas in the city centre. Extreme cases of limitation of natural factors are the artificial habitats of building interiors and sanitary installations (ventilation, drainage and sewage systems).

In general, cities are for wild animals an environment (complex of habitats) which has no history in terms of zoological evolution, and where natural structures and functions of the ecosystem are extremely reduced by anthropogenic factors. Expanding urban areas are for animals a new site with conditions (barriers and advantages) which are often new for them.

Urban Zoocoenoses

Urban development eliminates many, sometimes almost all, species from local fauna living in the region's original habitats. However, some species adapt to life in cities and even to drastically altered sites. Specific urban environments also create opportunities for the settling and spreading of alien elements which are geographically or ecologically different from native fauna.

In general, urban zoocoenoses (animal communities), in spite of all limitations, are much richer and more abundant than one could expect. In cities of Central Europe, 100–200 bird species and about 30 mammal species regularly occur. For example, in Berlin (West), during the last four decades 270 bird species (124 of them breeding in the urban area) and 50 mammal species (43 breeding) have been recorded (Sukopp, 1990). Recent zoological investigations in Warsaw (Pisarski, 1982; author's data) recorded over 2000 animal species. This number includes only certain taxonomic groups of invertebrates (mainly insects). The high diversity of these animals has not allowed us, so far, to list in any city their complete range of species. A total of 207 species of birds (142–144 breeding) have been recorded recently in Warsaw, and 41 species of mammals, 23 of them permanently residing in the city and fairly widespread (author's data).

In spite of high absolute numbers of species, urban zoocoenoses are generally characterized by significantly lower species diversity, as compared to animal communities of natural biotopes in non-urban areas. Studies carried out in Warsaw (see above) indicated the presence of approximately 40% of invertebrate species and two-thirds of breeding birds, in relation to the number of species found in the region of Mazovia, where Warsaw is situated. However, in the most urbanized area of the city centre, the respective proportions were lower and amounted to 15% for invertebrates and one-third for birds.

Declining species diversity of urban fauna is especially conspicuous when its quantitative structurer is analysed; this is highly disproportional in urban zoocoenoses. A few species, sometimes only one, become super-dominant, being much more abundant than other community components, and also compared to the population density of the same species in non-urban areas. For example, in the central areas of Warsaw (sample 53 km^2), two super-dominants – the feral pigeon and house sparrow – accounted for 73% of the total breeding bird population. In winter, when combined with the rook (*Corvus frugilegus*), this proportion rose to 82%. Similar examples concerning insect conununities in Warsaw were given by Trojan et al.(1982). They found that among 23 wasp (Sphecidae) species, 14 mosquito (Culicidae) species or 10 leafhopper (Auchenorrhyncha) species, single dominant species in the city centre couprised 90% or more. In natural zoocoenoses, quantitative proportions between species are much more balanced – proportions of particular dominants rarely exceed 10%.

Urban zoocoenoses often have higher population densities than non-urban conununities, which is connected with mass occurrence of dominant species. Studies in urban parks of Warsaw (Pisarski, 1982) found on average 2800 (or 730 per 100 leaves) insects and other invertebrate individuals in 1 m^3 of tree crowns. In 1 m^2 of lawns they found 940 individuals, and in 1 m^2 of soil about 40,000 individuals. The average density of bird population per 1 km^2 in highly urbanized areas of Warsaw (sample 53 km^2) was estimated at 1250 pairs (435 kg of biomass) in the breeding season and 3690 individuals (964 kg) in winter (author's data). The total biomass of urban bird communities may be many times higher than that in rural areas. This is linked in particular with large populations of feral pigeons, some corvid species (Corvidae) and mallards (*Anas platyrhynchos*), feeding on the abundant remains of human food. A classical example in this respect was given by Nourteva (1971), who found that in the city centre of Helsinki, the number of bird species was reduced to one half or one third in comparison to its surroundings, but total biomass of the bird community was as much as 20–30 times higher than in non-urban areas, reaching up to 250 kg km^{-2}.

A high proportion of exotic components is a very specific feature of urban fauna. In Warsaw (as well as in other European cities) both super-dominant and the 'most urban' vertebrate species – feral pigeon, house sparrow, domestic mouse (*Mus musculus*) and brown rat (*Rattus norvegicus*) – are of non-native origin. Among invertebrates, the best examples are common inhabitants of house interiors – cockroaches (*Blatta orientalis, B. germanica*) or pharaoh's ant (*Monomorium pharaonis*), which are alien species of southern origin. Examples from Warsaw (Pisarski, 1982) and other cities (Klausnitzer, 1987) show that, in comparison to non-urban areas, urban fauna includes a higher share of southern and xerothermic species, which find favourable conditions in the warm and dry urban mesoclimate. Also, the proportion of species (their number and abundance) demonstrating wide ecological valence – opportunists concerning their food and habitat needs – is increased. Specialized species and those located in higher positions of the food chain (typical raptors) are less represented. As regards the developmental strategy, 'r' type prevails over the 'K' type. Urban areas are more likely to have animal species which are less specialized in their ecological requirements, which produce more offspring, but with a relatively short longevity. Such a species adapts better to altered urban conditions.

The above examples show that specific features of urban zoocoenoses are: low species diversity, extremely high abundance of some species and some communities, disproportional structure and high proportion of exotic components and of less specialized species.

Synurbic Populations

The phenomenon of synurbization (i.e. adjustment of animal populations to specific conditions in cities) is a response of fauna to urban development. It has been described mainly for birds and mammals so far, but differences between urban and non-urban populations have also been studied for other animal groups, e.g. frogs (*Rana temporaria* and *R. arvalis*) by Vershinin (1990), or Egyptian mosquitos (*Aedes aegypti*) by Crovello and Hacker (1972).

In Warsaw, during the past few decades, at least a dozen bird species and two mammal species (rabbit, *Oryctolagus cuniculus* and striped field mouse, *Apodemus agrarius*) have settled in urbanized areas as new elements of fauna. The majority of them are spread out in numerous populations over the whole city. A similar number of new bird species settled in Moscow (Ilyichev et al., 1987) and in St Petersburg (Khrabriy, 1991). Studies carried out in 27 cities in Central and Eastern Europe (Luniak, 1990) confirmed this tendency in all cases investigated. Among birds, the most common examples of increasing synurbization in Central/Eastern Europe are: blackbirds (*Turdus merula*), mallards (*Anas platyrhynchos*), wood pigeons (*Columba palumbus*), hooded crows (*Corvus corone cornix*), magpies (*Pica pica*), black-headed and common gulls (*Larus ridibundus, L. canus*), jays (*Garrulus glandarius*), fieldfares (*Turdus pilaris*), and redwings (*Turdus iliacus*). Among mammals, one of the best-known examples of synurbization during the past decades is the red fox (*Vulpes vulpes*) in cities of Western Europe.

Synurbic populations show significant distinguishing features as compared with populations of the same species living in non-urban areas. Many studies and assessments concerning this topic (e.g. Tomiatojć, 1976; e.g. Andrzejewski et al., 1978; MacDonald and Newdick, 1982; Ilyichev et al., 1987; Luniak et al., 1990; Gliwicz et al., 1994) show that the most typical changes in the ecology and behaviour of birds and mammals living in cities include the following.

1. There is considerably higher density of population connected with reduction of individual (pair, family) territories. Density of urban blackbirds is 10–20 times higher than forest ones. For magpies (in Poland), density is 20–50 times in comparison to non-urban areas, for wood pigeons (Western Poland) 10–30 times higher, for hooded crows (Moscow) 20–60 times higher, for striped field mice (Warsaw) 5–10 times higher, and for the red fox (England) density is at least 10 times higher.

2. Migratory behaviour is reduced. For example, synurbic blackbirds and corvid birds often spend winters in their breeding places, whereas non-urban populations of these species migrate long-distance to winter quarters. A strong tendency to sedentary life is also a distinguishing feature of urban field mice and foxes.

3. The breeding season is prolonged. Urban blackbirds begin breeding 1–4 weeks earlier than forest ones in the same climatic zone, and their last fledglings leave nests 4 weeks later. Even cases of winter broods of some bird species were recorded in Central European cities. Field mice in Warsaw still show sexual activity in October, and even in winter, which is observed very rarely in rural population of the species.

4. There is greater longevity, e.g. in urban blackbirds, of 1–1.8 years. This is mainly due to better winter survival (favourable climate and food conditions in cities), a more sedentary life (migrations are exhausting and dangerous) and lower threat of predators (majority of raptor species avoid cities). There is evidence showing that natural selection in urban populations is weaker than in 'natural' ones.

5. Circadian activity is lengthened. Urban birds begin singing and feeding earlier in the morning and finish later in the evening. In blackbirds, even cases of nocturnal activity are known, which is never observed in forest populations. Feral pigeons, active at night, are commonly seen in cities. Field mice, which in natural habitats are strictly nocturnal, show diurnal activity in urban parks.

6. There are changes in nesting habits. Birds and mammals in cities use a great variety of human-made objects for shelters, nesting places and material for nests. The majority of bird populations in cities use buildings and other technical objects as nesting sites. A general tendency of urban birds is to locate their nests higher than is observed in non-urban populations. For blackbirds, the mean difference is about 1 m.

7. Changes are seen in feeding behaviour. Many bird and mammal species include anthropogenic foods in their diet. For some of them (feral pigeon, house sparrow, mallard, corvids in winter, collared dove, *Streptopelia decaocto*) this is the main source of food. Their behaviour is closely adjusted to human customs and it is aimed at finding or receiving food from man.

8. There is tameness toward people. Birds and mammals (e.g. squirrels, *Sciurus* sp.) living in cities have considerably reduced their distance of escape. In many cases they come to people (see paragraph above), and sometimes even sit on them. Such behaviour is not typical in 'natural' populations of the same species.

All of the above are examples of specific adjustments of synurbic populations. These changes seem to be within the range of natural plasticity of the species, and they allow these animals to exist successfully in cities. The problem of genetic differentiation between urban and non-urban populations has not yet been clearly solved. Some genetic differentiation is suggested by certain morphological, anatomical and physiological differences found in striped field mouse populations (Gliwicz, 1980), in field experiments on the introduction of blackbirds into cities (Graczyk, 1982), as well as in recent laboratory tests on the behaviour of urban and forest blackbirds (Walasz, 1990). Similar results were obtained in studies on invertebrate species (Crovello and Hacker, 1972; Owen, 1961). Whatever the conclusive answer for this essential question will be, the empirical picture is clear: Some animal species show surprising abilities for successful coexistence with urban development.

The Chance for Enriching Urban Wildlife

The growing tendency towards synurbization observed recently in an increasing number of bird and mammal species is an optimistic chance for improving diversity of urban wildlife. But this is only one side of the matter. The global ecological crisis affects fauna in cities as well as in the whole animal world. For example, in the area of united Berlin, out of 170 known breeding bird species, 97 (57%) are included in the Red List, of which 29 (17%) have become extinct, 28 (16%) are threatened by extinction, and 12 (7%) are highly endangered (Witt, 1991). Berlin is a typical example for other European cities.

On the other hand, synurbization of some species could cause practical problems when some populations (super-dominants, typical for urban zoocoenoses) grow to disproportional concentrations. Examples of such problems are roosting of starlings (*Sturnus vulgaris*) and corvids in cities. During winter roosting, the concentration of rooks, crows and jackdaws (*Corvus monedula*) in Warsaw is estimated to be about 250,000 birds (Luniak, 1990), and in Moscow, 800,000 crows (Ilyichev et al., 1987). Feral pigeons, gulls, starlings, and sometime mallards cause well known problems in many cities. A recent example of this kind is the Canadian goose (*Branta canadensis*) in some cities of North America. Hundreds of these birds pollute parks, golf courses and beaches.

Existing scientific knowledge and practical experience offer a wide range of possibilities for the management of animal communities in urban areas (i.e. controlled stimulation of synurbization processes) and even for making cities as places for the conservation of some endangered species. The general strategy for such measures should have as an aim the creation or protection of zoocoenoses which should be:

1. Diversified and include components important for improving the poor structure and functions of urban ecosystems;

2. Productive and abundant, but maintaining quantitative balance between their components;

3. Adequate to cope with urban conditions, with specific urban biotopes and with human and technical factors of the city system.

Managing urban zoocoenoses should be considered as part of a general ecological strategy concerning the whole system of the city because animal communities, and all wildlife (biocoenosis), are elements of the urban ecosystem. It means that managing zoocoenoses (stimulating synurbization) should support natural structure and functions of the ecosystem, with ecological and social needs of man in mind. A city is mainly a habitat of man and the management of urban wildlife should have that in view.

General recommendations derived from statements above are as follows.

1. For urban planning: to protect sites and natural elements of particular ecological value, to follow the natural structure of the landscape and that of habitats, to maintain (or to create) spatial continuity in ecosystems, e.g. in the form of 'green corridors'.

2. For operations of development and usage: to minimize destruction of natural habitats or particular elements (e.g. old trees); to introduce new elements favourable for wildlife (e.g. small ponds, nest-boxes).

3. For management of green areas: to create rich and diversified plant cover, thus increasing primary production and creating habitats for animals. Locally indigenous plant species and natural vegetation should be preferred. But plant cover should be able to cope with existing and expected characteristic urban conditions.

4. For social politics: to develop ecological education aimed at understanding nature and promoting friendly coexistence with wildlife. Public awareness and public involvement in the protection and management of wildlife is particularly important.

Practical measures for the application of the general recommendations above are given in several manuals (Baines, 1985; Emery, 1986; Sukopp and Werner, 1987; Adam and Dove, 1989; Barker and Graf, 1989).

With our current scientific knowledge and practical experience, it is possible that cities will be the home of a rich fauna and even play an important role for some elements of wildlife. Such a direction of urban development is in full accordance with the ecological and social needs of people living in cities. Recent years have brought a growing understanding of this by decision-makers and by a wide urban public. Growing activity in this field is indicated by a growing number of scientific meetings and publications (e.g. in Germany, Poland, USA, UK), specialized journals devoted to urban wildlife (e.g. published in UK, Italy, USA), the urban ecological park movements expanding quickly in countries like UK, Germany and The Netherlands. This activity is stimulated by the International Network for Urban Ecology – a working and coordination body of INTECOL and UNESCO. All this shows that progress is being made in the field discussed. However, there is still a long way to go before people in cities experience full enjoyment of a rich wildlife around them and the full chance of synurbization is given to animals.

Conclusions

1. The new term 'synurbization' has been introduced to name an increasing phenomenon of animals settling in cities connected with adjusting animal populations to specific urban conditions.

2. Global expansion of urban landscapes is for animals a new ecological space posing living conditions (barriers and advantages) which often are new for them.

3. Some urban habitats support relatively rich animal life. Specific features of urban zoocoenoses (animal communities) are: low species diversity, often high abundance, disproportional structure with few dominant species, high proportion of exotic components.

4. Some animal species show a high capacity to adjust to specific conditions of urban habitats. Among the best known differences in the ecology and behaviour of synurbic populations, in comparison to non-urban ones, are: higher density, more sedentary life, prolonged breeding season, better survival (higher longevity, changes prolonging circadian activity, changes in nesting and feeding habits), use of human-made objects and human foods, high tameness and adjustments to human behaviour.

5. Synurbization is an opportunity to enrich wildlife in cities – the main habitat of humans – and to encourage the coexistence of fauna with global expansion of urban development. Existing scientific knowledge and practical experience offer wide possibilities to stimulate synurbization in full accordance with ecological and social needs of man.

References

Adams, L.W. and Dove, L.E. (1989) *Wildlife Reserves and Corridors in the Urban Environment*. US Department of Interior, Washington, DC.

Andrzejewski, R., Babinska-Werka, J., Gliwicz, J. and Goszczynski, J. (1978) Synurbization processes in an urban population of *Apodemus agrarius*. I. Characteristics of population in urbanization gradient. *Acta Theriologica* 23, 341–358.

Baines, C. (1985) *How to Make a Wildlife Garden*. Hamish Hamilton, London.

Barker, G. and Graf, A. (1989) *Principles for Nature Conservation in Towns and Cities*. Nature Conservancy Council, Peterborough.

Crovello, T.J. and Hacker, C.S. (1972) Evolutionary strategies in life table characteristics among feral and urban strains of *Aedes aegypti* (L.). *Evolution* 26, 185–196.

Emery, M. (1986) *Promoting Nature in Cities and Towns*. Croom Helm, London.

Gliwicz, J. (1980) Ecological aspects of synurbization of striped field mouse *Apodemus agrarius* (Pall.). *Wiadomosci Ekologiczne* 26, 117–124.

Gliwicz, J., Goszczynski, J. and Luniak, M. (1994) Characteristic features of animal populations under synurbization – the case of blackbird and of the striped field mouse *Memorabilia Zoologicae* 49, 237–244.

Graczyk, R. (1982) Ecological and ethological aspects of synanthropization of birds. *Memorabilia Zoologicae* 37, 79-91.

Ilyichev, V.D., Butiev, V.T. and Konstantinov, V.M. (1987) *Birds of Moscow and Vicinity*. Izd. Nauka, Moscow.

Khrabriy, V.M. (1991) *Birds of Saint-Petersburg - Fauna, Conservation*. Izd. Zoologiceskiy Inst. AN Ros., St Petersburg.

Klausnitzer,B (1987) *Oekologie der Grossstadtfauna*. VEB G. Fischer Verlag, Jena.

Luniak, M. (1990) Avifauna of cities in Central and Eastern Europe - results of the international inquiry. In: Luniak, M. (ed.) *Urban Ecological Studies in Central and Eastern Europe*. Ossolineum, Wroclaw, pp. 132-149.

Luniak, M., Mulsow, R. and Walasz, K. (1990) Urbanization of the European blackbird - expansion and adaptations of urban population. In: Luniak, M. (ed.), *Urban Ecological Studies in Central and Eastern Europe*. Ossolineum, Wroclaw, pp. 187-199.

MacDonald, D.W. and Newdick, M.T. 1982. The distribution and ecology of foxes, *Vulpes vulpes* (L.) in urban areas. In: Bornkamm, R., Lee, J.A. and Seaward, M.R.D. (eds) *Urban Ecology*. Oxford University Press, Oxford, pp. 123-138.

Nourteva, P. (1971) The synanthropy of birds as an expression of the ecological cycle disorder caused by urbanization. *Annales Zoologici Fennici* 8, 547-553.

Numata, M. (1976) Methodology of urban ecosystem studies. In: Numata, M. (ed.) *Studies in Urban Ecosystems*. Tokyo University Press, Tokyo, pp. 1-14.

Owen, D. (1961) Industrial melanism in North American moths. *American Nature* 95, 227-233.

Philipson, J. (1989) Urban ecosystems: soils, soil fauna and productivity. In: *International Scientific Workshop on Soils and Soil Zoology in Urban Ecosystems as a Basis for Management and Use of Green/Open Areas*. UNESCO MAB Program, Berlin, pp. 101-123.

Pisarski, B. (1982) La faune de Varsovie, sa composition et son origine. In: Luniak, M. and Pisarski, B. (eds) *Animals in Urban Environment*. Ossolineum, Wroclaw, pp. 103-114.

Sukopp, H. (1981) Oekologische Charakteristika der Grossstadt. In: Klausnitzer, B. (ed.) *Tagungsber. I Leipziger Symp. Urbane Oekologie*. Karl Marx Unw. Leipzig, Leipzig, pp. 5-12.

Sukopp, H. (1990) *Stadtoekologie - das Beispiel Berlin*. Dietrich Reimer Verlag, Berlin.

Sukopp, H. and Werner, P. (1987) *Development of Flora and Fauna in Urban Areas*. Nature and Environment series No. 36. Council of Europe, Strasbourg.

Tomiatojć, L. (1976) The urban population of the wood pigeon *Columba palumbus* Linnaeus, 1758 in Europe - its origin, increase and distribution. *Acta Zoologica Crucov*. 21, 586-631.

Trojan, P., Górska, D. and Wegner, E. (1982) Processes of synanthropization of competitive animal associations. *Memorabilia zoological* 7, 125-135.

Vershinin, V.L. (1990) Features of amphibian populations of an industrial city. In: Luniak, M. (ed.) *Urban Ecological Studies in Central and Eastern Europe*. Ossolineum, Wroclaw, pp. 112-121.

Walasz, K. (1990) Experimental investigations on the differences between urban and forest blackbirds. *Acta Zoologica Cracoviensia* 33, 235-271.

Witt, K. (1991) Rote Liste des Brutvoegel in Berlin, 1. Fassung. *Berl. ornithol. Ber.* 1, 3-15.

Zimny, H. (1994) The city as an ecological system and its impact on environmental quality. *Memorabilia Zoologicae* 49, 21-125.

51 | Biodiversity Management in Peri-urban Environments in Switzerland

B. Schmid
Institut für Umweltwissenschaften, Universität Zürich, Winterthurerstrasse 190, CH-8057 Zürich, Switzerland

Introduction

To put the 1992 Convention on Biodiversity to work we may ask 'Why do we need biodiversity?' and 'What should we do about it?' In this chapter, I first present the problem and the research needs in peri-urban environments. I will then use an interdisciplinary case study to demonstrate the scientific diagnosis and finally suggest some strategies for therapy.

The major factors that change our global environment are the exponential growth of population and economy, leading to an unprecedented level of resource consumption by the human species (Daly, 1992). At present, the human population uses about one quarter of the primary production on earth. It is only because population growth is slow in the 'North' and economic growth is slow in the 'South' that the situation is not even worse.

The increased consumption by one species obviously leaves less resources for the remaining 5–50 million species or so. In contrast to severe alterations of the physical and chemical environment by humans, the loss of every species or variety is permanent and can not be corrected by the application of environmental technology. The level of biodiversity may be the best measure of the state of our planet.

Having identified the problem, what are the consequences? To answer this question requires research. The need for such research is particularly great in peri-urban environments. One reason for this is that peri-urban environments harbour some of the richest cultural landscapes at least in Switzerland and other parts of Europe. Further, however, peri-urban environments suffer from high human impacts such as pollution and habitat fragmentation and exert a high impact on humans. Here the proper management of biodiversity can not

© 1996 CAB INTERNATIONAL. *Biodiversity, Science and Development: Towards a New Partnership* (eds F. di Castri and T. Younès)

only have the greatest effect on our daily quality of life, it can also, for logistic reasons, best be implemented and monitored, and, not least, have a high educational value.

A Swiss Case Study on Biodiversity

For the above reasons a group of about 50 researchers in Switzerland decided two years ago to study peri-urban dry calcareous grasslands near Basel as a model system. These species-rich ecosystems are endangered because of high human impact, but at the same time there is high public motivation that sound environmental management should be introduced.

The guiding questions for the Swiss case study on biodiversity are the following:

1. How is biodiversity generated and maintained in nature?
2. Which anthropogenic factors reduce biodiversity?
3. How much biodiversity is needed to 'run nature' (climate, material cycles, energy flows, soil processes)?
4. Are we allowed to destroy biodiversity if we do not know the consequences?
5. How can we protect biodiversity?

The main threats to peri-urban areas in Switzerland are that habitat destruction and fragmentation lead to reduced biodiversity and that this reduced biodiversity negatively affects ecosystem functions and also the response to environmental change, for example the rising level of CO_2 (Fig. 51.1). Although rising CO_2 is a truly global problem, it is particularly relevant to study it in peri-urban areas because they themselves contribute over-proportionally to the so called greenhouse effect and because here this effect is added on to an already high impact load (Odum, 1989). The hypotheses depicted in Fig. 51.1 run as follows: fragmentation leads to reduced population sizes of animal and plant species, populations below some minimal size have a low survival probability leading to reduced diversity, systems with a lower diversity have a lower capacity to respond to environmental variation by internal buffering.

Habitat fragmentation in peri-urban environments is often caused by road building, condominium development, and intensive agriculture. Animals may not be able to cross barriers and therefore migration among and colonization of fragments is not possible. Both habitat loss and fragmentation have most severely reduced the calcareous grassland of the formerly typical peri-urban landscape in northwestern Switzerland. To test how fast biodiversity may be lost in these 'islands' we set up an experiment in which grassland fragments of different size are maintained by frequent mowing of the surrounding vegetation. Will there be increased fluctuations of plant and animal populations in the islands? Does the level of inbreeding increase and that of genetic diversity decrease? Will the rare and specialized species die out first? Only if we can provide answers to these scientific questions, that is only if we can diagnose

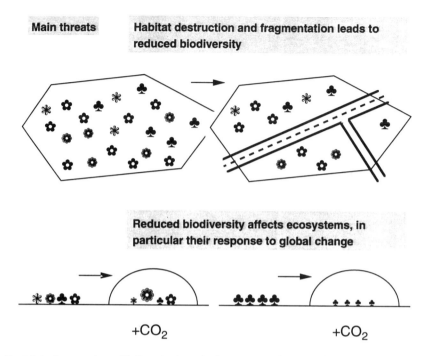

Fig. 51.1. Case study on biodiversity in peri-urban areas in Switzerland.

the illness, only then will it be possible to suggest a promising therapy in the form of a management plan.

The first results of the fragmentation experiment (J. Joshi, unpublished data) show that, as expected, the plant species number in the investigated grassland declines with a reduction in area. Several animal species declined within less than 6 months after fragmentation. Similarly, the fertility of the characteristic grass species was significantly reduced, possibly due to inbreeding. These early results already clearly indicate that reduced biodiversity is a likely consequence of habitat fragmentation.

In peri-urban areas the reduced biodiversity coincides with other anthropogenic factors leading to damaged ecosystems that are particularly vulnerable to the effects of global change. For example, ecosystem stability and productivity may be reduced, 'weedy' species may gain dominance, the food quality for animals may deteriorate, and soil processes may be disturbed. Whether reduced biodiversity can indeed have these effects can only be assessed by experimentally manipulating it and exposing manipulated plots to conditions expected under a global-change scenario. We are doing this in a biodiversity-by-CO_2 experiment using the same model system as in the habitat-fragmentation experiment. If our experiment does yield the expected scientific facts, they would provide strong support for plans of action to maintain biodiversity and to reduce CO_2.

Less than one year after the start of the experiment we found that CO_2 assimilation and carbon gain decreased with reduced biodiversity, i.e. from ecosystems containing 31 plant species to those containing only five species (P. Leadley, unpublished data). Whereas for these productivity measures the responses were parallel in ambient and elevated CO_2, the effects of reduced biodiversity became more pronounced under elevated CO_2 for some other measures. Also, depending on the level of biodiversity, particular species and genotypes showed fitness gains or losses under elevated CO_2, and in one case this could be traced back to changed interactions with symbiotic fungi in the soil (mycorrhizas) (A. Birrer, I. Sanders and T. Steinger, unpublished data). The first results of this experiment provide evidence that reduced biodiversity can disturb ecosystem functions under natural situations, a scientific fact of which we were completely ignorant at the start of the study.

Conclusions Rergarding Biodiversity Management

Our diagnosis of the problem is severe, because the series of events appears to trigger a vicious cycle of habitat change–reduced biodiversity–disturbed ecosystems–further habitat change–etc. that eventually may even disturb human culture. What therapy can we use to disrupt this vicious cycle before it is too late?

Clearly the problems should be tackled at the source, but the source can only be known from rigorous scientific study. The results of scientific study should be made available to the public; in particular, we should educate our children. In addition to the scientific research we need research on how to administer therapy: What are the perceptions and valuations of biodiversity? Which protective measures will be accepted? Do we have the necessary economic, political, and legal instruments? Management recommendations are the goal of the presented case study, but so far we have only had two years to develop the scientific basis.

No explicit goals of biodiversity management in peri-urban environments have been stated for Switzerland. For example, should ecosystems and landscapes be preserved at the level of 1990, or should they even be brought back to the level of 1950? The lack of agreement is not surprising considering the lack of scientific arguments. Nevertheless, I shall present some first views on management strategies in peri-urban environments in Switzerland. These views should mainly be considered as suggestions for further discussion:

1. Implement a 'rolling planning' of research and management to use scientific facts as soon as they are available.

2. Where these scientific facts are missing or contradictory, the opinion of, say 20 independent expert scientists, should be asked.

3. As a safety rule and to increase diversity: single, uniform strategies should always be avoided.

4. Management should use experimental designs to provide data for control and prediction of success.

In summary I would like to repeat that as long as we are almost completely ignorant about the causes and consequences of biodiversity, its general and rigorous protection is the only reasonable survival strategy. However, this argument by itself may be of little value if there are important reasons against protection. We do need to greatly intensify the scientific study and the development of management strategies to retard the continuing loss of biodiversity while we still have a good proportion of it. A prime area for both are peri-urban environments.

References

Daly, H.E. (1992) Steady-state economics: concepts, questions, policies. *GAIA* 1, 333–338.

Odum, E.P. (1989) *Ecology and Our Endangered Life-Support Systems*. Sinauer Associates, Sunderland, Massachusetts.

General Considerations on the Biodiversity of Urban and Peri-urban Environments

G. Vida
Department of Genetics, Eötvös University, Muzeum krt. 4/a, H-1088 Budapest, Hungary

Introduction

Urban and peri-urban environments can provide a rich source of examples of the biodiversity confusion. Unless it is clearly stated which level of biological organization (from genes to ecosystem, cf. Solbrig, 1991) and what kind of diversity measurements are referred to, contradictory examples can be cited to prove or disprove a certain statement.

Sound biodiversity measures should reflect at least three different attributes of the system studied (Vida, 1994):

1. The number of distinguished units (such as the number of allelic forms in genetic diversity or the number of species in species diversity);
2. The proportions (or frequencies) of these units;
3. The distances among these units.

Estimated biodiversity values always include (1), sometimes (1) and also (2), but only exceptionally consider (1), (2) and (3) together. The most preferred Shannon and Simpson indices incorporate (1) and (2) only.

Large cities can exhibit a very high species diversity of higher plants and animals if we include botanical gardens and zoos. However, this high biodiversity disappears as soon as we consider the frequency values, many species being represented by less than a dozen individuals.

If biodiversity qualifications are not based on precise calculations of (1), (2) and (3) attributes, they should at least be in accordance with the expectation of such a triple assessment.

Special Features of the Urban Environment

Urban environment shows the highest density of human population at the expense of thousands of other species, formerly present in the area (Murphy, 1988). This radical transformation has created several empty niches which invite spontaneous immigration (SI). In addition, humankind's need for a better environment resulted in intentional introduction (II) of several other species which are usually dependent upon our assistance.

Category SI can be regarded as the spontaneous urban biodiversity. It can be subdivided into four groups (in relation to our own species), such as:

1. Parasites (from viruses and bacteria to fleas and bugs);
2. Competitor 'pests' feeding on our food and other organic materials (several bacteria, fungi, house flies and other insects, rats, mice, etc.);
3. Tolerated neutralists (algae, mosses, lichens, weedy species, several lower invertebrates, some birds, etc.);
4. Supported 'mutualists' (songbirds, squirrels and other free-living pampered species).

Category II consists of species of pet animals, house and garden plants plus some specific collections like botanical gardens and zoos, city parks, and avenues. Many of these species cannot reproduce in the urban environment or at least require some contribution outside the cities. This part of the urban biodiversity is actually an 'extended' biodiversity of other biomes of the biosphere.

Genetic Diversity of Populations Living in Urban Environments

Spontaneous immigrants (SI) are often depleted in genetic diversity for two reasons. First, they colonize the urban environment with very few individuals, second, many pest species are frequently decimated as they are a nuisance to humans. Such populations are regarded by population biologists as r-strategists adapted to survive by very high reproductive rates. The frequent radical size reduction of these populations results in the so-called bottleneck effect, which prevents the maintenance of high genetic diversity (Huston, 1994).

The loss of gene forms can only be compensated by migration and mutation. In this way SI populations can evolve simple adaptations such as industrial melanism or resistance to pesticides and pollutants (*in situ* evolution).

Intentionally introduced (II) species, however, either do not form populations in the genetic sense or they are bred artificially according to the market's requests (*ex situ* evolution). Most of them are not viable without human provision.

The only species of the urban system which has very large populations with a suitable amount of genetic diversity for genetic adaptation is *Homo sapiens*.

Species Diversity of Urban and Peri-urban Environments

In order to assess urban species diversity we have to decide which species are to be included. If we count every living species the list might result in a misleading conclusion. In order to save biodiversity efficiently, we have to save urban biodiversity. Unfortunately, such action would protect mostly non-viable populations. Conservation genetics teaches us that for preservation and conservation of a species, viable and adaptable populations are needed (Loeschcke et al., 1994). Category II does not fulfil these criteria. Category SI does, but these are the least preferred species for humans.

Conclusion

Urban and peri-urban environments can be characterized by various levels of biodiversity depending on definitions. Genetic diversities of most populations are rather limited. Species diversity can be very high if we disregard frequencies of the species. This high species number, however, should not be a priority target in conservation strategy, for the more valuable II part is not viable alone in the long term.

Urban and peri-urban biodiversity has its greatest value in aesthetic and educational aspects through awaking people's interest in biodiversity issues.

References

Huston, M.A. (1994) *Biological Diversity. The Coexistence of Species on Changing Landscapes*. Cambridge University Press, Cambridge.

Loeschcke, V., Tomiuk, J. and Jain, S.K. (eds) (1994) *Conservation Genetics*. Birkhäuser Verlag, Basel.

Murphy, D.D. (1988) Challenges to biological diversity in urban areas. In: Wilson, E.O. (ed.) *Biodiversity*. National Academy Press, Washington, DC, pp. 71–76.

Solbrig, O.T. (ed.) (1991) *From Genes to Ecosystems: A Research Agenda for Biodiversity*. IUBS, Cambridge, Massachusetts.

Vida, G. (1994) Global issues of genetic diversity. In: Loeschcke, V., Tomiuk, J. and Jain, S.K. (eds), *Conservation Genetics*. Birkhäuser Verlag, Basel, pp. 9–19.

53 The Urban Dimension of Diversity

R. Pesci

FLACAM, Fundación CEPA, Calle 57 No. 393, 1900 La Plata, Argentina

Introduction

The International Forum 'Biodiversity, Science and Development. Towards a new partnership', is an exceptional occasion to bring to the forefront some of the serious problems concerning biodiversity conservation that exist in cities. In addition, there are other essential dimensions for the comprehension of these problems and the search for solutions, such as the case of cultural diversity.

The urban dimension of diversity reveals another universe of demands, as well as fascinating creative possibilities. This is the main subject of this chapter.

Is an Ecology of the City Possible?

Conflicts and dialectics as homeosis of urban systems

Today, three important forces have negative repercussions for our cities.

The impact of voracity
The city is highly consuming and does not recycle much of what it consumes. Italo Calvino (1972) called it 'Leonia', which is the city that will be covered with its own waste. It corresponds to the functioning of the Society of Lineal Flows which plunders the resources – including the human ones – returning to nature little of what it takes from it, and recycling almost nothing (ESAM, 1992) (Fig. 53.1).

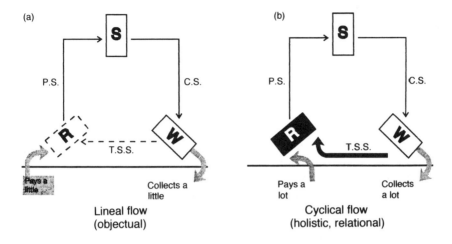

Fig. 53.1. Types of society. (a) the society of lineal flow (productive objectual); (b) the society of cyclical flow (a new humanism, holistic, relational). PS, productive system; S, society; CS, consumerism system; W, waste; R, resources; TSS, technological scientist system.

The impact of periphery expansion

This is also a manifestation of consumerism (of soil resources, in this case) since the old city is not recycled. Its empty spaces, its densities and social processes of production are not reinstated. Changes of scale in social systems, immigrations towards cities (the general process of urbanization crowds the population which has been thrown out of rural areas into poverty belts around the consolidated city) are phenomena which have not been well managed. In fact, this impact is a structural manifestation of modern society (of our own environment).

The historical city extends into the periphery, adding migrant populations, or throwing it out of its uninhabitable centres, destroying its rich urban–rural interface, without finding a better role to play than that of a formal and informal tertiary rank. These phenomena are all evidence of the city's lineal flow, which is dependent, a plunderer and consumer rather than a producer. Calvino (1972) named it 'Pentesilea': the city which is all 'urbis', where the 'civis' is difficult to find, or is almost entirely immersed in the great conflicts of a number of social struggles, including traffic flows, endless service systems and unmanageable security and control systems. This city is not civis, it is only suburb.

The impact of its obsolete planning systems

There is a tendency to freeze the planning system through artificial schemes, keeping urban areas as museum pieces and speculating on the soil retention or its mobilization just to increase its prize value. Calvino (1972) called it 'Zora' and it is the city which is intended to be managed artificially, without responding to its real evolving dynamics, and generating constant conflicts between an authentic – although complex – 'want to be' and a 'be able to be' which, in practice, is schematic, ignorant and obsolete.

Discontinued systems and plural societies

In fact, cities are in their current state not only because of blindness and stupidity. Human have made the three above-mentioned mistakes (which are used here as mere symbols of many others which are included in them) due to the rapid transformation of our artificial systems which made it possible for the city to triumph over rural areas. Among some of these transformations are: concentrated production, decentralized commercialization and distribution, integrated technology, and distant administration. But these cities happen to be made up of plural societies, which are new and migrant, do not know each other, and have different, and often opposite, social and territorial habits. The historical city, which managed to operate as a mature system (balancing inputs and outputs, recycling its resources, roles and spaces, with low entropy relative to its condition as living system which is in constant evolution) has given way to the current city, whose operation is much more erratic, probabilistic and even discontinued with high uncertainty.

As a matter of fact, such is the case with complex systems, with a high tendency towards entropy, shown through the impacts mentioned above and the permanent conflict among representative subcultures, each of which has different interests.

If a plural society does not agree to a cyclic flow behaviour, little can be done to succeed in improving the system, acknowledging opportunely its income and expenditure deficit and surplus. In fact, it is an evolving cyclical movement (like a helicoid, see Fig. 53.2; Pesci and Scudo, 1974), which is supplied in its condition of an open system, but it needs certain modelling and control in order to keep the biological, psychological, and functional balance among its thousands and thousands of parts (Fig. 53.2).

The dilemma is: control without diversity, and death of the system; diversity without control, and entropic dispersion; or 'on line', 'soft' control in constant readjustment. In other words, we must deal with cyclic control, feedback, plurality readjustment, and modelling, and the implicit contradictions.

The power of information-communication flows

In a homogeneous society (one culture, or several arranged microcultures), the decisions are agreed upon since they arise from constant communication which is nourished by identities. Matter and energy flows are cyclically conducted, for they are nutrients which are collectively shared and guarded. In a pluralistic society, composed of subcultures in search of their recognition which are willing to emerge from different levels of financial, religious, cultural or technological submersion, the decisions are counter-orders (and conflicts) which are made by powerful groups who make decisions for others.

The power of information flow governs the society, which passes from the equalitarian net of communities showing solidarity (a), to the hierarchic

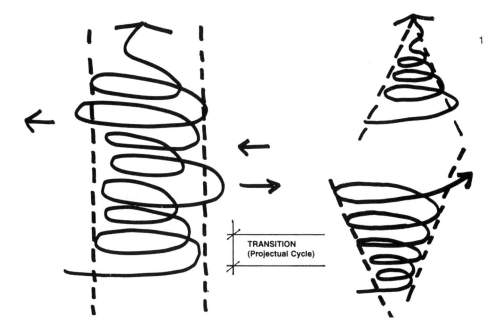

Fig. 53.2. Pattern of an holistic projectual process: the helicoid.

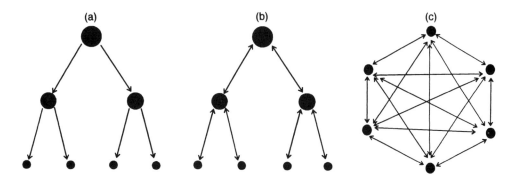

Fig. 53.3. Patterns of social information–communication.

community (b), to get to the dominant caste and mass anonymity (c) (CEPA, 1984) (Fig. 53.3).

The environment, however, cannot be improved without social control of its multiple and complex phenomena. It is necessary to recover the importance of localism, since it can be self-imposed and easily done. The global, planetary quality of modern society is not being denied, but it is not a question of being deceived by omnipotent statements. If there is not local action it will be impossible to improve the global condition. The environment where we live and belong is the first and unavoidable identity provider.

The environment (both human and natural, biotic and abiotic) will be improved only by a new culture based on ethics of solidarity, since all of us become involved with our culture, people and companions throughout our lifetime. Such awareness exists even today at the microcultural level (community identity), and it is the starting point for change in communication flow and decision-making: from the local, and increasingly towards the regional and global aspect.

The city is by definition the local experience: it is the civis and the social association. It is the possible society which Man was able to build and the fact that it is failing should act as a warning signal that the meaning of society has been misunderstood.

The city should be able to operate again as a balanced ecosystem. Or perhaps, today it should operate as a complex cultural biosphere, where a million phenomena combine to form a strange metabolism, but where modelling demonstrating both solidarity and structure must be possible. Otherwise, it is not maintainable and it becomes both a cruel cultural and ecological threat.

From Biodiversity to Diversity?

Need and beauty of diversity

We have learned to recognize both biologically and ecologically, the need and beauty of diversity. We know that in ecological processes the alimentary chains are kept if there are component species. We also know the magnificence of such complementary niches or roles, which prove the immeasurable genetic resources of nature. And we know that the morphogenetic canon of nature is amazing, and that a fly's beauty with its antennae and wings appears overwhelming to us when the magnifying glass shows us its architecture and engineering. We accept, and even admire nature's pluralism and heterogeneity of genetic and morphogenetic processes.

However, we are much more beguiled (and reactionary) when we refer to the human environment. Many people still believe in the virtues of pure races, with superior intelligence, people who are better than others, because of their religion, lineage or any other difference. Nevertheless, in Latin America, we know that the culture or race crucible – though difficult – is a fascinating process of increasing tolerance and beauty, on the condition that it is exercised in freedom and with true disposition to integrate.

Diversity is also beauty: it serves a relational, processional, heteronomous, realistic and also magical aesthetic vocation, such as my native country's literature, before the aesthetics of purism, aesthetically and artistically doubtful.

Fig. 53.4. Three historical patterns of urban configuration (sketch: Arq. Rubén Pesci).

Environmental geometry and city morphogenesis

Human beings have preferred a rationalistic control of our anthropocentric environment, and artifice has been deemed more prestigious than nature. The shape of our habitat appeared to be intelligent by differentiating rather than by integrating. Three geometries were analysed (Fig. 53.4): (i) genetic growth; (ii) preconceived order; and (iii) functional growth (cumulative).

We know today that this does not suffice. Genetic growth supports neither the great scale nor the pluricultural nature of our current urban centres. Preconceived order is necessary, but with wider, more systematic and not so schematic and simplistic patterns. Cumulative growth – our current suburb – is not sustainable.

However, we have increased our knowledge about deeper structures of environmental geometry (March and Steadman, 1974), which show us the systematic (more flexible and richer) wisdom of multiaxial, multiplural combinations, intersections, superposed levels and a whole range of generations – genetic processes – capable of organizing complex systems.

Interfaces as the shape of the complex order

We know, above all, that living systems interact as ecotones or interfaces, where two or more ecosystems touch, confront and exchange. Interfaces are the most interesting aspect of current environmental studies (Giacomini, 1983) due to such supplementary diversity, value: biodiversity in the homeosis of two or more ecosystems. For the same reason, an interface is also the component which has the greatest information capacity of the system; it contains information of all its constituents and permits a positive or negative interaction between data emission of each part.

In a vision of the biosphere (and we must compare the current city as

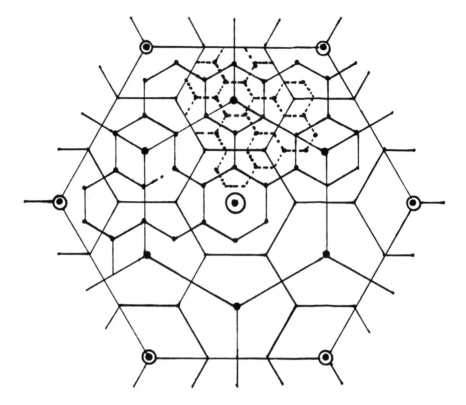

Fig. 53.5. Patterns of the distribution of cities on the territory (Christaller) that illustrates a theoretical model about how cities of different range are distributed on the territory, in order to set them up with efficient interfaces between themselves and each surrounding one.

regards its complexity with a biosphere, rather than with an ecosystem), interfaces are the essential shape or structure of system complexity. And that is why the urban environmental system – 'environmental' meaning interaction with the natural, manmade, socioeconomic, and cultural habitat – can be reinterpreted as a complex system where the interfaces are most evident.

Interfaces and shapes of cultural diversity

Mileto, the famous Hellenic city designed by Hipodamo and regarded as the first planned city, had only one culture. It had a cultural interface: the agora, the gymnasium, and the theatre; a productive interface: the harbour and its market; and a continuous border of a natural interface: the sea. Our historical western cities had neighbourhoods with strong identities which are still partly preserved. These neighbourhoods were characterized by their chapels, monuments, and microcultures. In fact, those cities were microcultural systems with cultural interfaces (the church, the club, the school, the square) and physical interfaces (a park, a river, a depression, a railway, or industrial area) which separated and joined them at the same time. They established a very positive interdependence with the country and small farms (Fig. 53.5). The interfaces showed (and supported) the diversity of the parts within the uniqueness.

Today's megacities, and even those which are not so big but suffering from voracity, periphery invasion, and poor planning, eliminate their interfaces and get into disorder. They reduce their natural accidents and borders, put streams into pipes and destroy hills, pollute their natural resources and drive away their biodiversity, and they build luxury 'ghettos' to protect their leading culture. The urban struggles in Los Angeles in 1992 showed this hard truth: they are not cities, they are only big neighbourhoods with opposing interfaces where thousands of subcultures in conflict come together.

Only respect for diversity – natural and cultural – and careful handling of the interfaces among them, will give the future of our cities a chance. It is known that in Los Angeles the society is almost organized in castes (Pesci, 1991), with privileges and aberrant urban discriminations, but little has been done to solve this. The 'lock out' was and is produced day after day in the violence which their own films clearly show.

The City and Diversity

The city as natural and cultural interface

The modern city and especially the one which is being formed for the future, either takes the informative–cultural wealth and human diversity crucible from its diversity (preserving the animal and vegetal biodiversity), or puts itself in danger of making those premonitory and dazzling hells of '1984', 'Metropolis', 'Blade Runner' or 'Brazil' come true. These are some of the literary and cinematographic works which have illustrated the horrors of prostitute, jailed, blood thirsty, and dependent cities.

The city was, on the other hand, created for arguable exchange: hinterland and city; market and international commerce; balanced income and expenditure; factory and consumption; everything was recycled: matter, energy, data; and other cities purchased and sold their surplus to each other. The city is a great cultural interface, and also a natural one; it is connected with the country and primary production which it transforms, exports, and commercializes.

The existence of the city is not negative. Those who want to make the cities rural (and return to them their ecology) are wrong, and willingly or unwillingly, they are supporting the American widespread urbanization model: cities with no diversity, no cultural exchange, individualism, anti 'civis', with all the foreseeable social and 'civilizing' consequences.

It would be good to urbanize the territory, that is to say, to take civility to the territory: the socioeconomic and cultural city interfaces throughout the whole territory[1], and make them cohabit in harmony on a well-handled physio-natural basis to preserve the evolving conservation of biodiversity. The city is diversity: tolerance, pluralism, integration, society. The city is an interface, composed of multiple interfaces.

The city and a diversified civilization

Defending the city then means defending diversity: a diversified city which is flooded with diversity (natural, cultural, physical, and social) must be a democratic city, managed as a society of cyclic flows, shaped as an open system, and a participating system as a communication net. In addition, big migration flows continue towards the cities and therefore, any other abstract or wilful model of 'civilization' is illusory.

The city is possible, but it will have to be different. It will have to be: open, and local-scaled (although it remains part of a huge entirety); multicultural (although each place allowing for microcultures); and widespread over the territory (but preserving and shaping the city in order to save it).

Jacobs (1989) found that successful maintainable cities are those which are regarded as 'city-regions', i.e. federated groups of cities which act as a complement (Po Valley and Ruhr Watershed are eloquent examples).

In fact, we encourage accepting and propitiating multifocality (many social interfaces) within each city, and as a system of cities inside a region.

Diversity and interfaces appraisal

In analysing cities, we return to the essential concept of interfaces. If we multiply positive interfaces, the city (civis) recovers tolerance, security and control within the multicultural pluralism which we ask of the city today. If we observe the biological value of physio-natural interfaces, we will recover or preserve the condition of natural resources such as water, air, soil, fauna and flora and, in general (as they coexist with interfaces), all of them mutually interactive.

If we protect the cultural and symbolic value of any interface (a city entrance, a woods transition, a beach or a meadow) we will create a consciousness about developing a culture based on solidarity (relational, in process) which will gradually replace our objective, sectoral, materialistic, and individualistic culture.

Principles to Recover and Propitiate Diversity in Urban Systems

It is not possible within the scope of this chapter to develop completely the theoretical and methodological principles which we have called upon to establish a better urban life (with diversity and biodiversity), from exposed criteria, out of long experience in urban and regional projects. Only a brief enumeration is possible here – an enumeration which permits a conceptual connection with what has been said so far and gives the possibility of continuing research on this fascinating topic.

Five principles have emerged from this praxis[2].

Multipolarity

Multipolarity is based on the idea of creating and reinforcing social interfaces as the focus of associated life – many towns which can be recognized inside the big city, or cities which are associated in regional leagues to increase their supply and demand of diversity and guarantee their support. Multipolarity permits: (i) growth, without gigantism; (ii) large social scale, with a local scale in each subsystem; (iii) different and complementary roles.

Interface system

Such a system underlines the importance of considering the physio-natural environment in terms of 'interfaces' (relational thinking) and only for this it proposes to rescue all the natural accidents, creating:

1. artificiality bumpers (which protect the neighbourhood scale);
2. diversity protectors (protection of fauna and flora, preservation of superficial watersheds and conservation of topography and its functioning);
3. a healthier and more beautiful urban landscape.

It also creates and recovers the value of constructed interfaces which can act similarly to the natural ones. What could these modern beneficial built interfaces be?

1. canals and their banks;
2. parks and boulevards;
3. terraces or great topographic unevenness used with social equipment;
4. great centres, such as universities, airports, harbours, technological parks, and others.

Urbanity and open spaces

There is no need to maintain (or create if they do not exist) microinterfaces within the urban area: points of social convergence, urbanity growth ('civis'), both constructed and natural attributes which are, in fact, 'open spaces' because they can be possessed by everyone. In their future vision, these heirs to the glorious tradition of agoras, forums and squares may be: (i) cultural centres; (ii) sports and leisure centres; (iii) neighbourhood associations; (iv) clubs; (v) recovery of squares and streets for social meeting.

Quality is related to the notion of 'open': plurality, diversity, and for this, interaction among varied groups in shared sites is essential.

Social participation

Social participation implies above all a legitimacy claim on any transformation action within the city. The city is either in favour of all its emerging micro and subcultures in its diversity, freedom and democracy, or it is sectarian, discriminating and antidemocratic. Therefore, all its citizens should feel like participants in its changes and major decisions.

We place emphasis on the expression 'to feel like participants' because participation may be direct (lets get to work, do it yourself) – which is often Utopian and dangerous, since the problems to be solved are not always widely known – or indirect, meaning consultive, respectful and inspectional through technical investigation and dialogue towards agreement, but where the final modelling is again technical.

Decisions must be legitimated, taking into account all diversity needs and aims of the social groups living in the city. In doing so, the basis of urban culture is recovered, which integrates differences and discriminations and praises diversity and its historical evolution, Fifty years ago, Lewis Mumford had already anticipated these matters (Mumford, 1966). The city was the first place which had visitors, foreigners, merchants, ambassadors, and scholars, as well as poor people and slaves in search of freedom and work.

This is not new and was very positive. It should happen again.

City production

The multifocal participating cities, with interfaces and open spaces, cannot go about their production as current cities do. Today's cities are monopolized by two extreme powers: speculative commercial production and marginalized class production, through illegal land possession. Both extremes are types of violence and anti-freedom. We must learn to agree on a production system, articulating both interests and priorities. But we must give rise to three new (and fairer) production techniques:

1. define and legitimate the great director lines socially;
2. but open to 'on line', continued, flexible actions and which can be conceived by each microculture;
3. incorporating both the social economy and the economic-financial capital, protected and promoted by the State through programmatic integrating agreements.

These five principles represent five new ways of bearing diversity pressure over old urban structures and obsolete production systems, and transforming the conflict pressure into the genesis of cultural, historical and social wealth. Five principles to stop 'civis' destruction, and thus, defend human diversity and biodiversity in the increasingly extended and aggressive urban territory.

Notes

1. As Ramon Folch said at the Forum on 'Our Own Solutions' organized by FLACAM at ECO'92. It is an idea which is supported by the new Utopia: the city-territory, which in Italy, for instance, dominated the cultural scene in the 1960s and 70s. CEPA greatly propitiated this model (CEPA, 1984).
2. The author refers to the 20 years of experience of the Foundation CEPA on environmental design, with special emphasis on urban systems.

References

Calvino, I. (1972) *Le Citta'Invisibile*. EINAUDI, Roma.
CEPA Foundation (1984) Proyecto de Ecologia Urbana del Sistema Urbano Pampeano. *Ambiente* (La Plata) 66, 45–56.
ESAM (1992) The society of cyclical flows. *A/mbiente* (La Plata) 73, 30–37.
Giacomini, V. (1983) Proyecto de Ecología Urbana de Roma. *Ambiente* (La Plata) 38-27.
Jacobs, J. (1989) *Las Ciudades y las Riquezas de las Naciones*. Altafulla, Barcelona.
March, L. and Steadman, P. (1974) *La Geometria dell' Ambiente*. G. Mazzotta, Roma.
Mumford, L. (1966) *La Cultura de las Ciudades*. Infinito, Buenas Aires.
Pesci, R. (1991) California, una mirada en los nuevos paradigmas ambientales. *Ambiente* (La Plata) 69, 10–14.
Pesci, R. and Scudo, G. (1974) Introducción a la proyectación ambiental. *Summarios* (Buenas Aires) 7, 3–6.

54 Biodiversity in the Twenty-First Century

J. de Rosnay

Director of Development and International Relations, Cité des Sciences et de l'Industrie, 30, avenue Corentin Cariou, 75019 Paris, France

When considering the importance of biodiversity, scientists often focus on questions concerning taxonomy, genetics, and classification of species. These are indeed important facets of biodiversity; however, they do not encompass all of the areas of life influenced by the presence or absence of biodiversity. A multitude of cultural and ethical issues inherent to biodiversity need to be addressed as well. Given that the consequences of biodiversity extend far past the scientific discipline of ecology to which it belongs, new strategies need to be developed to understand its complex nature if we aim to apply related scientific findings in an effective and knowledgeable way.

Biodiversity is a fundamental concept for a world entering the twenty first century. Its emergence corresponds to a conceptual shift in our reasoning, from an analytic to a systemic way of viewing the world. Biodiversity's range of influence is much more general than other fields of biological study, such as species biology or the biosphere. Its numerous manifestations can be found in inconspicuous places, for example in the technosphere (the sphere of manmade machines) and in the noosphere (the sphere of minds uniting everyone on the planet through various communication networks). Biodiversity includes not only living species, but the multitude of human inventions as well. It is concerned with the interaction of ideas, because this promotes variation which is necessary for evolution to continue. In summary, biodiversity appears to be the basis of a general law of dynamic stabilization of complex systems. This law is fundamental to homeostasis, a process familiar to biologists, and which is also applicable to the functioning of the planet as a whole,

A fundamental question arises, however, concerning the role of biodiversity in ecosystem functioning. If diversity is in fact a fundamental law of dynamic stabilization of complex systems, what degree of diversity is necessary in order for the system to establish itself? In other words, how many species ensure a stabilizing diversity? Does the presence of too many species cause a system to become saturated? Experiments carried out in ecotronics and other laboratories have demonstrated that the process is much more complex than we previously thought. A certain amount of diversity is required, but beyond a certain threshold, we sometimes observe results of decreasing output. Productivity which was already decreasing is in fact reduced. Therefore, there exists a kind of biodiversity optimum, not only for the number of species, but especially for the distribution of certain key species that play a determinant role in the dynamic interactions within ecosystems.

Answers to these fundamental questions about biodiversity cannot be obtained by relying on the classical analytical methods of science. We need new ways of thinking and new methodological tools. Among these new ways of thinking, the systemic approach and complexity sciences that have emerged over the past years may help to provide initial answers. Among the tools, computers offer a powerful simulating capacity making it possible to study the dynamics of complex systems. In a period of barely twenty years, the development of what is known today as 'complexity sciences' has resulted in a synthesis of analytic and systemic approaches. These sciences, for the most part, have arisen from the theory of chaos and self-organization. A great number of laboratories in the world are dedicated to this new approach as a means of demonstrating processes of auto-catalysis, auto-selection, dynamic stabilization and emerging properties. Computer simulation is now an indispensable tool for these studies, and over the past five years, computer science has made considerable progress in this respect. The calculating power of a microcomputer or a desktop computer today is equal to that of a Cray 1 computer seven years ago. Increased performance in computer graphics allows scientists to visualize complex structures, contracting time and dilating or concentrating space, transforming the computer into a kind of portable and experimental 'complexity' laboratory.

Two examples help to answer the question raised earlier concerning the existence of a diversity optimum and the stabilizing effect of species interaction. In Los Alamos, Doyne Farmer and his team have been working on the emergence of autocatalytic structures which are believed to be at the origin of living systems. Using a computer they simulated polymer interactions, represented by code chains, and they found that the emergence of an autocatalytic network would only come about if sufficient diversity in the interacting elements was maintained. Below a certain threshold nothing emerged, and above a certain threshold the system became saturated.

The second example is taken from genetic biology. Human beings have approximately 100,000 genes that produce proteins, enzymes, and metabolites which act to turn on or off certain genes. There is an extremely complex communication network linking these different genes, similar to a computer having a parallel multiprocessor. Even if the capacity of all these elements combined could potentially result in an infinite number of cellular species, in reality, we find that there are about 250 of them. Nature created attractors (in the chaos theory sense) which form pockets of stabilization, in turn causing the selection of certain elements over others. Some species require a high level of diversity whereas others require stabilization. Among other things, computers have become an essential tool for understanding complexity, teaching, encouraging research, linking professors and students, and increasing communication concerning complexity and system dynamics.

One of the underlying themes of the Forum which led to this book was partnership: the search for partners to address issues of biodiversity at an international level. It seems essential to me that a certain priority be given to financing and coordinating activities that both use and develop computer simulation techniques. Considerable sums of money finance computer studies that focus on classification, taxonomic and genetic sequence studies. These are unquestionably fields of great importance. However, long-term computer simulation of system complexity and dynamics is an essential element for understanding the emergence and maintenance of diversity.

In conclusion, biodiversity can be understood through methods of both analysis and synthesis. However, breaking down complexity into simple elements results in losing emerging properties along the way. On the other hand, reconstructing complexity from these elements means compromising the experimental grounding needed to confirm hypotheses. Computers require these previous analyses in order to create the mathematical equations necessary for simulation. Recreating interactions of complex systems, and thus of life *in vitro* – almost *in silico* – is a powerful tool not only for understanding but also for acting on complexity. Most importantly, computer-simulated experiments render it unnecessary to carry out large-scale experiments directly on our planet – experiments that could endanger the evolution of five billion human beings.

Biodiversity: Cultural and Ethical Aspects

M. Elmandjra
University Mohamed V, B.P. 53, Rabat, Morocco

Biodiversity, culture and ethics each raise complex conceptual issues of their own; to attempt to describe their interactions is no easy task especially if one wishes to keep in mind the theme of this Forum : 'Science and development, Towards a New Partnership'. The very concepts of 'diversity' and 'biodiversity' now have different meanings and content from the ones they had a few decades ago because of the progress made in scientific research and the increased awareness of the importance of ecological systems.

We now know that the first bacteria was born 3.4 billion years ago (Chambon, 1994). A census (phylogenetic tree) covers about 1,400,000 species as of today – the unknown species are estimated at between 5 and 50 million. The new quantitative information which the scientific community is gathering has qualitative impacts on the problematic of 'biodiversity'. It contributes to the understanding of the complexity of the relationships between living organisms and underlines the fact that this diversity is at the very basis of life.

Organic diversity is now accompanied by another type of diversity. We are moving from a society based on 'production' to one based on 'knowledge' – a knowledge which is producing an enormous quantity of information of great diversity. In the field of science alone, there are over 2,000,000 scientific articles published each year in about 60,000 scientific magazines – about one article every four minutes.

Scientific vocabulary is enriched by 40,000 new words every year. Diversity has thus become not only a condition for biological and ecological survival but also for the development of a society of knowledge where the process is much more important than the product. The process is in itself a source of diversity. With the help of innovation and creativity (added value), the 'process' has become an important source of industrial diversity.

This chapter will attempt to analyse the effects of culture and ethics on biodiversity without ignoring the reverse process, i.e. the key role of diversity in the areas of culture and ethics.

Three main topics will be dealt with succinctly: (i) the spiritual dimension of biodiversity; (ii) ethical implications of diversity; (iii) cultural diversity : a prerequisite for communication and world peace.

Spiritual Dimension of Diversity

Diversity is a very basic concept for the understanding of nature as well as of human behaviour. It is a physical and a sociocultural reality. It also has its spiritual and metaphysical facets. All of the Holy Books underline diversity as an essential element of Creation. I shall limit myself to a few quotations from the Holy Qur'an (in which we find 37 times the expression 'diversity'; translations by Ali, 1946).

> ... To each among you
> Have We prescribed a Law
> And an Open Way.
> If God had so willed,
> He would have made you
> A single people, but (His
> Plan is) to test you in what
> He hath given you: so strive
> As in a race in all virtues.
> The goal of you all is to God;
> It is He that will show you
> The truth of the matters
> in which you dispute. (V, 48)

> It is He who produceth
> Gardens, with trellises
> and without, and dates,
> And tilth with produce
> of all kinds, and olives
> And pomegranates,
> similar (in kind)
> and different (in variety):
> eat of their fruit
> in their season, but render
> The dues that are proper
> on the day that the harvest is
> gathered. But waste not
> By excess: for God
> Loveth not the wasters. (VI, 141)

> If it had been thy Lord's Will,
> They would all have believed,
> All who are on earth!

Wilt thou then compel mankind,
Against their will, to believe (X, 99)

And the things on this earth
which he has multiplied
in varying colours (and qualities)
Verily in this is a Sign
for men who celebrate
the praises of God (in gratitude) (XVI, 133)

And among His Signs
Is the creation of the heavens
And the Earth, and the variations
In your languages
And your colours : verily
in that are Signs
for those who know. (XXX, 22)

Seest thou not that God
Sends down rain from
the sky, and leads it
Through springs in the earth?
Then He causes to grow,

Therewith, produce of various
colours : then it withers;
Thou wilt see it grow yellow;
Then he makes it
Dry up and crumble away.
Truly, in this, is
A Message of remembrance to
Men of understanding. (XXXIX, 21)

This small selection of verses is meant to show the diverse ways in which the concept of diversity is used in the Qur'an and how it applies to human beings, nature, plants, and life in general. A remarkable verse X, 99 (see above), is the one in which it is said that if God had wanted to make all the people on earth believers he could have done so. It then goes on to criticize those who exert pressures in matters of belief, 'Wilt thou then compel mankind, Against their will, to believe'. This verse not only highlights the vital role of diversity in the Qur'an but also stresses the sense of tolerance and freedom which are essential conditions for diversity.

We also see how the Qur'an stresses cultural diversity (verse XXX, 22 above) and the 'variations in languages and colours' . . . 'If God had so willed, He would have made you a single people . . .' (verse V, 48 above). The reference to the spiritual dimension is also essential if we are to understand the shaping of values which affect our attitudes towards life and its diverse aspects. The spiritual concept of diversity leads to the understanding of 'unity'. Without a comprehension of unity we cannot understand the true sense of diversity.

Ethical Implications of Diversity

The following text of Chambon (1994) is left in its original form as it highlights beautifully the reasons why the living sciences of the last 50 years reveal to us a source of ethical values: the respect of the biological universe.

> Accroché à l'un des rameaux de la couronne de l'arbre phylogénétique, fruit plus qu'improbable d'une loterie cosmique, l'homme qui est le seul être vivant à pouvoir se représenter lui-même comme un autre, est aussi le seul à connaître ses racines Il me semble que cela lui impose des devoirs particuliers. Presque paradoxalement, s'il fallait trouver dans l'histoire du vivant que la biologie des cinquante dernières années nous la révèle une source de valeurs éthiques, celles-ci devraient sans doute prendre en compte le respect de l'univers biologique auquel nous sommes si profondément ancrés.[1]

The respect of the biological universe calls for a set of ethical norms including a high consideration for diversity. We cannot, however, separate the sociocultural systems of humanity from the biological universe of which they are a part. Yet, the whole game of 'power' consists of imposing one's own system of values and weakening, if not overtaking, the values of others, thereby reducing cultural diversity.

We are thus confronted with a delicate ethical problem because all research today is demonstrating the very close links between ecological systems and cultural systems in operational terms. Biodiversity is not an end in itself, it can no longer be thought of independently of the sociocultural environment which it sustains.

Diversity has positive as well as negative aspects. When we look at the diversity in the quality of life within and between countries, we are struck by huge inequalities. These inequalities make it very difficult to preserve biological diversity because of the rate of consumption and of pollution of the privileged minority on Earth.

How can one ethically give first priority to biodiversity as long as the problem of poverty has not been solved through the 'partnership of science and development'? About 1500 million people live in a state of absolute poverty, have no drinkable water or electricity and are illiterate or quasi-illiterate.

On the other hand, we find 20% of the world population earning 150 times more than the poorest 20%. The diversity in the situations found in the North and in the South is getting constantly worse.

There are more than 1100 million persons who earn less than a dollar per day! The bottom 20% of the poorest people in the world earn only 1.4% of the total world income in comparison with 82.7% in the case of the top 20% of the richest people.

The countries of the North which represent less than 22% of the world population consume 70% of the world's energy, 75% of its metals, 85% of its wood and two thirds of its food products. The North also accounts for over 90% of expenditures in R&D and 80% of the expenditures in education.

How long can the ethics of biodiversity endure the consequences of inequitable economic diversity within and between countries? I believe this to be a vital question to which a common answer must be found jointly by the North and the South on the basis of universal values to be agreed upon. The problem is one of equilibrium within diversity – an equilibrium which calls for an equitable redistribution and social justice. Without such a peaceful redistribution a socioecological explosion is inevitable in the medium term.

Cultural Diversity: A Prerequisite for Communication and World Peace

Younès (1994) highlights the cultural dimension when he states that in dealing with the theme 'Origins, Maintenance and loss of biodiversity', 'the major constraint is of cultural origin'.

Cultural implications are to be found in every step of life. Let us take something as simple as the title of this international forum as well as of the international programme which calls upon the collaboration of a number of intergovernmental and governmental organizations: *Diversitas*. Why choose a language which has a very clear cultural bias when dealing with a worldwide programme? What happens to cultural diversity? The word 'diversity' exists in every language. In fact the use of 'diversitas' can be interpreted as an unconscious form of ethnocentrism.

It is somehow contradictory to defend the principles of cultural diversity and cultural identity and at the same time claim 'universality' for one's own values. This is what the West is doing in its relations with the rest of the world. Such an ethnocentrism is based on the assumption that 'modernization' equals 'westernization' and 'westernization' signifies 'universalization'. These assumptions need to be corrected.

In a study entitled 'Agenda for Japan in the 1990s' carried out by NIRA (Nippon Institute for Research Advancement), the President of the Institute stressed in the Introduction the issue of cultural diversity (NIRA 1988):

> ... it is no longer appropriate to view the world in terms of military polarization, i.e., Pax Russo-Americana. Rather, it has become necessary to look at the world system differently, to put aside a long-sustained view of world order based on stratification under American rule. The new world order may be called the **Age of Diverse Civilizations**, based on the emergence of an age with multiple co-existing civilizations.

He then adds that 'Japan's modernization served as evidence that modernization is different from westernization'.

This last conclusion is very important because the ideological basis of colonialism, neocolonialism (and now postcolonialism) has always been and still is that one cannot go through the process of 'modernization' without going first through 'westernization'. There is no room left for diversity and for freedom of choice. We have to deal with prefabricated modernity and fast food democracy and human rights as defined by others.

When we speak of cultural diversity we speak of cultural identity, of cultural values, of preservation as well as of development of culture. This diversity does not cut off cultures from the 'universal' – on the contrary it is diversity which is the source of universality and which permits true communication and mutual understanding instead of one-way monologues. One of the dangers facing humanity is the absence of this cultural communication. We have a one-way communication – the one of the powerful who imposes his values by force if necessary while maintaining that they are of a 'universal' nature.

In 1978, during the First North–South Round Table (Rome), I stressed the fact that the main problem in North–South relations was first and foremost one of 'cultural communication' (Elmandjra, 1978). Cultural values are playing a greater and greater role in international relations. The big question is how to preserve diversity within harmony? It is one of the conditions for building world peace.

On 2 October 1986, in Tokyo, during a television programme of NHK with Jean-Jacques Servan Schreiber on the future of international cooperation, I maintained that future conflicts will have cultural causes and that we may witness such a type of conflict between the United States and Japan.

During the Gulf War, in an interview with *Der Spiegel* (11 February 1991) I qualified that war as the 'First Civilizational War'. In 1991 I published a book with that title (in Arabic and French). The weight I have always attached to the importance of culture and cultural values in development and in international relations has been a constant one. It is therefore interesting to quote the following excerpt by Samuel Huntington (1993), Director of the Institute for Strategic Studies at Harvard University:

> It is my hypothesis that the fundamental source of conflict in this new world will not be primarily ideological or primarily economic. The great divisions among humankind and the dominating source of conflict will be cultural ...
> The clash of civilizations will dominate global politics. The fault lines between civilizations will be the battle lines of the future.

Culture and science have become the main determinants of the international system. A 'partnership' between the two may help to solve many issues (Elmandjra, 1989). As the subtheme of this Forum is 'science and development towards a new partnership' one should recall René Maheu's (former Director General of UNESCO) definition of development: 'Le développement est la science devenue culture' ('Development is science which has become culture').

Ilya Prigogine has also very well described the link between science and culture by maintaining that the 'problems of a culture can influence the development of scientific theories'. He has also underlined the link between 'science', 'culture', 'diversity' and 'universality':

> la science s'ouvrira à l'universel lorsqu'elle cessera de nier, de se considérer étrangère aux préoccupations des sociétés, et sera enfin capable de dialoguer avec les hommes de toutes les cultures et pourra respecter leurs questions.[2]

The implication of this statement is that science is not yet 'open' to the 'universal'. Perhaps the big task of the 'science and development partnership' is to allow science to open itself to universality through a full respect of cultural diversity. It would be one of the best ways to contribute to better cultural communication and thus to the building of world peace and the respect of biodiversity.

Notes

1. 'Hanging from a branch at the crown of the phylogenetic tree, a more than unlikely fruit of a cosmic lottery, Man, the only living being capable of seeing himself to be like another, is also the only one to know his roots. It seems to me that this imposes particular obligations upon him. Almost paradoxically, if it were necessary to discover in the history of living things that the past fifty years of biology showed it to be a source of ethical values, these values should undoubtedly take into account respect for the biological universe in which we are so deeply rooted'
2. 'Science will open itself to the universal once it ceases to refute, to consider itself foreign to, societal preoccupations, and will at last be able to maintain dialogues with people from all cultures and respect their questions'.

References

Ali, A.Y. (1946) *The Holy Qur'an*, 2 Vols. Hafner Publishers, Boston, Massachusetts.

Chambon, P. (1994) Les obligations de l'homme génétique. *Le Monde*, 30 June 1994, Paris.

Elmandjra, M. (1978) *Political Facets of the North-South Dialogue*. North-South Round Table, Society for International Development (SID), Rome.

Elmandjra, M. (1989) Fusion de la science et de la culture: la clé du XXIème siècle. *Futuribles* 138 (décembre).

Huntington, S.P. (1993) The clash of civilizations. *Foreign Affairs*, Summer, 22–48.

NIRA (1988) Agenda for Japan in the 1990's. *Research Output* 1(1).

Younès, T. (1994) Editorial. *Biology International* 28.

56 The Value of Biological Diversity: Socio-political Perspectives

Crispin Tickell
Green College, Woodstock Road, Oxford OX2 6HG, UK

Introduction

The current debate about the diversity of life is relatively new. It has been created not by some new appreciation of the marvels of nature – would that it were – but by the evident destruction of the natural world as we and our ancestors have known it.

This chapter is about values: why we should worry, what we should do about it, and how we should do it.

Change and Its Impact

We are engaged in a process of extinguishing species and their ecosystems at something like a 1000 times the natural rate. It is comparable to the extinctions at the end of the Cretaceous period 65 million years ago when the long dominance of the dinosaurs came to an end.

How is it that one animal species – our own – could have had such destructive effects on others? Until the industrial revolution, the effects of human activities were local, or at worst regional, rather than global. All the great civilizations of the past have cleared land for cultivation, introduced plants and animals from elsewhere, and caused lasting change.

The consequences of the industrial revolution are still more serious. On the one hand there has been a huge growth in the human population; on the other there has been a huge growth in consumption of the world's resources and saturation of its sinks. Higher standards of living inevitably involve higher consumption and more waste.

© 1996 CAB INTERNATIONAL. *Biodiversity, Science and Development: Towards a New Partnership* (eds F. di Castri and T. Younès)

In broad terms there were around 10 million people at the end of the Ice Age, 1 billion in the lifetime of Thomas Malthus, 2 billion in 1930, and around 5.4 billion now. Short of catastrophe there will be around 8.5 billion in 2025. At the same time there has been a still deeper growth in urban populations, with all that implies for the resources surrounding cities. An observer from outer space, with a device for speeding up time, would see steadily increasing brown patches like freckles on the land surface of the Earth.

High consumption of resources in rich countries and heavy pressure on resources in poor ones have already changed its face. Earlier this year the UN Environment Programme published its 1993-4 Environmental Data Report which showed that 17% of the world's soils had been degraded by human activity. Demand for fresh water doubled between 1940 and 1980, and is likely to double again by 2000. Pollution of both fresh and salt water has increased. There is also the prospect of genetic change caused by increased ultraviolet B radiation in areas, both land and sea, unaccustomed to it.

There are comparable problems in the chemistry of the atmosphere. Acidification downwind of industry is a manageable problem if those concerned have the will to manage it. Depletion of the ozone layer, our protective blanket, has potential consequences for all forms of life. By adding carbon dioxide, methane and nitrous oxide to the atmosphere, we may be altering the global climate with unforeseeable local consequences. There could also be a significant rise in sea level.

Major uncertainties remain, but they are more about the magnitude and geographical distribution of change than about change itself.

The issue of biodiversity falls into four categories: ethical, aesthetic, direct economic, and indirect economic. Each has social and political implications.

On ethical grounds it is questionable whether we have the right to exterminate so many of our companions on the living planet whether they are of use to us or not. This is not a point which has caused Christianity much concern in the past. There are honourable exceptions; but most Christian thinkers have seen humans as separate from the rest of nature which they believe was for their plunder or delectation. But respect for life as such has always been a central tenet of Buddhism and Taoism, among other systems of belief.

The aesthetic aspects of nature usually go without saying, but they are very difficult to define. There is, I believe, a profound human instinct which causes people to feel linked to the natural world. Even the most hardened city dwellers need space and greenery in their work and play. The culture of every people is closely allied to its landscapes and their living inhabitants, and cannot be dissociated from them.

Ethical and aesthetic arguments are of enormous, indeed, primal, importance for the psychological health of any society, but they cannot have monetary values attached to them and are usually unpersuasive against short-term arguments of self interest.

Our direct economic interest in biodiversity is obvious. We need to maintain our own good health as well as that of the plants and animals, big and small, on which we depend for food. Our economies simply could not work without the raw materials available from living organisms. Take the simple case of medicine. More than three-quarters of the population in poor countries depend on plant-based drugs, whereas in industrialized countries about a quarter of prescription drugs contain at least one compound that is or once came from higher plants. Substances derived from the rosy periwinkle of Madagascar proved effective against childhood leukaemia and Hodgkin's disease, and the bark of a rare Pacific yew yielded drugs for use against ovarian cancer.

As well as conserving diversity at the level of species and ecosystems, we also need to cherish the genetic diversity that occurs within them. The wild relatives of useful strains are often lost when natural habitat is converted for other land uses. Without a large natural genetic reservoir, we make our food supplies vulnerable to disease as the Irish potato growers in the last century learned to their cost.

I suppose that monetary values might be placed on some of these direct benefits. The opportunity cost involved in trying to replace them would be enormous. The value of their potentialities is larger still. It is not easy to weigh up the unknown long-term potential value of a rainforest against the known short-term but rapidly diminishing value of a cattle ranch.

The same issues arise with still greater force over the indirect economic benefits provided by the diversity of life. At present we take as cost-free a broadly regular climatic system with ecosystems, terrestrial and marine, to match. We rely on forests and vegetation to produce soil, to hold it together and to regulate water supplies by preserving catchment basins, recharging groundwater and buffering extreme conditions. We rely upon soils to be fertile and to absorb and break down pollutants. We rely on coral reefs and mangrove forests as spawning grounds for fish and wetlands, and on deltas as shock absorbers for floods.

Likewise we need nutrients to be recycled, and wastes to be disposed of. We rely on the current balance of insects, bacteria and viruses.

There is no conceivable substitute for these natural services. Economic values tend to be based on scarcity, and so far there has been no permanent shortage of such commodities as fertile land, clean water, clean air or the supply of genetic resources. Yet we cannot continue to assume that this natural bounty will continue.

The Rise in Public Awareness

Until very recently most people were unaware of these issues most of the time. But public anxiety has greatly increased often in partial and muddled fashion. The UN Conference on the Environment in Stockholm in 1972 was followed by the creation of the UN Environment Programme. Conferences and publications multiplied. The subject took more and more space on the political and diplomatic agenda. Last came the UN Conference on Environment and Development in Rio de Janeiro in June 1992, which produced a Declaration, two Conventions and Agenda 21, or an agenda of environmental action for the next century. It was followed by the creation of a UN Commission on Sustainable Development.

The Convention on Biodiversity was the first of its kind. The idea of an international convention to protect it would have seemed bizarre only a few years ago. Yet the Convention was signed by 157 countries. It has now been ratified by more than the minimum necessary, and will shortly come into effect.

The various conventions and conclusions of Rio laid new obligations on individual governments. Among them were the submission of reports and strategies to explain how the good intentions of Rio were to be put into effect. My own government has produced mammoth papers on Sustainable Development, Climate Change, Forestry and Biodiversity within and beyond Britain, and created new machinery for increasing public awareness of local as well as global environmental problems. My government also launched a project of its own: the Darwin Initiative for the Survival of Species: it supports cooperation between Britain and countries rich in biodiversity but poor in resources, and already has some excellent projects to its credit.

Changing Minds

The Biodiversity Convention is an illustration of the way in which minds have changed. But we are still at the very beginning. Nowhere are our shortcomings clearer than in the conceptual as well as practical approach that we maintain to economics and economic management. Most economists still work within a framework in which long-term environmental considerations play a minor if not peripheral part. For example, measurement of wealth through Gross National Product or Gross Domestic Product is clearly defective: it does not treat natural resources like other tangible assets, and it does not include activities that increase or deplete them.

Economic growth, as enunciated by most politicians, is often cited as the only way out of our problems. But it takes no account of the impact of growth on the environment nor of its inevitable effect on natural resources.

Economists usually argue that environmental considerations must be given a monetary value for them to be visible in the decision-making process. It is true that the benefits of any development must be sufficient to outweigh the costs. But costs involve pricing, and prices should always tell the truth. In addition to the traditional costs of research, process, production and so on, prices should reflect the costs involved in replacing a resource or substituting for it; and the costs of the associated environmental problems.

Considerations of biodiversity create even bigger problems for economists. First there is the problem of irreversibility. For example, if we choose between building a dam and losing a species, blowing up the dam will not bring the species back. Nor will the species return when the dam has become filled with silt and is no longer functional. Discount rates cannot accommodate this point. Nor can they accommodate the second problem: that we are forced to make decisions without more than the sketchiest idea of their ecological consequences. It is an understatement to refer to this level of ignorance as mere uncertainty.

Thirdly, not all objects and processes are marketable. As we have seen, species are important in ways that have no direct or immediate effect on humans but are essential to the long-term health of the ecosystem of which they are a part. How should we value the ecosystem? A common method used by economists to assign value is to ask individuals how much they would be willing to pay for a particular item or service. Imagine asking someone how much payment they would require to accept an oxygen-poor atmosphere. More generally how can we value the totality of diversity, including the range of habitats within an ecological community? We have to value the things that count rather than the things that can be counted. That requires recognition of our place within a natural ecological community in which all species, including humans, contribute to its functioning. Nature would then be recognized as having an existence beyond that of a warehouse of raw materials to be protected because they were useful to our animal species.

How can we find means to reflect this multidimensional value of ecosystems in a way that conveys meaning? Perhaps a hint of the answer lies in the recent work of the UN Development Programme in devising a Human Development Index to replace the old misleading pecking order of wealth based on Gross Domestic Product. By adding indices of child mortality, life expectancy, access to health services, literacy, political plurality and so on, a new and very different order has emerged, often to the indignation of countries who had come high on the previous list. The same principle could be applied to establish an Index of Planetary Wellbeing or Health which could bring in the values of biodiversity – ethical, aesthetic, directly and indirectly economic – and require governments to show whether their countries were making a positive or negative contribution. That might sharpen some minds.

The Main Pressure Points

Obviously a new valuation of biodiversity will require a bottom up as well as top down approach. Most governments and politicians are sensitive to public opinion, albeit on short time scales. Public anxieties over the last twenty years have played an indispensable role in influencing governments and bringing about the current range of environmental agreements, including those of Rio.

For the future the prime responsibility of our political leaders is not just the establishment of a new system of values, through education and the like, but also to find the best means of giving such values practical effect. This needs identification of what might be called the main pressure points on biodiversity. We may have different lists of such pressure points but the main ones are fairly obvious.

Human population increase and its corollary higher resource consumption must come first. The two issues are now under discussion at Cairo. A number of countries are already on the threshold of ecological ruin. Africa is already in deep crisis. How can anyone in this moment of history expect Africans to give the priority we should like to wildlife? Productive cropland is becoming the most precious commodity on earth. In the past it has been easier and less costly to keep expanding into virgin lands rather than to rehabilitate exhausted land, but remaining virgin lands are mostly unsuitable for agriculture, and constitute refuges of biodiversity.

Projections of population increases vary widely. We could see a doubling by the middle of the next century. Of course Nature will take care of us sooner or later. Lack of resources, environmental degradation, famine, and disease will in the painful fashion known by our ancestors cut our species back. In all cases, the problem of population is fundamentally a cultural problem, and this has made it an issue of peculiar delicacy.

There was a healthy reaction from the world's scientists who met in New Delhi in October 1993 and published a statement signed by 58 academics. This made the fundamental point that human ability to deal successfully with the world's social, economic, and environmental problems would require the achievement of zero population growth within the lifetime of our children.

Next comes the problem of land abuse. Conservation is often seen as limited to setting aside islands or parks from which humans can generally be excluded. Such areas are very important. In the future they may act as reservoirs from which species can colonize other habitats as the opportunity arises. So far about 4.5% of the Earth's land area has been given some level of protection from development, although this will not preserve more than a fraction of the world's diversity, particularly as much of this takes the form of unguarded parks which look better on paper than on the ground.

Such areas account for only part of the story. One of the reasons for land degradation, particularly in industrial countries, has been the application of fertilizers and pesticides in quantities which have had the effect of devastating and sterilizing the soil. Neither agriculture nor industry nor even human settlements are exempt from the principles of conservation. Of course, human activity changes ecosystems in uncountable ways, but changing them need not mean destroying them. Respect for natural processes can even be profitable!

Pressure on forests and wilderness in that precious 4.5% of the world's land surface is of course pressure of people. They fall into two broad categories.

In the first are the invaders and destroyers coming from far away and cut off from their origins and familiar surroundings. Food tomorrow and next year are their priorities. Before condemning them too roundly, we should remember that our own forebears behaved likewise. Most are victims of circumstance. In devising new systems of value, we should look primarily at those behind them: governments with mega schemes of development or population transfer, with the usual apparatus of subsidies and inducements; and landlords, cattle ranchers, corporations and the rest with short-term commercial interests.

In the second category are the current inhabitants of forests and wildernesses. Their environment is their life. Not surprisingly their system of values is based on respect for it. The Amazon forests have, for example, been cultivated over thousands of years; but the system of cultivation is long-term, covers wide areas, and is geared to relatively small and stable populations. Thus it is hard to attach commercial value to the products of the forest, most of which are consumed on the spot. A lot of what passes for good sense in aid policy goes back to systems of knowledge and cultivation established in the past; for example the distribution of trees in and around human settlements in India, the return to old systems of potato growing and irrigation in the high Andes, and respect for seminomadic pastoralism in Africa. These are all systems which work provided that pressure of human numbers does not disrupt them.

A system to protect the commercial value of assets of this kind, current and potential, is very difficult. This was one of the conundrums faced by the drafters of the Biodiversity Convention. Here is another strong pressure point. Industrial interests in natural products have intensified due to advances in molecular biology and biological pharmacology, but it is too easy for people to pirate natural systems, and for pharmaceutical and other companies elsewhere to synthesize them (and later create substitutes for them). At the same time some recognition of intellectual property rights in such substances is essential.

The pharmaceutical industry's sales from drugs based on natural raw materials is estimated at around $200 billion per year. The potential value of the next generation of insect repellents, soaps, oils, food colourings, and cosmetics from the same sources could be greater still. Yet none of this comes back to those who have often selected, nurtured, improved and developed the varieties in question over hundreds if not thousands of years.

Whatever the questions of ownership, it is clear that conservation requires incentives and disincentives. People have to identify their interest in it. In industrial countries governments can regulate or manipulate the tax system accordingly. Elsewhere it is much more difficult. In most of the wildernesses of the world no one's writ runs. Vast quantities of hardwoods are exported illegally from Brazil without any but the recipients being the wiser. The ban on international trade in endangered species is only selectively observed.

In general, incentives are much more effective than disincentives. There are ways of limiting the tragedy of the commons. Thus in Zimbabwe local communities have responsibility for managing nature reserves and derive substantial profit from tourists and others in doing so. The absolute condition of success in such schemes is that those who manage conservation, whether of plants, elephants, or fish, should be the direct beneficiaries.

There is another often neglected pressure point, this time in industrial countries. Governments there have a tendency to regard environmental hazards, including destruction of biodiversity, as something which applies principally to others. They are more concerned about the fate of tropical rain forests in Brazil or Indonesia than that of their own boreal forests in Alaska and British Columbia. But three issues should be a direct worry to them: the effects of environmental change on human health, in particular through the rise of new or adapted pathogens; the consequences of biotechnology and

DNA transfer; and the development of biological weapons. To have any hope of coping with these problems requires the continued existence of the rich library of genetic resources afforded by biodiversity as we now know it.

Conclusions

It may be that I am optimistic in thinking that people and their governments will ever respond in sufficient measure to the kind of problems I have outlined. They may not change their minds in time. Economists and politicians may remain on the treadmill of economic growth. Our species may continue to proliferate with a rising tide of refugees washing across frontiers. Land, sea, and air may become increasingly polluted. The impoverishment of the Earth may become daily more evident, and our society and its leaders may become less and less able to cope. Let us remember that all previous civilizations have crashed. None over time has reached a well-regulated steady state with population, environment, and resources in balance.

Perhaps we need a catastrophe, not too big and not too small. Catastrophes wonderfully concentrate the mind. So far there have been few with sufficient impact. No one likes the spread of African bees, the worldwide decline in amphibians, the mass deaths of trees, the eruption of such new diseases as AIDS and of lethal bacteria resistant to antibiotics. The miners' canaries may be singing away lustily but we cannot yet understand their tune.

We remain astonishingly ignorant, above all about the degree of our dependence on others within the human ecosystem. An ecosystem can be likened to the structure of a Noah's ark. We can remove one, two or ten rivets without apparent damage. But at a certain point – it could be the eleventh or the thousandth rivet – the timbers fall apart and the ship founders. Such ships, big and small, have foundered countless times before in the history of the Earth.

I have a domesday day dream. If we perished more or less together – say over 50 years – what would become of the Earth? How long would it take for the cities to fall apart, for the Earth to regenerate, for the animals and plants we have selected for ourselves to find themselves a more normal place in Nature, for the waters and seas to become clean, for the chemistry of the air to return to what it was before we disturbed it? Life itself is so robust that the human experience could soon become no more than a tiny episode. Nature is not fragile. But we are.

57 The Ethical and Non-economic Rationale for the Conservation of Biodiversity

A. Campeau

IDRC/CRDI Secretariat, 250 Albert St, PO Box 8500, Ottawa, Canada K1G 3H9

That we should even have to address the reasons why we should conserve life, the magnificent diversity of other species, the 'web' in which we are but one strand, speaks volumes as to how we have evolved and our present ability to understand what may well be the greatest threat we have ever had to confront, if we are to survive on this planet.

The backdrop of the Biodiversity Convention was the largely intuitive but rising realization that the number and rate of anthropogenically generated extinctions – occurring as a consequence of human activities – is an evil, a threat that must be contained; equally, concern that genetically modified organisms – 'species' being 'created' or modified by biotechnology – may pose a threat to our species and other existing strands in the 'web'.

The rate of change to the *status quo* and the full impact of these changes cannot be assessed with the same relative scientific certainty as the depletion of the ozone layer or projected climate change. Unlike these threats, there is no 'critical load', no models to project the consequences. There is no quantitative definition conceivable for 'tolerable' loss of biodiversity. We have only very preliminary scientific information on the 'critical load' and mix of species necessary to the health of a particular ecosystem or, indeed, of the mix and vibrancy of healthy ecosystems, to a healthy biosphere.

Instinctively, intuitively, perhaps the loudening whisper of a survival instinct within many of us, is a deepening, more or less articulated sense that our survival as a species is interdependent on the coexistence of others.

Our own evolution as a species has concentrated those senses appropriate to our survival. But we have not evolved at the same dizzying pace at which we have acquired the ability to change the environment in which we live. The sight/sound world of *Homo sapiens* is woefully inadequate because many of the threats to our lives and future are not simple sensory events that can be incorporated into our 'caricatures' of the world. Indeed, many of them are not directly accessible to our senses at all.

The 'boiled frog' syndrome is directly applicable to us.

Like the frog, many people seem unable to detect the gradual but lethal trend in which population and economic growth – unsustainable consumption patterns – threaten to 'boil' civilization. They keep turning up the heat because they can not detect its rise! And our cultural evolution has not led to the establishment of governmental institutions that force politicians/decision-makers to pay attention to the long-term consequences of their actions.

It is against this backdrop that the opening words of the Preamble to the Convention must be appreciated. Contrasted and distinguished from the social, economic, scientific, educational, cultural, recreational, and aesthetic values of biological diversity, the contracting parties declare that they are 'conscious of the intrinsic value of biodiversity' – that they are 'conscious of the importance of biological diversity for evolution and for maintaining life-sustaining systems of the biosphere'. They note that 'it is vital to anticipate, prevent and attack the causes of significant reduction or loss of biological diversity at source'.

This recognition of the 'intrinsic value' of biodiversity is a rejection of the notion that life has value only to the extent that it can be exploited for economic/financial benefit – that life is sacred – something to be respected just as we respect the right to life of individuals of our own species – that when we lose respect for other forms of life, ultimately we lose respect for ourselves.

The sneering use of the term 'environmentalist' as an intended pejorative of a type of 'new age' fundamentalist is a remnant of a passing era in which any form of economic growth was regarded as being good in and by itself – rather than as only a means to an end; of an era that confused the concepts of 'standard of living' and 'quality of life' – one in which we strutted and stamped our way through the biosphere, defending our conduct on the basis of Descartes' 'hierarchy of species'. Why is it that the last 30 years in which change has accelerated to a dizzying pace has also seen the most dramatic increase in mental disorders, divorce and violence, of substance abuse? Have we condemned ourselves to a 'Prozac world'? I think not!

There is growing realization that disparities within the *status quo* represent one of the most significant elements of the threat posed to biodiversity. And if altruism, a sense of equity or compassion is insufficient to motivate us to address these disparities, then, hopefully, a realization that it is in our own enlightened self-interest to do so, will.

We stand on the edge of an even more profound period of introspection than that provoked by the consequences of our realization that we have lived beyond our economic means. There is growing awareness that we must not compound that error or, more significantly, that we cannot exceed the carrying capacity of the biosphere, we cannot continue to hack away at the other strands of the 'web of life' without its integrity being compromised.

Unconstrained economic growth as an end in itself, as a panacea to address the problems of the world, in a consumer society with rising material expectations, fanned by various forms of incentives and encouragements to increase rates of consumption that are already unsustainable, clearly spells disaster ahead.

We know this intuitively and the physical, scientific evidence is rapidly confirming what our survival instinct is beginning to shout.

The period of introspection to which I alluded is akin to the proverbial 'mid-life crisis', through which most of us pass, usually on or near a birthday that ends with a zero. Are we passing through a collective 'mid-life crisis' as we approach a year with three zeros?!

As we continue to peel away the layers of the onion, to confront the so-called 'values' we have come to accept without question, will we have the courage to challenge and change the arrogant certainty of past and present economic theories that ignore the carrying capacity of the biosphere and the myriad but interrelated ecosystems that comprise it?

The sense of futility, that there is nothing I can do to change how things are and how they will likely unfold, is the same lethargy that must be overcome if we are to break out of what some would define as a collective death-wish.

Does the search for meaning and purpose explain the explosion of interest and involvement in traditional and not so traditional religions – of interest in the spiritual – the belief of a significant majority of Americans polled by one of the major pollsters and reported in recent issues of *Time* and *Newsweek*, that angels do exist!

The cosmology of Matthew Fox, his books and videos, have attracted many. The Judeo-Christian God with no connection to physical things but rather one who had created all things but remains distinct and separate from them has been replaced by some with a 'God' that is to be found within ourselves and in every living thing.

New-age music, holistic medicine, books and videos on positive thinking, self-hypnosis, shamanism may be manifestations of another evolutionary step being taken by the vanguard of our species, who are starting to think in a different way – in longer time frames, far more holistically and with far more regard to ethical, non-economic values.

It is essential that the separate tracks down which science and religion, in their broadest definitions, have diverged since they first became bifurcated, be reunited. Decisions are made not simply on the basis of scientific and economic inputs but primarily on the basis of 'values', many of which we have blithely accepted for too long, unchallenged.

For those presently more motivated by 'the bottom line', appropriate economic incentives and disincentives, a critical re-examination of the use of fiscal policy, of subsidies, of national accounting methods, of the tangible and intangible benefits which biodiversity brings to a healthy economy, will come; and none too soon. But at the same time, I have no doubt that the ethical and non-economic rationale for biodiversity conservation will continue to struggle for articulation, that our previous, curious titillation with predictions of doom, will be overcome by a fuller appreciation of the intrinsic value of life, in its diversity and its religious/scientific sanctity.

Environmental Ethics and Biodiversity

Ph. Bourdeau
*Université Libre de Bruxelles,
26, avenue des Fleurs, 1150 Brussels, Belgium*

Sustainable development is the new paradigm, the new way to consider the whole of human society in its relationship to its enviroment. Progress toward sustainable development requires, *inter alia*, a profound understanding of the Earth's ecosystem. It must also be grounded in a philosophical attitude regarding the relationship between humans and nature, in order to guide human conduct.

This need for an environmental ethic arises from the increasing magnitude and seriousness of environmental problems which are due, on the one hand, to the increasing power and diversity of technology and, on the other, to the exponential growth of the human population, which is unequally distributed over the Earth's surface. Indeed the pressure exerted by humans is a function of the product of their numbers by the average consumption of resources per capita.

Humans are now able to affect the biosphere on a scale comparable to that of natural geophysical and geochemical agents and processes. Some of the changes induced, such as the loss of biological species, are irreversible. And yet, human life is submitted to the same ecological processes as the other parts of the biosphere.

It has become apparent that technological development alone is unlikely to enable humans to solve the problems generated. Changes in attitude and behaviour are needed to modify the patterns of resource depletion, pollution, and waste accumulation or, in other words, to strive for sustainable development in industrialized as well as in developing countries. This should reconcile high quality economic growth with the preservation of the environment.

In addition to external sanctions for doing what is needed, internal sanctions are also required, i.e. the adoption of an ethic or a set of principles and rules to govern human behaviour. Such an ethic can be viewed as an adaptive device which should enable the human species to thrive in the world. It is destined not only for individuals, but also for communities, governments, non-government organizations, businesses, etc. In order to be effective, it should be understood by and acceptable to peoples of all races, religions, philosophies, etc.

In order to define the desired ethic, one must consider the proper relationship between man and nature. Human beings are dependent on nature (of which they are both part and apart) but are also able to manage it. A review of religious and philosophical attitudes towards nature shows two main positions: anthropocentrism and ecocentrism.

This relationship has been seen quite differently through time and in the various regions of the world. Thus, for the classical Greek philosophers, the rational structure of the world was considered as simple and therefore they had no insight into the complexities of ecological systems. They were in fact the originators of the reductionist attitude that is still prevalent in many fields of science. Exceptions were Aristotle and Theophrastus. The former studied organisms and believed in teleology (everything in nature has a purpose); the latter examined relations between plants and their habitats and could therefore be considered as the first ecologist. It has been said that it is from Plato's ideas that the medieval attitude that 'love of nature detracts from love of God' was probably derived. The emperor and stoic philosopher Marcus Aurelius, on the other hand, stated that what is good and right in itself is to live in harmony with nature. In the Judeo-Christian tradition, Genesis 1:26-28 gives man dominion over the whole earth and enjoins him to subdue it, a mandate repeated in 9:1-2 where Noah and his sons are told to let the fear of them be upon all the animals. On the other hand, in 9:9 and in 10, God's covenant is established not only with humans but also with all animals. Descartes expressed the same idea of dominion in his often-quoted text on using science to make us 'masters and possessors' of nature.

Contemporary exponents of Christianity and Judaism have made the point that the term 'dominion' should be understood as 'custody' or 'stewardship'. John Paul II stated in 1990 that it is a moral duty for Christians 'to take care of the Earth so that it produces fruits'. In its statement at UN Conference on Environment and Development in 1992, the Holy See representative stressed the notion of responsible stewardship and solidarity, and of the need to respect the order of the 'cosmos'.

The idea of stewardship is also found in the Koran: Allah decided to create a vice regent on Earth and taught Adam the nature of all things. This notion is not accepted by Hinduism, for which all lives have equal value and the same right of existence. God and nature are one and the same. Duties are due to animals because of the belief in reincarnation. Hence the idea of 'ahimsa', non-violence to people and animals, which was also adopted by Buddhism.

Philosophical attitudes toward nature changed in the nineteenth century, starting with the wilderness movement in the USA. Transcendental philosophers (Emerson, Wordsworth, Thoreau) accepted an organic view of the world and sought communion with God through nature. Aldo Leopold developed an ethic which claimed the right to continued existence of natural resources, in a natural state. The contemporary deep ecology movement goes further and claims inherent rights for living and non-living components of the biosphere. This ecocentrism is at times opposed to various forms of anthropocentrism.

Although it is difficult to build a strong case for ecocentrism, pure anthropocentrism is not entirely satisfactory either. Anthropocentrism can be extended to encompass a broad concern for the environment, not only for its utilitarian value, but also for its aesthetic worth, or other concerns. Merging of the two approaches may be the most appropriate solution: primary human needs must be pre-eminent but not the secondary, often artificial needs. The latter may well come after the 'rights' of nature, i.e. the preservation of natural ecosystems and of biological diversity.

Once an ethical foundation is set, codes of environmental practice can be developed for individuals, communities, governments, businesses, and so on. Examples already exist, such as the codes advocated by various business groups or non-governmental organizations such as the International Chamber of Commerce or the Keidanren (Business Association of Japan). Some international agreements include reference to ethical principles. For example the UNCED Convention on Biological Diversity, mentioned above, is based on the remarkable acceptance of the intrinsic value of biological diversity.

The codes which have been proposed are tailored to apply ethical principles to their potential users, taking into account the state of scientific knowledge, the nature of the problems addressed, the socioeconomic circumstances, the legal and regulatory backgrounds, cultural situations, etc.

A code of environmental practice of general application, which has been developed by an international group of people following the 1989 Economic Summit Nations Conference on Bioethics (Berry, 1993) is detailed below as an example. It is founded on the ethic of stewardship of the living and non-living systems of the Earth, in order to maintain their sustainability for the present and the future, allowing development with equity. This ethic is based on the following premises:

1. The acceptance of the goal of sustainable development in all of its implications;
2. The need for scientific knowledge on ecological processes and the state of the environment, in order to inform the political decision-making process;
3. The full accounting of external and internal costs;
4. The recognition of the interdependence of environmental values, such as the demands of present and future generations and of nature itself.

A number of guidelines are formulated on the basis of the above-mentioned principles:

1. Safeguard ecological processes essential for the proper functioning of the biosphere;
2. Integrate natural resource conservation with socioeconomic development;
3. Improve the understanding of the environment through research, education, access, and dissemination of information;

4. Protection of global commons through international cooperation;
5. Avoid irreversible choices and apply the precautionary principles.

Detailed obligations have been derived to apply these guidelines to individuals, communities, governments, businesses, etc. They are as follows.

1. All environmental impacts should be fully assessed in advance for their effect on the community, posterity, and nature itself, as well as on individual interests.
2. Regular monitoring of the state of the environment should be undertaken and the data made available without restriction.
3. The accounting of activities involving environmental impact should incorporate social, cultural, and environmental costs, as well as commercial considerations.
4. The facilitation of technological transfer, should be guaranteed with justice to those who develop new technologies and equitable compassion towards those who need them.
5. Regulatory and mandatory restrictions should be controlled by cooperation rather than confrontation.
6. Minimum environmental standards must be effectively monitored and enforced, and regular review of environmental standards and practices should be undertaken by expert independent bodies.
7. Costs of environmental damage (fully assessed as in point 3 above) should be fully borne by their instigator.
8. Existing and future international conventions dealing with transboundary pollution or the management of shared natural resources should include:
   (a) the assumption of the responsibility of every State not to harm the health and environment of other nations;
   (b) liability and compensation for any damage caused by third parties; and
   (c) equal right of access to remedial measures by all parties concerned.
9. Industrial and domestic waste should be reduced as much as possible, if appropriate by taxation and penalties on refuse dumping. Waste transport should be minimized by adequate provision of recycling and treatment plants.
10. Appropriate sanctions should be imposed on the selling or export of technology or equipment that fails to meet the best practicable environmental option for any situation.
11. International agreement should be sought on the management of extranational resources ('Global Commons': atmosphere, deep-sea, regions covered by the Antarctic Treaty System, etc.)

The above is merely an example of a practical code for environmental behaviour which could be accepted by all. It could perhaps be used as a starting point for the elaboration of the Earth Charter, which the United Nations

Organization is supposed to adopt in 1995, as part of the celebration of the fiftieth anniversary of its foundation.

References

Berry, R.J. (1993) Environment concern. In: Berry, R.J. (ed.) *Environmental Dilemmas. Ethics and Decisions*. Chapman & Hall, London, pp. 242–264.

59 Ethics of Biodiversity Conservation

Gian Tommaso Scarascia Mugnozza
Università degli Studi della Tuscia, 01100 Viterbo, Italy

The ecological question which includes biodiversity cuts across all fields of knowledge and action, from experimental sciences to philosophy, politics, economy, religion, and ethics. The ethical, philosophical, religious anthropological, cultural, and legal implications of human responsibility towards the Earth have to be considered in a holistic and sustainable effort of protection and conservation of the Earth and its natural resources.

Up until the 1970s, ethics only concerned the relationship between human beings and between God and man. Since the 1970s, knowledge, but more importantly, prediction of the consequences caused by alteration of nature at local and international levels (as for instance, acid rains), or affecting the whole planet (such as the greenhouse effect and the hole in the ozone layer) gave rise to the need for ethical behaviour in the treatment of our environment. Indeed, the first journal on this subject (*Environment Ethics*) was published in 1979 in America.

Ethics concerning the position of man versus nature may be roughly divided into two groups: ecocentric ethics and anthropocentric ethics.

Ecocentric ethics judges as moral what may be useful to the ecosystem, and immoral what causes damage to it. Man has the same value as other species and there is no reason why his interests must be considered superior to that of other species. As a consequence, all natural entities have intrinsic value because they exist and not merely in relation to man's interest. Therefore, all human activities which decrease homeostasis and/or diversity of the natural community are always wrong, even if this decrement produces useful consequences to man and civilization. As an example, the most radical supporters of ecocentric ethics will consider agriculture as immoral because its practice entails the elimination of all harmful plants and animals, or a drastic reduction of their number.

The merit of this ecocentric ethics consists in stimulating both sensitivity

© 1996 CAB INTERNATIONAL. *Biodiversity, Science and Development: Towards a New Partnership* (eds F. di Castri and T. Younès)

towards nature and the awareness of the damage caused by man to terrestrial and marine ecosystems.

On the other hand, when radically applied to human behaviour, ecocentric ethics would stand as a significant obstacle to civilization's development. In addition, it would be very difficult to find acceptable solutions for hunger, disease and education problems of the developing countries, should the welfare of the environment have priority over that of man. If the same value attributed to man is attributed to other species, why should a child be saved from a fire before a rabbit? On the same principle, antibiotic substances should be forbidden, because this therapy is based on man's priority over pathogenic organisms. In my opinion, if we attribute the same value to all organisms, then man, as a natural entity, is bound to die. In this case, what is the meaning of ethics without man?

Anthropocentric ethics takes into account the nature of man: both as biological and cultural expression. Man is the only moral agent and, consequently, the only living being ethically responsible. According to the most radical view, the environment has no rights. Only the right of man over the environment exists. Environment, biodiversity, etc., must be conserved, not for their own benefit, but to man's advantage.

Anthropocentric activity is acceptable only if man considers himself as the caretaker of nature and not as its master. Therefore, moderate anthropocentric ethics should be the moral, the ethical guideline, that can direct and inspire the human approach to the environment.

Man has duties towards nature; that means that he is responsible towards nature, responsibility which increases with man's capacity to influence his environment, individually and collectively.

Man's benefit of nature is not only economic, but it must be extended to all the social, aesthetic and intellectual needs of man. Indeed, an environment kept in satisfactory condition supplies natural resources, enhances appreciation for the beauty of nature, stimulates the arts and promotes greater understanding of nature. Nature emerges as something which is beyond and which fully embraces human existence. The complexity and richness of the various forms of life, their concurrence with present and evolving diversity in the different ecological conditions existing on the planet: this is the natural reality with which human life is bound. Man's responsibility for the non-human natural environment is based fundamentally on the fact that the tremendous potential and the delicate equilibrium of the natural environment are ensured and safeguarded only to the extent to which man acknowledges that nature is not for man's benefit alone.

In addition the concept of humankind cannot be restricted to human beings now living on the planet, but must embrace future generations. If this concept is accepted, it is evident that the present influence of humans on the environment must not damage the Earth for future generations through ecosystem alteration and exhaustion of natural resources.

Moderate anthropocentric ethics (unlike the most radical ecocentric ones) do not condemn every modification of nature, but only those actions which are detrimental to the environment. Exploitation of wild ecosystems to man's benefit, avoiding their deterioration, cannot be forbidden. For example the Lombard water meadows, the Scottish moorlands, the Tuscan countryside, the Venetian lagoon and the European and Asian ricefields are examples of artificial ecosystems obtained by modifying, without damaging, the original ecosystems.

Although adhering to different religions and philosophies, there is increasing evidence that humans finally believe in corporate stewardship which could permit us to reach sustainability for the present and for the future, allowing for development with equity.

Consequently, it is moral, it is our responsibility, to adopt measures to protect and conserve biodiversity, because humans themselves are involved in the integrated system of relationships among all living beings on Earth, where a flow of energy, nutrient cycles, and homeostatic processes takes place. Through his understanding of these processes and relationships, man has freed himself, although slowly, from the determinism of some of these processes, and has developed a cultural approach (i.e., a system of concepts, values, instruments, as well as economic, artistic and ethical expressions, due to which he lives and works in a certain group, through which he influences the environment, causing alterations and adapting to them).

If biodiversity protection is of primary interest to man and belongs to human rights and values – because narrow are the ties between survival and progress of humankind, as well as between biodiversity and cultural diversity developed in the history of human societies – then the ethics of biodiversity conservation must become part of the ethos of humanity.

Central to the ethos of human culture are solidarity, non-violence, international tolerance; and biodiversity conservation, its evaluation and scientific and technological exploitation must become another issue central to the ethos of humanity.

This implies that man must be aware of such diversity if he wants to be able both to attribute to it a fair value and to propose an ethical behaviour coherent in its planning, managing, controlling, and conservation. Thus there is a need for strengthening scientific efforts towards identification, evaluation, and utilization of biodiversity. Human society has an interest, although it is not deterministically obliged, to work towards the safeguard of nature, its resources and its diversity. The likelihood of man to destroy nature and biodiversity is linked to poor management stemming from his ignorance.

Through scientific research and technology aimed at the exploration, identification, preservation, evaluation and utilization of animal, plant, and microorganism genetic resources, man is gradually increasing his knowledge of this invaluable natural treasure which is biodiversity.

Biodiversity protection is not only a complex biological problem from a scientific and technical standpoint, but requires also a clear identification of the 'right' and 'not-so-right', of the 'good' and 'not-so-good'. Therefore the problem becomes also a legal and economic one.

As discussed in earlier chapters, global, regional, and national biodiversity policies should be able to establish universally accepted rules of behaviour for inventorying and monitoring, management decisions, and evaluation of the intrinsic economic values of preservation and commercial exploitation of biodiversity. International laws and institutions should also be given clear and definite mandates, as well as the power of mobilizing financial resources and developing timely information, and the capacity to involve political, administrative, scientific and entrepreneurial organizations.

National strategies should be coherent with greater responsibility given to the advanced countries. Basic to all these problems, however, are environmental education and human resources development.

Primarily, the role performed by traditional societies, rural communities, especially by women, must be acknowledged, particularly in those areas in which plant and animal genetic resources are richer, as was pointed out in Chapters 8 and 30. The role of women has only recently been officially acknowledged at the Rio Conference.

As a matter of fact, all social classes must be provided with a sound and effective environmental education. It should correspond to appropriate national strategies, in the assumption that progress of humankind and protection of environmental and natural resources are indissoluble. This way of thinking must become part of our culture – part of those cultural values that have freed humans from natural determinism (predator versus prey) and selfish control of nature.

Whether a follower of ecocentric ethics or anthropocentric ethics, humans must act as caretakers of natural resources, in the interest of both current and future generations.

Principles and initiatives in the field of environmental education, now underway or being developed in all communities, will certainly improve the relationship between humankind and environment.

As a matter of fact, one of the main causes of environmental deterioration is ignorance, essentially linked to lack of access to knowledge. An increasing part of humanity has no access to education in whatever form. According to UNESCO, the number of illiterate people, particularly in developing countries, has increased from 600 million ten years ago to more than 800 million today. But environmental education must be implemented according to appropriate national strategies.

At present, people's environmental education is poor and will require a powerful effort. The approach in developed and developing countries, however, cannot be the same. In the former, one can observe a substantial homogeneity of behaviours. Also, a conscious awareness is spreading, mainly as a result of recently achieved higher cultural levels. There, the objectives will be a decrease of consumption, promotion of recycling and a sustainable use of renewable resources. In developing countries, on the other hand, there is necessarily a focus on basic, elementary aspects of social life: survival, human rights, solidarity, development, peace. Safeguard of the environment and environmental education will be considered as having marginal importance.

At the same time, however, environmental emergency is the most acute in developing countries, where insufficient human and economic resources are the main obstacle to environmental education. Thus, following also the resolutions adopted at the Rio Conference, international cooperation must implement urgent measures to remove those obstacles.

Environmental education must also be a vehicle for the creation and diffusion of realistic ecological awareness. It must also pursue another objective: to persuade human beings that development and environmental protection are indissoluble, inasmuch as individuals, communities, human activities, man's intellectual capacities and accomplishments can grow and progress only in a 'healthy environmental status'. 'Healthy' means appropriate and durable protection from devastations, which is a prerequisite for sustainable durable development, today as tomorrow.

Durable and sustainable development is a principle and a way of acting where ecology, economy, sociology, and culture must meet. Whence the opportunity that environmental education mould the consciousness of an ecological and economic interdependence – almost a cohesion – between biological, physical, anthropological, technological, scientific, social, and cultural problems. One must be convinced that man and nature are factors of a binomial whose product would be positive, if man were educated to live in the environment.

In order to achieve such results, effective environmental education will have to aim at a drastic change in attitudes, in ways of thinking and living, in economy, as well as in interpersonal, interracial and international relationships. Changes will also have to intervene in goals set for humanity, and therefore in procedures and instruments.

Psychological, cultural and structural obstacles hindering such revolution in attitudes must be overcome without delay and hesitation. Structures, thus far based on antinomies (individual vs. collectivity; selfishness vs. solidarity; cynicism vs. moral duty; south vs. north; national vs. global; isolation vs. interdependence; individual vs. system; saving vs. waste; specialization vs. interdisciplinarity; short-term vs. long-term prospective, etc.), will have to be replaced with the principle of singleness of humankind in future, in a development both more equitable and in accordance with nature and its biological and physical resources.

If this is the basic principle, the implementing strategy will inevitably be characterized by any effort towards a permanent and global education, so that environmental education becomes an integral part of general culture.

Family, school and the media must cooperate in promoting in young people the development of values and individual and social behaviours with respect to relationships between humans and their environment. Universities, academies, and research institutions face a similar duty, related to their responsibilities in training not only scientists and technicians in environmental fields, but also professionals and managers for the society at large. The same may be said for cultural and religious institutions, international organizations and agencies, professions and enterprises that utilize both human intelligence and natural resources.

Environmental education must become in all countries a clear priority, which will allow the same countries to strengthen their endogenous capacities to conserve and manage in a durable way the environment and its resources, and to ensure sustainable economic development and social, cultural and moral progress of national societies in the framework of the global human community.

Diversity between individuals, species and ecosystems is the raw material that permits human communities, today as much as yesterday, to adapt to gradual or abrupt environmental changes (for instance: earth heating, desertification, soil erosion, depletion of ozone layer, famine or malnutrition), to sustain demographic growth, to improve economic and social living conditions and to benefit from durable development.

Biological diversity is essential to the health of our planet, and a prerequisite, now as in the past, for the wealth and welfare of our societies, as well as for their cultural, social, and economic development.

The use of biodiversity, especially plant, animal and microorganism genetic resources, is indispensable for food, agricultural, and industrial production. The utilization of genetic resources in plant breeding has allowed in the United States, over the past 50 years, an average yearly increase of the added value of more than one billion dollars.

Nature's storehouse is truly huge. The still untapped potential (it is estimated that more than 200,000 plant species have not yet been identified and considered) represents an extremely valuable and virtually unlimited resource. For instance, an initial global inventory would include, with some approximation: 3000 tropical fruits, 10,000 grasses, 18,000 legumes (members of the family Leguminosae), 1500 edible nuts, 1500 edible mushrooms, 60,000 medical plants, 3000 species with purported contraceptive powers, 2000 plants with pesticidal properties, and 30,000 tropical trees.

In comparison, the present world agricultural and food production is still based on a very modest number of plant and animal species. For instance: of the 3000 tropical fruits, only four – banana, mango, pineapple, and papaya – are produced in any quantity on a global scale; of the 10,000 grasses, only seven – wheat, rice, maize, barley, sorghum, rye and oats – are employed globally; of the 18,000 legumes, only six – peas, beans, soybeans, peanuts, alfalfa, and clover – are used intensively, despite the fact that legumes are very nutritious plants.

The last issue I want to underline concerns the application of modern biotechnology methods in utilization of biodiversity. It is useful, or rather necessary, but it calls for a prior assessment of compatibility and fitness of processes and products to the safeguard of human rights and values. Of course this is true for all processes and products of modern technologies.

Modern technologies do not change the goals of agricultural and industrial productions, but make available new techniques of raw-material processing and transformation and, therefore, a continuously renewed range of products.

Genetic engineering, and the development of new or improved organisms through the utilization of alien DNA (e.g. plant types resistant to parasites or adverse climatic conditions; microorganisms able to modify or intensify the output of products useful to humans, as in the case of additives in food processing; microbial destruction of pollutants), may greatly benefit from a better understanding and use of the up until now scarcely explored biodiversity.

However, the introduction and utilization of new products require a systematic, preliminary approach, where the behaviour of the new introduction, and its medium-term interactions with life in all its expressions (social, economic, ethical), must be evaluated, and its compatibility with human rights verified.

As a matter of fact, the risk of subjection, of limiting human freedom, whether at the individual or collective level, is evident. It is quite difficult to separate, in theory as well as in practice, the attainment of new forms of knowledge from their technical and industrial applications. Through biotechnologies, genetic engineering techniques, medical applications, the potential of computer science and the use of data banks, the exploitation of energy and space, and environmental misuse, an excessively 'technicized' society may violate the human rights of both today and tomorrow's generations, both individually and collectively. We feel therefore the need to protect humans from any possible form of subordination and slavery.

It has been amply shown that, although the flow of technological innovation grows ever faster and the elapsed time between the arising of new knowledge and the implementation of innovation in production becomes shorter and shorter, a long time is still necessary for man to accept innovation and to become aware of the real compatibility between technology, human rights and humankind's cultural and social values. Yet, refusing or delaying the implementation and dissemination of new technology may result in severe financial damage to enterprises and to national economies.

By promoting a profound and critical consciousness in public opinion, free from any emotional or instrumental conditioning, it is possible to favour a prompt, economically convenient diffusion of scientific innovation and new technology. Such public awareness will originate through a network of relationship, debates, confrontations, and interactions, whose basic prerequisite is represented by a series of local, national, and international initiatives considered as indispensable for the diffusion in all human communities of an effective and appropriate environmental education.

Informing public opinion must conform to criteria of objectivity and responsibility, so as to avoid that biased representation of reality and human interest be engendered. Information disseminated must enable man to understand the evaluations of risks and benefits stemming from technological innovations, and foster the acceptance of those risks, of the new technological processes and products, when compatible with human rights and values.

The role of universities, academies, and research institutions at large is, however, instrumental to test the compatibility of new technologies with human rights. It will be the task of such institutions, in cooperation with representatives of technical and productive activities, to promote interdisciplinary studies and to entrust them to scientific and technical teams working together with human and sociojuridical groups, in order to ascertain the ties that bind human values to technologies, and to attain, in the end, the awareness of a balanced and compatible relationship between human beings and the technological society.

Furthermore, in case of the use of biodiversity in modern biotechnologies, particular care shall be exerted to ensure that processes and products obtained through the use of biological materials (and therefore usually considered more life-compatible) truly represent innovations useful and beneficial to man. For instance, the release of organisms (plants or animals) modified with use of genetic variability and the wealth of new characteristics present in biological diversity and apt to allow man to increase his welfare and control of nature, must be preceded and accompanied by a continuous orientation of scientific research towards a better understanding of their long-term, far-reaching effects. In this way, man's action develops according to the principle of 'moderate anthropocentric ethics' and strives to guarantee, at one time, compatibility with human rights and values, conservation of biodiversity, safeguard of natural resources and of natural and human environment.

In conclusion, with such knowledge and scientific and technical guarantees, and within a definite sociojuridical framework, policy-makers, opinion leaders, religious leaders, mass-media, cultural associations and workers' and entrepreneur's organizations will be able to guarantee all worried users and consumers, afraid of technological innovation, that the new processes and products, also due to future improvements made possible by new scientific knowledge and technical developments, are compatible with human rights and cultural values and, therefore, wholly legitimate.

Strategies for rational use of biodiversity, basic element of living and non-living natural resources, for a more equitable and prosperous global society, for the solution of the many problems involving the entire human family and the environment in which we live, need also to be based on a coherent, moral, ethical vision of the world.

This forum is important (and is a great merit of the organizers) because now is the real time to develop hypotheses, which must become opinions, and opinions which must influence decisions. Among the scientific, technical, economical, juridical, social, political, and cultural elements of the decision power, non-quantifiable factors, such as ethical values and principles, must also enter and be taken into account.

60 Reformation Towards a Nature-oriented Culture

Wakako Hironaka
Member, House of Councillors of Japan, 2-1-1 Nagata-cho, Chiyoda-ku, Tokyo 100, Japan

It is a great honour and a privilege for me to be given the opportunity to participate in the IUBS General Assembly to speak in front of such a distinguished audience. I am not a biologist, not a scientist, not a scholar, but a politician concerned with the future of the people of my country, Japan, and her relationship with the rest of the world. Thus, I am interested in such global issues as environment, poverty elimination, population, and migration. This chapter will consider the cultural ethics of Japan in relation to biodiversity.

For those who are not familiar with or have never visited Japan, I would like to give a brief description of my country. In Japan, as in other countries, the nation's culture and national character have been largely moulded by geography and climate.

Japan is an island nation off the east coast of China. Four major islands, plus several thousand smaller islands, stretch over 3000 km from northern Hokkaido to the southern tropical island of Okinawa. The land is mountainous, and covered by green forests which occupy 67% of the land even now. The range of mountains are mostly volcanic with swiftly flowing rivers.

Although there are some differences between the north and south, Japan has four distinct seasons, consisting of gentle spring, hot humid summers, beautiful autumn, and dry clear winter days. But occasionally, we face natural catastrophes including typhoons, floods, earthquakes, and drought, which bring people's minds in awe towards mother nature.

The original inhabitants of the islands date back to perhaps 12,000 years, and they were basically hunters and gatherers believing in nature-worshipping, called Shintoism. Roughly two thousand years ago, an agricultural civilization was brought into the Japanese archipelago by those who made their way from the Asian continent through the Korean peninsula. Japan is located far enough from China to have escaped involvement in continental upheavals and yet close enough to have benefited from their culture. Buddhist

© 1996 CAB INTERNATIONAL. *Biodiversity, Science and Development: Towards a New Partnership* (eds F. di Castri and T. Younès)

belief of momentary beauty in nature was brought in by the immigrants, and greatly affected the lifestyle and customs of the Japanese inhabitants. But the old belief of Shintoism remained to coexist with the Buddhist teachings, and the original animistic philosophy continued to teach the people that all living creatures were to participate equally in a 'seamless' web.

Historically, Japanese people treated nature not as something that is to be mastered but as an entity of infinite depth. Not only have the seasonal changes played a major role in creating haiku poetry, scenic gardens, and other aesthetic customs of the upper class families, but the Japanese people of all social levels attempted to bring nature into the proximity of their daily lives. People wore kimonos with patterns designed to represent natural objects and phenomena. They gathered to view the blooming of cherry blossoms, the full moon, and autumn leaves. When the world outside was covered with snow, a painting of the Adonis flower pushing out of the frost was hung in the alcove. When summer arrived, it was replaced by a picture of a waterfall against a background of fresh green trees, and a wind-bell was fastened to the deep overhang. Japanese people of the past were careful to harmonize with mother nature indoors and out through the medium of art. Thus, traditional houses and gardens were shaped to blend in everyday activities with the surrounding environment. As a result, this concept created architectural devices such as the paper sliding doors, called 'shoji' so that one can be sensitive to the changes in light, wind and temperature outside.

As has often been pointed out, Japan during the Edo Era (1603–1867) was moving towards its peak as an agricultural society. The country was strictly closed off from the outer world under the isolation policy of the Tokugawa Government. Using our limited natural resources and energy, the Japanese people of Edo lived materially simple lives and developed an environment friendly culture. I might add that the population of Japan during the Edo, period for nearly 300 years, was kept at around 30 million; that was perhaps what the natural environment could sustain. However, the Meiji Restoration in 1868 led to the development of the modern pursuit of efficiency and convenience, and Japanese society was overridden by the capitalistic concern of industrialization imported from the West without a proper adoption of the western way of observing, managing, and caring for nature through science and technology.

Often environmental concern has given way to economic development, and this trend escalated especially after World War II. The population, has grown fourfold since the Edo period. Now, a population of 120 million people, one-quarter live in and around Tokyo. Their daily life is certainly enriched by such modern conveniences as electric appliances and automobiles, but at the same time it is hectic from long commuting hours and poor living conditions. People are now beginning to wonder what price they have paid for their economic development and affluence.

Ever since the beginning of human civilization, starting with Mesopotamia, Egypt, the Indus River Basin, and China, human beings have continuously abused their natural environment at a regional level. However, it is in the last

50 years that this exploitation has been occurring on a global scale. This is mainly due to the massive scale of industrialization; mass production, mass consumption, and mass waste, during the rapid industrialization in developed nations, and the ever-increasing population and poverty of the South. Japan experienced serious pollution and health problems, and our natural environment has been destroyed. Fortunately, we were able to overcome most of these problems, by investment in pollution prevention technology and facilities, and by strict regulations. Japan's experience demonstrates that prevention is more cost effective than environmental clean-up or other counter-measures.

Now, the people of Japan are beginning to look back, and to re-examine our traditional life-styles and nature-oriented philosophies that once played an important role in enhancing the quality of life. We acknowledge that historically, two different religions, Shintoism and Buddhism, did not dominate one another but coexisted to bring forth a harmonious blending of people with the environment: that means our present human-centred values which have pushed older values out of the mainstream need to be questioned. By recalling the traditional concepts neglected by modernization, people are reminded that we live as a part of nature from which we receive our sustenance.

The Brundtland Commission predicted a five- to tenfold increase in industrial development as the population doubles over the next 50 years. We are not certain of the carrying capacity of the planet, but a number of experts estimate that if our present production and consumption pattern continues, the earth may not be able to sustain our human existence even with the help of science and technology. Whether we are endangering our own existence or not, the Earth will still continue to rotate and the fundamental laws of nature will not change. Therefore, it is our way of viewing nature and our life-styles that should change.

As much as we are all aware that we cannot go back to the Edo period, we must change our society into a sustainable one, integrating environmental issues into our economic and social policies while making the best use of science and technology.

This is where political leadership comes in. The Environment Agency of Japan, recently passed the New Basic Environment Law which sets forth a government-wide effort, consisting of three fundamental goals: the preservation of a robust natural environment for present and future generations, the transformation of our society into one which allows sustainable development and the promotion of international cooperation on global environmental issues. Now, the implementation of this law, our 'Environmental Constitution' requires political will and leadership support.

Sustainable development will require a radical transformation in our values, and most importantly will depend on how seriously we translate these values into our daily life-styles; by every country, every community, every household and every individual. This transformation will rely on a value system of being aware that the planet Earth is not ours alone and that we share it with other living things.

Biodiversity and Quality of Life 61

G. Hauser
Histologisch-Embryologisches Institut, Schwarzspanierstr. 17,
1090 Vienna, Austria

The awareness of problems of an ethical nature and of the necessity to deal with them appears to have evolved with the development of Man's awareness of his own self, his mental and physical capacities. These problems became of increasing importance as human communities developed in size and complexity. Codes of practice to deal with them had already evolved by the time of the great Greek civilizations, probably earlier, and still affect our behaviour today. For example, Hippocrates lived on the island of Kos in the sixth century BC and his ethical views still govern modern medicine. They have provided guidelines for the traditional doctor–patient relationship, but have been severely tested by the coming of modern biological advances which have made possible a variety of medical applications (Table 61.1).

The rationale for the use of these applications is that they improve the quality of life for the affected individual, the parents or the family. These new applications include techniques for visualizing a fetus *in utero* and obtaining specimens of fetal tissue or products for examination. Fetal visualization is carried out by echosonography or fetoscopy, whereas fetal tissue is obtained by amniocentesis or chorionic villus biopsy. This material can be cultured for examination of fetal chromosomes or to obtain material for biochemical assay or for DNA analysis. Prenatal visualization allows the detection of gross life-threatening defects and preparations can be made for immediate postnatal surgery. If the fetus has an untreatable defect, then termination of the pregnancy may be offered and the procedures for doing so have improved greatly over the last few decades.

Postnatal genetic screening can be carried out at the population level for some of the more common genetic errors where treatment is possible, and at the family level for the rarer disorders. Screening of potential patients is possible for carriers of a number of severe, but less common inherited diseases. Once the disease has been diagnosed after birth, there is considerable interest

Table 61.1. Modern techniques in medicine that may be applied in order to improve the quality of life for the individual or family.

Biosampling	Amniocentesis, chorionic villus biopsy
Prenatal diagnosis	As for trisomy 21 and other chromosomal aberrations, neural tube and numerous other defects.
Genetic screening	Newborn as in phenylketonuria (enzyme deficiency-diet control possible, mental retardation if not treated) Potential parents Tay-Sach's disease (enzyme deficiency, mental deficiency, no diet control)
Major malformation screening	Echosonography
Neonatal surgery	e.g. Heart defects
Gene therapy	Adenosine deaminase deficiency, implantation of cells to switch on production of the enzyme (bone marrow cells of relatives)
In vitro fertilization	Success rate 30% (varies from unit to unit from 0 to 30%) problem of multiple pregnancies
Organ transplantation	Kidney transplant in early renal failure
Prodromal diagnosis	e.g. Huntington's chorea onset in middle age, polyposis coli.
Euthanasia	Rarely permitted officially.

in the potential for gene therapy. At a still later stage, prodromal diagnosis can help with early commencement of treatment and with reproductive decisions. For those couples who are infertile, an *in vitro* fertilization is a possibility, and this technique is also available for those who know that they carry a gene for a severe defect and prefer to accept donor gametes rather than run the risk of producing a severely affected child of their own. The application of each of these techniques brings its own ethical problems ranging from the acceptability of pregnancy termination or of submitting a child to a series of surgical interventions to the legal rights of donors of gametes (Roberts and Chester, 1991).

In the majority of these cases resulting from application of modern techniques, the ethical problem can be largely resolved by consultation between doctor and affected individual, or in the case of the fetus, between the doctor and the parents. Here, of course, arises the problem of whether the decision should be that of the mother alone. This consultation involves the explanation of relative risks, although in many instances these alone are insufficient and other factors, e.g. religion and other cultural considerations, may be of overriding importance. For example, no calculation of risk will make any difference to a Jehovah's witness who would prefer their child to die rather than to receive a blood transfusion. Difficult though these cases are, even more difficult are the bioethical problems which have arisen from either people's ignorance of – or lack of interest in – the effects of their actions, or from mere overestimation of their powers to restore nature's balance.

One cannot attempt to justify to an Amazonian hunter–gathering tribe the destruction of their traditional habitat and food supply by the development of a new mine and the roads serving it. One cannot even begin to attempt to justify to the koalas of New South Wales the destruction of their eucalyptus stands in order to accommodate new human housing development. The other side of the coin is that sometimes it is only through the tragic intervention of humans that biology appreciates that a particular species is a keystone species in the ecosystem, as for example, the sea otter of the kelp forests of the Pacific coast of Alaska.

These examples have touched on the quality of life of the human individual, the family, the population and the quality of function of the ecosystem. Let us turn now to a conspicuous variable in intraspecific biodiversity – the variable of sex. I suppose that one of the key pointers to the quality of life is the desire to continue living. Table 61.2 gives details of suicides in Austria (Hauser et al., 1995). It shows that in both 1981 and 1991, the total number of male suicides exceeded very clearly female suicides. Fig. 61.1 presents the data diagrammatically. Along the y-axis the ration of males to females is plotted: 100 equals equality, 300 means that there are three times as many males committing suicide as females. The figures for 1981 show a male excess at all ages, threefold in males of all age groups up to the fifties. The figures for 1991 show a considerable increase in the male disproportion, especially at younger ages where it is fivefold. There appears to be a general tendency for earlier occurrence. Over the ten-year period, the maximum number of male suicides shifted from the age class 45–59 years in 1981, to 30–44 years in 1991, and in females from the age class 60–74 years to 45–59 years.

This change is not uniform for all age groups. If instead of using absolute numbers one calculates in each age group the percentages of suicide in relation to the total number of deaths in that age group, a lower proportion of female suicides occurs in the younger age groups in 1991, whereas the percentages in the older age groups remain fairly constant. In males, on the other hand, suicide contributed to a greater proportion of deaths in the two oldest age groups, and especially in the age group 15–29 years. These secular changes therefore influence the sex ratio of deaths by suicide. Over the ten-year period, the tendency of more suicide in younger males appears to be part of a general phenomenon, as an increase in the incidence of suicide in younger males was occurring in many of the developed countries (Denmark, Finland, France, Hungary, Japan, Sweden, Switzerland; Gibbs, 1968).

The French sociologist Emile Durkheim (1952) demonstrated at the beginning of this century a relationship between suicide rates, social class, religion, and affluence. One may argue, in agreement with the Indian physicist and

Table 61.2. Suicides in Austria.

Age classes	Male 1981		Male 1991		Female 1981		Female 1991		Sex ratio 1981	Sex ratio 1991
	N	%	N	%	N	%	N	%		
0–14	9	0.90	2	0.35	3	0.44	0	0.00	300.00	∞
15–29	283	19.80	263	22.50	90	19.36	56	16.76	314.44	469.6
30–44	350	17.50	311	17.33	115	12.29	92	11.23	304.35	338.0
45–59	368	6.06	252	4.98	134	4.18	113	4.69	274.63	223.0
60–74	278	1.87	262	2.23	153	1.24	111	1.25	181.70	236.0
75+	149	0.79	209	1.14	100	0.32	98	0.31	149.00	213.3
Total suicides	1437		1299		595		470			
Total deaths	44,235		38,639		48,458		44,789			

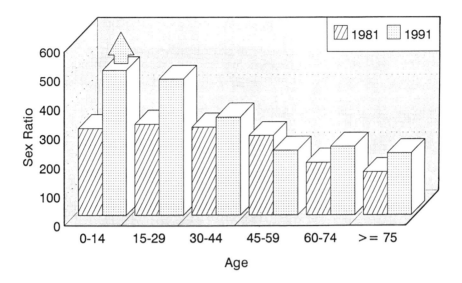

Fig. 61.1. Sex ratio and suicide in Austria.

philosopher Vandana Shiva (1993), that the modern world has built its ideas of nature and culture on the model of the industrial factory, where a forest is rather judged for the financial value of its timber than for the variety of its life forms and its life-supporting capacities. Both of these writers draw attention to the relationship between human behaviour and environment.

One interpretation of these suicide figures is that more and more people, and especially young people, are no longer satisfied with the quality of their lives and this seems to affect males in particular. I do not think that Austria is unique in this respect.

It is a long way to argue from suicide to loss of biodiversity, but may they not be parts of the same phenomenon? A human response to a change in values?

References

Barron, S.L. and Roberts, D.F. (1995) *Issues in Fetal Medicine*. MacMillan, London.
Durkheim, E. (1952) *Suicide. A Study in Sociology*. Routledge and Kegan, London.
Gibbs, J.P. (ed.) (1968) *Suicide*. Harper & Row, New York, Evanston, London.
Hauser, G., Seidler, H., Wytek, R. and Wilfing, H. (1995) Geschlechtsproportionen der Österreichischen Bevölkerung-Biologische und soziodemographische Überlegungen. *Zeitschrift für Morphologie und Anthropologie* (in press).
Roberts, D.F. and Chester, R. (eds) (1991) *Molecular Genetics in Medicine. Advances Applications and Ethical Implications*. Macmillan, London.
Shiva, V. (1993) *Monocultures of the Mind*. Third World Network, Penang.

Index

Aframomum melegueta 511
Agricultural subsidies 275-276
Agroecosystems
    in Canada 341-346
    in India 116-126
    legal aspects 314
    management 312-322
    microbial loss 135-136
    shifting agriculture 116-126
Air pollution 334
Alfalfa 470
Algae 155, 410, 458-459, 490-491
All-Taxon Biodiversity Inventory (ATBI) 8
Alnus nepalensis 124
Among-stand diversity 328-329
Anas platyrhynchos 569, 570, 572
Ancient civilizations 347-350, 507, 558
Ancistrocladus 527
Animal breeding 313, 315, 386-401, 414-419
Antarctica 133
Anthropogenic forests 349-351
Apodemus 570, 571
Aquaculture
    biodiversity loss 412-415
    biotechnology 456-465
    conservation opportunities 404-406, 416-420
    demands 406-407
    freshwater introductions 446-450
    gene banks 439-445
    genetic engineering 415-416
    marine ranching 424-434
    species farmed 410-412
    in Taiwan 452-454
    trends 403-404
Aquilaria agallocha 479
Arboreta 316, 547
Argentina 35-37, 540-541
Aromatic plants 510
Artefacts 83-87
Artemisia annua 475, 476
Ascidia aperta 464, 465
Atlantic Ocean 141, 143, 145-146, 155
Atractoscion nobilis 424, 432
Atropa belladonna 476
Australia 209, 447, 458, 463-464
Azadirachta indica 509
Azov Sea 154, 429

Baccharis punctulata 541
Bacteria
    distribution 131-135
    drug production 461-463, 464-465, 490-491
    genetics 59-60
    root transformation 478
Bambusa 124

Bangladesh 412
Barley 358
Beaver 157, 204
Bees 82
Belgium 545
Belize 176, 258
Benin 513
Biological collections 172
Biotechnology
    companies 462–465
    defence 500–505
    development 495–496, 522–528
    marine 456–465
    training 532
Birds 53, 83, 541, 568–572
Black Sea 154, 429
Boiga irregularis 155–157
Botanical gardens 316, 319, 496–497, 547, 556
Botswana 529
Bower birds 83
Bradybaena 198
Branta canadensis 572
Brazil
    anthropogenic forests 350
    biotechnology 308
    cattle ranching 217
    city ecosystems 545
    fish gene banks 444
    forest destruction 221, 254
    forest restoration 559, 563
    malaria 475
Bromus biebersteinii 470
Butterflies 209, 258, 259, 283, 292

Cambodia 185
Cameroon 447, 473, 527, 529
Canada
    agroecosystems 341–346
    aquatic invaders 155
    fish gene banks 441–445
    forest management 324–337, 545
    insect outbreaks 53
Cancer 177, 475, 478, 527
Carassius auratus 448
Carbon dioxide 162–164, 169, 459, 577–579
Carica papaya 508, 510
Carp 411, 414, 417, 432, 447

Cassia 509
Castanopsis cuspidata 560
Catharanthus 475–476, 477, 508, 510, 513, 528
Cattle
    breeds 313, 320–321, 397–399
    history 386–387
    ranching 216–217, 221
Centella asiatica 508, 528
Centre National de la Recherche Scientifique (CNRS) 6
Chanos chanos 412
Chile 559, 563
China
    aquaculture 409, 412, 444, 454
    biotechnology 308
    flora 185
    medicinal plants 474–475, 507
Christmas Island 158
Chrysanthemum cinerariafolium 529
Cichlids 447–449
Cinchona 490, 508
Cities
    aesthetics 552, 607
    animal populations 566–574, 582
    birds 541, 568–572
    dynamics 585–589
    habitat types 540–541, 545–546
    horticultural planning 552–553
    human populations 544–545
    interfaces 589–592
    morphogenesis 589
    periphery expansion 585
    phytosociology 554–555
    zoos 554, 582
Climatic trends 205–206, 334–335
Coastal ecosystems 224–241, 288–289
Cochliomyia hominivorax 175
Cockroaches 569
Cocoa 352–353
Cod 424, 425, 432
Codium fragile 155
Coffee 509
Coleoptera 173
Coleus blumei 477
Colombia 444
Combretum micranthum 511
Convention on Biodiversity xi–xii, 1, 15, 18–19, 28, 503–505, 519–523, 530, 609

Copepods 143, 174
Coptis japonica 476, 477
Coral reefs 105, 233, 412, 416, 463
Corsica 141
Cortaderia selloana 541
Corvids 569–572
Cosmetics 480, 528
Costa Rica 34, 168, 180–183, 219, 442, 474
Cranes 95–97
Crayfish 429
Croton tiglium 527
Cryopreservation 417–418, 444
Cynodon dactylon 120
Cytotoxic compounds 460–462

Darwin 12, 46, 47, 67, 387
Dendracalamus hamiltonii 123, 124
Desert ecosystems 158
Digitalis lanata 477–478
Dioscorea deltoidea 477
Diplopods 202
Disease resistance 470–471
DIVERSITAS programme xii, 2, 5, 9, 14, 20, 24, 28–29, 151, 167, 169, 225, 232, 237–238
DNA fingerprinting 94–102
DNA transfer 59–65
Dreissena polymorpha 155
Drimycarpus racemosus 123
Drosophila melanogaster 92–93
Drugs (*see* Medicinal plants)
Dyes 510

Echinacea pallida 477
Echinocarpus dasycarpus 123
Echinoderms 464–465
Economic aspects
    fisheries 442
    global markets 256–258, 267–279
    pharmaceuticals 524–529
Ecosystem cycles 48–52
Ecotourism 295
Ectomycorrhizas 133, 135, 136
Ecuador 529
Education 293–294
Eels 453, 454
Egypt 409, 412, 473, 507, 529

Endangered species 97, 370, 371, 375, 428
Enzymes 460, 478, 479, 493–494
Escherichia coli 59, 457
Estonia 417
Ethics 602–603, 607, 614–629
Ethiopia 320, 358, 440
Eucalyptus 209
Euphorbia hirta 511
Eusideroxylon zwageri 255
Extinctions 27, 67, 155–158, 204, 266, 288, 348, 606
Extreme environments 131, 132–133

Fagus crenata 562
Fallowing 122–123
Family planning 39
Fire 333
Fish (*see also* Aquaculture and under individual names) 403–468
Flagship species 224, 234
Flemingia vestita 123
Floras 173–174, 185–186, 284, 287
Food additives 478–479
Food and Agriculture Organization (FAO) 13, 319, 345, 360–361, 403, 410, 412, 439, 485–486
Foraminifera 143
Forests
    among-stand diversity 328–329
    boreal 133, 135, 324–337
    Canadian 324–337
    management 176, 321, 322, 324–337
    Mayan 349–354
    microorganisms 133, 135
    restoration 558–563
    Russian 201–202, 205
    temperate 133, 135, 201–202, 205, 324–337
    tropical 133, 158, 176, 194, 254–255, 347–354
France
    biotechnology 308
    embryo storage 317
    government policy 19–22, 26
    livestock farming 386, 388, 392–393, 395–399
    perfume industry 524
    urban ecosystems 551–552, 554–556
Freshwater ecosystems 134, 441, 446–450
Frogs 570

Fruit 510
Fungi 130–136, 313, 318, 358, 470, 490–491

Gambusia affinis 447
Gecarcoidea natilis 158
Gene banks
    fish 417, 439–445
    general 309, 316
    plants 316, 342, 357, 361, 365
Gene transfer 59–65, 306, 309
General Agreement of Tariffs and Trade (GATT) 258, 270–271, 314, 454
Genetic engineering 415–416, 419, 457–458
Genetic markers 432
Genome organization 69–77
Germany 308, 568, 572, 573
Germplasm banks 36, 309
Ghana 176, 412, 473, 513
Global Biodiversity Assessment Project (GBA) 33
Global markets 256–258, 270–271
Glycyrrhiza glabra 477
Grasslands 321, 342–343, 577–579
Great Barrier Reef 463, 464
Green Revolution 307–308
Groundnut 470
Growth hormones 457–458
Grus 95–97
Guam 155–157
Guatemala 470
Gulls 570, 572

Harpagophytum procumbens 529
Hatchery enhancement programmes 424–434
Hibiscus sabdariffa 529
Hirudinea 202
Homo erectus 83
Human resources development 531–532
Hungary 417
Hunter-gatherers 351

INBIO Programme 34, 168, 182, 474
India
    agroecosystems 116–126

ancient pharmacopoeias 507
aquaculture 412
aromatic plants 479
biotechnology 308
fish gene banks 441, 444
Indicators 327, 328, 335–336, 371
Indonesia 38–41, 255, 308, 412, 418, 440
Insecticides 529
Insects 53, 173, 199, 569
Intellectual property rights 482, 523
International Biosphere Programme (IBP) 13
International Board for Plant Genetic Resources (IBPGR) 361, 362
International Council of Scientific Unions (ICSU) xii, xiii, xiv, 12, 13, 14, 23–25, 170
International Marine Biodiversity Programme 237
International Plant Genetic Resources Institute (IPGRI) 319, 362, 404, 444
International Union of Biological Sciences (IUBS) xii, xiv, 6, 9, 14, 20, 23, 26, 29
International Union for the Conservation of Nature (IUCN) 13, 18, 25, 235, 290
International Whaling Commission 21
Introductions 405, 413–414, 429–431, 446–450
Invading species 153–157
Iran 470
Iraq 473
Ireland 449
Israel 155, 458
Italy 433, 547, 551, 574
Ivory Coast 412, 475, 513

Jacaranda caucana 527
Japan
    biotechnology 308
    cosmetic and perfume industry 524, 528
    cultural ethics 603, 630–632
    fisheries 424, 454
    flora 185, 186–188
    marine algae 458
    native forests 558–563

Kangaroo rat 158
Kenya 255, 444, 475, 521, 529
Keystone species 123-124, 158, 283, 321
Korea 454

Languages 249, 251
Laos 185
Lates niloticus 429, 448-449
Legislation 256, 275, 314, 482-484
Lesotho 529
Libya 175
Lichens 132, 133, 136, 175
Limnothrissa miodon 429
Linum flavum 477
Lithospermum 476, 477, 528
Littorina littorea 155
Livestock farming 385-401
Lobaria pulmonaria 175
Lolium multiflorum 470
Lycopersicon 363, 364, 470-471

Madagascar 475, 529
Maize 358
Malaria 447, 474-475, 490, 508
Malawi 412, 414
Malaysia 176, 185-186, 308, 473, 559, 563
Marine ecosystems
    coastal 224-241
    deep-ocean 104-106
    mammals 21, 234
    microorganisms 134
    ranching 424-434
    sediments 139-146
    South Africa 288-289
    toxins 461, 462
Mauritius 473
Maya forests 349-354
Maytenus buchananni 521
Medicinal plants (*see also* individual species names)
    history 488-490, 507
    main species 508-511
    outlook 515-516
    overculling problems 513-514
    political aspects 514
    production and processing 514
    screening 472-473, 519
    South Africa 295
    tonnage 510
Mediterranean Sea 155
Memes 82-83
Metroxylon sagu 117
Mexico
    biotechnology 308
    corn varieties 470
    defaunation 158
    economic aspects 256-258
    forests 349-355
    land use changes 216-217
    languages 249, 251
    legislation 255-256
    population distribution 249
    protected areas 252-255
    species richness 247-248
    tuna fishing 270
Microbial Resources Centres (MIRCENs) 14, 34
Microorganisms 130-137
Mites 150
Mixed cropping 122
Mnemiopsis leidyi 429
Molluscicides 529
Molluscs 143, 155, 198, 417
Monomorium pharaonis 569
Morinda citrifolia 477
Mozambique 529
Mullet 412, 432
Museums 172-175
Mus musculus 569
Mutagenesis 59-65
Mutualism 149-150
Mycorrhizas 133, 135, 136
Myriapods 198

Namibia 529
National parks 36, 286-289, 294, 497, 545
Nematodes 140-143, 470
Netherlands 573
New Zealand 447
Nicotiana tabacum 476, 477
Nigeria 444, 475, 529
North Atlantic Free Trade Agreement (NAFTA) 258
North Sea 143-145
Norway 425, 441

Norwegian Sea 143, 145
Nothofagus 559, 563

Oil pollution 234
Oils 510
Okoubaka aubrevillei 513
Oligochaetes 202
Oncorhynchus 199-200, 424
Orchards 316
Organization for Economic Cooperation and Development (OECD) 501
Ornamental fish 448
Oysters 230-232, 236, 417

Pagrus major 424
Palaeontological data 174, 204
Panama 448
Panax ginseng 477
Papua New Guinea 259, 449
Patents 482-483, 523
Pausinystalia yohimbe 509, 513, 529
Penaeus japonicus 433
Perfumes 479, 510, 524
Peri-urban ecosystems 576-583
Persea thunbergii 560
Peru 258, 471
Pesticides 322, 333
Peumus baldus 529
Pharmaceutical industry
    companies 471, 473, 475, 477-481
    drugs from microorganisms 490-491
    environmental protection 492-495
    growth and economics 524-531
    intellectual property rights 482-486
    partnerships with developing countries 442, 473-476
    plant screening 472-473
    production biotechnologies 476-481
Pharmacopoeias 508, 512
Philippines 410, 414, 416-418, 441, 464, 473
Phytolacca dodecandra 529
Pigeons 569, 570, 571
Pigs 313, 386
Pinus 205, 541
Plant breeding 307, 313, 315, 358-366, 470-471

Plant Patent Act 482
Plumbago zeylanica 476
Poecilia reticulata 447
Poland 568-571, 573
Politics 514, 533-534, 610
Pollachius virens 425
Pollution 234, 429
Polychaetes 140
Population growth 39, 249-250, 371, 374, 513, 606, 611
Potato 358, 363
Poultry 313
Prairies 342-343
Predatory birds 541
Protected areas 217-220, 235, 252-255, 286-289
Pseudooryx 199
Public awareness 372, 374, 417, 608-609
Puccinia striiformis 470
Puerto Rico 254
Pygeum africanum 509, 510, 513

Quercus 205, 541, 560, 562

Rabbit 111-112, 317, 570
Radioactivity 495
Rauwolfia 475, 476, 477, 508, 510, 529
Red Sea 134, 155
Reintroductions 294
Religions 600-601, 616, 618, 632
Reptiles 155-157
Rhopilema nomadica 155
Rice 109-111, 358, 470, 471
Ricinus communis 529
Roads 332
Root transformation 478
Rubber 471
Russia 197-206, 417, 462, 463, 570
Rwanda 529

Salix humboldtiana 541
Salmonids 414, 424, 428, 442-443, 447, 457
Saudi Arabia 308
Schistosomiasis 175-176, 529
Sciaenops ocellatus 424

Scientific Committee on Problems of the Environment (SCOPE) xii, 14, 20, 24, 29, 153–154, 159
Seaweeds 458–459
Seeds 316, 342, 361, 365, 444
Senegal 175–176
Sheep 387
Shifting agriculture 116–126
Shorea robusta 89
Singapore 473, 474, 544, 545
Social issues 268–271
Solanum khasianum 477
Somalia 444
South Africa 282–297, 529
Spain 541
Sparrow 569, 571
Species identification 27
Spices 510, 511
Sponges 460–461, 462
Sri Lanka 544
Streptopelia decaocto 571
Strophanthus 509, 510, 529
Sturgeon 424, 428, 429
Sturnus vulgaris 572
Subsistence farming 358
Sudan 447, 473, 529
Suicide rates 635–637
Sweden 308
Switzerland 553, 576–579
Syria 473
Sysygium cuminii 123

Tabebuia seratifolia 527
Tabernathe iboga 509
Tagging 426, 432
Taiwan 452–454
Tanzania 529
Taxonomy 177, 181
Taxus 177, 478
Tea 509
Technology transfer 514–515
Tessaria integrifolia 541
Thailand 185, 412, 474
Thalictrum minus 477
Tilapia 412–414, 416–418, 429, 447, 453
Tilletia 470
Timber management 331–334

Tissue culture 476–481
Togo 513
Tomato 363, 364, 470–471
TOTAL Foundation xiv, 9, 26
Tourism 295
Traditonal livestock breeds 320–321, 385–387, 397–399
Traditional plant varieties 307, 315, 320, 349–354, 358–359, 361
Training 293–294, 532
Transgenic fish 415–416, 457
Transport vehicles 84–89
Trapping 334
Trigonella polycerata 120
Tropical forest ecosystems
    cattle ranching 216–217, 254
    logging 254, 255
    mammals 158
    protected areas 217–219
    traditional usage 213–216, 347–354
Tuna 442
Tunicates 464–465
Turdus merula 570–572
Turkey 470
Tylophora indica 476
Typhula 470

UK 175, 308, 386–387, 417, 573
United Nations Development Programme (UNDP) 264
United Nations Educational, Scientific and Culture Organization (UNESCO) xii–xiv, 9, 12–17, 20, 24, 29, 170, 214, 518, 573
United Nations Environment Programme (UNEP) xiv, 1, 13, 20, 30–34, 264, 608
UPOV Convention 482–483
Urban ecosystems (*see* Cities)
Urocystis 470
USA
    animal breeding 391
    biotechnology 308
    cereals 358, 440, 470
    coastal ecosystems 228, 236
    ecosystem management 374
    endangered species 155, 157, 219
    fish extinction 442

USA *contd*
    marine ranching 425, 432
    medical plant screening 472–473
    protected areas 254
    salmon 265
    sport fishing 429
    urban ecosystems 544, 573

Venezuela 444, 545
Vietnam 185, 199, 412, 475
Village ecosystems 116–119
Viruses 440, 459, 470, 471, 490–491
Voacanga 509, 510
Vulpes vulpes 570, 571

Wars 444

Water fluxes 164
Water supply 125–126
Weeds 120
Whales 21
Wheat 317–318, 470
Withania somnifera 477
Within-stand diversity 330
World Intellectual Property Organization (WIPO) 482

Zaire 513, 529
Zimbabwe 429, 612
Zoos 317, 496–497, 554
Zostera japonica 155

DISCARDED

JUL - 1 2025